THE WORLD IN SPACE

A Survey of Space Activities and Issues

Prepared for UNISPACE 82

United Nations

Ralph Chipman, editor

Prentice-Hall, Inc.   *Englewood Cliffs, New Jersey 07632*

Library of Congress Cataloging in Publication Data

Main entry under title:

The World in space .

    Revised background papers prepared for the Second
United Nations Conference on the Exploration and
Peaceful Uses of Outer Space, to be held Aug. 9–21,
1982 in Vienna, Austria.
    Includes bibliographical references and index.
    1. Space sciences. 2. Outer space--Exploration.
3. Astronautics. I. United Nations. II. United
Nations Conference on the Exploration and Peaceful
Uses of Outer Space (2nd : 1982 : Vienna, Austria)
QB500.W67   1982   333.7′0999   82-3794
ISBN 0-13-967745-3   AACR2

Editorial/production supervision and interior design
   by *Kathryn Gollin Marshak*
Case design adapted by *Debra Watson*
Jacket design by *Diane Saxe*
Jacket photograph courtesy of *NASA*
Manufacturing buyer: *Gordon Osbourne*

Printed in the United States of America

10 9 8 7 6 5 4 3 2 1

ISBN 0-13-967745-3

Prentice-Hall International, Inc., *London*
Prentice-Hall of Australia Pty. Limited, *Sydney*
Prentice-Hall Canada, Inc., *Toronto*
Prentice-Hall of India Private Limited, *New Delhi*
Prentice-Hall of Japan, Inc., *Tokyo*
Prentice-Hall of Southeast Asia Pte. Ltd., *Singapore*
Whitehall Books Limited, *Wellington, New Zealand*

CONTENTS

FOREWORD

Man's interest in space extends back to prehistoric times. The sun, the moon, and the stars appear in old rock paintings. Astronomical observatories were built in many parts of the world long before the invention of the telescope. Thousands of years ago man learned to navigate with the help of various celestial objects. Through observations on lengths of shadows cast by a vertical stick at two places separated by a known distance, ancient Egyptians deduced that the earth was round and that its radius was about 6000 km. The beginnings of physics, the most fundamental of modern sciences, are closely associated with attempts to understand the orbits of planets. Heavenly objects were always a part of man's global consciousness. They intrigued him, made him curious, and oftentimes, made him afraid, particularly when the sun or the moon was blocked out during an eclipse. Without a detailed understanding of planetary motions he learned to forecast eclipses and conjunctions. He looked for causal connections between otherwise unexplained events on earth (and differences in the state and progress of individuals) and associated star clusters and configurations of planets. This was the foundation of the still flourishing industry of astrology.

Through the development of telescopes and other instruments and the increasing understanding of the structure of matter and physical laws, the science of astronomy became one of the most popular and productive branches of the human quest for knowledge and understanding. It is no wonder, therefore, that when the arrival of the space age made it possible to look at the heavens from outside the thick blanket of the atmosphere, the astronomers were thrilled with the prospect of opening so many new windows. Thus during the last 25 years the progress in the science of astronomy has been phenomenal. However, the exploration has just begun and the future is likely to be far richer than the past.

Astronomy, until recently, has been nothing but a remote-sensing of the heavens. The transport systems of the space age have now provided the opportunity of looking at the planets, the sun, and the moon not only from outside the atmosphere, but also at close range. The moon and some of the planets have actually been visited and their material analyzed. Many *in-situ* studies of planetary phenomena have brought new understanding about the interaction between the sun and its satellites. In spite of the great impact of these voyages of discovery, it is already obvious that the adventure has barely begun and that many lifetimes of intense activity would be required even to approach the degree of knowledge to which man does and should aspire.

While astronomy and space science had progressed significantly even before the arrival of the space age, the earth-bound uses of space technology have come as a revolutionary new innovation. Man-made objects in the sky can provide active mirrors for radio signals to put large parts of the earth in instant communication with each other without the need for proximity or developed terrestrial networks. This has revolutionized the mode and intensity of communication for all countries and has a very special significance for those parts of the world where the painstaking and expensive activity of establishing terrestrial communication systems has not yet taken off. The new communication systems can be developed with special configurations to address problems of education and development in a manner which was impossible in an earlier age. Indeed, the new possibilities demand that there be a radical rethinking about the institutional arrangements involving communication, extension work, education, and development. Some of this has started, but much innovation lies ahead. On a global scale, distance stands abolished; going via synchronous satellite, the cost and quality of the signal are independent of the physical separation on the ground. Should this method of communication become widely accessible and appropriately reciprocal, the concept of neighborhoods would be completely altered. If origination of such communication is confined to a few centers of influence and power, the new technology could lead to increased domination and homogenization. On the other hand, it is in fact easier and less expensive than before to provide initiative and voice to groups who were never heard from before. New technology has given birth to new possibilities. But these possibilities will become practical only through positive institutional and organizational efforts, in addition to a great number of technical innovations on the ground.

Observation of the earth from space has provided powerful tools to monitor the weather and other hydrometeorological phenomena on a continuous basis and, what is more important, simultaneously over large parts of the earth. It also enables the survey of earth features related to geological structures, crops, forests, ocean currents, man-made pollution, etc.—an activity which is clearly crucial to sustaining and improving the quality of life of the ever-growing human population. That such techniques can be used effectively has already been demonstrated, although the operational forms of this activity have yet to emerge.

There are a great many other uses for which space technology is eminently suited. Artificial satellites can be and are being used for ship-borne communication, navigation, and geodesy. Very soon they will also be routinely used for air traffic control and for search and rescue missions all over the earth.

As the multifarious applications of space technology come into being, there will be a need to develop new and appropriate organizational structures to realize its full promise, as well as new means of regulating some of its activities. This has already begun through various international organs. There is a distinct possibility that this new technology with its very special attribute of unifying the world may yet be used predominantly by countries which are already developed industrially. There are, however, many elements of this technology and many aspects of its usage which could be adopted and taken over by even the least industrially advanced countries. There is an urgent need to examine the manner in which this could be done and to enhance the capabilities of various countries so that they should not only begin to derive benefits, they should also begin to contribute to its development.

Beyond the adventure of space science, the understanding of the universe at large, and the evolution of the solar system, man, and life itself, beyond the practical applications of space systems which encompass the whole globe, in communication, meteorology, earth observation, etc., the coming of the space age marks, in my mind, the infusion of a new ethical input into man's progress toward his distant destiny. The nature of this activity, the questions it deals with, and the global character so intrinsic to its very substance should influence the very consciousness of man. It should establish a new relatedness of all things living or inert, near or far, and especially of all humans who inhabit this small, beautiful, lonely planet, emphasizing the inescapable fact that happiness and misery, or affluence and deprivation, cannot for all time stay separate on this earth. I believe that man's entry into the space era provides, on the one hand, a set of desirable and ethical goals, and on the other the means to fulfill them. Both these aspects should govern the conduct of future space activities. I hope they will, and in particular, that the deliberations of the United Nations Conference on the Exploration and Peaceful Uses of Outer Space will help to create such awareness.

Yash Pal
Secretary-General
Second United Nations Conference
on the Exploration
and Peaceful Uses of Outer Space

PREFACE

The United Nations has been a forum for international discussions concerning space activities since the first man-made earth satellite was launched in 1957. Efforts were made at that time to ensure that space would be used exclusively for peaceful and scientific purposes. In 1959, the Committee on the Peaceful Uses of Outer Space, with a Scientific and Technical Sub-Committee and a Legal Sub-Committee, was established as the focal point of United Nations activities in this field. A fundamental principle that has guided United Nations activities was declared by the General Assembly in 1961: "The exploration and use of outer space should be only for the betterment of mankind and to the benefit of States irrespective of the stage of their economic or scientific development."

In 1968, the United Nations convened the first Conference on the Exploration and Peaceful Uses of Outer Space in order to assess the practical benefits that could be derived from space activities and to find practical means to enable all countries to share in these benefits. The results of the Conference were summarized by the then Secretary-General of the United Nations, U Thant:

Discussions in the Conference indicated that these practical applications could assist materially in alleviating some of the economic and social problems created by the explosive growth of population, the serious shortage of food, the spread of disease—problems of great concern to a vast majority of mankind. At the same time, the Conference highlighted the fact that, because of their complex, expensive and specialized nature, these tools of outer space are known only to a few nations. The developments in space science and technology have thus far benefited most of those countries which are already far ahead in the economic and social time-table of the world.

Space activities have developed rapidly since 1968, in terms of both the technology and the applications throughout the world. Well over one hundred countries now use satellites for international or domestic communication and have used data from meteorological and remote sensing satellites for environmental surveys. A rapidly growing number of countries are acquiring their own satellites. Within the current decade these satellites will be broadcasting television programs directly to home receivers and will be routinely making detailed surveys of natural resources. The unrealized potential of space technology for economic and social development, particularly in developing countries, is still very great.

In light of these developments and prospects, the United Nations decided to convene the Second United Nations Conference on the Exploration and Peaceful Uses of Outer Space, or UNISPACE 82 as it is commonly called, from 9–21 August 1982 in Vienna, Austria. The purpose of the Conference was to review recent developments in space and to consider how international cooperation in space activities might be improved.

The member countries of the United Nations felt that there was a need to review and assess developments in space science and technology. Many countries also felt that there was a need for greater international participation in space activities through the United Nations. At the Conference, representatives from these countries discussed their experiences and the actual and potential impact of space technology, and assessed the effectiveness of existing programs for developing and using space technology.

The Conference examined the particular requirements of the developing countries with respect to education and training, equipment, methods for applying space techniques to economic and social problems, and the organizational problems related to the use of new information systems. New technology that will provide new information and services in the coming decades was discussed, as were the implications of these developments for national development and international co-operation. Finally, member countries generally felt that efforts must be made to increase awareness in the general public and in managers and planners of the capabilities and implications of space technology and applications.

As part of the preparations for the Conference, the United Nations

Outer Space Affairs Division, with the assistance of international scientific organizations and individual scientists, compiled a series of Background Papers to assist countries in their preparations for the Conference. A total of 177 distinguished scientists from 28 countries, including the foremost international experts in many different disciplines, contributed to these documents which have been distributed to the 157 member countries of the United Nations to assist them in their preparations for the Conference.

In order to make the Background Papers available to a wider audience, they are now being published in book form. The material has been slightly revised to incorporate new information that has become available since publication of the UNISPACE documents.

The topics covered correspond to the agenda for the Conference and to the division of the work of the Conference among three main committees. Part I includes four chapters which constitute a general review of the state of space science and technology, an assessment of applications, and a look at the developments that are likely to occur in the next decade. Parts II and III include five chapters which study several of the key issues to be dealt with by the Conference and consider the social and economic implications of the use of space technology. Part IV includes three chapters which survey the activities of international organizations concerned with space programs.

Such a book, compiled from independently prepared reports, each with the participation of numerous experts from many countries, can both benefit from and suffer from a wide diversity of viewpoints. All aspects of the peaceful uses of space are considered, and the material has been extensively reviewed for accuracy. The different attitudes toward space science and technology expressed or implied in different chapters reflect the diversity of opinions in the international community. This book, therefore, is not intended to express the official opinions of the United Nations, but should provide the reader with a comprehensive and authoritative review of space activities and international cooperation in space.

Part I
SPACE SCIENCE,
TECHNOLOGY,
AND APPLICATIONS

Chapter 1

CURRENT AND FUTURE

STATE OF SPACE SCIENCE

ABSTRACT

This chapter is one of four general review chapters. It was prepared with the assistance of an international team of experts organized by the Committee on Space Research (COSPAR) of the International Council of Scientific Unions (ICSU) (Sections 1.1 through 1.7) and by the International Academy of Astronautics (IAA) of the International Astronautical Federation (IAF) (Section 1.8).

The purpose of this chapter is to review man's knowledge of outer space and to survey the contribution that space activities can make to other scientific fields. Where appropriate, historical knowledge is reviewed and likely future directions of space research are projected. Sections 1.1 through 1.4 review man's knowledge of outer space, from earth's upper atmosphere through the solar system to the most distant observable regions of the universe. Sections 1.5 through 1.7 review the contribution that scientific experiments using space vehicles can make to man's understanding of geodynamics and geodesy, materials science, and biology and medicine. Finally, Section 1.8 examines the search for extraterrestrial intelligence (SETI). Scientific studies from space of the earth's surface and lower atmosphere are examined in Chapter 3, and the technology for acquiring some of the knowledge reviewed in this chapter is reviewed in Chapter 2.

INTRODUCTION

At a time when space research has been an experimental science for almost 25 years, it is appropriate to provide an overall review of the subject. Not only does this volume attempt that impossible task, but it also tries to point out some directions in which space research may proceed in the future.

Every space experiment consists of designing and building sophisticated instrumentation to be launched into space by a rocket. The instrumental package either follows a parabolic trajectory in the earth's gravitational field, for about ten minutes, or goes into orbit around the earth, as an artificial satellite, encircling the globe in an hour and a half or more. The observations made in space are encoded and transmitted to the ground over a radio telemetry link. The signals are then decoded and analyzed, often using a computer. They are studied and interpreted by research scientists.

A synthesis of the results obtained in this way is presented in the various sections of this chapter. The discussion is pitched at a level for the metaphorical "intelligent layman," although it must be admitted that he must not be distracted by the manner in which scientists communicate their results to others.

Some appreciation and knowledge have to be assumed of basic physics. This is this science of "how things tick," not only in the environment where man finds himself but also in nature generally. At the heart of physics is the concept of energy. Energy may take several forms, such as kinetic energy (the energy of directed motion of a moving object), heat (the energy of the random motion of atoms and molecules), or waves (the energy of particle motion in a sound wave, or in a wave on the surface of a liquid, or the energy associated with electromagnetic radiation). Visible light, with a wavelength between 400 nm (for blue) and 700 nm (for red) and with a corresponding frequency between 10^{15} Hz and 4×10^{14} Hz, constitutes radiation in a small part of the electromagnetic spectrum. Electromagnetic radiation occurs over a much wider range of wavelengths, from radio waves with a wavelength of 100 km at a frequency of 3 kHz in the VLF band, or of 3 m at 100 MHz in the VHF band, to the microwave part of the spectrum, with a wavelength of 1 mm at a frequency of 300 GHz, and to the infrared. Beyond the visible, to higher frequencies, are ultraviolet radiation, where the wavelength is 100 nm and the frequency 3×10^{15} Hz, X-rays, and gamma rays. Gamma rays are the highest energy photons, that is, "packets" of electromagnetic energy, observed in the universe.

Much of the discussion in this chapter is concerned with the interaction between radiation in the electromagnetic spectrum and matter, that is, atoms in which negatively charged electrons move around a positively charged nucleus.

The number of protons in the nucleus is equal to the number of encircling electrons; there is a comparable number of electrically neutral neutrons in the nucleus too. If the photon has sufficient energy, an electron can be removed from its atom in this interaction; in this case, an ionized gas, or plasma, results. More than 99 percent of the matter present in the universe is believed to be in the plasma state.

The style of this chapter is similar to that of this introduction; words and phrases between a pair of commas or in parentheses are often solely explanatory. Further explanations are provided in the appendices. Appendix 1.1 discusses some important physical quantities and the units in which they are measured; the reader may find it helpful to start with this background information. Appendix 1.2 lists the chemical symbols and Appendix 1.3 lists the acronyms and abbreviations used in this chapter.

1.1 THE IONIZED ATMOSPHERES OF THE EARTH AND PLANETS

1.1.1 The Earth's Ionosphere

The ionosphere is the part of the earth's upper atmosphere that reflects radio waves. It has this property because it is partly ionized; it contains free electrons and ions that are produced by photodissociation, that is, by the action of ultraviolet light and X-rays, mainly from the sun, on the tenuous air of the upper atmosphere. Even by day, only a small fraction of the air is ionized, roughly one part in 10^8 at 100 km height and one part in 10^3 at 300 km height. It is only at heights above 1,000 km that a substantial proportion of the air is ionized. Nevertheless, the mixture of negatively charged electrons and positively charged ions, a plasma, has certain electrical properties that can bend radio waves, or even reflect them completely. This property makes possible long-distance radio communication, and it was Marconi's transatlantic signaling in 1901 that established the existence of a reflecting layer in the upper atmosphere. However, it was the systematic experiments of pioneers such as Sir Edward Appleton, in the years around 1925, that really began the scientific study of the ionosphere. Today the ionosphere is studied by radio experiments involving the reflection or scattering of radio waves and also by experiments involving the transmission of radio signals between spacecraft and earth. Experiments using rockets and satellites have given detailed information about the ionosphere not obtainable from the ground. Nowadays, sophisticated radar systems are used to probe the ionosphere in detail at a few sites, while simpler, extensively used ionosondes sound the ionosphere regularly (usually hourly), at over a hundred locations worldwide, in order to produce information for radio communications users and scientists.

Ionospheric science, which remains a very active subject internationally, has the dual objectives of measuring and predicting the ionosphere's variations for the benefit of radio communications engineers and of using knowledge of the ionosphere to help understand the physics and chemistry of the upper atmosphere, an important part of man's environment.

     1.1.1.1 Regions of the ionosphere. The important parts of the ionosphere extend from about 70 km height up to several hundred kilometers. This height range is divided into regions, conventionally lettered D, E, F, in a way originated by Appleton (Figure 1.1). Within each region, a graph of electron density versus height, the electron density profile, displays one or more layers, although some layers are more accurately described as ledges. The reflecting properties of a layer depend on its maximum (or peak) electron density which is related to the critical (or penetration) frequency. Radio waves incident vertically on any layer will penetrate it if their radio frequency exceeds the critical frequency; otherwise they will be reflected. The upper part (F2) of the F layer, the most important for communications because it normally has the greatest electron density, is found at a height that is usually between 250 and 400 km, depending on time and place.

     In spite of the development of satellites, the intensive use of the ionosphere for communications seems likely to continue indefinitely. The ionosphere normally reflects the medium- and high-frequency (MF and HF) waves used for broadcasting; it can bend or scatter very high-frequency (VHF) waves to some extent, but it has little effect on microwaves, which travel straight through it.

     The various layers shown in Figure 1.1 exist because the spectrum of ionizing radiation from the sun is complex, and so is the chemistry of the atmosphere. As a result, the rates at which the plasma is produced by radiation and destroyed by the recombination of ions and electrons vary with height and time in a complicated way. At greater heights, the ionosphere merges into the magnetosphere, the outermost part of the earth's atmosphere, which contains a very tenuous plasma permeated by the earth's magnetic field. Charged particles from the magnetosphere are at times precipitated into the upper atmosphere at high latitudes, causing the spectacular polar aurora (or northern lights), and creating an additional source of ionization in the high latitude ionosphere.

     Figure 1.2 shows the chemical composition of the daytime ionosphere from 100 to 400 km. As previously mentioned, the charged particles (electrons e^- and the ions O^+, N^+, NO^+, O_2^+, N_2^+) are vastly outnumbered by the neutral gas atoms and molecules (O, N_2, O_2). Whereas in the E region the ions are mostly molecular (NO^+, O_2^+), at greater heights the ions are virtually all atomic oxygen: the O^+ and e^- curves in Figure 1.2 merge above about 200 km. The transition between molecular ions and atomic ions takes place around 170 km by day, the level of the F1 layer; an analysis of the chemical situation shows that this circumstance can account for the fact that the F1 layer (or rather ledge, as in Figure 1.1) is only evident by day and under certain conditions. At night the transition is higher, above 200 km, and there is no F1 ledge. At heights well above the F2 peak (not

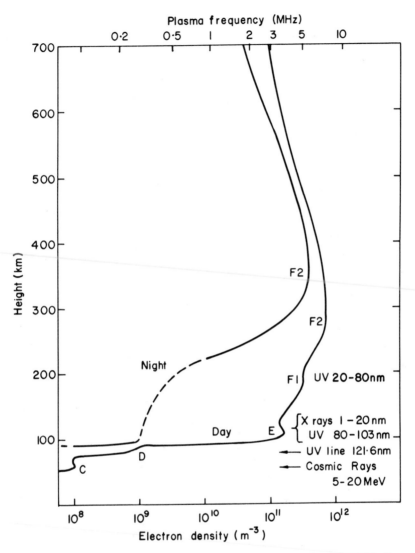

Figure 1.1 The distribution of electron density with height in the ionosphere. The F region lies above 150 km, the E region between 90 and 150 km, and the D region between 65 and 90 km. Shown at the right are solar radiations mainly responsible for producing the ionized layers within these regions. The diagram is compiled from rocket and satellite data obtained in the Eastern U.S.A. in 1968 (above 100 km) and radio data obtained in Great Britain (below 100 km); it may be taken to represent typical summer conditions at middle latitudes. The top scale, plasma frequency, approximately corresponds to the highest radio frequency that would be reflected at any height if incident vertically on the ionosphere. (From W. C. Bain and H. Rishbeth, *Radio & Electronic Engineer,* 45, 3, 1975, published by Institution of Electronic & Radio Engineers, London.)

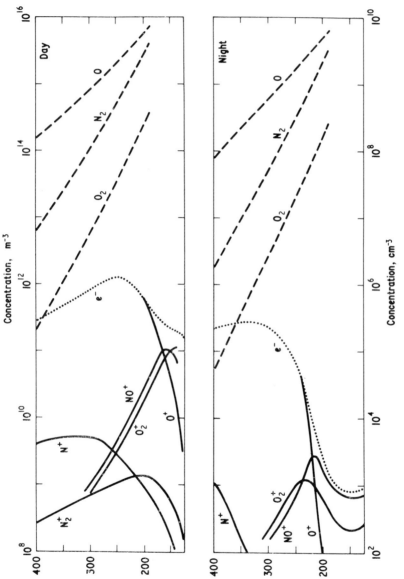

Figure 1.2 The chemistry of the ionosphere between 125 and 400 km. The full lines represent the concentrations of ions obtained from rocket experiments in France in February 1969; the dotted line shows the electron density obtained at the same time by radar. (The O^+ curve merges with the electron e^- curve above about 200 km.) The dashed lines show the concentration of the principal neutral gases, obtained from a model based on satellite data. (From H. Rishbeth, P. Bauer and W. B. Hanson, *Planetary & Space Science*, 20, 1287, 1972.)

shown in Figure 1.2), O^+ ions give place to the lightest ions, H^+ and some He^+. The chemistry of the F region is now well enough understood for accurate measurements of the various constituents, obtained from satellite-borne instruments, to be used to determine the rates of certain chemical reactions. In other words, the ionosphere can be used as a chemical laboratory as well as a radio laboratory.

An international reference ionosphere model of electron and ion densities and temperatures is now being developed by the International Union of Radio Science (URSI) and the Committee on Space Research (COSPAR).

As the solar ultraviolet radiation is absorbed during the ionization of the atmosphere, its energy is passed to the atmospheric molecules and atoms; it also heats the ionospheric electrons. The electron temperature is determined by the heat input, heat loss by collisions, and heat conduction. Elastic and inelastic collisions of the electrons with atmospheric particles cause the electrons to lose heat. Heat conduction occurs primarily by electrons moving along the lines of force of the earth's magnetic field. The physical and chemical processes occurring in the ionosphere are sometimes strongly dependent on the electron temperature. Therefore, the electron temperature plays an important role in the heat budget as well as in various physical and chemical processes in the ionosphere.

1.1.1.2 Ionospheric theory.

As in other branches of science, progress in ionospheric science comes from a combination of experimental and theoretical work. Many kinds of equations occur in the theory. Correlation analysis and statistics are useful aids in the treatment of ionospheric data to find out the relationships between different phenomena and to shed light on their causes. The theory of random processes is used to study the small-scale irregular structure of the ionosphere, which has a considerable influence on the propagation of radio waves. Furthermore, wave theory has been greatly developed to account for the propagation of many kinds of waves (radio waves, plasma waves, and atmospheric waves) in the upper atmosphere.

The three most important equations of ionospheric theory are:

(a) *The continuity equation,* expressing the conservation of mass or of particle concentrations:

(Rate of change of concentration) = (production) − (loss) + (transport)

where transport includes diffusion and motion produced by winds and electric fields.

(b) *The equation of motion,* expressing the conservation of momentum:

(Rate of change of momentum) = (applied force) − (friction) + (transport)

where transport includes viscosity, friction results from collisions with other kinds of particles, and the equation includes effects of the earth's rotation on the motions.

(c) *The energy equation,* expressing the conservation of energy or heat content:

(Rate of change of energy) = (heat gain) − (heat loss) + (transport)

where transport includes conduction and convection, and gain and loss include the absorption and emission of radiation. These equations may be applied to the electrons in the ionosphere and also to any of the numerous kinds of ions and neutral particles. The transport terms, which represent the effects of a flow of particles in (a), of momentum in (b), and of energy in (c), render the equations mathematically complicated so that simple solutions can only be obtained by making drastic and usually unwarranted simplifications. It is better to solve the equations numerically with a computer. With the large data storage capacity needed for the solution of complicated equations, the progress in this kind of ionospheric theory has depended very much on the development of computers and techniques for using them.

Given solutions to these equations, there are two basic ways of using them in ionospheric theory, which may be termed analytic and synthetic. In the analytic approach, experimental data (such as observed variations of electron concentration) are inserted in the equations and suitable analyses carried out to yield values of ionospheric parameters, such as rates of production or loss coefficients. In practical situations, these parameters cannot be found accurately, even in the clear-cut situation in which a solar eclipse cuts off the ionizing radiation and causes the ionosphere to decay. In the synthetic approach, the equations are solved to yield theoretical values of concentration, velocity, and temperature. Examination of the solutions often provides considerable insight into how the ionosphere actually works. Both approaches have contributed to the present understanding of the complexities of the F region, an understanding which, though still incomplete, represents a satisfying combination of experiment and theory.

Though a vast amount of knowledge about the upper atmosphere has been gained in the last few decades, there is still much to learn, particularly about the chains of cause and effect that link the solar wind, the aurora, and the winds in the ionosphere. Many experiments with rockets and satellites, and the building of large radar systems, are of particular help in this regard. Studies of the ionosphere give much information on the vast heat engine that is the upper atmosphere.

1.1.1.3 The F region.

The F1 layer varies in a regular way with the sun's elevation above the horizon and also with the 11-year sunspot cycle, as the sun's output of ionizing radiation varies. However, apart from exhibiting a general correlation with the sunspot cycle, the F2 layer does not behave in a regular way. Some of its anomalies [(1), (2), and (3) in the following list] are evident in a contour map (Figure 1.3) which shows how the F2 layer critical fre-

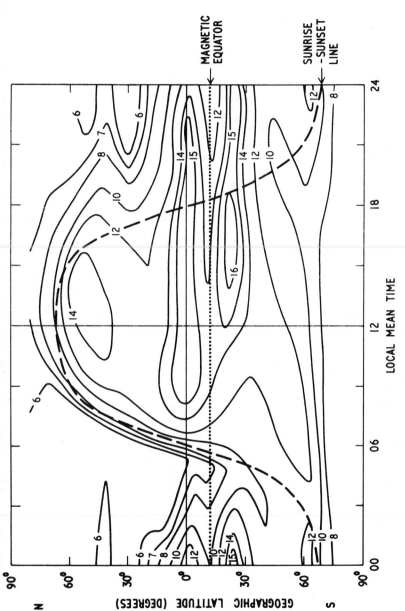

Figure 1.3 Contour map of the F2 layer critical (or penetration) frequency (MHz) for December 1957, for the longitude of North and South America. Some features are symmetrical about the magnetic equator, which is about 12° south of the geographic equator. At middle latitudes, the daytime electron density is greater in the northern (winter) hemisphere than in the southern (summer) hemisphere. These features can be explained in terms of the chemistry and motions of the ionosphere. (Map drawn from data supplied by Central Radio Propagation Laboratory, U.S.A. Department of Commerce.)

11

quency varied with latitude and local time in December 1957, near the maximum of the solar cycle:

1. At noon, the maximum electron density is greater at northern middle latitudes (winter) than at the corresponding southern latitudes (summer). This anomaly is believed to be due to changes in the neutral chemical composition that affect the rates of production and loss of ionization.
2. This maximum varies regularly with the time of day in winter (with a maximum around noon) but not so regularly in summer.
3. By day there is a minimum electron density at the magnetic (not geographic) equator, with greater values of electron density to the north and south of the magnetic equator.
4. During magnetic storms, when the earth's magnetic field is disturbed as a consequence of charged particles emitted from disturbances on the sun, the electron density may be abnormally low or sometimes abnormally high, with corresponding effects on radio communications.
5. At auroral latitudes the F2 layer tends to be more variable and irregular than elsewhere.

Though (5) is associated with the precipitation of charged particles from the magnetosphere, the others all depend, to some extent at least, on motions of the plasma and the neutral air.

Motions in the ionosphere are conventionally termed winds if they refer to the neutral air and drifts if they refer to the plasma of ions and electrons. In the F region, winds are caused by:

1. Daytime heating of the F region by the ionizing radiation, which sets up pressure gradients causing winds to blow from the dayside to the nightside of the earth.
2. Tidal effects produced by heating in the lower atmosphere, transmitted up to the F region in the form of waves of 24-hour, 12-hour, and other periods.
3. Heating at high latitudes by charged particles and electric currents fed from the magnetosphere, largely associated with auroral displays.
4. Drifts at very high latitudes arising from the flow of the solar wind past the earth's magnetic field.

In the F region winds and drifts interact with one another in a rather complicated way, determined by the collisions between ions and neutral gas particles and by the influence of the earth's magnetic field on the ions and electrons. The details of this interaction cannot be explored here, interesting as they are; nor can the very complicated links between plasma emitted by the sun (the solar wind), the earth's magnetosphere, and the earth's ionosphere that are involved in (3) and (4).

Winds and associated drifts affect the F region electron density, either directly because they modify the distribution of ions and electrons or indirectly because they can affect the chemical composition of the air, and hence the production and recombination processes of the plasma. Such chemical changes take place on a seasonal basis as a result of (1), and on time scales of hours and days as a result of disturbances (3). Suffice it to say that some explanations exist for all the anomalies (1) through (4) in the first list above in terms of (1) through (4) in the second list above.

1.1.1.4 The E region.

The E region is part of the lower ionosphere which also contains the D and the F1 regions, neighboring on it below and above, respectively. The altitude 90 km is usually accepted as the boundary between the E and D regions. The upper boundary of the E region is not so well defined and is usually taken as being at about 150 km. From the height distribution (or vertical profile) of the electron density during the daytime shown in Figure 1.1, it can be seen that between 90 and 100 km there is a strong increase of electron density. Sometimes, but not often, the electron concentration above the E region is lower than in the region itself. Thus a valley occurs between the E and the F1 regions, as in Figure 1.1. In this case, there is a maximum electron density near 110 km.

The bright line of the solar spectrum, termed hydrogen Lyman-β and having a wavelength of 102.6 nm, plays an important role in E layer formation. This line can ionize oxygen but not nitrogen. Radiation at slightly shorter wavelengths ionizes molecular nitrogen forming N_2^+ ions; these are quickly transformed into nitric oxide ions (NO^+) through chemical reactions with the neutral atmosphere. As a result, the chemical composition of the ions (the ion composition) of the daytime E region is about 70 percent NO^+ and about 30 percent O_2^+.

After sunset the vertical distribution of electrons in the E region changes drastically. The absolute value of the electron density decreases at all heights, by as much as a factor of 100. This is much more than the diurnal variation of the F2 maximum, where it is a factor between 3 and 10.

The nighttime profile is, as shown by the dashed line in Figure 1.1, highly variable. These are maxima and minima which often are only 5 to 10 km apart. For quiet geomagnetic conditions, the electron density at the maximum of the nighttime E layer is about 10^9 electrons/m³. Modern photochemical theory of the ionosphere explains the presence of such comparatively large values during the night by the ionization of the atmosphere by solar ultraviolet radiation scattered by the geocorona (the outermost part of the earth's gaseous cover at several radii from its surface) finding its way to the night side of the atmosphere.

In some cases, however, electron densities as high as 3×10^{10} electrons/m³ are observed in the E region during the night. Such cases are often associated with disturbances of the geomagnetic field; charged particles with energies of several kiloelectronvolts (keV), precipitating from the earth's magnetosphere produce additional ionization in the E region. As far as the ionic chemical composition is concerned, the nighttime E region does not differ substantially from the daytime one; the proportion of NO^+ ions may be about 80 to 95 percent, the rest being mainly O_2^+ ions. Sometimes in the vertical profile of the E region one or more narrow layers appears. The half-width of such a layer may be as small as 1 to 2 km, with the concentration within this sporadic-E layer differing by a factor of 10 or more from that outside. These layers are observed both by day and night. The ion composition of these sporadic-E layers is predominantly of the alkali metals and of iron and silicon; these elements typify the chemical composition of meteoric matter. Thus it is suggested that the destruction of micrometeors in the upper atmosphere may be the source of the major ions, with NO^+ and O_2^+ abundances being reduced to a few percent of the total.

Sporadic-E layer formation is explained as follows. Neutral atmospheric winds in the earth's magnetic field produce a vertical motion of charged species. If there are winds blowing in different directions at relatively close levels, termed wind shear, then at altitudes where the direction changes, maxima and minima of ionization will be formed, and a layer type structure will appear in the vertical profile. This mechanism operates most effectively for atomic ions having long lifetimes between their formation and their disappearance through recombination, because it takes dynamic processes some time to move ions and electrons and to form the layer. After sunset, the lifetimes of ions and electrons increase because of the general decrease of electron and ion densities, so the night offers more favorable conditions for the formation of these layer-type structures.

A sporadic-E layer that occurs over a sufficiently large area will reflect radiowaves of frequencies used for radio communications. This makes possible communications at frequencies and along paths not normally useable. Further work on the morphology and physics of the formation of sporadic-E layers is needed before they can be used effectively for radio communication purposes.

1.1.1.5 Equatorial electrojet.

Observations of an abnormally large amplitude of the diurnal variation of the horizontal component of the earth's magnetic field at Huancayo, Peru (near the dip equator) in 1937 led to the discovery of the phenomenon now known as the equatorial electrojet. This is a narrow band of intense electric currents flowing in the E region near the dip equator. The study of the spatial distribution of these currents is now being carried out with ground-based magnetometers in South America, India, and Africa. Backscatter radars at Jicamarca, Peru, and Thumba, India, are being used to study the height distribution of electron density irregularities and their movements, and hence the fine structure of the equatorial electrojet currents. Rocket-borne magnetometers are being flown from the Peruvian coast and from Thumba to study the vertical distribution of the electrojet currents. Langmuir probes and resonance probes on rockets study the electron density profile, the spectrum of the electron density irregularities and their production. Satellite-borne magnetometers have also been used to study the global distribution of these currents.

The normal daytime electrojet is mainly an eastward flow of electric current. Starting with a minimum value in the morning, the current increases steadily and attains its peak value around noon. Thereafter, it decreases slowly and changes its direction during the night. On some days, the behavior of the electrojet currents departs from normal, and the currents reverse for a period of a few hours; this is referred to as the counter electrojet. During magnetically disturbed periods, such reversals may take place on time scales as small as a few tens of minutes.

Many investigations have shown that the electrojet current system is closely related to the worldwide quiet day ionospheric current system; however, the former is much stronger than the latter. This can be explained by the enhanced conductivity near the magnetic equator due to the inhibition of vertical currents. Some features of the electrojet current system at the equator indicate that it is different from the current system elsewhere. For example, the electrojet is confined to a very narrow range of altitude, its vertical half-width being about 14 km. The current density can be as large as 10 amp/km². The north–south extent of the electrojet is a few hundred kilometers on either side of the magnetic dip equator. Many models of the equatorial electrojet have been developed, but they are based on simplifying assumptions. From theoretical considerations, the peak of the electrojet currents should be at 103 km, as illustrated in Figure 1.4. The large discrepancy between the observed and calculated altitude of the electrojet is not yet explained.

The electric fields driving the electrojet current can also give rise to electron density irregularities in the E region through various plasma instability mechanisms. Electron density irregularities are also produced through the interaction of gravity waves with the ionospheric plasma and through neutral turbulence. The presence of these irregularities is likely to affect the electrical conductivity of the plasma, and thereby also the electrojet currents.

The electrojet currents are mainly driven by electric fields generated through the interaction of large-scale winds, such as tidal motions, and medium- and small-scale winds, such as gravity waves, with the ionospheric plasma in presence of the earth's magnetic field. Short-term variations in the electrojet currents are sometimes related to geomagnetic variations observed in the auroral regions, indicating that a part of the equatorial electrojet is connected with the auroral electrojet.

1.1.1.6 The D region.

The D region is the lowest part of the ionosphere, normally defined as the region between 65 and 90 km above the earth's surface. This region is of interest and importance both to the atmospheric physicist and to the user of radio communications and navigation systems. Studies of the D region became of practical importance when it was realized that high-frequency radio waves, which are reflected from the E and F regions, may suffer significant absorption at altitudes below 90 km, and that low- and very low-frequency waves are reflected in the D region. Communication in the high

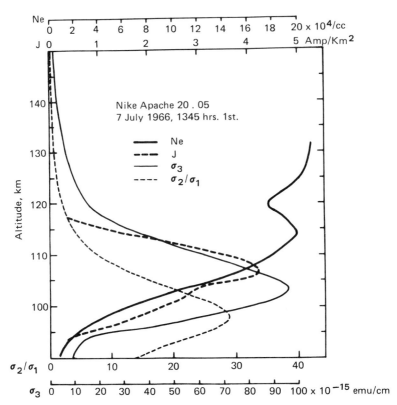

Figure 1.4 Electron density (Ne) profile measured with a rocket-borne Langmuir probe and a resonance probe, electrojet current density (J) measured by a rocket-borne magnetometer, and calculated electrojet current (represented by σ_3). (Satya Prakash, in *Planet. Space Sci.,* 20, 47, Pergamon Press, 1972.)

frequency band (HF, 3 to 30 MHz), using radio signals reflected from the iono-sphere, can be inexpensive and efficient over long distances. Reliability is, however, often limited by ionospheric absorption, particularly in high latitudes and in middle latitudes in winter.

The low-frequency (LF, 30 to 300 kHz) and very low-frequency (VLF, 3 to 30 kHz) bands are widely used for navigation and timing systems. Examples are the Omega, Decca, and Consol systems. These long waves are reflected low in the D region and are less susceptible to absorption than HF waves. However, since the height of reflection determines the travel time of a radio wave from trans-mitter to receiver, fluctuations of this height will limit the accuracy of timing and positioning. In order to correct for these variations, it is important to map the normal diurnal and seasonal variations in the D region on a global scale and to measure and predict the effects of different types of disturbances in the lower ionosphere. In general, a D region disturbance will intensify HF absorption and lower the reflection height of LF and VLF waves, as illustrated in Figure 1.5.

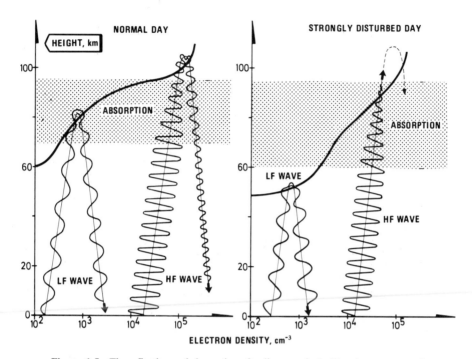

Figure 1.5 The reflection and absorption of radio waves in the D region on a normal and a strongly disturbed day. The solid curves are typical electron density profiles. The disturbance, caused by an enhanced intensity of ionizing radiation, lowers the reflection height of LF waves and completely absorbs the HF waves.

The physics of the D region (Figure 1.6) differs from the physics of the E and F regions in several important respects. Under normal conditions, the D region ionization densities are small. The ionization is formed by X-rays and extreme ultraviolet radiation from the sun and by galactic cosmic rays (Figure 1.1). An interesting feature is the selective ionization of certain minor constituents. For example, the strong hydrogen Lyman-α line (wavelength 121.6 nm) ionizes nitric oxide, NO, in the D region. This mechanism dominates near 80 km, even though nitric oxide is only present in concentrations of one part per 10 million. Nitric oxide is a long-lived constituent produced by photochemical processes mainly in the D and F regions, and it is therefore important to understand the transport mechanisms by which NO is brought into the D region. Such transport is probably partly responsible for the enhanced D region ionization observed at mid-latitudes in winter, thereby causing the high and variable radio wave absorption termed the winter anomaly. In high latitudes, energetic charged particles from the sun and magnetosphere precipitate into the D region, causing strong and variable ionization.

The primary ions formed by solar electromagnetic and particle radiation react with neutral molecules, such as water vapor, to form large complex positive and negative cluster ions. The ion chemistry resulting from this interaction is a

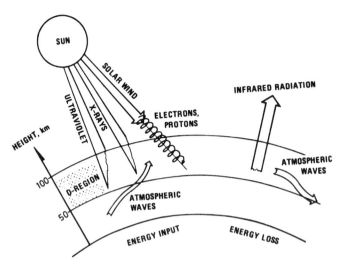

Figure 1.6 The energy input and loss processes that affect the D region shown schematically. Ionizing radiation from the sun and atmospheric waves from the lower atmosphere deposit energy in the D region, whereas atmospheric waves and infrared radiation carry energy away.

special feature of the D region. Because some ion reactions depend on temperature, the state of the neutral atmosphere is important for the ionization balance.

The D region is not easily accessible to instruments; the air there is too dense for satellites and too tenuous for balloons. *In-situ* measurements can only be made using instrumented sounding rockets, but these can be relatively small and inexpensive. Such rockets have provided a wealth of information about the D region, some using simple techniques to measure basic parameters such as ionization densities and others carrying very sophisticated mass spectrometers to give detailed information on the neutral and ionic composition.

Rocket experiments cannot, however, give adequate coverage in time and space for global synoptic studies of the D region. For such purposes ground-based techniques using radio waves reflected from or absorbed in the D region are invaluable (Figure 1.7). The simplest instrument is the riometer (relative ionospheric opacity meter) which measures the intensity of cosmic radio noise penetrating the ionosphere from space. Frequencies between 27 and 40 MHz are commonly used. A D region disturbance will cause absorption of the radio noise and a deviation from the quiet-day curve will be recorded. Absorption of HF waves reflected from the E and F regions is a sensitive indicator of the state of the D region, but it is more difficult to interpret. Two interesting methods should be mentioned: the partial reflection and wave-interaction techniques. The first measures the amplitude of HF waves scattered from irregularities of ionization density; the second measures the nonlinear interaction of two HF waves trav-

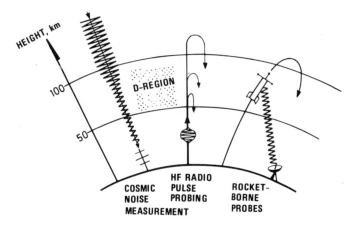

Figure 1.7 Three important methods for studying the D region. From left to right, these are: riometer measurements of the absorption of cosmic radio noise, probing by means of HF radio pulses that are absorbed and scattered from the region, and probing by sounding rockets that transmit in-flight measurements to the ground.

eling through the same region. These techniques are in use for routine observations of D region electron density distributions at a small number of stations throughout the world.

Although satellites cannot survive in the dense air in the D region, remote probing from satellites carrying infrared spectrometers has recently provided information on the thermal structure of the D region.

Although the D region is only weakly ionized, there is a strong coupling between the ionized and neutral constituents, and, through complex photochemical processes, between the incoming electromagnetic and particle radiations and the atmospheric gases. Studies of these coupling processes are important for understanding the energy balance in the upper atmosphere. How do the weather systems in the troposphere, that is below 15 km, influence the upper atmosphere, and is it possible that the upper atmosphere and its interaction with solar electromagnetic and particle radiation can influence weather and climate? These are important and difficult questions, and the answers must be sought through international synoptic atmospheric studies, including the study of the D region as an essential element. The Middle Atmosphere Programme (planned for the period 1982 to 1985) is an example of such a co-operative effort in which intensive studies of the atmosphere between 10 and 100 km will be made. Many experimental techniques will be used, including ground-based, rocket, and satellite probing of the D region. The time is past when single measurements could add substantially to knowledge. What is now needed is mapping of the global "weather" system in the upper atmosphere, using all available resources in a comprehensive effort to understand the basic mechanisms governing ionization densities, composition, temperature, and winds. Such understanding should ultimately lead to models adequate for reliable predictions of short- and long-term changes in the ionosphere, and perhaps even aid in the prediction of climatic changes at the earth's surface.

1.1.2 Planetary Ionospheres

Every body in the solar system responds strongly to processes that occur on the sun. Both shortwave electromagnetic radiation and charged particles from the sun interact with the outermost regions of the gaseous envelope of a planet, its upper atmosphere. This natural barrier prevents the most energetic part of the solar spectrum, the ultraviolet and X-radiation, as well as energetic charged particles and the solar wind, from reaching the planetary surface. Molecules in the upper atmosphere dissociate into atoms and both are ionized to form the planet's ionosphere.

1.1.2.1 Venus and Mars.   Among the terrestrial planets, both Venus and Mars possess ionospheres. But in contrast to the earth whose atmosphere contains much oxygen, the photochemical processes occurring in the upper atmospheres of earth's neighboring planets are principally controlled by carbon dioxide which dominates below about 200 km altitude. Above this level the dissociation products of CO_2, CO, and O, occur; from about 400 to 500 km upward, helium (He) and hydrogen (H) become most abundant.

The radio occultation technique, that is the cut-off of the telemetry signal as a spacecraft passes behind the planet, contributes much to the study of the ionospheres of the other planets. Compared with the earth's, both the Venusian and Martian ionospheres are less dense and closer to the planet. The mass-spectrometry and retarding potential analyzer measurements aboard the PIONEER Venus orbiter and VIKING probes give information on the chemical composition of these ionospheres; developments of theoretical models help the evaluation of the principal processes responsible for ion production and loss and for the dynamics of the ionosphere.

The main peak of the dayside ionosphere of Mars occurs at 135 to 140 km in height, and the maximum electron density amounts to 2×10^{11} electrons/m^3. This is about an order of magnitude less than the F2 maximum of the earth's ionosphere. A minor maximum of 7×10^{10} electrons/m^3 occurs at 110 km. There the principal ion is molecular oxygen, O_2^+; it is produced in a charge exchange reaction of a carbon dioxide ion with an atom of oxygen.

The ionosphere of Venus (Figure 1.8) is mainly composed of O_2^+, whereas O^+ ions prevail above about 200 km. The abundances of other ions in the lower ionosphere (NO^+, CO^+, CO_2^+) are less than 10 percent of the total. The dayside maximum, with an electron density of 4×10^{11} electrons/m^3 at a height of 145 km resembles the F1 layer in the terrestrial ionosphere. There is an extended nightside ionosphere, with its main peak at nearly the same height, and several lower local maxima that exhibit noticeable variations and a slightly increased NO^+ content.

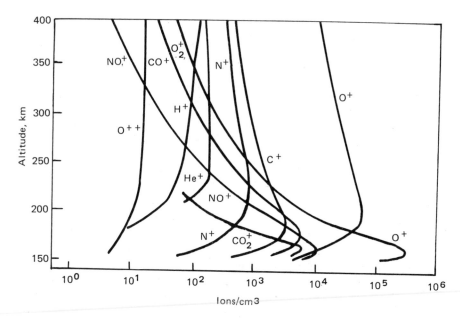

Figure 1.8 Ion composition of the dayside ionosphere of Venus according to PIONEER Venus Orbiter measurements on orbit 12. The position of the main O_2^+ peak is evident at an altitude of approximately 160 km. (S. J. Bauer et al., "Venus Ionosphere: Photochemical and Thermal Diffusion Control of Ion Composition," *Science,* 205, 109–111, Fig. 1, 6 July 1979. Copyright 1979 by the American Association for the Advancement of Science.)

Thus the principal difference between the ionosphere of Mars or Venus and that of the earth is the absence of the F2 maximum which exists because of the interplay between the processes of ionization, recombination, and diffusion. Because O^+ ions react faster with CO_2 on Mars and Venus than with N_2 on earth, the accumulation of O^+ ions into an F2-type layer is prevented on Mars and Venus.

If the planet has either a very weak intrinsic magnetic field (Mars) or none at all (Venus), the solar wind interacts with the ionosphere directly. This leads to an interaction which is very different from that which occurs when the planet has an intrinsic magnetic field and hence a magnetosphere. A boundary, termed an ionopause, is formed where the pressure of the solar wind plasma (which amounts to about 10^{-9} mb at the Venus orbit) is balanced by the thermal pressure of the ionospheric plasma. The position of the boundary varies from 250 to 1,800 km above the surface of Venus, depending on the solar wind, and the dayside ionosphere above the main peak exhibits dramatic variations.

Theoretical treatments of the problem consider electric currents being induced and flowing in the ionopause. This results in an induced magnetic field. Because the time constant for magnetic diffusion is much longer than the characteristic time scale of changes of the interplanetary magnetic field direction, diffusion of this magnetic field into the undisturbed ionosphere should be negligible.

Ionospheric and plasma measurements made aboard VENERA, MARS, PIONEER, VENUS, and VIKING orbiters in the vicinities of Mars and Venus have investigated the heating and thermalization of ions, the formation of a bow shock, and convective motions in the ionosphere. Such processes can explain the recently discovered, unexpectedly high electron and ion temperatures in the ionosphere of Venus, namely, 5,000 and 1,000°K respectively; these are much greater than the upper atmospheric neutral gas (exospheric) temperature. Even more of a surprise is the fact that very high electron and ion temperatures occur in the dark hemisphere, when the neutral temperature is as low as 100°K. Thus a mechanism of permanent energy input to the nightside must be invoked; this appears to be due to soft electron (with energies of tens of electronvolts) precipitation and/or to dynamic exchanges (including O^+ ion transport from the dayside) and to electromagnetic interactions. Large (up to 1 km/s) horizontal drift velocities of ions, in the nightside, are measured. As far as the dayside is concerned, besides the direct ultraviolet absorption, the solar wind might serve as an additional energy source via whistler waves. The analogy with the interaction between the solar wind and a comet can be drawn here.

1.1.2.2 Jupiter and Saturn.

Jupiter possesses a rather extensive ionosphere, though the maximum number density of electrons does not exceed 5×10^{11} electrons/m³. According to VOYAGER radio occultation measurements made during morning and evening conditions, and obtained at the equator and at 67° latitude, respectively, the height of the main electron density peak is 700 and 2,500 km above the upper cloud deck, the maximum electron density being 2×10^{10} electrons/m³ and 2×10^{11} electrons/m³ respectively. The plasma temperature, determined from the measured scale height assuming that the electrons and protons are distributed according to diffusive equilibrium, is between 1,200 and 1,600°K.

The ionosphere of Saturn, measured using the same technique during the PIONEER Saturn flyby, is known to be large in extent but not very dense. The peak of electron density under morning and evening conditions near the equatorial plane is 1.1×10^{10} electrons/m³ at 1,800 km above the "surface" of the mean equatorial radius (61,000 km), and a minor maximum of 9×10^9 electrons/m³ at 1,200 km (Figure 1.9). The relatively low electron density is explained by the ionized particles interacting and recombining with solid particles (possibly ice grains) populating Saturn's rings. This idea is confirmed by the increased electron density in the gap between the C and D rings, the so-called Guerin division. Saturn's moon, Titan, is the largest satellite in the solar system; it is just larger than the planet Mercury. It was believed that Titan could be formed entirely of ice, or of ice with up to 15 percent of rock; however, VOYAGER-1 observations made in November 1980 indicate that its surface is of liquid nitrogen. Titan possesses an atmosphere, part of which could be ionized forming an ionosphere.

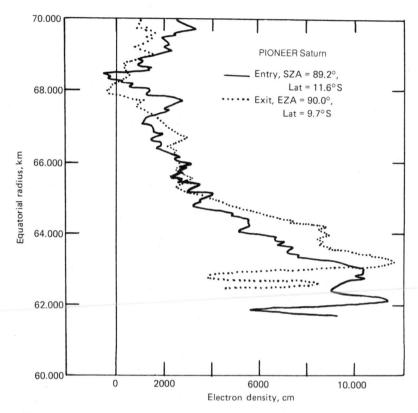

Figure 1.9 Electron density profiles in the ionosphere of Saturn according to PIONEER Saturn radio-occultation measurements near the equatorial plane between 60,000 and 70,000 km. (A. J. Kliore, in "Structure of the Ionosphere and Atmosphere of Saturn . . . ," *J.G.R.*, 85, no. All, November 1980, 5862.)

The ionospheres of both Jupiter and Saturn appear to consist mainly of atomic hydrogen (H^+) ions. The energy sources responsible for the thermal structure of their upper atmospheres and for the formation of their ionospheres are solar ultraviolet radiation, soft electrons of magnetospheric origin, joule heating, and gravity waves. The loss mechanisms are mainly photochemical processes, turbulent (eddy) diffusion, and outflow of H^+ ions along magnetic field lines.

It is interesting to note that ionospheres are also found on two moons, the earth's moon and Io which is the innermost of the four Galilean satellites of Jupiter. The very weak ionosphere of the moon (up to 10^9 electrons/m^3) is close to the lunar surface; it appears to consist mainly of argon ions ($^{40}Ar^+$). Lines in the optical emission spectrum of Io are identified as corresponding to sulphur ions and sodium ions. These ions are probably produced by energetic electrons from the powerful radiation belts of Jupiter hitting atoms that are ejected by Io's volcanoes; Io seems to be the most volcanically active body in the solar system.

1.2 SOLAR SYSTEM PLASMAS AND FIELDS

1.2.1 The Earth's Magnetosphere

The magnetosphere is the region of space surrounding the earth in which the earth's magnetic field has a controlling influence on the behavior of the plasma (ionized gas) which it contains. The word magnetosphere was coined in 1959. It refers to a region very much larger than the earth itself, extending from the ionosphere (the ionized part of the upper atmosphere beginning at an altitude of about 65 km) out to about 60,000 km (10 earth radii) on the sunward side of the earth and, in the other direction, to perhaps several hundred or even thousands of earth radii in a comet-like tail extending away from the sun (see Figure 1.10). The behavior of the plasma is determined by the electric and magnetic fields present; the origin of these fields may be either external or internal to the plasma.

The phenomenon of geomagnetism has been known since ancient times through its effect on magnetized rocks (lodestones) and compass needles. The fact that the earth has a magnetic field similar to that of a simple bar magnet was first recognized by William Gilbert, physician to Queen Elizabeth I. In 1716, Edmund Halley connected this concept with the behavior of the polar aurora, which often has a rayed structure forming a pattern similar to that expected for such a terrestrial dipole magnetic field. He suggested that the auroral emissions result from "magnetic particles" which move along the magnetic field lines. Fifteen years later, de Mairan proposed that solar gas was somehow directed into the polar regions where the auroral phenomenon normally occurs. The associated phenomenon of geomagnetic disturbances was first noted in 1724; the relationship between solar and geomagnetic activity was established in the second half of the nineteenth century. Despite extensive ground-based observations (for example, the International Polar Years of 1882/83 and 1932/33) and extensive theoretical work, progress in understanding the problems of geomagnetic activity and the aurora was relatively slow until the International Geophysical Year in 1957 and the launching of the first earth satellite.

For many years, ionospheric physicists were concerned mainly with problems of ionospheric effects on radio propagation as observed from the ground. There was a natural tendency to ignore questions associated with the extension of the ionosphere into space in the regions above 300 km altitude, which were not directly observable from the ground. In the early 1950s, however, L. R. O. Storey caused a significant change in this attitude with his discovery that the behavior of very low-frequency radio signals called "whistlers," which are associated with lightning flashes, can only be interpreted if the radio wave propagates along geomagnetic field lines out to large distances (25,000 km or 4 earth radii) and if a relatively dense plasma exists in these regions to permit propagation to occur. This plasma, the plasmasphere, constitutes the upper extension of the ionosphere into the magnetosphere. There is an intimate connection between

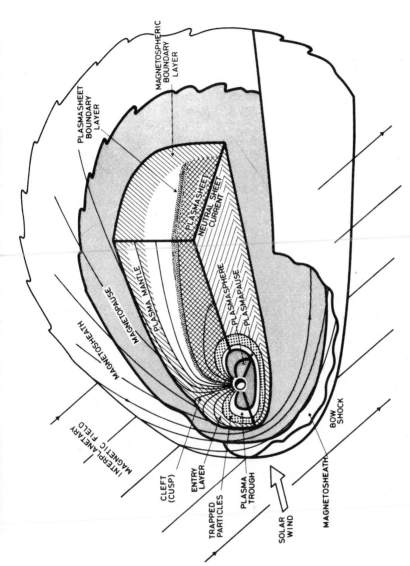

Figure 1.10 The "anatomy" of the earth's magnetosphere. The earth's magnetic field is not only a "trap" for charged particles (electrons and ions) but also a "shield" preventing solar wind plasma and cosmic ray particles having free access to the entire terrestrial atmosphere. Many physical processes operate within the many different regions of the magnetosphere. The terms are explained in the text. (Adapted from W. J. Heilika, *EOS*, 54, no. 8 (1973), 764.)

the ionosphere and the magnetosphere and similar auroral, geomagnetic, and ionospheric phenomena occur in magnetically conjugate regions in both hemispheres of the earth.

1.2.1.1 Discoveries made using spacecraft.

The first major discovery of the space era was made in 1958, by James Van Allen and his colleagues, using Geiger-tube observations from the satellite EXPLORER-1. They found that the earth's magnetic field contained trapped energetic charged particles (radiation), which on more detailed investigation proved to be the source of the largest geomagnetic disturbances at low latitudes (geomagnetic storms) and also of the excitation mechanisms for the auroral emissions in the upper atmosphere. New phenomena have been discovered and new problems have emerged. For example, the earth has been found to be a strong source of radio emissions (terrestrial kilometric radiation), evidently generated in the polar regions of the magnetosphere by the energetic particles which produce the aurora, and large-scale reconfigurations of the whole magnetosphere (substorms) occur, which are in many respects similar to eruptions observed on the sun.

It is now believed that energetic magnetospheric particles mainly have their origin in the sun's outer atmosphere which expands away in all directions at very high speeds (about 400 km/s) forming the solar wind. In the early 1950s, studies of comet tails, which always point away from the sun, had resulted in the suggestion that the only explanation for their behavior was the presence of a continuous flux of charged particles flowing more or less radially outward from the sun. More recently, this electrically neutral plasma, termed the solar wind, has been observed directly by numerous satellites and interplanetary probes. It is comprised mainly of protons and electrons having kinetic energies typically a fraction of a kilovolt.

The extension of the geomagnetic field into space is limited by its interaction with the solar wind. As the solar wind plasma encounters the geomagnetic field; its protons and electrons are deflected in opposite directions giving rise to a vast current system. This current produces a magnetic field that acts to shield the solar wind plasma from the geomagnetic field, and it forms the boundary between them. These fields add to the earth's magnetic field on the inside of this boundary and distort it from its dipolar shape. The region where the geomagnetic field persists is called the magnetosphere. The boundary at which the interaction between the geomagnetic field and the solar wind occurs and where these currents flow is called the magnetopause (see Figure 1.10).

Because of the supersonic velocity with which the solar wind flows past the region of the earth as it encounters the earth's magnetic field, a detached bow shock exists "upstream" of the earth. The region between the bow shock and the magnetopause is called the magnetosheath. This region is dominated by the behavior of the shocked solar wind.

In the "inner" magnetosphere the geomagnetic field remains essentially dipolar. Charged particles "trapped" in this dipolar region form the radiation belts and provide the source of the aurora. Their study, which must necessarily be undertaken mainly with the aid of earth satellites, has become important not only for the light it sheds on terrestrial phenomena such as those described but also because we now know that charged particle acceleration in very low density plasma is a widespread cosmic phenomenon, occurring in all planetary magnetospheres, on the sun in the form of solar flares, in the interstellar medium, in supernova remnants, in pulsars, X-ray sources, and galactic nuclei.

Charged particles, such as those making up the energetic Van Allen radiation belts, circle around and bounce along magnetic field lines as they orbit the earth (see Figure 1.11). Like the Van Allen radiation, many more lower energy particles are also trapped in the inner magnetosphere. Detailed studies of the magnetic field in this region have shown that these trapped particles produce a ring current that persists at all times and which tends to weaken the total magnetic field in the inner magnetosphere. Sometimes, in association with solar flare events, this ring current becomes enhanced and produces the magnetic storm signature seen in the earth's surface magnetic field. This takes the form of a 1 percent decrease in the 0.3 gauss field; the field perturbation is typically 300 nT (or gamma: 1 gamma is equivalent to 10^{-5} G).

Although more detailed studies of the energies of the ring current particles have to be performed, it is known that during magnetic storms many particles having kinetic energies of tens of kilovolts (keV) populate this inner magnetosphere region. Since solar wind particles have energies typically not exceeding 1 keV, these current systems provide evidence that charged particles must be energized considerably within the magnetosphere. This important observation has yet to be explained in quantitative detail.

A third magnetospheric current system of major importance flows across the center of the anti-solar, or tail, region of the magnetosphere. A cross section of the tail (Figure 1.10) shows that it is comprised of three regions; two lobes, where the magnetic field is directed toward (in the northern hemisphere) and away (in the southern) from the earth, are separated by a third region, the plasma sheet, where plasma flows across the center of the tail. It is in this plasma sheet region that the tail currents flow across the center of the tail and then close on the magnetopause or in the magnetosheath. These tail currents are responsible for the shape of the tail magnetic field and play a major role in magnetospheric substorm dynamics. The magnetospheric substorm is characterized by sudden changes in the structure of the plasma sheet and the tail magnetic field. Dramatic changes in the high latitude ionosphere and surface magnetic field are also noted during magnetospheric substorms.

Although these three major magnetospheric current systems persist at all times, each current responds in different complex ways to the solar wind, other magnetospheric phenomena, and the interplanetary magnetic field (the magnetic field associated with the solar wind). Variations in the strength and size of the magnetopause current system can be determined by equating the kinetic

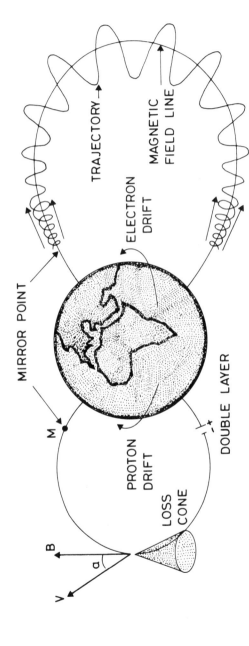

Figure 1.11 Trajectory of a trapped charged particle oscillating between two conjugate mirror points along geomagnetic field lines. Azimuthal, i.e., longitudinal, drifts of electrons and ions are due to magnetic as well as electric forces. When the pitch angle (α) of a particle is too small, i.e., is within the loss cone, the charged particle is not mirrored, but it penetrates into the denser layer of the terrestrial atmosphere where it ionizes or excites neutral atoms. Localized field-aligned electric potential drops or "double layers" are sometimes present in the magnetosphere. The resulting parallel electric fields accelerate ionospheric ions upward and magnetospheric electrons downward; their mirror points are also reduced. (Adapted from R. Gendrin, *La Recherche*, 39 (1973), 961.)

pressure of the solar wind to the energy density (or pressure) of the total geomagnetic field (comprised of the main field and the fields due to the three magnetospheric current systems). The pressure of the solar wind is dependent on its density and speed. The solar wind density has been observed to vary from less than 10^5 to over 10^8 particles/m^3. The speed of the solar wind is somewhat less variable, ranging from 200 to 800 km/s. In response to this variability in the solar wind, the position of the magnetopause, which is typically found (in the direction toward the sun) at about 10 earth radii from the center of the earth, is observed to vary from less than 6 earth radii to more than 15 earth radii. The strengths of the magnetopause currents and their associated magnetic fields vary by about an order of magnitude between these locations, increasing when the solar wind compresses the magnetosphere.

A major conclusion concerning our present understanding of the magnetosphere is that it "breathes," that is, it is not static, but rather has current systems (and associated magnetic fields) which change in response to the variability of the solar wind and of the interplanetary magnetic field. Both the field topology of the magnetosphere and the charged particle populations that it contains therefore exhibit pronounced temporal variations. Changes in the magnetospheric magnetic field with time induce electric fields in the magnetosphere that play an important role in the energization of charged particles.

1.2.1.2 Electric fields in the magnetosphere.

The electric field is a quantity whose significance in magnetospheric physics has only recently been fully realized. One reason for this is that in the early stages of space research the limited observational data were interpreted in terms of highly idealized theoretical models of the magnetospheric plasma. These models underestimate the role of electric fields. However, as a result of increasingly sophisticated satellite-borne measurements, the significance of magnetospheric electric fields is increasingly evident, especially in the context of the polar auroras. Another reason is the fact that direct measurements of the electric field are exceptionally difficult technically, and in the outer magnetosphere the first electric field measurements have only recently been achieved.

The matter filling the magnetosphere is a plasma, that is, it consists of electrons and positive ions rather than neutral atoms and molecules. Unlike an ordinary gas, where collisions between the constituent atoms or molecules are of decisive importance, the tenuous magnetospheric plasma is "collisionless." The motions of the electrons and ions forming the magnetospheric plasma are controlled almost exclusively, except at the lowest altitudes, by electric and magnetic fields. Thus the electric field is intimately involved in the dynamics of the magnetospheric plasma. Furthermore, electric fields are the only means by which charged particles of the magnetospheric plasma can be energized. Although the energization may often be caused by the small-scale electric fields of plasma waves (see Section 1.2.5.4), direct energization by macroscopic electric fields is also important.

While essentially the whole universe consists of magnetized plasma, the magnetosphere is the only cosmic plasma system readily accessible to *in-situ* studies. What is being learned in the magnetosphere about the complex interplay between electric fields and cosmic plasmas may therefore have far-reaching significance in addition to helping to solve the age-old problem of polar auroras.

The most important source of the magnetospheric electric field is the solar wind. Its motion in the presence of an interplanetary magnetic field constitutes an electric dynamo. Thus the magnetosphere is immersed in an interplanetary electric field with a strength of the order of 1 to 2 V/km, which corresponds to a voltage difference of several hundred thousand volts across the entire magnetosphere.

Over an area as large as the magnetospheric cross section facing the solar wind, the solar wind kinetic energy represents a power supply of the order of 10 terawatts (1 terawatt = 1 million million watts). A small fraction of this power enters into the magnetosphere, where it feeds magnetic storms and auroral displays and it energizes charged particles trapped in the geomagnetic field. How much power actually enters depends in a characteristic way on the direction of the interplanetary magnetic field. Thus a southward pointing interplanetary magnetic field, which implies a large electric field directed from the dawnside to the duskside of the magnetosphere, favors the penetration of the electric fields, energy, and plasma from the solar wind into the magnetosphere. Indeed, the overall response of the magnetosphere in this respect is much like that of a rectifier in which the "conduction direction" is from dawn to dusk. Although this overall characteristic of the power input can be described in such simple terms, it must be emphasized that it is the result of extremely complicated plasma dynamic processes.

Another source of magnetospheric electric fields is the rotation of the earth in the presence of its own magnetic field. The atmosphere is carried along in this motion, and its electrically conducting uppermost part, the ionosphere, acts like the conductor of an electric dynamo. Thus the entire ionosphere, when viewed from outer space, appears positive near the magnetic poles and negative at the magnetic equator. Its total voltage is nearly a hundred thousand volts. The "load" of this terrestrial dynamo is formed by the various plasmas that populate the magnetosphere. At low latitudes the response of the plasma to the dynamo field is a simple one, namely, co-rotation with the ionosphere. At high latitudes the ionospheric dynamo is connected to volumes of plasma which are too large to co-rotate and which by virtue of their low density and high temperature have peculiar electrical properties. These plasmas are subject to electric fields imposed externally from the solar wind as well as internally from the rotating ionosphere. Together they determine the large-scale electric field distribution in the outer magnetosphere.

Ionospheric winds, which cause additional electric dynamo fields superposed on those due to the rotation of the ionosphere as a whole, constitute a third source of electric fields.

As a consequence of the solar wind and magnetosphere interaction, large currents (millions of amperes) flow inside the magnetosphere, especially to and from the auroral regions and across the extended tail of the magnetosphere. Energy from the solar wind can be temporarily stored as magnetic energy associated with an enhanced cross-tail electric current. When this energy is released, there occurs a violent phenomenon, the magnetospheric substorm, characterized by rapid changes in the geomagnetic field and intense auroral displays. The changing magnetic field causes large electric induction fields at all levels in the magnetosphere. At ground level these magnetic variations cause earth currents and corresponding surface potential differences that can disturb power networks and telephone and signal systems. In the magnetosphere there occur dramatic relocations of magnetospheric plasma and heating. Spacecraft that become immersed in substorm-heated plasma can become electrically charged to many thousand volts, and in a number of cases this has led to destructive effects. The mechanism responsible for the extremely rapid motion and heating of magnetospheric plasma is still not understood, but large-scale electric induction fields in the magnetosphere are likely to play a key role.

Magnetospheric electric fields were first studied, and are best known, at the low-altitude end of the magnetosphere, that is, in the ionosphere. At sounding rocket altitudes, two main techniques have been used to determine the electric field; these are releases of artificial ion clouds (whose motions reveal the electric field) and direct measurements by electric double probes. The most extensive mapping of ionospheric electric fields has been made by means of double probes and plasma drift meters on low-altitude polar orbiting satellites. Additional information on the large-scale electric field distributions has been obtained from balloon-borne probe measurements and from ground-based radar measurements.

As a result, the overall character of the large-scale distribution of electric fields in the ionosphere, and their dependence on interplanetary conditions, is fairly well known. The main characteristic is a general electric field across the polar cap, at latitudes greater than about 70° magnetic latitude, directed from the duskside to the dawnside of the earth, and an oppositely directed field at lower latitudes. The electric equipotentials form a characteristic pattern of loops. The largest field strengths occur at auroral latitudes and reach 100 V/km or more.

The overall characteristics of the magnetospheric electric field are obtained by extrapolating the ionospheric electric field upward along geomagnetic field lines. These have been confirmed by satellite measurements of geomagnetically trapped particles moving under the influence of the electric field. The spatial distribution of electric potential inferred from such charged particle measurements is shown in Figure 1.12.

Direct measurements of the electric field high in the magnetosphere are difficult. Two main methods are used. One relies on electric probes of spherical or cylindrical geometry extended on booms from the spacecraft; the other derives

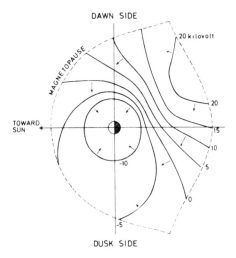

DAWN SIDE

MAGNETOPAUSE

20 kilovolt

20

TOWARD
SUN

15
10
-10
5
0
-5

DUSK SIDE

Figure 1.12 Distributions of electric potentials in the equatorial plane of the magnetosphere as deduced by McIlwain from observations of charged particles drifting under the combined influence of forces due to both electric and magnetic fields. The solid lines show lines of constant electric potential. The electric field, whose direction is shown by a few sample arrows, everywhere forms a right angle with the local equipotential line and is directed from higher to lower potential. The strength of the field is inversely proportional to the local separation between the equipotentials. The region of plasma corotating with the earth is characterized by closed equipotential contours and corresponding inward-directed electric fields. The open equipotentials outside the corotating region are inclined, reflecting an electric field with both a dawn-to-dusk and a sunward component. (© 1974 by D. Reidel Publishing Company, Dordrecht, Holland.)

the electric field from the motion of an electron beam ejected from and returning to the spacecraft. The results, which are still being analyzed, have already produced several results of great importance. Measurements aboard the S3–3 satellite prove that the electric field above the auroral regions has components not only transverse to but also parallel to the geomagnetic field. This follows from the observation of high-altitude transverse electric fields that are incompatible with electrically equipotential magnetic field lines. One consequence of this is that charged particles, notably electrons, which move most easily along geomagnetic field lines, can be directly accelerated by the electric field and hurled down into the atmosphere to cause auroral displays. At the same time, positive ions of ionospheric origin are ejected into space. This explains why the outer magnetosphere, until recently believed to consist exclusively of hydrogen plasma from the sun, is partly populated by a plasma of terrestrial origin and consisting largely of oxygen ions. The GEOS and ISEE satellites measured the electric fields in the low and moderate latitude regions of the outer magnetosphere. Figure 1.13 shows an example where the above-mentioned general dawn-to-dusk electric field is clearly evident at both spacecraft. One particularly interesting result from the ISEE–1 data is the detection of a (dawn-to-dusk) tangential electric field at the magnetopause itself.

These direct measurements show that the electric field in the magnetosphere is much more variable than previously envisaged. The variations of the field strength are comparable to its average value, a few volts per km and occur on time scales of typically 0.5 to 5 min. Much stronger electric fields occur during

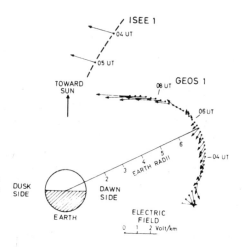

Figure 1.13 Electric fields measured by the satellites ISEE-1 and GEOS-1 on November 20, 1977. The strength and direction of the electric field at a number of points along the orbits are shown by arrows. The scale at the bottom of the figure shows how the arrow length corresponds to the electric field strength. As the satellites moved along their orbits the measurements at different positions were taken at different times (in universal time, UT). (From "Quasistatic Electric Field Measurements on the GEOS-1 and GEOS-2 Satellites," in *Quantitative Modeling of Magnetospheric Processes,* p. 287, 1979, copyrighted by the American Geophysical Union.)

magnetic substorm events when transient field strengths of the order of 20 to 30 V/km are often recorded. The duration of such transient electric fields, as observed at the satellite position, is of the order of a minute. Frequently the electric field shows regular oscillations, with an amplitude as large as 10 V/km, reflecting large-scale oscillations of magnetospheric plasma.

1.2.2 Plasma Waves in the Earth's Magnetosphere

1.2.2.1 Plasma waves.
A plasma of electrons and ions permeated by a magnetic field can support the propagation of a wide variety of plasma waves. The waves are conventionally classified by relating the wave frequency to certain characteristic frequencies peculiar to the plasma. These fundamental frequencies are the cyclotron and plasma frequencies. The electron and ion cyclotron (or gyro) frequencies are those at which electrons and ions respectively rotate in the ambient magnetic field in which the plasma is immersed. Ions, being much heavier than electrons, rotate at a correspondingly slower rate, and the stronger the magnetic field, the higher are all the cyclotron frequencies. The electron plasma frequency is the frequency at which the electrons, if moved bodily with respect to the ions, would oscillate in endeavoring to reestablish equilibrium. It is independent of the magnetic field and is related to the number density of electrons; the higher is the density, the larger the plasma frequency. An analogous plasma frequency exists for each ion species, again its value being much lower because of the larger mass of an ion than that of an electron.

1.2.2.2 Magnetospheric parameters.
Within the earth's magnetosphere, the magnetic field strength decreases rapidly on moving away from the earth. The electron gyrofrequency in the polar ionosphere is about 1.6 MHz and in the equatorial ionosphere it is 800 kHz. At distances of $10R_E$, where R_E is an earth radius (6,370 km), the magnetic field is so weak that it cannot withstand the

pressure of the bombarding solar wind plasma. The electron cyclotron frequency at such distances on the sunward side of the earth is as low as 1 kHz. Much lower gyrofrequencies exist in the magnetotail, especially in the equatorial plane at large distances.

The plasma density and, therefore, the plasma frequency also decrease as the distance above the ionosphere increases. The decrease occurs only relatively slowly within the plasmasphere, which is the region of co-rotating plasma bounded by the plasmapause lying on magnetic field lines crossing the equator at approximately $4R_E$. The plasma frequency in the ionosphere is usually in the range 2 to 10 MHz, whereas just inside the plasmapause it may be as low as several hundred kilohertz. At the plasmapause, the density drops sharply so that the plasma frequency decreases to a few tens of kilohertz in less than one earth radius.

The values of the cyclotron and plasma frequencies quoted above are meant merely as guides to the orders of magnitude involved, so as to enable the plasma waves which will be described to be visualized in the appropriate context. The adjectives "high" or "low" frequency applied to plasma waves convey little information in the absence of comparative values of the cyclotron and plasma frequencies. From the above-mentioned values it may be deduced that in most of the magnetospheric regions the plasma frequency is greater than the cyclotron frequency. Important exceptions are the nighttime plasma trough and the auroral and polar regions.

1.2.2.3 Magnetospheric plasma waves.

Figure 1.14 is a synopsis of the various plasma waves observed in the different magnetospheric regions. At first sight it appears prohibitive, but on categorizing the wave types according to the characteristic frequencies of the plasma in the different regions, a clear picture emerges. Where the word turbulence appears, either it has been impossible to identify the wave mode, or it is genuinely a case of turbulence; such phenomena are not considered here. Another phenomenon evident in Figure 1.14, but which has not yet been unambiguously fitted into the plasma wave picture, is that of magnetic noise bursts seen in the magnetotail and which may be whistler mode noise.

Figure 1.15 attempts to classify the various types of plasma waves that can exist in a plasma consisting of electrons (e) and two ion species, hydrogen ions (H^+, or protons), and helium ions (He^+). The presence of the magnetic field not only fixes the cyclotron (c) frequency f_{ce}, f_{cH^+}, f_{cHe^+}, but it also imposes certain restrictions on how the various waves propagate. Hence Figure 1.15 is divided into two planes: one that includes wave modes which can propagate with their wave normals parallel to the magnetic field and one that lists those that can have perpendicular wave normals. The wave types which lie in the latter plane can exist only in a noncold plasma and are found to be heavily damped if their wave normals deviate too far from the perpendicular. They are classified as electrostatic waves because their electric fields are so large compared to their

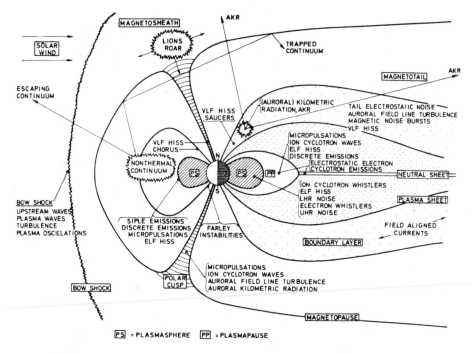

Figure 1.14 Regions of plasma wave occurrence in the non-midnight meridian cross section of the earth's magnetosphere. (From "Magnetospheric Plasma Waves" by S.D. Shawhan, in *Solar System Plasma Physics–a twentieth anniversary review,* (ed. C.F. Kennel, L.J. Lanzerotti, and E.N. Parker), North Holland, 1978.)

magnetic fields. The waves listed in the parallel plane are electromagnetic modes which can propagate in both cold and noncold plasmas. These waves can also propagate perpendicular to the magnetic field, although the frequency ranges in which this is possible differ considerably in certain cases from those ranges available to parallel wave normals.

It is convenient to picture the various wave modes listed in Figure 1.15 as falling into three categories. First, the localized waves are unable to propagate very far from their source regions without being highly damped. All the modes listed in the perpendicular plane fall in this class. In the second category are the confined waves which, although they are able to propagate over large distances as compared with localized waves, are restricted to the earth's magnetosphere. These are the modes listed in the parallel plane which have high-frequency limits to their frequency ranges, that is, the ion-cyclotron and electron-cyclotron waves. Clearly, as these propagate away from the earth they move into regions of lower gyrofrequencies, and thus their allowed frequency range of propagation becomes more and more limited as f_{ce} and f_{cH^+} decrease. The third category, the escaping modes, contains the remaining modes in the parallel plane, namely the O (ordinary) and X (extraordinary) waves which have no high-frequency limit to

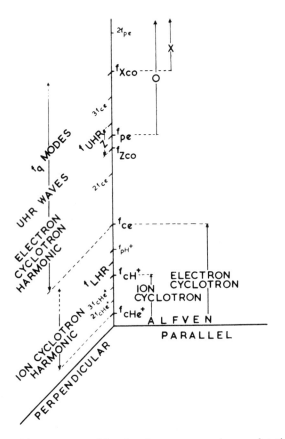

Figure 1.15 Diagram summarizing the plasma wave modes as a function of frequency with respect to the major characteristic frequencies of the plasma, for wave normals parallel and perpendicular to the magnetic field.

their propagation. These can escape from the earth's magnetosphere to propagate freely through interplanetary space and beyond. It is by observing such waves that an extraterrestrial observer would identify the earth as a radio source.

1. *Localized modes:* each of the modes listed in the perpendicular plane is considered in turn, starting at the lowest frequencies.

   (a) Ion cyclotron harmonic waves (ICHW). These can exist in frequency bands lying between harmonics of the ion-cyclotron frequency, f_c ion. Therefore, in Figure 1.15, they may propagate between f_{cHe^+} and $2f_{cHe^+}$, between $2f_{cHe^+}$ and $3f_{cHe^+}$, and so on, up to and above the lower hybrid frequency f_{LHR}, which is the frequency above which ions have little effect. However, the ICHW suffer large damping at the gyrofrequency harmonics themselves, that is, at nf_{cHe^+}, where n is an integer, and hence propagation from one frequency band to another is inhibited. The S3–3 satellite has observed ICHW in the auroral zones associated with the harmonics of the proton gyrofrequency, protons being the ions that dominate in that region. The waves are apparently produced by intense field-aligned electric currents and are believed to be instrumental in increasing the temperature of the cold ions such that they escape the auroral ionosphere to permeate the outer magnetosphere during active periods. Ion cyclotron harmonic waves are also observed by GEOS–1 in the dayside plasmatrough under quiet geomagnetic conditions. Linked to harmonics of the He^+ gyrofrequency, they occur when the cold He^+ concentration is relatively high. In this case, it appears that energetic protons are the energy source, and there are also indications that the waves preferentially heat the cold He^+ ions.

(b) Ion acoustic waves, which are not listed in Figure 1.15, can propagate perpendicular to the magnetic field at frequencies below the proton plasma frequency $f_{pH}{}^+$. They may be produced in the ionosphere by electrons streaming, with respect to the ions (and neutrals), with a velocity greater than the ion-acoustic velocity.

(c) Electron cyclotron harmonic waves (ECHW). These are the electron equivalent of ICHW and they can exist in frequency bands between harmonics of the electron gyrofrequency and have thus been termed $(n + \frac{1}{2})f_{ce}$ waves. They are observed to be most intense at and beyond the dayside plasmapause and appear to be confined to within a few degrees of the magnetic equator, possibly allowing for the first time an accurate determination of the latter's position. The energy source for these waves lies in the low-energy electron distribution; certain components of this electron distribution are tightly confined to the geomagnetic equator, that is, in the regions where the waves are observed. It is believed that ECHW are responsible for diffuse auroral precipitation of 1 to 10 keV particles. Weaker ECHW are observed at latitudes removed from the equator and are also seen occasionally in the tail plasma sheet (see Figure 1.14).

(d) Upper-hybrid resonance waves (UHRW) and f_q modes. These belong to the same family as ECHW but are considered separately here because their frequency is comparable with that of the O and X escaping waves. They are unable to propagate to any great distance, but it is being suggested that they may couple linearly or nonlinearly to O and X waves, which is an important concept within the field of planetary radio astronomy. UHRW are one of the most intense wave types observed within the earth's magnetosphere; they are located in the same regions as ECHW and are believed to be similarly produced. They have direct access to the cold plasma Z mode and this is the reason why the latter, although an electromagnetic mode, is included in the perpendicular plane in Figure 1.15. Relatively weak line emissions are at times observable from GEOS-1 at frequencies ~ 45 kHz; these are the f_q modes, which have the characteristic that their group velocity is zero. Hence, once the waves are excited at the f_q frequencies, the energy cannot propagate away. f_q emissions are also members of the ECHW family.

2. *Confined modes:* as in (1), these waves are considered in turn, starting with the lowest frequency mode, namely Alfvèn waves.

   (a) Alfvèn waves. These are hydromagnetic (or magnetohydrodynamic) waves whose oscillation periods are of the order of many minutes; the magnetic field lines resonate similarly to a vibrating stretched string such as a violin string. The oscillation frequency is related to the length of the magnetic field line and also to the plasma "load" that is moved during the oscillation. The longest period waves are associated with the longest field lines, namely, those extending out to the magnetopause, and are believed to be excited by surface (Kelvin-Helmholtz) waves on the magnetopause, induced by the solar wind. Shorter period waves are linked to intense electric currents, the auroral electrojet, in the midnight sector, and possibly at other times to surface waves on the plasmapause. Because their characteristics are related to the plasma load, they are being increasingly used to determine, from the ground, estimates of plasma density at large distances.

   (b) Ion cyclotron waves (ICW). These are the natural electromagnetic extension upward in frequency of Alfvèn waves. They occur at frequencies approaching the ion cyclotron frequencies at which it is no longer sufficient to consider the plasma as a fluid. They are believed to be produced by energetic protons; conversely, once produced they could be responsible for the loss of ring current protons. GEOS-1 has observed them to be commonly located outside the plasmapause on the dayside, and when ICHW are observed, ICW are observed simultaneously. A low-frequency noise called lion's roar is thought to be produced by ion cyclotron resonance when 10 keV protons stream through the magnetosheath, particularly during geomagnetic storm periods; ion cyclotron waves associated also with streaming ions are observed in the polar cusp region. Ion cyclotron whistlers have a completely different origin. Their energy source is in lightning flashes at the earth's surface and the electron cyclotron whistler waves, which propagate upward through the ionosphere, are converted to ion cyclotron whistlers. The latter, which are observed on low-altitude satellites, have been used to provide information on the ion species along the propagation path.

   (c) Electron cyclotron waves (ECW). As mentioned above, electron cyclotron whistlers originate in lightning flashes. They are observed on the ground, and their propagation paths extend out to the equatorial plasmapause and beyond. Whistlers are therefore probably the most intensively studied of magnetospheric wave phenomena, since they can yield valuable information on the plasma density in these distant regions. To explain multihop whistlers echoing between hemispheres, it is necessary to invoke field-aligned density ducts, and ground-based goniometer (direction-finding) techniques are now yielding information on duct movements and hence electric fields within the plasmasphere. The most widely observed electron cyclotron emissions produced within the magnetosphere are chorus and hiss (auroral and plasmaspheric). Chorus signals are primarily rising frequency emissions that are believed to be produced at the geomagnetic equator by Doppler-shifted cyclotron resonance. Due to the relative motion of electrons moving in one direction and whistler mode waves moving in the opposite direction along a geomagnetic field line, the electrons "see" the wave frequencies Doppler-shifted upward to the cyclotron frequency and yield some of their energy to amplify the waves. The reciprocal process can occur in that, once the waves exist, they can cause electrons to be precipitated into the atmosphere to produce aurora. Auroral hiss is believed to be produced by Cerenkov emission from low-energy precipitating electrons; VLF "saucers" indicated in Figure 1.14 may be similarly explained.

(d) The lower-hybrid frequency f_{LHR} is the frequency below which electron whistlers are allowed to propagate at all wave normal directions with respect to the magnetic field. Noise seen at the LHR in auroral regions may therefore be due in part to reflected and dispersed electron whistler mode waves propagating perpendicular to the magnetic field. At much lower frequencies, electron cyclotron waves co-exist with ion cyclotron waves; both naturally become Alfvèn waves at even lower frequencies.

(e) Because whistlers provide such a wealth of information on the distant plasmasphere, man-made ground-based transmitters, for example, at Siple Station in Antarctica, operating at whistler-mode frequencies have been used for controlled experiments. It has been found that these man-made signals can stimulate emissions similar to chorus, and a great deal of interest is focused on the possibility that even 50 and 60 Hz powerline radiation from industrialized regions can cause wave stimulation and charged particle precipitation from the magnetosphere.

3. *Escaping radiations:* at present only two types of radiation are known to escape the earth's magnetosphere and to be observable at large distances. They are terrestrial nonthermal-continuum or myriametric radiation (TMR) and terrestrial or auroral kilometric radiation (TKR or AKR); myriametric and kilometric refer to the free-space wavelengths (10^4/m and 10^3/m respectively) at which the radiations are predominantly observed.

(a) TMR. This is radiation in the frequency range 100 Hz to 100 kHz predominantly observed in the ordinary (O) magnetoionic mode. It appears to have its source in the intense UHR waves observed on the dayside plasmapause, and there is evidence accumulating from the GEOS and ISEE spacecraft that it is beamed away from the magnetic equator in a manner compatible with mode-coupling of Z-mode waves to O-mode waves at $f = f_{pe}$. Waves with frequencies less than the magnetosheath and solar wind plasma frequencies are trapped within the plasmatrough as indicated in Figure 1.14 only the higher frequencies can completely escape.

(b) TKR. This is very intense radiation, in the frequency range 50 to 500 kHz, having its source in the auroral zones. It is associated with electron precipitation and auroral arcs, but there is no accepted generation theory at present. Some controversy surrounds its mode of propagation, most results implying that it is in the extraordinary (X) magnetoionic mode. However, recent observations on the EXOS–B satellite indicate that it is propagating in the O-mode. Because of its similarity to Jovian decametric radiation, a knowledge of its source mechanism is important for planetary radio astronomy in general.

1.2.3 The Magnetospheric Plasma Population

1.2.3.1 Low-energy plasma.   Until very recently, conventional knowledge has centered on the understanding of the earth's plasma environment as being basically made up of two charged particle populations, those of solar origin and those of terrestrial origin. In the past few years, however, there have been hints of a third plasma population having a composition that indicates an origin in the earth's ionosphere, but having energies much higher than those expected from an origin near the earth. This hybrid plasma population is found throughout the magnetosphere and displays energies ranging from tens of electronvolts up to tens of kiloelectronvolts.

In the 1960s and early 1970s, ground-based whistler measurements in combination with direct satellite measurements furnished a fairly comprehensive picture of the density distribution of the cold plasma population in the earth's magnetosphere. Compilations of these measurements led to a general understanding of the morphology of the plasmasphere together with some information on the dynamics of the plasma associated with co-rotation and varying convective electric field drifts. There were, however, no direct comprehensive measurements of the low-energy plasma temperature, pitch angle, or density distributions along the magnetic field line. This left open many questions about the mechanics of plasma interchange between the ionosphere and plasmasphere.

As a result, theoretical speculations led the field. For example, concerning filling of the plasmasphere, it was suggested in 1971 that supersonic polar wind flow out of conjugate ionospheric regions might fill the magnetic flux tubes.

Because of the strong influence of this outward flow on the density profile of the topside ionosphere, the nature of this filling process is also a key element in the physical linkage between the location of the marked plasmapause density gradient in the magnetosphere and the light ion trough density gradients in the topside ionosphere. To test this model, the presence of polar wind flows inside the plasmapause should be sought; shock fronts, which display density and temperature changes along the magnetic field line, and regions of enhanced density near the equatorial regions of magnetic flux tubes in the outer plasmasphere would also be expected. The rates of refilling magnetic flux tubes with plasma have also been studied from whistler observations.

Plasma temperatures of 5 to 10 eV were observed in the outer plasmasphere using OGO–5 and also in the dayside plasmatrough using IMP–6. The ATS–6 satellite, placed in geosynchronous orbit in 1974, carried an instrument that could study the differential angular and energy distributions of plasma between 1 eV and 80 keV. The local time distribution of encounters of ATS–6 with regions of 1 to 30 eV plasma with a density between 10^6 and 10^7 ions/m³, exhibiting a maximum in the afternoon-dusk sector, indicate the presence of dayside filling of the plasmatrough and the location of the bulge region of the plasmasphere, confirming expectations. Also in agreement with earlier convection models, ATS–6 data show a predominant region of eastward flow on the dawnside of the magnetosphere, whereas the flow around dusk is predominantly westward. New and unexpected information was obtained on plasma temperatures in the range of 10 to 25 eV (about 100,000 to 300,000°K).

Instruments specifically designed for low-energy plasma investigations have recently been launched on GEOS–1 and 2 and ISEE–1. There are two such experiments on the GEOS–1 and 2 satellites, namely, the Ion Composition Experiment and the Suprathermal Plasma Analyzers. Initial measurements made by the Ion Composition Experiment confirm ATS–6 observations of both magnetic field-aligned and conical distributions (with peaks 20° off the geomagnetic field line). In general agreement with earlier OGO results, the dominant ions are found to be H^+, He^+, and O^+. However, there is also a surprising amount of D^+ or $^4He^{++}$ (with an atomic mass to charge ratio M/Q of 2), and $^{16}O^{++}$ ($M/Q = 8$). These multiply charged ions in the 2 to 100 eV energy range are unexpected features of the thermal plasma and suggest new physical processes.

The new phenomena observed by ATS–6 and GEOS–1 and 2 have been confirmed by an instrument aboard ISEE–1. Ion energy spectra often reveal plasmas with temperatures of, for example, 0.5 eV, 6 eV, and 12 eV simultaneously. As the spacecraft moves in toward the earth, the proportion of the hotter components decreases in favor of the cooler plasma. Ions with energies up to 100 eV have been measured by ISEE–1 throughout the magnetosphere, that is, in the plasmasphere, plasmatrough, entry layer, and magnetotail, with a variety of pitch angle and energy distributions.

Recent measurements show the need to search for energization mechanisms that apparently act on the ionospheric plasma as it is transported into the magnetosphere. The relative proportions of field-aligned, conical, and trapped low-energy plasma distributions, and their influence on the generation of resonant wave phenomena, are important in this regard. Future studies of the hybrid plasma regime, and its relationship both to the cold ionospheric plasma and to the hot magnetospheric plasmas, will require direct differential measurements of energy spectrum, pitch angle, and ion composition from a spacecraft with carefully controlled potential and surface characteristics. Such a capability is now being planned for the origin of plasmas in the earth's neighborhood (OPEN) mission.

1.2.3.2 High-energy plasma.

Essentially all of the mass, momentum, and energy transfer from the solar wind to the magnetosphere is via hot (that is, high energy) plasmas with energies of a few tens to 100 keV. Although *in-situ* observations of the earth's magnetosphere have been carried out for over two decades, several major regions of the magnetosphere were not discovered until deep-space probes were available with advanced instrumentation. Some of these major regions (with years of discovery) are the plasma sheet (1967), cusp (1971), magnetospheric boundary layer (1972), and plasma sheet boundary layer (1976) (see Figure 1.14). Quantitative measurements of three-dimensional velocity distributions of charged particles and of gradient-scale lengths were not available until the ISEE satellites were launched in late 1977. Several research efforts are currently underway to use detailed velocity distributions along with plasma wave and magnetic field data to test predictions of collisionless plasma theory. These studies may be crucial for future progress in fusion research and other areas of pure and applied plasma physics.

The recently discovered plasma sheet boundary layer and the magnetospheric boundary layer are transition layers between regions of open magnetic field line topology and closed field line topology. These are the regions through which transport occurs from plasma source regions to energy storage regions. The transport of mass, momentum, or energy between various regions has been primarily determined by analyses of hot plasma data.

The high-latitude boundary layer differs from the other boundary layer regions in that it does not separate an open from a closed field line region. The magnetic topology still differs, however, in that it separates field lines that have one "foot" at the earth from the purely interplanetary field lines. The principal source of the plasma sheet may be the low-latitude boundary layer which directly transports magnetosheath plasma into the plasma sheet instead of the indirect transport mechanism involving the high-latitude boundary layer. During periods of enhanced geomagnetic activity, however, the plasma sheet may receive a much greater input of plasma from the high-latitude boundary layer.

A solar wind magnetosphere coupling function has been found which is closely correlated with the rate of total energy consumption by the magnetosphere. This coupling function has been interpreted variously as the power generated by the solar wind magnetosphere MHD dynamo, or as that portion of the solar wind electromagnetic power flux that penetrates the magnetosphere. Activity in the auroral ionosphere has also been found to correlate with this coupling function. Thus, a substantial portion of the power transferred from the solar wind passes directly toward earth without being stored in the magnetotail.

Energization of the magnetosphere is known to be correlated with a change in the orientation of the interplanetary magnetic field from northward to southward. A widely accepted view is that this correlation is evidence for a magnetic merging process taking place, in which field lines embedded in the solar wind plasma are pressed against the magnetopause and merge with the earth's geomagnetic field. The merged field lines are then swept over the polar cap into the tail lobes. In this model, magnetic flux is "eroded" from the dayside and accumulates in the tail lobes. This stored magnetic energy in the magnetotail can be released in an explosive manner, involving hot plasma energization with the onset of reconnection at neutral lines in the tail, and causing a substorm. Particle energization near the sunward magnetospheric boundary is expected in merging models. This has been observed only very recently; clear cases of energization are observed only rarely. In these few isolated cases, dayside merging (if it occurs at all) is usually considered to take place in highly time- and space-variable merging patches.

Plasma source regions have higher densities and lower temperatures than the storage regions, whereas the transport regions usually have intermediate densities and temperatures. The transport regions, however, have shorter characteristic spatial and temporal structures, and the hot plasmas there often exhibit anisotropies and gradients. Except for the tail lobes, the transport regions are all boundary layers (or extensions thereof) and they all have enhanced electrostatic wave emission and electric currents. Indeed, this statement applies to all transport regions if the tail lobes are classified as regions where magnetic flux is stored.

Density, temperature, and velocity are examples of plasma parameters that are derived by taking sums over the charged particle distribution functions. The most direct link to plasma theory, however, is by evaluating energy spectra and velocity distributions. These distributions can be constructed from the raw data collected by three-dimensional electrostatic analyzers aboard the ISEE spacecraft. Color-coded energy–time spectrograms have been constructed which quickly and effectively summarize the plasma data and permit rapid identifications of the plasma regime, energy spectra, and flow behavior. As various analysis groups develop and implement such new display formats of previously unavailable types of data (for example, three-dimensional, high-time resolution, or ion composition data sets), the possibilities for new (and unexpected) discoveries are almost unlimited.

1.2.3.3 Energetic charged particles.   Electrons and ions having energies above 100 keV are found to be stably trapped in the earth's magnetosphere in addition to the lower-energy particles which are much more abundant. The density of these high-energy particles is extremely low, yet they are of substantial technological significance because of the damaging radiation effects they cause. Consideration of the high-energy charged particle population must be a part of every spacecraft design and mission plan in order to ensure its successful operation and the desired lifetime.

The two dominant sources and some minor sources of energetic charged particles in the magnetosphere are described below:

1. *Particles accelerated in situ:* The majority of the energetic particles in the magnetosphere represent the high-energy tail of the energy distribution of ions and electrons accelerated *in situ* by magnetospheric processes. Given the copious quantities of hot plasma (with energies below 100 keV) in the magnetosphere discussed in the previous section, the acceleration of some of the particles to high energy is expected on the basis of general plasma physics principles which appear to operate in plasmas of all physical scales. Thus the ultimate source of these high-energy particles becomes a question of the source of the thermal (keV) plasma. *In-situ* acceleration is responsible for the outer zone energetic particles and for inner zone particles with energies below about 1 MeV for electrons and below about 10 MeV for protons. Such particles are transported toward the earth throughout the magnetosphere by a process known as radial diffusion.

2. *Cosmic ray albedo neutron decay (CRAND) protons and electrons:* The upper atmosphere of the earth is continuously bombarded by galactic cosmic rays with energies ranging from hundreds of megaelectronvolts up to many gigaelectronvolts. The interaction of a cosmic ray with an atom of nitrogen or oxygen in the earth's upper atmosphere may result in an energetic neutron moving outward into the earth's magnetosphere. The free neutron is radioactive and decays to an electron, a proton, and an antineutrino with a half-life of approximately 10 min. If the decay occurs within the magnetosphere, the electron and proton are trapped. The nature of the neutron decay is such that the proton receives almost all of the kinetic energy of the parent neutron. As a result, this CRAND source generates electrons with energies below about 1 MeV and provides protons with energies from 10 MeV to at least 1,000 MeV. The ability of the earth's magnetic field to contain protons of such high energy is limited, and this consideration restricts stable trapping to the equatorial regions of the magnetosphere within approximately one earth radius of the earth's surface. The CRAND source is significant only for the inner zone. It should be noted that the upper energy limit for trapping has not been determined experimentally, since particle intensities at the highest energies are very weak.

3. *Minor sources:* There is evidence that other sources supply trace quantities

of high-energy particles to the earth's magnetosphere. These sources include the direct capture of energetic ions and electrons, energized in Jupiter's magnetosphere and in solar charged particle events, which escape from Jupiter's magnetosphere and the sun, respectively, and reach the vicinity of the earth. A very small percentage of these particles may then be trapped in the earth's magnetosphere. In addition, the so-called "anomalous galactic cosmic rays," believed to be singly charged ions, can be trapped in the magnetosphere if stripped of their atomic electrons by collisions with the residual atmosphere.

The composition of the energetic ions is important because it can reveal the ultimate source(s) of these particles. It is well known that the bulk of the energetic radiation belt ions are protons, with the helium ion abundance ranging from a few percent down to a few hundredths of 1 percent. This relative abundance is higher near the equator and following large geomagnetic disturbances. Only a single measurement has been reported which distinguishes among the heavier elements (carbon, nitrogen, oxygen) and nuclei of still higher mass. The sum of the carbon, nitrogen, and oxygen abundances was found to be a few percent of the abundance of helium, with the relative abundances of the three elements being comparable with the abundances found in the sun. The observation strongly suggests that these energetic ions initially came from the solar wind and were energized by magnetospheric processes.

The abundances of these high-energy particles exhibit temporal variations with time scales varying from seconds to centuries. The very energetic protons found in the inner zone have lifetimes measured in thousands of years. In the outer zone, abrupt changes are seen in response to strong magnetic storms, with slow variations occurring during the time interval of the order of months between impulsive events of magnetic activity. Near the outer boundaries of the outer zone, changes on time scales shorter than a day are the rule. A rough analogy with ocean dynamics may be appropriate: The outer regions of the magnetosphere respond readily to external disturbances, as does the surface of the ocean, whereas the center of the inner zone is well shielded, as is the deep ocean.

1.2.4 The Solar Wind

In 1951, Ludwig Biermann deduced from observations of comet tails that ions and electrons should continuously flow away from the sun. This prediction raised lively interest in the experimental investigation of interplanetary space. Consequently, even the very first scientific satellites with high apogee orbits were equipped with instruments suitable for the detection of the still hypothetical interplanetary medium. In 1960, Konstantin Gringauz reported the discovery of ion fluxes with about the expected properties with his instrument on LUNIK-2. Much more elaborate measurements yielding quantitative details such as density, temperature, and velocity of the flow of charged particles were carried out aboard MARINER-2 in 1963. From that time on, essentially every highly eccentric earth-orbiting scientific satellite and all space probes carried instruments for the investigation of this phenomenon which had been named "solar wind" by Eugene Parker in 1958.

1.2.4.1 Properties.   We now know that the solar wind is a highly ionized gas, a so-called plasma, mainly consisting of protons and electrons. Usually, also an admixture of fully ionized helium ("α-particles") with a relative number density up to 20 percent is observed. The heavier ions in the solar wind, amounting to less than 1 percent in density are, in contrast to the light ions, not fully ionized.

Near the orbit of the earth this plasma is observed to flow almost radially away from the sun with speeds between roughly 200 and 800 km/s, the number density being of the order of 10^7 particles/m^3. Even though the plasma temperature which ranges from 10^3 to 10^6 °K, may appear high, the thermal velocity of the particles is always smaller than the plasma bulk velocity. This means that the plasma flow is supersonic. This fact is important whenever the plasma interacts with such obstacles as planetary bodies and their magnetic fields. In these cases, shock waves comparable with those generated by supersonic aircraft are formed.

The plasma carries along a magnetic field which is about one-ten-thousandth of that observed near the earth's surface. This field is believed to be "frozen in" to the plasma because the electrical conductivity of the plasma is so high that currents (which are the ultimate source of the magnetic field) continue to flow essentially unweakened for very long times. The magnetic field plays a very important role in determining the propagation of high-energy particles in the plasma; wave modes and instabilities which do not exist in common gases or fluids can also occur.

1.2.4.2 Understanding.   Even though no *in-situ* measurements in the region of generation of the supersonic plasma flow near the sun can be made at present, the basic principles of plasma acceleration, which first appeared mysterious, are probably understood now. The plasma is accelerated to supersonic velocities against the sun's gravitational forces due to the high temperature of the solar corona if a sufficiently large radial energy transport in the plasma is assumed. The acceleration of the gas as described by this theory resembles very much the processes going on in the nozzle of a burning rocket where the thermal energy of a hot gas is transformed into a supersonic bulk motion.

Since the average solar wind properties near 1 astronomical unit (AU), the sun–earth distance, are known and the reasons for plasma acceleration in the outer solar corona are basically understood, it might appear that most of the problems were resolved in this field of space research. Unfortunately, this is not true at all for several reasons:

1. The solar wind most likely fills the whole heliosphere to a radial distance of at least 100 AU. Only a very small fraction of this space is presently accessi-

ble to *in-situ* research. In particular, the important region of plasma acceleration near the sun, and the processes going on at large distances from the sun where the solar wind must somehow interact with the interstellar medium, have not yet been explored. Beyond the ecliptic plane (above about 15° latitude) no *in-situ* measurements have yet been made at any distance from the sun.

2. The solar wind is a highly variable medium, in part due to nonstationary processes near the sun, and in part due to interactions between plasma streams of different speeds in interplanetary space. Therefore, large-scale dynamic processes and deviations from spherical symmetry, which are not included in the basic theories, must be taken into account. Fast solar wind streams persisting over several solar rotations are embedded in a slower flowing plasma. The fast streams interact with the slow solar wind mainly by generating compressed and heated regions at their leading edges. In order to understand such dynamic processes, many space probes spread over the huge volume of the heliosphere, and measuring simultaneously with suitable instruments, would be required. If the situation in solar wind investigations were to be compared with meteorology, for example, knowledge of the solar wind is now at that stage when, in meteorology, the thermometer and barometer had just been invented.

3. In addition to the large-scale processes, the small-scale features of the plasma are important but are still insufficiently understood. Proton velocity distributions measured at different solar distances in the same plasma stream by the two HELIOS spacecraft show that the thermal behavior of the solar wind described by means solely of temperatures is very often not satisfactory. Deviations from Maxwellian distributions are believed to be due to the lack of thermal equilibrium and to wave-particle interactions which probably play the decisive role in the energy transport in the solar wind which is important but not specified in Parker's theory.

4. Better developed diagnostic tools are required. For example, the rate of loss of the sun's angular momentum, which is important for theories of star formation and development, cannot yet be measured because the accuracy of presently available solar wind instruments is limited. The situation is similar with respect to other measurements of astrophysical importance, for example, the mass composition of the solar wind. Such measurements have so far only been possible under very special flow conditions and therefore do not permit general conclusions to be drawn. Further limitations, for example, in the determination of accurate velocity distributions with high time resolution, which are necessary to resolve some open questions in plasma physics, result from both the restricted capabilities of present-day instrumentation and the low data transmission rates available from the spacecraft. As a result of the efforts in instrument development being made by a widespread scientific community, and also

of the progress in spacecraft technology, many of these problems will, it is hoped, be overcome in the near future.

5. The difficulties of investigation are increased, and the present state of our knowledge is correspondingly restricted, if the interaction of the insufficiently well-understood solar wind with the complex system of a planetary atmosphere or magnetosphere is considered. Apart from the fact that the outer planets (beyond Saturn) have not yet been reached by any spacecraft and that spectacular types of interaction, such as those between the solar wind and comets, are still unexplored, it must be said that even solar–terrestrial relations, modulated by the solar wind, are still far from being understood.

1.2.4.3 Large-scale interplanetary magnetic field.

Interplanetary magnetic field lines are attached to the sun at one end, and their form is determined primarily by the motions of the solar wind and by the rotation of the sun. The solar wind, in effect, carries the magnetic field lines with it as it moves radially away from the sun, while the sun carries the base of each line along a circle as it rotates. The result is that the magnetic field lines are spirals; this geometry has been confirmed by direct measurements from several spacecraft between 0.3 and 10 AU. The interplanetary magnetic field is more complicated than that of the earth, because it is related to the magnetic field of the sun, which is complex and highly variable. The interplanetary magnetic field is actually simpler than the photospheric field because most solar magnetic field lines extend only to small distances (about two solar radii) from the sun and then return to the sun. The interplanetary field is believed to be a manifestation of the "large-scale" solar magnetic field. Thus, *in-situ* measurements of the interplanetary magnetic field can provide a description of the sun's global magnetic field, which in turn is related to the dynamo that powers solar activity.

If the sun's main magnetic field were a simple dipole, like that due to a bar magnet, and if the dipole axis were perpendicular to the ecliptic plane, then during the last solar cycle the interplanetary magnetic field direction would be "away from" the sun (positive polarity) north of the ecliptic plane and "toward the sun" south of the ecliptic plane, the surface of transition from north to south being a plane (actually a current sheet) coinciding with the ecliptic plane. If the sun's magnetic dipole axis were tilted slightly (say 10°), the current sheet would be tilted 10° with respect to the ecliptic plane, and this plane would rotate with the sun about the solar rotation axis. Thus, an observer in the ecliptic plane would see field lines directed away from the sun during one-half of a solar rotation period and directed toward the sun during the other half. Such a pattern, called a two-sector pattern (one positive sector and one negative sector), is often observed. A two-sector pattern would also be produced by a highly tilted solar magnetic dipole, for example, a dipole whose axis lies in the solar equatorial plane, in which case the current sheet (sector boundary or neutral sheet) would be perpendicular to the ecliptic plane. The issue of whether or not the current sheet lies close to the ecliptic plane can be settled by sending a spacecraft far above the ecliptic plane.

The sector pattern is not constant with time. Sometimes, for example, four sectors are observed. This might be due to a "warp" or "undulation" in the surface associated with a slightly tilted dipole. Alternatively, it might be due to a much more complicated solar field. Little is known about the variations of the three-dimensional configuration of the sector boundary surface. As solar activity increases, the sector pattern is disrupted, presumably by ejecta from active regions. Relatively little is known about the complex polarity patterns that are measured at these times.

Even in the absence of transient ejecta, there are often distinct streams of fast plasma. Some of these streams persist for many solar rotations. The magnetic field polarity is usually constant throughout a stream, and sector boundaries always occur in slow flows between streams. Most large co-rotating streams originate in coronal holes, which are regions in the sun's high atmosphere, the corona, where (in contrast to the remainder of the corona) X-rays are not emitted. The polarity of the magnetic fields in the streams is related to that in the coronal holes. Models of the coronal magnetic fields indicate that most magnetic field lines in coronal holes are "open," that is, they extend from the sun to the distant interplanetary medium. There is some evidence that fast plasma may occasionally originate in open magnetic field regions that are not associated with coronal holes.

As the fast plasma in streams co-rotating with the sun advances into the slow plasma ahead, there is an interaction between the two. The magnetic field and plasma are compressed; the plasma is heated in this interaction region. Information about the interaction propagates by means of magnetohydrodynamic sound waves both into the slow flow and into the fast flow. At larger distances from the sun (2 AU), these waves steepen into shock waves.

In the co-rotating streams themselves, there are fluctuations of a different sort; these are Alfvèn waves, which propagate by virtue of an effective tension in the magnetic field lines. These waves are observed to propagate away from the sun. It has been suggested that they originate in stream interaction regions, but there is also evidence that they may be produced at the sun itself. The solar origin hypothesis is supported by observations of the wave amplitudes by Helios to 0.3 AU and by other spacecraft out to several astronomical units.

The geometrical form of Alfvèn waves is not fully understood. Present theories assume for simplicity that they are plane waves or simple wave patterns like those produced on a water surface by a moving boat or a falling pebble, and most observations are interpreted in terms of these theories. Recent observations suggest that the waves may rather be analogous to waves on the open sea, and that a statistical description may be more appropriate. This field is expected to be an active area of research during the next decade.

Measurements of the interplanetary magnetic field are now made with very high time resolution, corresponding to a spatial scale of the order of 15 km, i.e., 10^{-7} AU. Most phenomena taking place on this scale are of importance for understanding fundamental plasma processes. The solar wind is a particularly simple magnetized plasma, free of the boundary effects and contamination which complicate the analysis of plasmas in the laboratory. Moreover, the interplanetary plasma and magnetic field can be measured very accurately and in great detail. Thus, the solar wind is, in some senses, a laboratory in which plasma processes can be studied *in situ*.

The study of the structure of interplanetary shocks is a rich subject that is only beginning to be explored. Very many different shock structures are possible, depending on the local parameters at the time a shock is observed. There is a correspondingly large set of physical processes related to the different types of shocks and shock structures.

1.2.5 Planetary Magnetospheres

Magnetospheres have been discovered surrounding other planets, notably Mercury (discovered by MARINER-10 in 1971), Jupiter (detected first by its associated radio emissions and later by direct *in-situ* measurements from PIONEER-10 in 1972), and Saturn (discovered in 1980 by PIONEER-11). It seems clear that Venus has no significant magnetic field of internal origin and therefore no magnetosphere, whereas the situation at Mars remains unresolved, although any magnetosphere that exists is quite small. Magnetospheres are also to be expected in the cases of Uranus and Neptune, but this will not be confirmed (or otherwise) until the planets are encountered by VOYAGER-2 in 1986 and 1989, respectively. The concept has also been generalized to include "magnetospheres" of neutron stars (pulsars) and some X-ray sources where strong magnetic fields and associated particle acceleration phenomena are of primary importance.

1.2.5.1 The magnetosphere of Jupiter.   Jupiter is not an ordinary planet. It is the largest planet in the solar system with a mass more than twice the combined mass of all the others. Jupiter radiates substantially more heat energy than it receives from the sun; thus, as judged by its energy budget, it is a weak star rather than a planet. Nearly three decades ago, it was discovered that Jupiter was a strong emitter of radio noise and the brightest radio source in the sky. This radio emission comes from the inner part of a large and active magnetosphere, a magnetosphere so huge that, if it could be seen, it would appear from earth to be the largest object in the sky and to vary from two to nearly four times the lunar (or solar) diameter (Figure 1.16). In contrast, the earth's magnetosphere has a diameter only slightly larger than the body of the planet Jupiter itself. Hence the earth's magnetosphere, if seen from Jupiter, would appear in the sky to be no larger than the planet Jupiter appears to us from earth, a mere dot in the sky. Orbiting around Jupiter are the largest number (16 at the last count) and the most

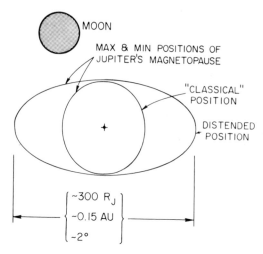

Figure 1.16 Relative angular sizes of the moon and Jupiter's magnetosphere as seen from the earth, when Jupiter is at its closest approach to the earth (four times the distance from the sun to the earth). The magnetosphere is taken in cross section perpendicular to the earth–Jupiter line. Principally because of variations in solar wind pressure, the size of Jupiter's magnetosphere varies with time between an earth-like "classical" radius, which would be expected if Jupiter's magnetosphere had negligible internal plasma pressure, and an equatorial radius distended by internal plasma forces. The distended radius is larger than the classical radius in the equatorial plane by roughly a factor of two. The magnetopause need not have dawn–dusk symmetry. The earth's magnetosphere (which is roughly 10 to 15 times larger than the earth) would be about half the size of the degree symbol in the 2° scale indication of the figure if viewed from Jupiter.

interesting collection of moons (satellites) of any of the planets. For example, the innermost of the four satellites discovered by Galileo, a satellite called Io, appears to be the most volcanic body known in the solar system. During the 1979 VOYAGER spacecraft encounters with Jupiter, eight volcanoes were observed on Io, all erupting simultaneously. Jupiter, its satellites, and its magnetosphere are also studied because they form a miniature solar system that provides clues to the origin and evolution of a full-scale solar system.

In the region within about 20 Jupiter radii (R_j), the magnetic field appears roughly to be that of a dipole (that is, similar to the field produced by a bar magnet), except that close to the surface of Jupiter the field structure is more complex, involving magnetic anomalies. Beyond about 20 R_j, the magnetic field is increasingly stretched out; the stretching is best seen on the nightside of Jupiter (see Figure 1.17) where the solar wind pressure does not oppose the stretching forces. The stretching is caused by a relatively thin, hot sheet of ionized gas (plasma) usually referred to as the magnetodisc. Electric currents flow within the magnetodisc plasma and cause the otherwise dipole-like field to be stretched.

What is the source of the energy that drives the wide variety of Jovian

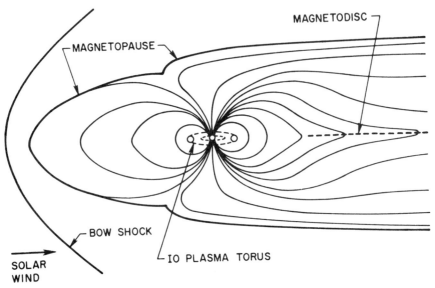

Figure 1.17 A cross-sectional view of Jupiter's magnetosphere in the plane in which the north magnetic pole is up and the Sun is to the left. The magnetic field lines are shown drawn into a magnetodisc structure on the nightside by the presence of a thin sheet of hot plasma that, a week or two earlier, was part of the Io plasma torus. The magnetodisc is a thin, well-formed structure only on the nightside. Jupiter and the Io torus have been doubled in relative size to make them more clearly visible. The earth's magnetosphere would be approximately the size of the circle near the center that represents Jupiter. The magnetosphere is flattened in the north–south direction and extended in the east–west direction; a three-dimensional model would show it to be more like a discus than a bullet.

magnetospheric phenomena? Where does the plasma in the magnetodisc and elsewhere in the magnetosphere come from? While it is not yet definitively established, it appears that the energy is extracted from the rotational energy of Jupiter itself. That is, Jupiter's rapid spin (its present rate is one rotation in just less than 10 hours) is being imperceptibly slowed by its interaction with its magnetosphere. Even more remarkable, the satellite Io is the primary source of the plasma that populates the magnetosphere and the magnetodisc. Io is also the primary source of the particles that form Jupiter's huge radiation belt. The development and dynamics of Jupiter's magnetosphere are therefore unlike those of the earth's magnetosphere and radiation belt, for which the solar wind provides much of the plasma and all of the energy. It appears that, at Jupiter, solar wind provides neither the plasma nor the energy.

The origin and distribution of plasma in the Jupiter/Io magnetosphere system is rather simple in concept. Because of its volcanoes, Io is covered with condensed volcanic material, principally sulphur dioxide (SO_2) frost. Some of this material is knocked off Io by energetic particle bombardment (that is, sputtered from Io) with sufficient velocity that it escapes from Io. This escaping material is then ionized by additional particle bombardment so that a plasma torus forms in the vicinity of Io's orbit. The torus is large and relatively massive for a magnetospheric plasma feature. Approximately one ton/s of material (mainly SO_2) is sputtered off Io and injected into the plasma torus. The Io torus contains roughly one million tons of SO_2 debris plus other material such as sodium and potassium. By contrast, all the plasma in the earth's magnetosphere above about 1,000 km altitude amounts to only about 10 tons, or a hundred thousandth of the mass of the Io torus. This plasma is transported throughout Jupiter's magnetosphere and accelerated to create the most extensive and the most energetic radiation belt found so far within the solar system.

There is fair to good agreement among scientists on the above outline of the origin and distribution of Jovian magnetospheric plasma. However, there is no corresponding consensus on the cause-and-effect chain of events that lead from the Io plasma torus to the observed magnetodisc and radiation belt. Our understanding of the Jovian magnetosphere is still in a primitive state; we can expect, during the next decade or so, significant progress that will at least modify our current theoretical ideas of Jupiter's magnetosphere. The following theoretical description should be read with this caveat in mind.

The torus drives a large-scale magnetospheric convection pattern that both transports torus plasma throughout the magnetosphere and causes a small fraction of the plasma to be accelerated to form Jupiter's energetic radiation belt. The torus can drive the convective motion because it is significantly heavier on one side than the other. This heavy portion of the torus is referred to as the active sector, which is related to an extensive magnetic anomaly in Jupiter's northern hemisphere. Because of the association with a magnetic anomaly, this particular theoretical development is referred to as the "magnetic anomaly model."

The region between the torus and Jupiter's ionosphere contains enough plasma to form a good electrical connection between them. The rotation of Jupiter causes currents to flow that, much as in an electric motor, exert forces on the torus that make it tend to rotate rigidly with Jupiter. The torus approaches within a few percent of total co-rotation with Jupiter. The rapidly spinning torus feels a centrifugal force that tends to fling it outward. The outward centrifugal stress on the co-rotating torus is a maximum in the active sector because the torus is heaviest there, so the plasma in the active sector moves away from Jupiter. However, the torus is part of a complex electrical circuit, and motion of one section of the torus causes motion of other portions of the torus as well as of other plasma throughout the magnetosphere. The outward motion of the active sector of the torus sets up an electric current pattern, and this results in large-scale circulation of plasma throughout the magnetosphere, which takes the form of a two-cell convection pattern that co-rotates with Jupiter.

The torus is so massive that its convective motion in the centrifugal force field releases approximately 10^{14} W of power throughout the magnetosphere. The radio emission from Jupiter has a time variation that is obviously controlled by the ten-hour planetary-spin period. No earth-type magnetospheric model has yet been proposed that accounts, even qualitatively, for the observed ten-hour periodicities. Yet, because of our detailed familiarity with the earth's magnetosphere, it is perhaps natural that significant effort should be expended on applying this fund of knowledge to Jupiter's magnetosphere. However, if Jupiter has a pulsar-type magnetosphere, much of terrestrial magnetospheric physics is not applicable to Jupiter's magnetosphere. The magnetic anomaly model thus far represents the only attempt to explain Jovian magnetospheric phenomena from something like a pulsar magnetosphere point of view.

Two differences between Jupiter and pulsar magnetospheric models can be recognized: Jupiter's surface magnetic field is significantly weaker than that assumed for pulsars, and Jupiter's pulsation period is significantly longer than that of the observed pulsars. If these differences can be taken into account, we shall have the opportunity to study the physical behavior of a body within our solar system that exhibits weak but genuine pulsar-like behavior and to use the knowledge gained from such studies to understand the physical principles governing astrophysical pulsars.

The discoveries and new insights made possible by the two PIONEER and the two VOYAGER spacecraft fly-bys of Jupiter, particularly the discoveries of volcanoes on Io and of the Io plasma torus, have opened up new frontiers of research in magnetospheric physics. The interval between the present time and the time when data from the GALILEO spacecraft (scheduled to be put into orbit around Jupiter before the end of the decade) is digested should be one in which theories are developed, embellished, and made ready for future testing. With its remarkable range of newly discovered properties, as well as its potential application to astrophysics, the magnetosphere of Jupiter will be an exciting area of research.

1.2.5.2 The magnetosphere of Mercury.

So little was known of Mercury before the historical voyage of MARINER-10 that the mission was virtually man's first look at the innermost planet of the solar system. MARINER-10 made three close fly-bys of the planet on 29 March and 21 September 1974 and on 16 March 1975. On its first encounter with Mercury, MARINER-10 unexpectedly discovered the global, intrinsic magnetic field and the modest-sized magnetosphere of Mercury. The trajectory of the first fly-by was on the darkside, near the equator, with a closest approach distance from the surface of 703 km. The second encounter was targeted so as to optimize imaging coverage of the south polar regions and passed too far from Mercury (50,000 km) to provide any direct observations of its magnetosphere. The third encounter was similar to the first, being a very close approach toward the darkside, but was carefully modified to pass near the north polar region at a distance from the surface of only 327 km.

The observational data of Mercury's magnetosphere was obtained by three scientific instruments on MARINER-10: a magnetometer, a plasma spectrometer, and a charged particle telescope. Mercury, the smallest of the planets except possibly Pluto, has an equatorial radius of 2,439 km (compared with 6,378 km for earth). It has a detached bow shock wave, a magnetopause, and a tail current sheet similar to those of earth. The magnetopause crossing data indicate that the size of Mercury's magnetosphere is very small (see Figure 1.18). Mercury itself occupies a much larger fraction of its magnetosphere than does earth. There is no evidence for the permanent existence of a trapped charged particle radiation belt. The observational data from MARINER–10 have been used to carry out quantitative studies of Mercury's magnetosphere. Determining the characteristics of the intrinsic planetary magnetic field is difficult because the magnetospheric magnetic field of Mercury includes a substantial contribution attributable to external field sources. Even at the closest approach to the planet, the magnetic field observations were made in regions not very distant from the effects of the electric currents which flow in the magnetopause and the tail current sheet. Therefore, a proper representation of the external field becomes very important for the success of a quantitative model of the magnetosphere of Mercury. The most recent result is a three-dimensional model magnetosphere that includes a tail current sheet and the confinement of all planetary field lines within a modest-sized magnetosphere. The size of the magnetosphere agrees well with the magnetopause crossings directly observed from MARINER–10.

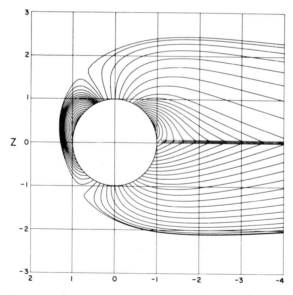

Figure 1.18 Mercury's magnetospheric magnetic field lines in the noon–midnight meridian plane calculated from a model study of observational data obtained by MARINER–10. Field lines are tangential to the magnetopause at the boundary. The X-axis is directed from Mercury to the sun, and the Z-axis is normal to the orbital plane of Mercury and directed northward. (Y.C. Wang, in *J. Geophys, Res., 82,* 1029, March 1977, copyrighted by the American Geophysical Union.)

The model, based on all magnetic field data from MARINER-10, concludes that the planet has a dipole moment that is 3,400 times weaker than the earth's dipole, tilted 2.3° from the normal to the planetary orbital plane. Its polarity is identical to the earth's. However, the planetary intrinsic field of Mercury is very much distorted from that of a simple centered dipole. This indicates that the average radius of the currents flowing inside the planet is a significant fraction of the planetary radius. This is consistent with our present understanding of the internal structure of Mercury, namely, that the planet has a large dense solid core with a radius about three-quarters of the planetary radius.

Results from the first encounter of MARINER-10 indicate that the distance to the magnetopause is 1.32 ☿ (radii of Mercury) or greater. This distance may change by ± 10 percent due to the varying dynamic pressure of the solar wind. Across the tail current sheet, the magnetospheric magnetic field undergoes a sharp change of direction (see Figure 1.18).

The interior plasma features in the magnetosphere of Mercury are similar to those of the earth's magnetosphere. The flux of low-energy electrons along the field lines connected with the polar cap region of Mercury is very small. Hot electrons in the kiloelectronvolt range are observed along the field lines connected with the inner edge of the tail current sheet. The remainder of the open field lines region was observed to have electrons of energy 100 to 200 eV. Observations by the charged particle telescope showed intense transient bursts of energetic electrons along field lines connected with the inner edge of the tail current sheet. This indicates that a local acceleration process must be active in the magnetospheric tail.

The external magnetic field, which represents a significant fraction of the magnetospheric magnetic field of Mercury, is induced by electric currents flowing in the tail current sheet and the magnetopause. The location of the two source surfaces and the intensity of the source currents vary all the time in response to the time-varying conditions of the solar wind. As a result, the solar wind conditions produce a direct temporal effect on the magnetospheric magnetic field of Mercury. In the region where the gradient of the external field is large, relatively large fluctuations in the magnetic field were observed by MARINER-10.

1.2.5.3 The magnetospheres of Venus and Mars.

Contrary to the planets described previously, neither Mars nor Venus has an intrinsic magnetic field, although each has a well-developed ionized gas envelope (ionosphere). This is why the ionized cavity surrounding each of these planets is sometimes called a pseudo-magnetosphere. The usage of this term can be avoided if we use an extended definition for a magnetosphere by considering it to be the limited region of space where the presence of the planet perturbs the direction and changes the value of the interplanetary magnetic field downstream of the bow shock. Such a general definition is then applicable to all known planets, even those not having an intrinsic magnetic field.

Since there are fewer spacecraft studying magnetospheres of the other planets than those studying the earth's magnetosphere, it is clear that there remain many unsolved problems in the physics of nonterrestrial magnetospheres. Studies of the magnetosphere of Venus are now at a stage comparable with studies of the earth's magnetosphere 15 years ago. Our knowledge of the Martian magnetosphere is at an even earlier stage.

Magnetic field and plasma measurements near Venus were carried out for the first time by VENERA–4 on 17 October 1967, at a height of 200 km above the surface of the planet; MARINER–5 approached the planet the next day at a minimal distance of 300 km, while MARINER–10 passed at a distance of several thousands of kilometers. VENERA–9 and –10 explored Venus in 1975–1976, and PIONEER Venus Orbiter has studied the planet since the beginning of 1978. The first magnetic field measurements were carried out by S. S. Dolginov and his colleagues, with VENERA–4. Their results showed that the intrinsic magnetic field of Venus is weak, less than 10nT (or gamma), and that it cannot create an obstacle for the solar wind as the geomagnetic field does.

The solar wind interacting with the ionosphere of Venus forms a bow shock that is relatively much closer to the planet than in the case of the earth. This difference arises from the fact that the obstacle creating the bow shock is not the intrinsic magnetic field of Venus but the small magnetic field produced by currents induced in the ionosphere by the solar wind flow itself. The interplanetary magnetic field moving with the solar wind plasma cannot penetrate into the electrically conducting ionosphere and accumulates in front of the planet forming a magnetic barrier (magnetic obstacle) to the solar wind.

The electron density profile observed in the dayside ionosphere of Venus exhibits a sharp upper boundary where the electron density decreases rapidly. It is at a height of 260 km and is called the ionopause. It is a typical feature of the ionosphere of Venus observed both by American and Soviet spacecraft. Until 1979, it was generally believed that the shielding electric currents producing the magnetic obstacle were distributed throughout the dayside ionosphere of Venus. However, from the magnetic field measurements of PIONEER Venus Orbiter, C. T. Russell and his colleagues have shown that these currents are mainly confined in a relatively narrow (~ 90 km) layer of the upper ionosphere of Venus. These currents are closed in the solar wind downstream from the bow shock, in a transition layer. The direction of these electric currents depends on the direction of the interplanetary magnetic field component perpendicular to the sun–Venus line. The dynamic pressure of the solar wind is usually balanced by the magnetic field pressure in the magnetic barrier; it is balanced by the kinetic pressure of the ionospheric plasma at the ionopause of Venus.

Magnetic and plasma measurements on board VENERA–9 and VENERA–10 show that Venus has an extended plasma-magnetic tail (Figure 1.19) similar to the magnetotail of the earth. This magnetotail has a diameter slightly larger than the diameter of the planet itself. The magnetotail of Venus is composed of two bunches of oppositely directed magnetic field lines; the orientation of the layer separating these bunches depends on the direction of the interplane-

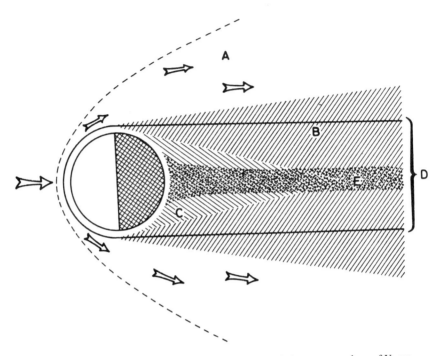

Figure 1.19 Schematic diagram of the bow shock and the magnetosphere of Venus from the data of VENERA-9 and VENERA-10. The arrows show the direction of the solar wind plasma stream. Region A is the transition region downstream from the bow shock. Region B is the boundary layer. Region C is the corpuscular umbra. The solid line shows the boundary of the magnetosphere (region D). Region E is the plasma sheet inside which is located the neutral layer that separates oppositely directed field lines.

tary magnetic field. Immediately behind the nightside ionosphere there is a cone-shaped region C of corpuscular shadow where the ion fluxes are essentially weaker than in the transition layer downstream from the shock. The apex of this cone is at about 5 Venus radii. Electron fluxes with energies ranging from ten to a few hundred electronvolts are recorded inside the corpuscular shadow of Venus. These fluxes are variable in time and form the main contribution to the fluxes observed in the nightside ionosphere of Venus.

A general view of Venus' magnetosphere and the shock wave ahead of it is shown in Figure 1.20. A rapid increase in the magnetic field intensity and a decrease in the plasma fluxes are observed at the separation between the transition layer and the magnetospheric boundary layer of Venus in a rather similar way as for the earth's magnetosphere. Figure 1.20 also shows a simplified current system which can produce a magnetic field distribution similar to that measured in the dayside and nightside of Venus.

The existence of a bow shock in front of the planet Mars was first discovered by magnetic field measurements from MARINER-4 in 1964. Other

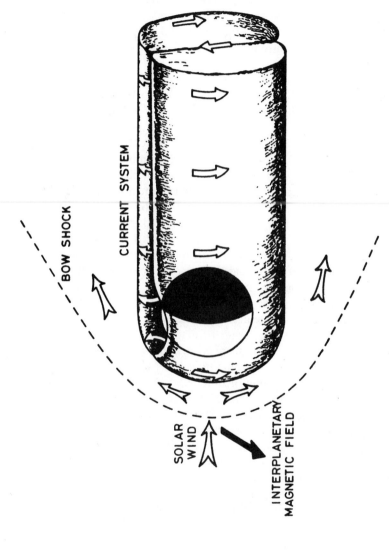

Figure 1.20 A possible simplified current system that forms the magnetic barrier ahead of Venus and the magnetic tail behind it. The orientation of the θ current structure in the magnetospheric tail changes with the changing orientation of the interplanetary magnetic field component perpendicular to the sun–Venus line.

information on the magnetosphere of Mars was obtained with MARS-2 and -3 in 1971-1972 and MARS-5 in 1974, which were equipped with magnetometers and instruments for plasma measurements. These three spacecraft did not approach the planet closer than 1,100 km. Furthermore, the instruments for plasma and magnetic field measurements were not switched on each time that the satellites passed close to the planet; these instruments were operated intermittently. This mode of operation limited the spatial and temporal resolution of the data. For these reasons, many problems concerning the Martian magnetosphere remained unsolved.

Nevertheless, the magnetospheric boundary of Mars could be determined in the same way as for the other planets, including the earth, from the increase in the magnetic field intensity, and from the decrease of plasma fluxes when the spacecraft penetrates from the transition layer (downstream from the bow shock) into the magnetosphere itself. From all crossings of the magnetosphere of Mars, it can be concluded that its shape very much resembles that of the earth's magnetosphere. From the altitudinal distribution of the electron density, it can be concluded that the ionopause is usually absent in the dayside ionosphere of Mars. This suggests that the solar wind interaction with Mars differs from that with Venus; indeed, the Martian ionosphere is shielded from the solar wind by the intrinsic magnetic field of Mars. The presence of such a magnetic field is confirmed by the fact that the diameter of the Martian magnetotail is 3.2 $R\sigma$ (radii of Mars), larger than the planetary optical shadow of 2 $R\sigma$. By contrast, the magnetotail of Venus has a diameter of 2.2 R_V.

Evidence for the presence of magnetic fields near a planet is not proof of the existence of an intrinsic planetary magnetic field. Indeed, in the case of Venus, a magnetosphere-like structure is formed with a magnetotail as a consequence of current flowing in the ionosphere around the planet. Recent observations with the PIONEER Venus Orbiter now under way will help us to understand many more details concerning the interaction between the ionosphere of Venus and the solar wind. For the future, the degree of ionization of the ions trapped in the Venusian atmosphere is not known; measurements that can determine the energy and also the mass of the ions are required. Several processes occurring in the Martian magnetosphere are not understood. An ion energy-mass spectrometer on a spacecraft passing through the Martian magnetotail would give new information on the interaction between the solar wind and Mars. The precise determination of the magnetic moment of Mars requires a low-altitude satellite, equipped with a sensitive magnetometer, orbiting around Mars.

1.2.5.4 The magnetosphere of Saturn.

In September 1979, the PIONEER-11 spacecraft discovered and penetrated the magnetosphere of Saturn. Saturn is the second largest planet in the solar system; its radius is 60,000 $\pm$ 500 km. The planet rotates with a period of 10 hours and 14 min, and it revolves around the sun with a period of 29.46 years. PIONEER Saturn made photometric and polarization measurements of four of Saturn's moons

(Iapetus, Rhea, Dione, and Tethys) and passed within 356,000 km of Titan. It may also have discovered a previously unknown moon of Saturn. Earth's moon is well outside the terrestrial magnetosphere and has no effect on it. However, many of Saturn's moons are inside the zone of trapped radiation and strongly absorb the high-energy trapped charged particles. These phenomena will be studied further using VOYAGER–1 observations made in November 1980.

Before the PIONEER encounter, five rings of Saturn had been tentatively identified. From the planet outward, they are the D-, C-, B-, A-, and E-rings. The spacecraft discovered two new rings. One of these, called the F-ring, lies just outside the A-ring. The gap between the F-ring and the A-ring has been tentatively designated the Pioneer division. The other new ring has been called the G-ring, and it lies well outside the F-ring.

The magnetic moment of Saturn is 530 times larger than the dipole moment of the earth. The polarity of the field is opposite to that of the earth. The field is largely dipolar. It has a surprisingly high degree of axial symmetry; the dipole axis of the field is tilted less than $1°$ from the rotational axis of Saturn. An offset of the magnetic center from the center of Saturn is no more than $0.04\ R_S$, principally in the polar direction.

The magnetic field of Saturn has created a magnetosphere intermediate in size between the magnetospheres of earth and Jupiter, with trapped particle intensities comparable to earth's. Saturn has a detached bow shock wave and a magnetopause similar to those of the earth. The overall form of the magnetosphere is simple and compact. The size of Saturn's magnetosphere is very responsive to changes in the solar wind dynamic pressure. The outer magnetosphere, out to the magnetopause at $7.5\ R_S$, contains lower energy plasma; its flow appears to be consistent with co-rotation with Saturn's magnetic field. The ion species measured in this region preclude the solar wind or Saturn's ionosphere as being the main source of Saturn's magnetospheric plasma. The tentative identification of O^+ and OH^+ as the dominant ions indicates that the low-energy particles are probably produced by photodissociation of ice on Saturn's rings or moons. The fluxes of trapped particles in the outer magnetosphere and their angular distributions are strongly influenced by the time-varying solar wind.

Rings of particulate material, and several small moons near the rings, strongly affect Saturn's trapped radiation. The outer atmosphere is terminated sharply by a sudden drop in both the proton and electron fluxes in a region between 6.5 and $4\ R_S$. This reduction in charged particle flux is attributable to the effective particle absorption by Dione, Tethys, Enceladus, and the E-ring. The inner region, inside $4\ R_S$, contains higher-energy particles; the particle fluxes and energies increase inward and the spectra become much harder and more complex. Particle absorption features near $2.5\ R_S$ and the imaging photopolarimeter have indicated the existence of a small Saturn moon with a diameter of about 200 km. Inside $2.3\ R_S$, the outer edge of the A-ring, there is a sharp cutoff of all trapped particles. A nearly complete absence of radiation belt particles on magnetic flux tubes intercepted by the Saturn rings leaves a shielded region close to the planet. This shielding prevents the further buildup of electron intensities at lower altitudes, which otherwise would have been present and would have made Saturn a strong radio source observable from the earth.

The radio occultation observations of Saturn indicate that Saturn has an ionosphere composed of ionized atomic hydrogen with a temperature of about 1,250° K in its upper regions. The ionosphere has two peaks in electron density, at 2,800 km and at 2,200 km.

After its encounter with Saturn, PIONEER Saturn headed out of the solar system, traveling in the direction in which the solar system moves with respect to the local stars in our galaxy. Other spacecraft are following along the trail blazed by PIONEER Saturn. VOYAGER-1 passed by Jupiter in March 1979 and Saturn in November 1980. VOYAGER-2 has also passed beyond Jupiter and encountered Saturn in August 1981, with the further possibility of traveling on to an encounter with Uranus in 1986.

1.2.6 Perspectives and Future Projects

The preceding paragraphs have presented a remarkably different picture of both the earth's plasma environment and the solar system than would have been presented less than a quarter of a century ago. In place of the presumed vacuum of outer space, the entire solar system is filled with a sparse atmosphere of fully ionized gas (plasma) rushing away from the sun at speeds up to several hundred kilometers per second, the ever-present solar wind. Further, the interaction of the solar wind plasma with planetary environments produces at each planet a unique mixture of turbulent plasmas and energetic particles, the planetary magnetospheres. The relative sizes of known planetary magnetospheres are illustrated in Figure 1.21. The dramatic change in our perception of planetary environments is demonstrated in Figure 1.21 where, for example, the magnetosphere of Jupiter is seen to be fully ten times the size of the sun.

The realization that similar plasma environments are found throughout the universe and that the bulk of the matter in the universe is in the plasma state has extended directly the applicability of solar system plasma studies to our efforts to uncover the mysteries of the cosmos and man's position in it. Figure 1.21 also shows two examples of cosmological objects to which magnetospheric concepts have been applied to explain certain of their observable features.

In the 23 years since the beginning of the space age, the field of solar system plasmas has evolved from a period of discovery and exploration to the creation of an important branch of science concerned with fundamental scientific problems extending from laboratory physics through large-scale astrophysics. What is to be done now? Are there any practical benefits to the continuation of these studies? The answers to these questions are best illustrated by considering a specific subset of solar system plasma studies, namely, solar–terrestrial relations. It is in the area of solar–terrestrial relations that social benefits can be identified, while the research results are applicable to solar system plasmas in

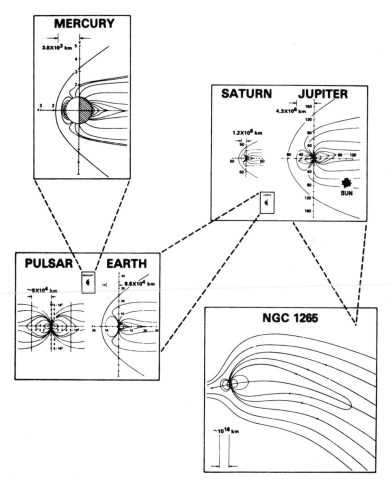

Figure 1.21 Magnetospheric plasma environments are common throughout the universe. Their sizes range from Mercury's small magnetosphere to those of galactic dimensions; Jupiter's magnetosphere is larger than the sun. The coordinates are labeled in units of the radius of the parent body; typical values of the minimum (front side) magnetospheric dimensions are given in kilometers. (NASA photograph.)

general. Figure 1.22 indicates the scope of worldwide efforts in solar–terrestrial studies.

1.2.6.1 Physics of solar–terrestrial relations.

The power flux contained in the quiet solar wind, with a density of 5×10^6 particles/m^3 and velocity of 300 km/s, is rather low, only 0.1 mW/m^2. This number is drastically increased for fast solar streams for which it can reach values as large as 3 mW/m^2. One of the interesting properties of this power flux is that it illuminates an area much larger than the earth's surface area, for the area of the magnetospheric

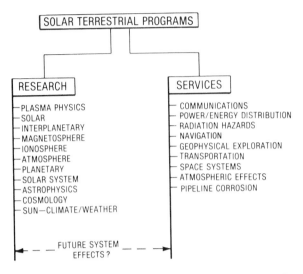

SCOPE OF THE WORLDWIDE SOLAR TERRESTRIAL EFFORT

Figure 1.22 Diagram illustrating the scope of the worldwide solar–terrestrial effort, both from scientific and practical viewpoints.

cavity as seen from the sun is of the order of 200 times the earth's cross-sectional area. Obviously, not all the impinging solar wind energy is transferred to the earth's magnetosphere, the transmission factor being rather low (less than a few percent) and, in fact, poorly known. However, two factors play a major role in amplifying the effects of this small amount of energy captured by the earth:

1. Because there is some time (a few hours) during which the energy is accumulated in the tail of the magnetosphere before being suddenly released in a rather short time (a fraction of an hour) the power available in the atmosphere is an order of magnitude greater than it would be otherwise.
2. The energy release is mainly concentrated in latitudinally narrow regions of the earth's surface, the auroral zones, whose area is of the order of 1 percent of the earth's total surface area.

Consequently, the power flux may reach non-negligible values (10 to 100 mW/m^2, or about 10^{-5} to 10^{-4} of the solar constant of $1.37 kW/m^2$). Since auroras occur on the nightside of the earth, where there is no energy input due to solar radiation, this sudden and localized input of energy to the atmosphere during magnetospheric substorms causes considerable heating of the high-latitude atmosphere at around 100 km altitude. Such events occur strongly about two or three days after the causative solar disturbance, the propagation time between the sun and the earth being three days for particles propagating at about 500 km/s. Associated with the charged particle precipitation, a system of strong ionospheric currents, flowing at altitudes between 100 and 150 km, is created. This is the auroral electrojet; its intensity can reach millions of amperes. These currents produce intense magnetic perturbations which are readily detected on the ground primarily at high latitudes. By induction in the resistive crust of the solid earth, high potential differences are created (up to 5 V/km). These may have dangerous consequences on conductive man-made systems of large dimensions such as pipelines, power lines, and telephone cables, and may they upset geological surveys. The effects of geomagnetic storms on power distribution systems are most spectacular. Many large storms (March 1940, February 1958, May 1969, August 1972) were responsible for temporary blackouts and transformer station breakdowns in the northeastern part of America. Because of the grounding of power transformers, large unexpected DC currents (up to 100 A) may flow in the windings, thereby inducing a half-cycle saturation in the cores which, in turn, leads to a voltage dropout (up to 50 percent). Heating and subsequent destruction of the insulators may also be the cause of voluntary removal of the transformers from the supply network. All these effects have been scientifically and technically assessed. It seems that with technological improvements, good training of the operators, and continuous geomagnetic monitoring, problems in this area will be overcome in the future.

More and more geological magnetic surveys (ground-based, airborne, or shipbased) are being conducted in a search for the magnetic anomalies in subsurface rocks that are indicative of petroleum and mineral resources. Since these surveys are influenced by earth-induced currents, it is crucial that when an expensive survey party is working in the field the ambient geomagnetic conditions are well understood. For such projects, it is important that predictions of magnetic activity should be improved.

1.2.6.2 Controlled experiments and environmental pollution.

In order to understand better the processes by which mass, energy, and momentum are exchanged between the different regions of the earth's environment, it has become more and more common to perform "controlled" or "active" experiments. These experiments consist of the injection into specific magnetospheric regions of a known stimulus whose characteristics can be varied and whose effects are carefully recorded. These stimuli can be waves sent out by powerful ground or space-borne radio transmitters. The transmitter power may be very high (up to 2 MW on the ground or 15 kW in space) and the frequencies can cover from the VLF (3 to 30 kHz) to the HF (3 to 30 MHz) waveband.

Electron or ion beams (with currents up to 1 A, at a voltage up to ∼ 40 kV) can be injected into the magnetosphere from rocket-borne or spacecraft charged particle guns. Large quantities of natural constituents, such as water, or of exotic species like barium, lithium, and xenon, can also be injected into the ionosphere or magnetosphere. Many of these experiments have shed interesting light on basic processes occurring within the earth's environment. In order to study the consequences of potentially environmentally detrimental activities in space (PEDAS), a panel has been created under the auspices of COSPAR.

It is important to realize that, through the generation and distribution of large quantities of electric power, we already, unintentionally, disturb our magnetospheric environment. Harmonics, up to 3–5 kHz, of the 50 or 60 Hz waves radiated by high-voltage power lines are detectable within the whole magnetosphere, and especially on those geomagnetic field lines linked to the most industrial countries. Such radiation may play a role in triggering VLF emissions and in causing charged particle precipitation, and therefore in modifying the radiation belt content. It is urgent to consider seriously both the short-term and the long-term effects that human activities in science or industry, not to mention in military fields, have on our environment.

1.2.6.3 Solar activity and the weather. A number of observations seem to point toward the existence of a connection between the solar wind and troposphere, although many scientists doubt the reality of such connections. For example, a number of parameters related to weather and climate are claimed to exhibit an 11- or 22-year periodicity, suggestive of a solar cycle dependence. On shorter time scales, statistical studies have revealed a 10 percent decrease in the vorticity area index, an indicator of well-developed cyclone activity in the troposphere, one day after the earth passes through a sector boundary when the polarity of the radial component of the interplanetary magnetic field changes from away to toward the sun or vice versa. The accuracy of the meteorological forecasts, which do not take account of this vorticity change, seems to decrease following a sector boundary passage. Furthermore, it has been found that the vertical electric field and the air–earth current through the atmosphere, measured on a mountain 3 km above sea level, increase by more than 20 percent during the two days after a solar wind sector boundary passes the earth.

It is not known whether events on the sun are really linked with the earth's climate or whether such a link, if it really exists, occurs via solar electromagnetic radiation, the solar wind, galactic cosmic rays, or atmospheric electricity. Research in this field has accelerated in the last few years as a consequence of the growing interest in climatological research. A period of co-ordinated multidisciplinary research, involving solar physics, magnetospheric, ionospheric, and atmospheric sciences, is necessary to further understanding in this area. Attention also has to be paid to finding quantitatively satisfactory physical mechanisms.

1.2.6.4 The future for solar–terrestrial physics. Some of the processes by which changes in the solar–terrestrial system occur and which may have a non-negligible influence on man's immediate environment have been mentioned. It is indeed fortunate that solar–terrestrial physics is now a mature field of research encompassing solar, interplanetary, magnetospheric, ionospheric, and atmospheric physics. Past and continuing studies of the individual components of the solar–terrestrial system have shown that they are highly interactive and that an understanding of the system as a whole requires a co-ordinated program of observations and theoretical studies dealing simultaneously with each of the key elements of the overall solar–terrestrial system. This realization not only has allowed a definition of the most appropriate problems to be attacked in order to understand solar–terrestrial system behavior but also has allowed the identification of areas in which applications of potential direct benefit to systems used by society can be foreseen.

Previously, the solar–terrestrial environment was explored and studied as a system of independent component parts—the sun, the interplanetary region, the magnetosphere, the ionosphere, and the upper atmosphere. From these earlier explorations, it now is known that this environment is a complex system composed of highly interactive parts whose total behavior differs significantly from a simple linear sum of the individual components. Whilst previous programs have advanced understanding of these components individually, an understanding of the solar–terrestrial environment as a whole requires a planned program of simultaneous observations and theoretical studies keyed to a global assessment of the production, transfer, storage, and dissipation of energy throughout this system. It is this current understanding of the various solar–terrestrial components plus the long-awaited availability of the required instrumentation that have allowed a comprehensive and quantitative study of the solar–terrestrial environment as a whole to be defined and planned.

The final goal of this program is to construct physical models of the solar–terrestrial system capable of leading to cause-and-effect predictions. To begin this task, initial objectives of the future solar–terrestrial program are as follows:

1. To trace the flow of matter and energy through the system from input by the sun to ultimate deposition into the atmosphere.
2. To understand how the individual parts of the closely coupled, highly time-dependent, solar–terrestrial system work together.
3. To understand the physical processes controlling the origins, entry, transport, storage, acceleration, and loss of plasma in the solar–terrestrial system.
4. To assess the importance to the terrestrial environment of variations in atmospheric energy deposition caused by solar–terrestrial plasma processes.

A key element in obtaining an understanding of the solar–terrestrial system is the simultaneous observation and study of the individual system components. This will require a co-ordinated ground-based and *in-situ* observing program strongly coupled to theoretical studies and modeling efforts. Since many nations are now formulating their solar–terrestrial programs for 1985 to 1995, it is necessary to continue the excellent international co-operation that has marked the field in the past in order to fulfill these ambitious, yet achievable, goals.

Today's technologically complex social systems have forced man to extend his environmental horizon from the lowest parts of the atmosphere of the earth to the surface of the sun. Winds and storms that rage in the plasmas and magnetic fields existing from the top of our atmosphere to the solar surface have effects on man which in their own right can be as potentially hazardous as the atmospheric storms which sweep across the earth's surface.

A wide range of solar–terrestrial variations can produce effects in a variety of areas including communications, transport, power and energy, space systems, geophysical exploration, navigation, and possibly, weather and climate. Magnetic storms have caused extensive power blackouts in cities and regions of the United States and Canada. These same magnetic storms can shorten expected lifetimes of the power transformers feeding energy to cities and factories. Geophysical exploration studies are curtailed during magnetic storms. Vital radio communication links are destroyed during solar and magnetospheric storms. Large solar eruptions can produce significant radiation hazards to both astronauts and airline passengers. Airline crews and passengers on a polar flight during the August 1972 solar flares could have received a dose exceeding the United States government minimum levels and exceeding that estimated for the Three Mile Island nuclear reactor accident. Solar activity also has direct impact on our upper atmosphere; it not only affects our space missions and communications systems, but it also may have effects on the earth's weather and climate.

The solar–terrestrial impacts on future terrestrial systems are not known. Whether solar–terrestrial indices may be used to improve weather and climate predictions is also not known. Thus, a better physical understanding of the solar–terrestrial system must be obtained. An international solar–terrestrial program aimed at improving understanding of the solar–terrestrial environment will produce vast improvements in the solar–terrestrial services program of great benefit to man wherever he may be.

1.3 SPACE ASTRONOMY AND ASTROPHYSICS

1.3.1 Introduction

1.3.1.1 Astronomical observations. The space age has provided scientists with unique opportunities to increase man's knowledge of the universe. By placing instruments on rockets, balloons, and spacecraft, they study many of the particle and electromagnetic fluxes radiated by astronomical sources and absorbed by the earth's atmosphere. These objects include not only the nearby sun and the stars and condensed objects of the Milky Way, man's own galaxy, but the interstellar matter between these stars, and a myriad of distant galaxies. These galaxies or "island universes," each containing hundreds of billions (about 10^{11}) of stars, occur throughout space; therefore, they can be observed far back in astronomical history. Some of the most remarkable, energetic, and little understood objects in the heavens reside at the edge of the visible universe.

The increase in man's knowledge of the nature, origin, and evolution of the universe and of galaxies, stars, and the solar system has been so remarkable in recent years that today's ideas and understanding scarcely resemble those of a generation ago. Much of this increase in knowledge has been due to observations that were impossible before the advent of space exploration.

1.3.1.2 Electromagnetic radiation.

Most of the information about the universe is obtained through the electromagnetic spectrum. As shown in Figure 1.23, each region of this spectrum is characterized by a different frequency and wavelength of the electromagnetic wave or photon, all of which travel in space at the speed of light (3×10^8 m/s). Also shown in Figure 1.23 are regions where the waves penetrate the atmosphere and reach the ground. Ordinary light waves, with a wavelength of almost 1μ (10^{-6}m), pass through such a "window"; in this way, optical astronomy has been performed for millenia. Fortunately, most ordinary stars and galaxies radiate intensely in this visible region of the electromagnetic spectrum, so optical astronomy was well established before the space age. Since the 1940s, radio astronomy, the observation of radio-frequency electromagnetic waves which also reach the earth's surface, has become an important astronomical technique. It has provided a very different view of the sun, the galaxy, and external galaxies from that obtained using ground-based optical astronomy.

It is also apparent from Figure 1.23 that vast regions of the spectrum are only accessible to instruments mounted on a space-borne vehicle. Between the radio and optical regions is the infrared part of the spectrum. Because of the molecular composition of the atmosphere, absorption in this spectral region is highly structured, and external radiation can reach the ground only in certain narrow regions. Here new results obtained from specially implemented ground-based telescopes have already shown the wealth of new understanding likely to be obtained from space observations. Radiation at all wavelengths shorter than the violet, that is, the ultraviolet (UV), extreme ultraviolet (EUV), X-ray, and gamma ray, must be observed from above the atmosphere. It is in these regions where some of the most remarkable advances in astronomy have occurred in the last 20 years because of space observations.

Even in the optical region, the atmosphere limits man's ability to observe astronomical structures. This is because the turbulent upper atmosphere bends, or defocuses, the incoming light rays in a random manner. Thus the angular resolution, that is, the ability to distinguish separate structures, even with the largest ground-based telescopes, is restricted. This limitation, called "seeing," will be overcome in the future with large optical telescopes in earth orbit.

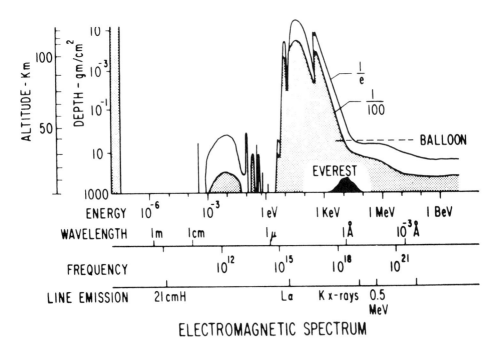

ATMOSPHERIC ATTENUATION

ELECTROMAGNETIC SPECTRUM

Figure 1.23 Diagram illustrating the effect of the attenuation, by molecules in the atmosphere, of the electromagnetic spectrum, from radio frequencies to gamma rays. The line shows the altitude at which the intensity of radiation from an astronomical object is reduced to 1/2.7 (or 1/e) of its value at the top of the atmosphere, at different wavelengths (or photon energies in electronvolts). The frequency, in hertz, is $3 \times 10^8/\lambda$, in meters. The shaded area shows the altitude at which the intensity is a hundredth of its original value; to be effective an observing instrument has to be higher than this level. The mass of atmosphere above that level, termed the atmospheric depth, is also shown in grams per square centimeter ($10\,kg/cm^2$). (Courtesy of University of California at San Diego.)

1.3.1.3 Particle radiation.   In addition to the electromagnetic spectrum, two kinds of corpuscular radiation are incident on the earth. The first of these is cosmic radiation, the nuclei of atoms stripped of their electrons and moving with nearly the speed of light. Some of these originate in flare events on the sun; at higher energies, they definitely come from outside the solar system. Cosmic rays are believed to be accelerated in the interstellar medium after injection from various sources, such as supernova explosions. Although secondary particles generated by cosmic rays interacting in the atmosphere do reach the ground, and even penetrate far underground, direct measurements of the astrophysically produced cosmic rays can only be obtained from above the earth. The second form of particulate matter consists of dust, meteorites, and similar small solid bodies which are part of the solar system and the interplanetary medium. Although occasionally the larger of such bodies reach the ground as meteorites, the majority "burn up" in the upper atmosphere and reach the earth's surface only as fine dust.

1.3.2 The Sun

1.3.2.1 The sun as a star.   Observations of the sun, and the interpretation of these observations, occupy special positions between the field of astrophysics and fields relating to the earth's environment. The sun is the source of energy for the entire planetary system, including the earth, and many important aspects of man's life have direct relationships with events on the sun. From the viewpoint of astrophysics, the sun is the only star for which structural features can be detected directly, and it thus serves as a unique source of ideas about what may be occurring elsewhere in the universe. Furthermore, many phenomena occurring on the sun are phenomena of plasmas (ionized gases) and magnetic fields, as also studied in earth-based laboratories. The sun provides a means of observing such processes on a different scale, with density, temperature, and magnetic field regimes unavailable to the laboratory worker.

The sun, when viewed with an unaided eye through a hazy atmosphere, is seen to be a sphere with a diameter of about 1.4 million km; its surface temperature is about 6,000°K. The surface, or photosphere, is marked by "sunspots," cooler regions which contain magnetic fields whose strength may be many thousands of gauss. These are shown in (A) of Figure 1.24. Parts (B) and (C) show additional features of the same sun observable in the light of red hydrogen line emission (Hα) and in the violet, respectively. Part (D) shows the complex detail that may be seen in hydrogen alpha emission at higher magnification. The finest features visible here are at the resolution limit attainable from ground-based telescopes.

The sun, like other stars, is powered by nuclear reactions in its hot (10 to 15 million°K) interior. These reactions have been occurring for at least 5 billion years, and it is expected that this energy generation will continue at its present level for billions of years more. The sunspot and surface magnetic fields are believed to be generated by material motions under the photosphere; they are manifestations of solar activity. The number and magnetic polarities of sunspots vary over a 22-year cycle; the associated magnetic fields emerge and then disappear. Also, more dynamic phenomena occur with time scales ranging from minutes to hours; these are solar flares, surges, and prominences. Although these variations were observed initially in the optical and radio bands, space observations at shorter wavelengths, such as EUV and X-ray, have led to new understandings of the basic mechanisms involved.

Above the sun's photosphere is a thin region, the chromosphere, which creates the brilliantly colored emissions seen during a total eclipse. The visible flare phenomena occur in this layer. At higher altitudes still, the temperature in-

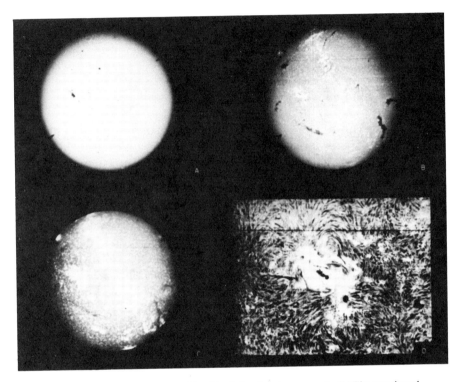

Figure 1.24 (a) The sun viewed in white light, showing sunspots. (b) The sun viewed in hydrogen α, the red Lyman α light of hydrogen (656 nm), showing active regions. (c) The sun viewed in violet light. (d) Hydrogen α observations of spatial structure in the vicinity of a pair of sunspots, showing an active solar flare. (Courtesy of Hale Observatories.)

creases to over 1 million°K in the solar corona. This extends millions of kilometers into space, merging into the solar wind and the interplanetary medium, and coupling the plasma output of the sun to the earth's environment. Throughout the solar chromosphere and corona are extensions into space of the photospheric magnetic fields. Other nearby stars, observed in soft X-rays, have recently been shown also to have coronas similar to the sun's.

1.3.2.2 Space observations. Space observations from rockets, balloons, and especially from satellites provide the most important source of new information regarding solar behavior. This is because the most direct measurements of high temperatures and dynamical phenomena are made at UV and X-ray wavelengths. Two types of space observations are made. First, the integrated output of the whole sun is measured; this is most useful for hard X-ray, gamma ray, and charged particle observations. Particular information about flares and other powerful explosive phenomena are obtained on short time scales. Second, images of the sun are formed for soft X-rays, EUV radiation, and, of course, visible light. These reveal the fundamental and slowly varying structures on the sun's surface and in the solar atmosphere.

The first astrophysical observations in the UV, EUV, and X-ray regions of the spectrum were of the whole sun, made from rockets and balloons. Space observations are now extremely sophisticated, measuring total emission fluxes and images throughout the spectrum. Understanding of the dynamic and high-temperature features of the sun and of their influence on the earth began with the first space observations; however, ground-based studies continue to be an important complementary source of information. Among the early space-borne solar instruments were those of the USSR COSMOS series of satellites and of the United States SOLRAD spacecraft and Orbiting Solar Observatories (OSO). Spectroscopic, photometric, and broadband X-ray observations were made of the whole sun. Later OSOs also carried imaging instrumentation, and in 1973 and 1974 SKYLAB (the Apollo Telescope Mount) provided the fundamental set of high-resolution images of the sun in the UV, EUV, and soft X-ray ranges.

1.3.2.3 The chromosphere.

Images made at various wavelengths in the UV and EUV regions have allowed a detailed determination of the temperature structure in the solar atmosphere to be made. Figure 1.25 shows the appearance of the sun in the light of a number of species of ionized atoms (hydrogen, nitrogen, oxygen, neon, magnesium, silicon, iron, and calcium). From data such as these the temperature and density distributions as a function of height over both the quiet sun and regions of solar activity can be obtained. This information has provided new knowledge of the radiation transfer and gas dynamics in the solar atmosphere and has influenced ideas on stellar atmospheres. Such detailed knowledge is totally inaccessible on other stars.

One of the most important problems in solar physics is that of coronal heating, that is, of the mechanisms by which the turbulent motions of the photosphere cause the outermost atmosphere of the sun to be heated to a million°K to form the corona. Space observations have provided diagnostic data that rule out an earlier explanation of coronal heating, namely, that of acoustic waves generated in the photospheric turbulence, OSO-8 and SKYLAB observations have indicated a mechanism involving magnetic fields. Observations of stellar X-ray emissions have clearly shown that the coronas in other stars are far more common than could be explained by the acoustic wave hypothesis.

1.3.2.4 The corona.

The most remarkable new discoveries concerning the sun have occurred from exploring the X-ray region. Indeed, the detection of solar X-rays from the quiet sun was a direct verification of the high temperature of the solar corona. In addition, X-rays have revealed structures and indicated dynamic processes which were, at best, only suspected previously. Figure 1.26 is a composite photograph of the sun, with the corona being taken from the ground during an eclipse and with the disc of the sun being in soft X-rays.

The SKYLAB observatory has provided the best continuous sequence of

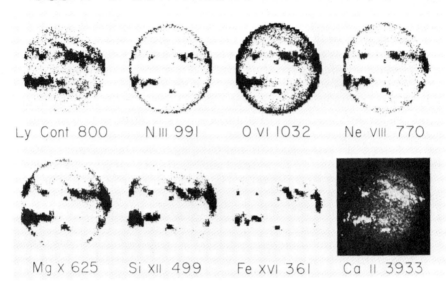

Ly Cont 800   N III 991   O VI 1032   Ne VIII 770

Mg X 625   Si XII 499   Fe XVI 361   Ca II 3933

Figure 1.25 Solar images in the light of hydrogen (Lyman continuum radiation), doubly ionized nitrogen, five times ionized oxygen, seven times ionized neon, nine times ionized magnesium, eleven times ionized silicon, fifteen times ionized iron, and singly ionized calcium; each type of radiation is characteristic of a certain temperature and height in the chromosphere. The wavelengths are given in Ångstrom units, Å (10^{-10} m; note that 10 Å = 1 nm). (Reproduced, with permission, from the *Annual Review of Astronomy and Astrophysics,* Vol. 9, © 1971, by Annual Reviews Inc.)

images of the sun at soft X-ray wavelengths, showing the "natural" emission of the solar corona. Discoveries made using SKYLAB include coronal holes, coronal transients, bright points, and the magnetic loops that appear ubiquitously in regions of solar activity. Coronal holes consist of regions in which the magnetic field extends radially outward, linking the solar surface to the interplanetary medium and to the rest of the solar system. These are the sources of the high-speed solar wind streams, which cause geomagnetic disturbances at the earth. Elsewhere in the corona the magnetic field is closed to another part of the sun's surface, and the whole corona consists of a variety of magnetic loops. The magnetic field that defines the loops may confine hot coronal plasma which is visible in soft X-ray images. The loops first emerge in the form of newly born active regions and subsequently grow as the active regions develop. Later on, as the active region decays, the loop structures "thin out"; a few loops remain and create the basic loop structure of the "quiet" solar corona. Particularly interesting are those loops that interconnect individual active regions across large distances, even across the solar equator.

The solar active regions, with looped magnetic field structures, give rise to solar flares, filament eruptions, charged particle acceleration, and other kinds of

Figure 1.26 A white light coronograph picture of the sun, taken during an eclipse, superimposed on a soft X-ray picture. Related features, such as coronal streamers and coronal holes, are seen; these delineate magnetic field regions that are either closed (loops) or open to interplanetary space. (VanSpeybroeck, Krieger and Vaiana, in *Nature,* 227, 818, 1970.)

solar activity. The SKYLAB observations showed that the emergence of magnetic flux from the solar interior was a much more common phenomena than had been thought. Hundreds of small regions of enhanced magnetic flux emerge every day over the whole solar surface, but only very few of them develop into active regions. All of the regions reveal themselves as bright points in the X-ray pictures; together they bring as much magnetic flux into the solar atmosphere as do the active regions themselves.

SKYLAB also discovered the existence of coronal transients, enormous clouds of plasma expanding from outbursts near the solar surface and propagating through the outer corona into interplanetary space. As the SKYLAB coronograph demonstrated, these coronal transients originate in disrupted filaments (magnetic field-aligned regions of high plasma density seen as eruptive prominences on the limb), both in active regions and outside them. Apparently, an instability opens the originally closed magnetic field line, and large amounts of plasma, including the filament, are ejected into the corona. As the opened field lines reconnect, strong heating is produced, which may give rise to a solar flare. Very powerful "two-ribbon" flares occur if the filament disrupts inside a well-developed active region; in such a case, the associated transient in the corona carries enormous amounts of energy, up to 10^{25} J.

1.3.2.5 Solar flares. The most remarkable explosive phenomenon on the sun is the solar flare. As shown in part (D) of Figure 1.24, a sudden brightening appears above a sunspot region in the light of hydrogen alpha; the brightening occurs in a few minutes and lasts up to an hour. These flares have a wide range of sizes and associated phenomena, and they also cause terrestrial effects such as sudden ionospheric disturbances. They have total energy output in the range from 10^{22} to 10^{25} J.

Figure 1.27 shows the typical sequence of events during a solar flare observed at many wavelengths. The sudden increase in the intensity of hydrogen alpha radiation, the impulsive phase, is associated with bursts of radiation in the microwave region, at UV, and at all X-ray wavelengths. During this time, X-ray observations show that the plasma is abruptly heated to about 20 million° K, and it is during this time that electrons and protons are accelerated to relativistic energies. Some of these particles reach the denser regions of the solar atmosphere or photosphere, and on occasion produce very hard X-rays, to hundreds of kilo-electronvolts, and bursts of gamma rays due to nuclear reactions.

Following the impulsive phase is the gradual phase, where the heated and dynamically driven plasma leaves the solar atmosphere and moves into the corona and interplanetary medium. It thus gives rise to radio emissions at lower frequencies, softer X-rays, and various other effects in the corona. Not shown in Figure 1.27 are the accelerated charged particles or solar cosmic rays, injected into the interplanetary medium. These are detected near the earth some one to three days after the flare, depending on the configuration of the interplanetary magnetic field between the solar flare region and the earth.

1.3.2.6 Solar observations: present and future. The most recent solar observations are from the Solar Maximum Mission (SMM) launched on 14 February 1980 and carrying seven different experiments from the United States, the United Kingdom, the Netherlands, and the Federal Republic of Germany. The instruments consist of a co-ordinated set of telescopes and whole-sun sensors devoted to the study of solar activity, and especially solar flares. A coronagraph images the solar corona at distances between 1.6 and 6 solar radii. This instrument can distinguish between hot (coronal) ejecta and cold plasma stemming from the remnants of a disrupted filament. The coronagraph shows that the cold ejecta can be seen at very high altitudes in the corona as they escape into interplanetary space. An ultraviolet polarimeter and spectrometer provide images (to a resolution of one arc second) of solar atmospheric structures as viewed in particular spectral lines. The X-ray polychromator studies in detail, and for the first time, the instant at which the line emissions of the very hot plasmas are formed in solar flares. As shown by the pioneering observations of USSR instruments on INTERCOSMOS satellites, these lines, including those from 24 times ionized atoms, provide essential information about the conditions in solar flares.

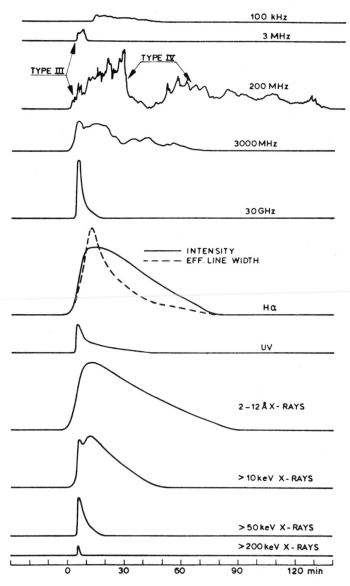

Figure 1.27 Solar flare associated radio frequency, microwave, hydrogen Lyman α, UV, and X-ray variations; microwaves and hard X-rays have their maximum emission during the impulsive phase of the flare, whereas the hydrogen α, radio, and soft X-rays are associated with the heated plasma and its ejection into interplanetary space. (Z. Svestka, "Solar Flares," courtesy of D. Reidel Publishing Co., Dordrecht, Holland.)

For higher energies, the SMM has a gamma ray spectrometer, which detects solar cosmic rays at their acceleration site near the flare proper, a hard X-ray burst spectrometer for observing the hard X-ray bursts that mark the most intense energy release in a solar flare, and a hard X-ray imaging spectrometer. A preliminary result suggests that energetic electrons are located in the magnetic loop structures and that they stream downward into the solar atmosphere as do the electrons that produce the terrestrial aurora borealis.

The Solar Maximum Mission also carries the first accurate instrument for continuous monitoring of the solar constant, the total flux of solar energy input into the terrestrial environment, and its variations as sunspot activity waxes and wanes. Such observations are of direct and practical significance for studies of the problems of energy and climate on the earth.

The development of space techniques for solar observations has only just begun. The exploitation of the natural advantages of space platforms for optical observation to obtain high spectral resolution at all wavelengths has not yet been accomplished. The solar optical telescope (SOT), a Space Shuttle-launched SPACELAB facility, heads the list of future experiments. This telescope will, for the first time, permit observations to be made with a reasonably large telescope (1.25 m diameter) at the diffraction limit set by the instrument itself rather than by the atmospheric "seeing" conditions. These observations will resolve fine structures on the sun down to size scales of 100 km, with uniform "seeing" conditions that will greatly increase knowledge of solar atmospheric structure.

The International Solar Polar Mission, a joint European Space Agency/National Aeronautics and Space Administration program, will carry out the first explorations of the interplanetary medium beyond the ecliptic plane. The spacecraft will carry solar imaging experiments as well as the particle and field sensors necessary to characterize the interplanetary medium. At some later time, the Star Probe spacecraft will actually approach the solar surface, reaching as close as four solar radii, thus effecting a link between remote sensing and *in-situ* measurements of the outer solar atmosphere. The two HELIOS spacecraft of the Federal Republic of Germany have already performed the first reconnaissance of the solar wind as close as 0.25 AU from the sun.

Finally, future observational tools will include the very high spatial resolution instruments necessary to answer many questions, especially those relating to solar activity where compact magnetic field structures seem to be the seat of the solar flare instability. Among these is the Pinhole/Occulter Facility, which will utilize two separate spacecraft to obtain a very long focal length instrument, with an angular resolution, for X-radiation, of a small fraction of an arc second.

1.3.3 Astronomy, Astrophysics, and Space Research

1.3.3.1 Stellar evolution.   The Milky Way galaxy consists of about 10 billion stars, arranged in a saucer-shaped configuration. The sun is a rather ordinary star; it is situated toward the edge of the galaxy, being about 30,000 light years from the galactic center. One light year is the distance traveled by light in one year at its speed of 300,000 km/s. The Milky Way is the galactic disc observed edge-on. Because of dust and other interstellar matter, objects only a small fraction of the distance from the sun to the galactic center can be seen from the earth. It is mainly through radio and infrared studies, and by comparison with external galaxies, that the spiral structure of the galaxy has been inferred. Optical observations of extragalactic objects are limited to those in a direction away from the plane of the galaxy, but they extend far out toward the edge of the universe.

Ground-based studies have given a remarkable understanding of the origin, evolution, and present state of stars. Stars "condense" from interstellar gas due to gravitational attraction; at a certain stage of condensation, they ignite their nuclear "furnaces." They then live (burn), decay, and die, much like animals. Depending on their mass, they may explode as supernovae, sending forth their newly produced chemical elements to follow a further evolutionary sequence, or they may collapse to form white dwarfs, neutron stars, or even "black holes." Several combinations of these processes may occur during the birth, life, and death of a star.

Man's knowledge of stellar evolution is best displayed using the Hertz-prung-Russell (H–R) diagram, shown in Figure 1.28. The "main sequence" is the temperature-luminosity relation that stars of various masses have during most of their lifetime. Massive stars (with a mass 15 times that of the sun, $M_\odot$) in the upper left of the diagram are at high temperatures (about 30,000°K) and evolve rapidly. Low-mass stars, like the sun, are cooler (down to 3,000°K) and are found at the lower right of the diagram; these live for thousands of millions of years. Degenerate stars, such as white dwarfs, are also hot, but because of their small size they radiate very slowly and hardly fit on the H–R diagram. Neutron stars, which are still more condensed, are not even shown in Figure 1.28.

Beyond the Milky Way are uncounted thousands of millions of other galaxies. Based on their optical appearance, these are classified into spiral, barred spiral, and elliptical galaxies. Figure 1.29 shows the giant elliptical galaxy NGC 5128 located at a distance of about 5 million light years in the constellation Centauris. This shows an unusual dark dust band believed to be the residue of evolved stars. Like many of the galaxies, CGC 5128 emits radiation at radio frequencies strongly and has also been recently observed to radiate X-rays and gamma rays. These active galaxies may be related to the mysterious quasars. Because the universe is expanding, distant objects appear to be receding. Many of the most interesting objects are at such extremely large distances that they are apparently moving away with nearly the speed of light. At these distances the radiation observed now was emitted a long time ago; thus the early evolution of the universe can be studied. The understanding of the origin and evolution of the universe, and its relation to fundamental ideas of gravitation, space–time, general relativity, and elementary particle physics, forms the subject called "cosmology."

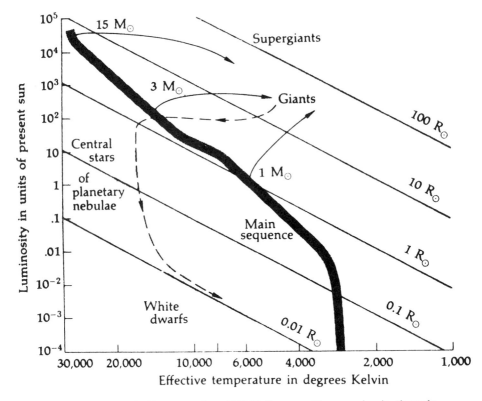

Figure 1.28 The Hertzprung-Russell (H-R) diagram, with arrows showing the evolution of main sequence stars of different masses, expressed in terms of the sun's luminosity, mass ($M_\odot$) and radius ($R_\odot$). (John C. Brandt and Stephen P. Maran, *New Horizons in Astronomy* (Second Edition), W.H. Freeman and Company. Copyright © 1979.)

1.3.3.2 Observations from space.

Space observations have revealed new and important phenomena on all scales in the universe, from solar to stellar, and from galactic to cosmological. Many of these would not or could not have been predicted from ground-based work; discoveries made over the past 20 years have substantially modified man's view of the universe.

Results in the ultraviolet spectral regions have been obtained from the United States Orbiting Astronomical Observatories (OAO), including COPERNICUS, the Astronomical Netherlands Satellite (ANS), and, most recently and importantly, from the joint United States/ESA International Ultraviolet Explorer (IUE). Rocket and balloon observations, in addition to making important contributions to the exploratory UV and EUV work, have thus far been the principal source of nonground-based infrared studies. However, it is in the area of high-energy astronomy (X-rays, gamma rays, and cosmic rays) that the most unexpected and remarkable new discoveries have been made. Exploratory obser-

Figure 1.29 The giant elliptical galaxy NGC 5128 in Centaurus, which has an unusual dust band which absorbs the light from its stars. This galaxy is also a powerful radio, infrared, X-ray, and gamma-ray emitter. (Courtesy of Hale Observatories.)

vations were carried out by rockets, balloons, and modest instruments on early United States and USSR spacecraft. The United States Small Astronomy Satellites, which include UHURU, SAS–2, and SAS–3, and later the European COS–B, established X-ray and gamma-ray astronomy as forefront fields of research by the mid-1970s for studies of stellar and galactic evolution, cosmology, and fundamental physical processes. Recently, the three spacecraft of the United States High-Energy Astronomical Observatory (HEAO) series have raised the capabilities of X-ray astronomy to equal those of ground-based optical astronomy, and they have continued the flow of significant new discoveries and understanding of astrophysical processes throughout the universe.

1.3.3.3 Galactic ultraviolet and infrared observations. Thousands of stars and many galaxies have now been observed in the ultraviolet (UV) region of the spectrum. While many of these observations have only confirmed the emissions expected from hot stars in the UV, the new data have led to a much more detailed knowledge of stellar atmospheres. The COPERNICUS satellite established the existence of a hot, thin component of the interstellar medium in the galactic plane. It explored the interstellar medium in many directions and provided new information on deuterium abundances. Now the IUE, with a 0.45 m diameter mirror and a high resolution spectrometer, is producing the most important new data.

Figure 1.30 shows typical observations, which are impossible from the ground, of a class of degenerate stars fitted with refined model atmosphere calculations. This work has required the re-evaluation of temperature distributions and elemental abundances, particularly of metals, in these objects. Other important results from UV observations include the discovery of hot, compact

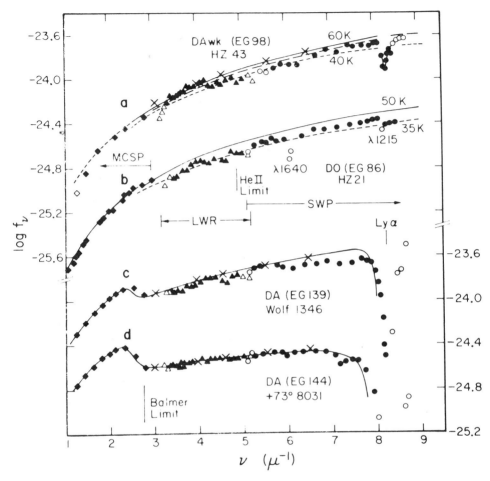

Figure 1.30 Ultraviolet spectra of various degenerate stars; the intensity is plotted, on an inverse logarithmic scale, against the wave-number ν, which is the reciprocal of the wavelength in μ, 10^{-6}m. Ground based telescopes can only observe radiation to the left of the Balmer limit. (J.L. Greenstein and J.B. Oke, *Astrophysical Journal,* 1979 Ap. J. (Letters) 229, L141 (May 1, 1979).)

stars and circumstellar envelopes of coronas in a wide variety of stellar objects. Stellar winds (mass outflows from stars) are far more common than had previously been thought. These latter results have been particularly important in understanding the gas dynamics of binary stars, that is, two stars in orbit about each other. The strength of the UV emissions in cool stars may be associated with rotation rates and, by implication, with stellar magnetism. Also, the Milky Way galaxy and many others apparently possess much material from coronas.

The impact of infrared (IR) observations on astronomy from ground-based work indicates its potential from space, where the full range of the spectrum is available. Exploratory work from rockets and balloons has provided a map of hundreds of possible sources and has indicated excess emissions at the longer wavelengths, beyond that predicted by a star radiating only as a "black body." Star-like cool objects, with temperatures as low as 800°K, have also been discovered. It is in the IR that the lower-right part of the H–R diagram (Figure 1.28) can best be investigated. Dust shells containing silicates and ices are common in many stellar structures. Figure 1.31 shows Orion Nebula, which is a nearby galactic region where stars are in the process of condensation. The cool protostars can best be studied in the infrared. The earliest phases of stellar evolution are thereby studied. Because of interstellar absorption or "reddening," it is only in the infrared, radio, and X-ray bands that "maps" of the galactic center region can be made.

Space observations in the EUV (about 80 to 5 nm) also have significance for astrophysics. Again, because of absorption by the interstellar gases, large distances in the galaxy cannot be studied. Early exploratory work with an instrument on the joint United States/USSR APOLLO–SOYUZ has already shown an unexpectedly large emission from a binary system known as HZ43, apparently due to a very high temperature (about 100,000°K) region.

1.3.3.4 Galactic X-ray and gamma-ray astronomy.

X-ray astronomy was born with the discovery in the early 1960s that emissions in the 1 to 10 keV range from cosmic sources were detectable with rather modest instruments placed above the earth's atmosphere. Since then, with the OSOs, UHURU,

Figure 1.31   A nebula in Serpens, containing protostars, where the early phases of star formation can be studied by infrared techniques. (Courtesy of Hale Observatories.)

SAS-3, ARIEL, HEAO-1 and -2, and numerous other satellites, over 1,000 sources on the celestial sphere have now been found. Hundreds have been identified with known optical or radio objects and have been studied in great detail. X-ray observations provide information on extremely hot gases (at 10^7 to 10^8 °K), on nonthermal particle populations, on the behavior of plasma in dynamical magnetic field regions, and on processes occurring in extremely dense and unusual states of matter. Figure 1.32 is a map in galactic coordinates (the galactic center is at the map's center), showing that the X-ray sources are concentrated in the galactic plane. There are also many sources at high galactic latitude; these are mainly extragalactic (beyond the Milky Way galaxy). The strongest of the galactic objects radiate about 10^{30} to 10^{31} J/s (or watts) in X-rays alone; this is about 10,000 (10^4) times the total energy output of the sun, and comparable with the most luminous stars in the H–R diagram (Figure 1.28). The map also identifies some of the most important and well-known X-ray sources.

X-ray emission from binary systems, with a compact component (a white dwarf neutron star, or "black hole") produced by accretion from the mass outflowing from the larger stellar component, is also observed. Many binary stellar systems show highly variable X-ray emissions at low luminosity, about 10^{23} to 10^{25} W, believed to be associated with the unstable transfer of mass between the two objects. Supernovae remnants, having turbulent gases at a temperature about a million° K also radiate at X-ray wavelengths. Many otherwise thermal sources have, like solar flares, a "nonthermal" component of X-rays, indicating populations of accelerated electrons. There is a class of sources known as "bursters," which occasionally or quasi-periodically produce a burst of X-ray radiation. Some of these seem to be located in globular clusters of normal stars.

Observations made from the Einstein Observatory (HEAO-2) instrument, which has a sensitivity about a million times greater than those of the first exploratory instruments, show that many stars have coronas like the sun, but with much greater powers in the X-ray emissions. Even Alpha Centauri, our nearest stellar neighbor 4.3 light years away, has a corona about ten times more intense than the sun's.

Gamma-ray emission is unique to astronomy in that it is the specific radiation associated with the occurrence of nuclear processes rather than being primarily associated with interactions among the electrons of atoms. Even though observations in this spectral region are still in an exploratory phase, a number of discoveries have been made. Many X-ray sources, such as the Crab Nebula (Figure 1.33), Cen A (shown in Figure 1.29), and the Galactic Center have been shown to have high energy emissions extending to tens and even hundreds of megaelectronvolts. Gamma-ray bursts, of short duration (1 to 10 s) and relatively intense (about 10^{-16} J/m² at the earth), and pulses of gamma rays, with variable temporal and spectral characteristics, are totally unexpected phenomena. About 100 of these events have been observed over the past decade. Their origin is believed to be galactic; their association with known objects is

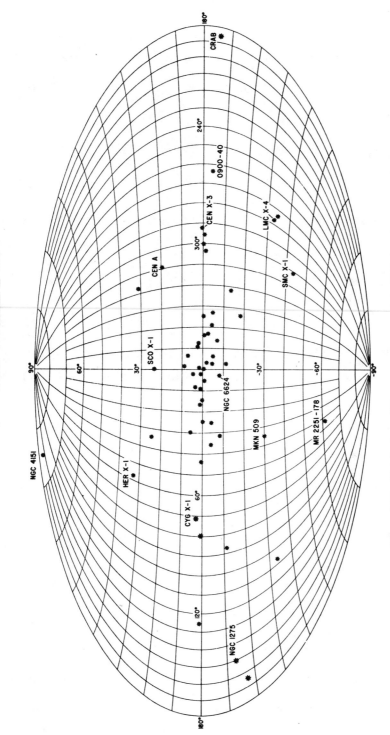

Figure 1.32 Distribution of X-ray sources mapped, in 1975, in galactic coordinates; over a thousand sources have now been catalogued. The stronger sources are clustered on the galactic plane; most sources at high latitudes are, in fact, extragalactic. (Courtesy of the University of California at San Diego.)

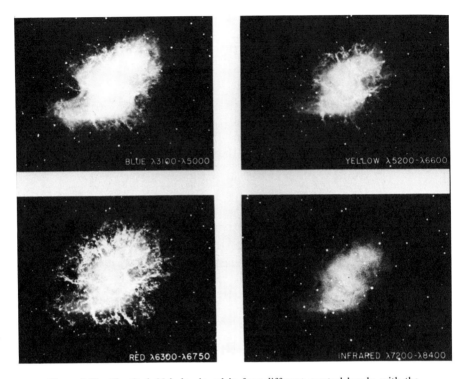

Figure 1.33 The Crab Nebula viewed in four different spectral bands, with the wavelengths shown in Å; it is also a source of radio and X-ray emissions. The pulsar NP 0532, the neutron star remaining after the supernova explosion of 1054 A.D., is near the center of the nebular region. This pulsar emits radiation at all wavelengths of the electromagnetic spectrum. (Courtesy of Hale Observatories.)

only speculative so far. One gamma-ray burst source has been identified as a supernova remnant in the Small Magellanic Cloud.

Steady gamma-ray emission at 100 MeV from the disc of the galaxy has been discovered and mapped by SAS–2 and COS–B. This is believed to be caused by cosmic rays interacting with interstellar matter. It thus provides a connection between cosmic-ray acceleration and propagation, radio astronomy, and conventional astrophysics. In addition, there are about a dozen localized sources of 100 MeV gamma rays in the galactic plane. Once again, although several of these have been associated with radio pulsars in supernovae remnants, the majority are of unknown origin.

Cosmic rays, coupled to galactic magnetic fields, are important for studies of the structure and dynamics of the galaxy. This is because their pressure is comparable with that of the interstellar gas. Cosmic rays interacting in the interstellar medium of the galaxy are expected to produce a number of gamma-ray lines due to such nuclear processes as positron annihilation at 0.51 MeV and the de-excitation of C^{14} at 4.41 MeV. Already balloon and HEAO observations have detected the annihilation line from the galactic center, and there are indications of other lines. Gamma-ray spectroscopy has enormous potential as a technique in astrophysics; it can only be carried out with a sophisticated instrument in space.

1.3.3.5 Supernovae, neutron stars, and black holes.

High-energy astronomy has been particularly successful in discovering and understanding the processes associated with compact stellar objects. White dwarfs are stars which have collapsed into a degenerate electron gas; neutron stars have undergone a further collapse into degenerate nuclear material. A white dwarf with a mass of the sun has a diameter about that of the earth; a neutron star may have a diameter of only about 10 km. More massive collapsed objects may have a sufficient gravitational attraction at their surface such that no electromagnetic radiation can escape from them. This is the explanation of the term black hole.

Figure 1.33 shows the Crab Nebula, visible in the optical band as an expanding gas shell after the original supernova explosion in 1054 A.D. Supernovae are among the most violent events in the universe and are thought to occur in the galaxy at a rate of about three per century. The Crab Nebula is now 5 light years across and contains a magnetic field. It produces unusual radio, optical, and X-ray emissions due to the presence of extremely high-energy particles. Embedded in the nebular region is the collapsed neutron star left over from the explosion. This can have an intense magnetic field, about 10^{12} G. The neutron star in the Crab Nebula is rotating at the rate of about 30 times per second; it is the seat of the intense pulsed emissions observed over the entire electromagnetic spectrum, that is, it is a pulsar.

Although the discovery of radio pulsars indicated the existence of isolated rotating neutron stars, it is observations of their pulsed X-ray emissions that determine their properties in dynamic stellar systems. Figure 1.34 is an artist's concept of a binary stellar configuration involving a compact object moving in a close orbit around a more massive ordinary star. Matter outflowing from the ordinary star is formed into a swirling accretion disc and then eventually falls into the gravitational "hole" formed by the compact object, producing X-rays. If this object is a rotating neutron star, the X-ray emission may be pulsed at the rotation period of the neutron star. For example, the Her X-1 system has a 1.24 s rotation period and a 1.7 day orbital period. Alternatively, if the compact object is a black hole, such as the candidate object in Cyg X-1, the X-ray emission is highly variable on millisecond (ms) time scales.

Attempts to understand the X-ray emission from binary systems have led to new knowledge of binary stars, their evolution, the gas dynamics associated with the mass transfer between them, processes occurring in the atmosphere of neutron stars, and the structure of neutron stars. The very idea that black holes may indeed exist in nature has resulted in new developments in general relativity and in understanding physical processes near the "event horizon."

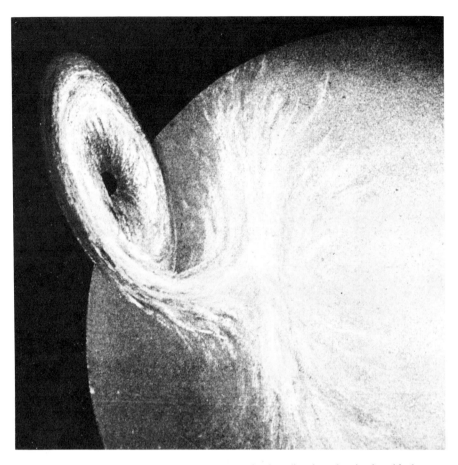

Figure 1.34 Diagram illustrating mass accretion in a disc-shaped region by a black hole toward the upper left. X-rays are formed when stellar material hits the surface of the compact object. (Victor J. Kelly © National Geographic Society.)

1.3.3.6 Ordinary galaxies. Progess in understanding the structure, composition, and evolution of ordinary galaxies, like the one containing the sun, has been particularly remarkable in the past few decades. Space observations in spectral regions unavailable from the ground have contributed much to this knowledge. The ability to observe to a redshift $z = 1$, where galaxies with an age half that of the Milky Way are observed, will aid understanding of stellar evolution. The Space Telescope will have this ability. Observations in the IR, UV, and X-ray regions of the spectrum have already made important contributions. COPERNICUS and IUE have observed such UV lines as 1,550 Å (155 nm) from C IV (three-time ionized carbon) in this and other nearby galaxies. They have confirmed the general existence of hot galactic coronas, a phenomenon predicted by Lyman Spitzer, Jr. some 25 years ago. Ultraviolet studies have also allowed the study of hot stars in other galaxies. Infrared observations indicate the presence and, indeed, the distribution of such interstellar dust components as graphite, ice, and silicates. In the far infrared and submillimeter regions of the spectrum, the detection of many more molecular species in both this and nearby galaxies can be expected.

X-ray observations have confirmed that nearby galaxies have emissions similar to those in this galaxy. A pulsating X-ray binary system has been discovered in the Large Magellanic Cloud; the distribution of bright sources in M31, the nearby Andromeda Nebula, has been mapped. Soft X-ray studies seem to indicate that the sun is in a region of hot gas in the galaxy.

1.3.3.7 Active galaxies.

The evidence for explosive, dynamic, and rapidly varying phenomena associated with the nuclei of galaxies, or with compact galaxies, is incontrovertible. These galaxies include powerful radio emitters, such as those in Centaurus and Virgo, the Seyfert galaxies, and quasistellar objects. The genesis and evolution of these and a number of other strange extragalactic phenomena are at present unknown.

Seyfert galaxies are characterized by having a very bright compact center and strong emission lines. NGC 4151 is particularly well studied. Over 15 years ago, the Stratoscope balloon-borne telescope showed its nucleus to be very compact, less than 0.2 s of arc across. X-ray and gamma-ray observations have indicated an exceedingly hard and time-varying spectrum, observed up to several hundred kiloelectronvolts. UV observations detect and study important atomic emission lines. They extend knowledge of the nonthermal continuum emission, and, therefore, of the structure and composition of the compact emission region. In addition to Cen A, 3C273, and NGC 4151, many other active and compact galaxies have now been detected and studied at X-ray wavelengths with HEAO–1 and the Einstein Observatory satellite (HEAO–2). Most of these galaxies have their greatest luminosity at X-ray wavelengths. Many of these objects are observed at high redshifts, that is, they are at the very edge of the universe.

1.3.3.8 Clusters of galaxies.

Galaxies are not usually found in isolation, but instead are grouped into a gravitationally bound dynamic configuration, known as a cluster; this contains many tens of galaxies. The observation that X-ray emission from the intracluster medium is a common feature of many clusters is an important discovery. The presence of the X-rays implies an intracluster gas at a temperature of about 100 million degrees (10^8 °K). Either this gas is flowing out of the galaxies or else it has been left over from their formation; it is heated by the motion of the galaxies. This discovery has provided insight into the "missing mass" problem, that is, the problem that the observed mass in the actual galaxies is insufficient to bind the clusters together gravitationally. The large contribution of the intracluster gas to the mass of the universe has significance in cosmology. Ultraviolet observations of distant clusters may eventually provide important information needed to understand the temperatures, densities, and distribution of hot gases in clusters.

1.3.3.9 Cosmology. The accidental discovery of the omnipresent 3°K background radiation can be understood on the "big-bang" model of the origin of the universe. Here, ordinary astrophysics, fundamental particle physics, and general relativity are all required. One of the parameters that determine the structure of the big bang is the rate of expansion of the universe, measured by the "Hubble constant." Another is the mean mass density which permits a determination of the deceleration parameter to be made. The degree of anisotropy of the universe, as determined by the distribution and motions of the distant galaxies and of the diffuse background radiation, is also an important parameter.

Space observations have contributed much to observational cosmology. Balloon work has verified the spectrum of the 3°K microwave radiation and shown that it has the blackbody characteristics expected from the remnants of the original explosion. X-ray and gamma-ray studies have shown the existence of a generally bright sky background, which has a complex spectrum and which presumably originates at very large (cosmological) distances. It is not yet known whether this can be explained simply by a superposition of X-ray emitters in the galaxies or whether it is a more fundamental phenomenon. Certainly X-ray and gamma-ray observations of the most luminous, the hottest, or the most non-thermal objects in the distant universe will contribute to the development of cosmology.

Of particular interest is the apparent lack of anti-matter in the universe. It is possible that the laws of physics that govern the fundamental particle interactions in the first few microseconds of the universe are nonsymmetrical. Otherwise, it would be expected that a quantity of anti-matter comparable to the quantity of matter in the universe must have been produced; if so, it must be somewhere. Gamma-ray observations could provide information that may solve this fundamental problem.

1.3.4 The Future of Astronomy from Space

The potential of astronomy from space is clearly enormous. Even in its infancy, the subject has led to many discoveries. New observational tools are under development by members of various nations of the earth. All nations recognize the necessity of using the space environment to obtain knowledge so that man may understand the origin and evolution of the universe.

1.3.4.1 Proposed infrared, optical, and ultraviolet observations. The Space Telescope, a joint United States/ESA venture, has already been described as being the single most important step in furthering an understanding of the stars, galaxies, and cosmology in the "classical" spectral regions. The 2.4 m diameter diffraction limited telescope will provide images of about 0.15

arc sec diameter. It will have five instruments at the focal plane and, with Space Shuttle refurbishments, should have a lifespan of 20 years. The Space Telescope is expected to become operational in 1984.

The Space Telescope is an important new tool designed to attack the most important problems in cosmology. It should be able to observe five times further into the distant universe than is possible using the largest ground-based optical telescopes. The volume of the universe studied will therefore be 100 times greater than that known at present. Observations will be made covering the entire region from near infrared to the far ultraviolet. The angular resolution of the Space Telescope will be about 0.15 arc sec, some five times better than the best ground-based instruments. Furthermore, this resolution will be available on a regular basis, not under unusual conditions of "seeing," and independently of the weather.

Figure 1.35 shows the effect of this improved resolution on a distant galaxy. This figure is based on a computerized simulation of the barred spiral galaxy NGC 2523. If the ground-based observations were equivalent, for example, to $z = 0.27$, the Space Telescope would provide an image in an equivalent time to that at $z = 0.05$. The additional detail to be expected from the Space Telescope observations is evident. Furthermore, spectroscopic investigations can be accomplished over each pixel (square element) of the image.

The focal-plane instruments of the Space Telescope include a faint-object spectrograph, a high-resolution spectrograph, a wide-field planetary camera, and a high-speed photometer, all supplied by United States investigators, plus a faint object camera provided by ESA. Other focal-plane instruments can be installed in orbit as technical developments make them available. The Space Telescope is thus a versatile observatory that can make fundamental contributions over a very long period.

IUE is a joint United States/ESA satellite which is expected to continue making observations well into the mid-1980s. A joint United States/Netherlands program is the Infrared Astronomy Satellite (IRAS). This has a 0.5 m diameter mirror and an IR sensor assembly, all cooled to liquid helium temperature; it will perform a sky survey in four broad spectral regions. It is expected to detect and locate 10^5 to 10^6 sources after launch in 1982. A United States project, Cosmic Background Explorer (COBE), is designed to explore the spectral distribution and the isotropy of the $3°K$ radiation in the 100 to 1,000 μ (0.1 to 1 mm) wavelength region. The ESA astrometry mission, HIPPARCOS, is intended to provide new data of a different nature. Accurate and extensive measurements of the positions and proper motions of nearby stars are needed to determine accurate distance scales and hence to study galactic dynamics. The positions of a hundred thousand stars will be measured to an accuracy of 0.002 arc s ($''$), and their motions to an accuracy of 0.005 $''$/year.

In addition, there are many investigations planned from balloons and rockets, and from the United States/ESA Space Shuttle-launched SPACELAB. These include the shuttle infrared telescope facility (SIRTF), which is a 1-m dia-

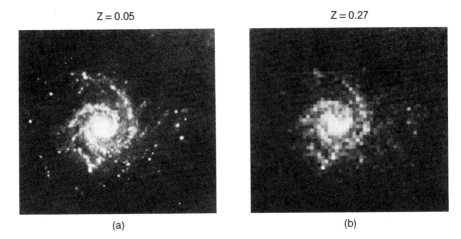

Z = 0.05   Z = 0.27

(a)   (b)

Figure 1.35 (a) Photograph of the barred spiral galaxy NGC 2523 taken by an excellent ground-based telescope. (b) Simulated (degraded) photograph of this galaxy, with poorer resolution. A very distant galaxy, which appears from the ground as in (b), would be recorded using the Space Telescope as in (a). The improvement in resolution obtainable by an instrument in space over ground-based telescopes, where atmospheric "seeing" limits the resolution, is strikingly shown. (Source: U.S. Government Printing Office.)

meter cryogenically cooled telescope with focal-plane instruments for imaging and for making photometric and spectroscopic observations. This will provide specific observations of celestial objects, therefore complementing the IRAS survey. In Europe, an Infrared Space Observatory (ISO), which is expected to have a similar performance, is being planned. Under discussion in the United States is Starlab, a 1-m class UV/optical telescope with a wide field of view of about 0.5°. With its extension to shorter wavelengths at high resolution, it will complement the Space Telescope.

1.3.4.2 Proposed X-ray, gamma-ray, and cosmic-ray observations.
Following the United States HEAO spacecraft, the first major observations will be made by the ESA EXOSAT, designed to study temporal variations and to survey sources in the X-ray sky, after launch in 1982. The Gamma Ray Observatory (GRO), under development in the United States for a Space Shuttle launch in 1985, also has major European involvement. This 10,000-kg spacecraft contains five major instruments to cover the range from 30 to 1,000 MeV with high sensitivity. Among its important objectives are high-resolution gamma-ray spectroscopic studies, galactic surveys to about 0.1°, and the detection of weak gamma-ray bursts. France and the USSR are developing GAMMA-1, to be launched on a USSR spacecraft. This will complement the GRO and will make observations at about the same time.

Another major United States endeavor, which is being planned under the direction of a worldwide consortium, is the advanced X-ray telescope facility

(AXTF). With a 1- to 1.5-m diameter grazing incidence telescope, this will have a number of instruments at the focal plane, including both high- and low-resolution imaging and crystal spectroscopy. This should have enough sensitivity to detect X-ray objects well beyond the distance limit of the Space Telescope. The United States is also planning an X-ray timing explorer whose objective is to observe temporal and spectral variations of strong X-ray sources. These data should provide much understanding of binary stellar system dynamics, accretion phenomena in compact objects, and neutron star structure. To be included in the Spacelab is the large area modular array (LAMAR), a grazing incidence X-ray "collector" designed for high sensitivity rather than for high spatial resolution.

Many other single instruments or joint efforts are underway, each of which will investigate a specific problem. Many of these, such as cosmic ray electron measurements, X-ray polarimeters, hard X-ray imaging devices, and EUV cameras form part of the early United States/ESA SPACELAB program.

1.4 SPACE OBSERVATIONS OF OBJECTS IN THE SOLAR SYSTEM

1.4.1 Introduction

1.4.1.1 Historical background.   The study of the solar system, particularly of the motions of the planets, caused the scientific revolution of the seventeenth century and ushered in the modern world. However, the different motions of stars and of the sun and moon across the sky of the constellations in the course of the year had been studied by the earliest civilizations. It was soon recognized that there were five wandering "stars," or planets, which moved in a strange way among the constellations of the zodiac, that the sun traversed during the year. An important advance was made when the morning star and evening star were proved to be the same planet, Venus. Precise observations were made of these motions by all the ancient civilizations (Chinese, Babylonian, Greek, and Incan) as they had an importance for agriculture and navigation and for predicting eclipses. Even peoples of whom no written records exist, the builders of the great stone circles such as Stonehenge and Karnak, appear to have correctly observed the motions of the sun and moon.

The Ptolemaic system, that is, the geocentric model of these planetary motions in the solar system, was accepted for a thousand years. In 1543, Copernicus, a Polish canon, substituted a sun-centered, or heliocentric, model for this earth-centered one. Eventually this theory led to a great public controversy because the Christian leaders had invested the former geocentric system with theological significance. The wealth of new observations of planetary motions by Tycho Brahe, astronomer to the great Danish King Christian IV, enabled Kepler to discover three laws of these motions. Galileo, championing the Copernican system, discovered the law of inertia, thus laying the basis for dynamics, the first and the most fundamental branch of physics. Born in the year that Galileo died, 1642, Newton completed Galileo's work by formulating the three laws of motion which explained Kepler's laws by the universal law of gravitation.

The discovery of the telescope in Holland around 1600 led to greatly improved observations of the moon and planets, as well as to Galileo's discoveries, in 1609 and 1610, of sunspots, the four major "Galilean" satellites of Jupiter, the rings of Saturn, and, quite importantly for his controversies, the fact that the moon resembled the earth in having topography. The increasing need for a method of finding longitude at sea led to the foundation of the Royal Observatory at Greenwich by King Charles I and of the Paris Observatory by King Louis XIV.

For the next two centuries, the major question was whether or not the detailed motions of the moon and planets could be exactly explained by Newtonian theory; this engaged the attention of the foremost mathematicians of the time, such as Laplace, Lagrange, and Adams. The ephemerides for navigation, the standards of time based at first on the earth's rotation, solar time, and later on ephemeris time provided by the motions of bodies in the solar system, were the practical outcomes of this work. The determination of the masses and other properties of the planets, the discovery in 1780 by Herschel of a new planet, Uranus, the discoveries of Neptune and of Pluto, of tidal phenomena, and of irregular variations in the length of the day were the scientific outcomes of this work. In particular, the observations of the perturbation of the orbit of Uranus led Adams at Cambridge and Leverrier in Paris to predict the existence of a new planet, and the race to discover it, won by the Berlin Observatory, was perhaps the first example of national pride intruding into scientific research. It certainly excited great public interest in a scientific discovery.

Three small peculiarities of these motions remained unexplained by Newtonian theory. One is a small deceleration in the longitude of the moon, which was finally shown to arise from tidal friction, evidently of the ocean tides, and which lengthens the day at a rate of 2 ms per century. Superimposed are irregular variations in the longitude of the moon, ascribed to a varying rate of rotation of the earth and reaching 3 ms; these are either due to motions in the earth's core, associated with the processes of generating the geomagnetic field, or possibly due to changes in the atmospheric circulation. Another peculiarity is the unexplained part of the precession of the long axis of Mercury's orbit, by 43 s of arc per century; this was explained by Einstein's general theory of relativity in 1915.

1.4.1.2 The planets and their formation.

The inner planets are Mercury and Venus, closer to the sun than the earth is, and the outer planets are Mars, Jupiter, Saturn, Uranus, Neptune, and Pluto, in order of increasing distance from the sun. Consequently, Mercury and Venus never subtend a greater angular distance in the sky from the sun, their elongation, than 24° and 40°, respectively; they may appear either as morning or evening stars. The outer planets move eastward among the stars, except when they are near inferior conjunction when they move westward, in "retrogression." This phenomenon was explained by Copernicus, using the heliocentric theory, more simply than on the Ptolemaic model. A more physically interesting division is between the planets, Mercury, Mars, and Venus, which are of comparable size and density to the earth, and thus termed the terrestrial planets, and those whose radius is an order of magnitude greater than the earth's radius and whose density is much lower than that of terrestrial rocks, the "major planets." The orbits of the planets lie close to the ecliptic, the plane of the earth's orbit around the sun, and near to the sun's equator. The planets rotate in the same sense as they move in their orbits, anti-clockwise looking down from the north, with their axes only making small angles, $21\frac{1}{2}°$ for the earth, with the pole of the ecliptic. Here Venus and Uranus are exceptions; in addition, Uranus' axis makes an angle of 95° with the pole of the ecliptic.

All these facts suggest that the solar system was formed by the contraction of an interstellar dust cloud, many of which are observed to be 100,000 AU in extent, into a thin disc of gas and dust flattened by rotation. In this disc the sun forms, and the planets form by condensation first into grains, then planetesimals, and finally planets. Understanding the origin of the solar system is a primary goal of the study of the planets. The development of rockets, sophisticated guidance systems, and long-distance transmission of information by radio led to observations of the solar system using space techniques. These have greatly increased the amount of information now available.

It is, however, important to emphasize that the record of early solar system events has been lost in many cases. On the earth no vestige of the original crust is known, or likely to be found. Even though the rocks of the moon's crust are older than those of the other terrestrial planets, the moon was molten early in its life. The isotope anomalies found recently in the Allende meteorite strongly suggest that the primeval solar system nebula received matter contributed from different nucleosynthesis events not very long before the condensation of the early solar system nebula. It is thus reasonable to suppose that such an event, a supernova explosion, triggered the gravitational condensation of the solar system.

Indeed, from astronomical photographs of blue stars, that is, stars at an early stage of their development, in certain of the great interstellar dust clouds, it has been concluded that a shock wave from a supernova explosion is likely to trigger the process of condensation to form a solar system. Such an interstellar dust cloud would, by the principle of the conservation of angular momentum, result in an increase in its spin rate sufficient to form the flattened disc needed to explain the closeness of the orbits of planets to a single plane. While the formation of tiny grains, or planetesimals, from which the planets would eventually grow is somewhat easy to visualize, the process of their coalescence to form larger objects is not well understood. It seems possible that the process of formation of the planets may involve an intermediate stage, in which the solar system consists of a very large number of moon-sized objects. It is easiest to explain the departures of the axes of rotation of the planets from the pole of the ecliptic by supposing that the existing planets were subject to large impacts by comparatively large objects in the final stages of their formation.

In this century, astronomy turned away from the solar system to the study of our galaxy and beyond. Had it not been for amateur astronomers, who devotedly observed the markings of the surfaces of the planets, the motions of their satellites, asteroids, and comets, and who have the discoveries of many new bodies to their credit, the historical record would have had very unfortunate gaps. The detailed observation of the moon was also continued largely by amateur astronomers; this led to the settlement of the controversy between the volcanic and meteoritic impact theories of crater formation in favor of the latter. When space technology ushered in a new epoch in the study of the moon and planets, it was accomplished by a renewed interest in ground-based astronomy, in spectroscopy, radio noise observations, and later radar study of the surfaces of the moon, Venus, and Mercury.

1.4.2 The Terrestrial Planets and the Moon

To explore space, either with or without spacecraft, is first to explore the terrestrial or earth-like planets (Figure 1.36), for these are the earth's nearest neighbors in the solar system. The broadest definition of the adjective "terrestrial" is to call all "large" bodies with solid surfaces "terrestrial." These include the inner planets (Mercury, Venus, earth, and Mars), the earth's moon, and the four Galilean satellites of Jupiter (Io, Europa, Ganymede, and Callisto). The larger satellites of Saturn, Uranus, and Neptune, as well as the planet Pluto, may eventually be found to be "terrestrial" too.

Knowledge of the terrestrial planets comes from two sources: earth-based observations and observations made from spacecraft. The former provide a framework of physical properties; the latter show each planet to be a unique place, which becomes more familiar and less strange as more observations are made.

1.4.2.1 Pre-space age knowledge.   The major terrestrial planets have all been known since prehistoric times; they were all envisioned as symbols of gods or goddesses in the ancient religions. As such, their positions in space, and especially the moon's position, were carefully watched and noted. Events on earth were synchronized with or measured against events in the heavens; thus the terrestrial planets helped to guide the development of calendars.

Pre-space age studies of Mercury included measurements of the planet's radius (as seen in transit across the sun) and its mass (obtained from the perturbations of cometary orbits). From these observations, the most important fact about Mercury, namely, its high density (5.5 to 6 $\times$ 10^3kg/m^3, or 5.5 to 6 g/cm^3) was found. Studies of faint albedo features on the planet's surface suggested synchronous rotation, that is, the period of rotation about its spin axis being equal to its period of revolution about the sun, 88 days. Mercury was known to be hot,

Figure 1.36 Earth, the quintessential terrestrial body; a METEOSAT photograph of Europe and Africa taken from geostationary orbit on 7 March 1978 at 12 U.T. Knowledge gained about the earth influences man's perception of other planets, and knowledge gained through the study of other members of the solar system leads to a fuller understanding of the earth and its environment. (ESA photograph.)

above $600°K$, and to lack an atmosphere. Little else was known because observations of objects close to the sun are difficult because of the sunlight scattered into a telescope.

Venus is the terrestrial planet most similar to the earth in size and distance from the sun. However, its surface is hidden from view by a perpetually cloudy atmosphere; a retrograde rotation of the atmosphere, with a four-day period, was inferred from repetitive changes in subtle markings visible only in ultraviolet light. That Venus had an atmosphere composed principally of carbon dioxide was inferred from spectroscopic analyses.

The moon, the earth's only natural satellite, is sufficiently close and large to have been well studied from earth prior to the space age (Figure 1.37) Telescopes revealed craters with diameters ranging from several thousand kilometers down to the limits of resolution. These craters were believed to be formed by the impact of interplanetary bodies. The moon was seen to have two types of terrain, one bright in albedo and heavily cratered and the other dark and lightly cratered. Although known as "terrae" (highlands) and "maria" (seas) respectively, the

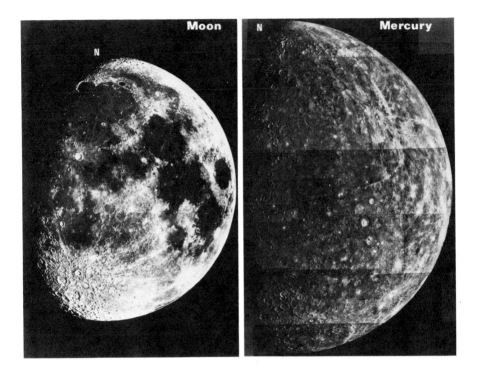

Figure 1.37 (left) The moon, Earth's nearest neighbor. Studied intensively for hundreds of years and the only extra-terrestrial body to be visited by man, the moon remains enigmatic. (right) Mercury, the "test" planet; a MARINER-10 photo-mosaic. By virtue of its position nearest the sun, Mercury is crucial in constraining and testing theoretical models for the formation of the solar system and its constituent members. Its surface records a lunar-like history, but its internal properties, revealed in the high planetary density and surface tectonic landforms, suggest a dramatically nonlunar interior. (NASA photograph.)

bright and dark areas were not actual land masses and oceans; they were known to be old and young surfaces, presumably of igneous rock. Many questions existed prior to the space age, among them being whether the craters were due to impacts or to volcanic activity, whether the surface was covered by an unconsolidated dust layer or was solid rock, and whether the moon was still internally active or was dead. In order to answer these and other questions, robots, and eventually man, traveled to the moon.

Telescopic observations of Mars have fascinated man for centuries by showing a planet with many changing features. The "annual" growth and retreat of the Martian polar caps, the appearance of clouds of many colors, shapes, and sizes, and both real and imagined changes in the color of the surface on a seasonal basis all led to conclusions that Mars was, perhaps, the most earthlike of the planets, even to the extent of perhaps harboring life. Light and dark areas were seen and mapped. These changed only slightly in shape but varied in intensity and color, suggesting to some observers the growth and decay of primitive life forms. Even canals were imagined, although these were not believed by the middle of the twentieth century. The red color of Mars was thought to result from the oxidation of iron-bearing silicate rocks, in much the same way as metallic iron rusts to red iron oxides. The temperature of Mars was known to vary seasonally from an equatorial maximum of 300°K (27°C) to a polar minimum during the winter of 150°K (− 123°C). Its atmosphere was examined by spectroscopic techniques and interpreted to be mainly of carbon dioxide, but the atmospheric pressure was unknown. In summary, Mars was an intensely observed, but poorly understood, planet.

The largest satellites of Jupiter were discovered by Galileo, and so they are called the Galilean satellites. The sizes of these "moons" of Jupiter (Io, Europa, Ganymede, and Callisto) were determined by measuring their small, but visible, disks and timing the movements of their shadows across the face of Jupiter. They were found to be large; Ganymede and Callisto were shown to be roughly the size of the planet Mercury, and Io and Europa the same size as the earth's moon.

Studies of the terrestrial planets have undergone a revolution since the beginning of the space age, both from the advent of close-up studies and from earth-based studies fostered by direct exploration of the planets. In the following sections the present state of knowledge, derived from space observations, of each planet is summarized, and major questions that remain unanswered are outlined.

1.4.2.2 Mercury.

In 1973 and 1974, the United States spacecraft MARINER–10 flew past Mercury three times, returning the first and, for the foreseeable future, only close-up observations of that planet (Figure 1.37). It appears to be a very moon-like place, densely cratered in some regions, sparsely cratered in others, and with a low overall brightness (or albedo) indicative of iron and magnesium-bearing silicate materials. Enormous scarps, hundreds of kilometers long and more than 1 km high, are seen on Mercury's surface. These suggest compressional tectonic action, perhaps caused by the shrinking of the planet as it cooled, early in its history. Measurements indicate that those portions of the planet in shadow reach temperatures even less than 100°K (− 175°C), while the sunward side reaches temperatures in excess of 700°K (430°C). A small, intrinsic magnetic field is observed. Although less than one-thousandth of the earth's field, this discovery is among the most important in planetary exploration, for it demonstrates the existence of the magnetic field of the only terrestrial planet other than the earth; it seriously stretches internal dynamo models for the generation of such magnetic fields. Finally, it is now known through earth-based radar studies that Mercury rotates about its axis, not once in one revolution of the sun, and thus always keeping its same face toward the sun, but rather rotates three times for every two revolutions. This 3/2 spin-orbit resonance is believed to reflect the evolution of Mercury's orbit as the sun was forming.

A great deal is not known about Mercury, for MARINER-10 observed only one-half of the surface and carried no geophysical or geochemical instrumentation. Mercury is important in that it provides a test case for examining the mechanics, thermal evolution, and chemical composition of bodies formed in the near-sun environment. Given the right observations, Mercury could be used to test theories of planetary formation and evolution and to constrain models of solar nebula composition. Such observations, including the surface chemical composition, the distribution and size of subsurface gravity anomalies, and the planetary heat flow, would require both a low-altitude orbiter and a lander. Owing to thermal problems, such spacecraft are today almost impossible to construct, and a fuller exploration of Mercury will no doubt be deferred.

1.4.2.3 Venus.

Venus has been the object of successful exploration by both the United States (MARINER-2, -5,-10 and PIONEER Venus) and the USSR (VENERA-8,-9,-10,-11, and -12). One of the first major discoveries, however, was made by earth-based radar, which showed Venus to rotate very slowly in a retrograde direction, that is, in a direction opposite to that of the earth, Mercury, the moon, Mars, and all the other planets except Uranus. Early spacecraft studies addressed principally the atmosphere; the atmosphere is carbon dioxide, at almost a hundred times the atmospheric pressure at the earth's surface, with trace amounts of noble gases. The surface temperature is about 740°K (470°C). Recent observations show that the atmosphere of Venus has a considerably greater amount of noble gases than does the earth's. Considering the potential sources for, and methods of, outgassing these volatile elements, these observations have brought into question nearly all models for planetary formation.

Observations of the surface of Venus are limited by its optically opaque clouds. Earth-based radar observations, at very low resolution (tens of kilometers) and covering only small fractions of the surface (~ 10 to 15 percent) show features that are interpreted to be volcanoes, craters, mountains, and rift valleys (like those in East Africa). Confirmation of many of these interpretations has come from recent measurements obtained by the radar altimeter aboard the PIONEER Venus Orbiter. Having viewed the topography of over 80 percent of the planet's surface, PIONEER data reveal Venus to have two large, continent-sized highland regions (called Ishtar and Aphrodite) and several smaller elevated areas. Mountain ranges, volcanoes, many rift valleys, and, perhaps, even craters have been observed (Figure 1.38). VENERA lander observations of radioactive elements (uranium, thorium, and potassium) and photographs of the different landing sites, combined with PIONEER Venus gravity and altimetry data, provide scientists with sufficient information to conclude that Venus has had a very active geological history. The history may, in many ways, be much like the earth's, and Venus may perhaps still be dynamic, although in a less demonstrative way than by showing earth-like plate tectonic behavior.

Many questions remain about Venus, the most important being the age and

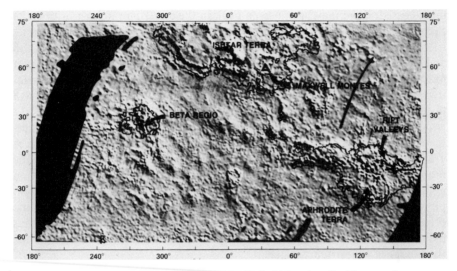

Figure 1.38 Venus observed, beneath the veil of a thick carbon dioxide atmosphere, by the PIONEER Venus radar shaded relief image, reveals a surface of craters, mountains, canyons, and volcanoes. (NASA photograph.)

chemical composition of its land masses. Higher-resolution photographs, using radar to penetrate the clouds, are needed to refine views of Venusian surface processes, especially to distinguish between impact and volcanic craters. As with Mercury, seismometry and heat flow measurements are necessary to infer the interior properties of the planet. Because of the high surface temperatures, the absence of day/night temperature differences, high surface pressures, and caustic materials in the atmosphere (the clouds contain appreciable amounts of sulphuric acid), long-lived landers are not within the current state of technology. Short-lived landers, balloon-based landers or "floaters," orbiters, and sample returns from Venus are more likely than these within the next several decades.

1.4.2.4 The moon. More is known about the moon than about any other planet in the solar system except the earth. Nearly 400 kg of lunar material has been returned to earth from eight separate locations. A network of four seismometers operated continuously throughout most of the 1970s, returning information about the moon's interior. The outward flow of heat, measured at two sites, is an appreciable fraction of that of the earth.

We now know that the moon was formed of materials depleted of volatile and siderophile (iron) elements; it was apparently born at high temperatures. It formed a crust rich in calcium-bearing alumino-silicates very quickly, in less than 200 million years, after its birth about 4.55 billion years ago. This crust is preserved as the light-colored highlands. The moon was subjected to intense bombardment by interplanetary objects. Some of the objects created craters as much as 1,000 km across. About 4 billion years ago, this torrential bombardment ceased; there may have been a peak in the cratering rate just prior to its cessation. The lunar maria were formed by the eruption of vast sheets of basaltic (rich in iron and magnesium-bearing silicate minerals) lava between about 3.8 and 2.6 billion years ago. Vulcanism may have occurred earlier, but evidence of it has been severely modified by the heavy cratering. The moon has been a relatively inactive planet for most of its history; only occasional impacts have disturbed an otherwise sedentary existence.

The largest problem remaining in lunar science is that of placing the information derived from the returned samples within a global lunar framework. A great deal is known about eight small regions on the earth-facing side of the moon, and something is known about its interior. But there is little information about regional variations, and the far side of the moon has scarcely been explored. It is likely that the moon will be the measuring stick against which other planets are compared. Without a global framework, these comparisons may not reveal the most important differences, and also the similarities, between the family of terrestrial planets.

1.4.2.5 Mars. As noted earlier, Mars has occupied a prominent position in studies of the planets. As a potential abode of life, it has received the greatest attention in space exploration except for the moon. No less than 11 spacecraft have sought to provide information on the planet and its surface. These have included early fly-by missions, orbiters, and most recently landers. From these spacecraft have come new views of a fascinating world.

The surface of Mars displays several major geologic provinces (Figure 1.39). The principal division is along a great circle inclined about 35° to the present equator. To the "north" of this line are low-lying plains relatively devoid of large impact craters; to the "south" is higher standing, heavily cratered terrain. Mars has two major volcanic fields: The Tharsis Montes contains over a dozen enormous volcanoes, the largest, Olympus Mars, being 550 km in diameter and 26 km high. The Elysium Montes are smaller and older. A gigantic trough system, the Valles Marineris, extends over 4,000 km. Fault scarps and massive landslides attest to the tectonic origin and large-scale modification of the troughs. Channels believed to be produced by catastrophic floods and valley networks probably formed by running water are found throughout the heavily cratered terrain. Layered deposits in the polar regions, sand dunes, dust storms, and water and ice clouds continue to change the Martian surface.

The Martian atmosphere is principally composed of carbon dioxide, with about 1 to 2 percent of nitrogen and argon. The isotopic abundances of gases suggest that Mars once had a more appreciable atmosphere than its current average surface pressure of 6 mb (millibar) would suggest.

The surface materials examined by the VIKING landers were found to resemble iron-rich clays formed by the chemical weathering of basaltic lavas. A small proportion of a magnetic mineral was found in the soil. The Martian soil is chemically active and strongly oxidizing. No organic molecules were detected;

Figure 1.39 Mars, the most fully explored planet outside the Earth–moon system, observed from VIKING. Fly-bys, orbiters, and landers have produced a wealth of information about Mars' atmosphere, geology, geophysics, and chemistry, but its biology remains a problem. (NASA photograph.)

the experiments conducted were unable to reveal any irrefutable evidence for biological activity. Mars appears to be lifeless.

A great deal still remains to be discovered about Mars. Its interior, as that of most of the other planets, remains a mystery. Studies of the absolute ages and chemical constitution of dozens of radically different places on the planet will be needed in order to decipher its complex history. Recent studies have found areas where the environment appears to be much more conducive to the development of life than others. These need to be explored fully.

1.4.2.6 The Galilean satellites of Jupiter. The two VOYAGER spacecraft have provided a first view of four new planets whose total surface area rivals that of Mercury, the moon, and Mars combined. Not only are the Galilean satellites large, but each has a surface that records a history unique from its neighbors and different from those found in the inner solar system.

Io, about the size of the moon and the nearest satellite to Jupiter, is covered with sulphur and sulphur-bearing condensates and sublimates. Io is the first planet other than the earth to show active vulcanism. At least eight major eruptions, blasting material over 100 km high, were occurring simultaneously during the VOYAGER fly-by (Figure 1.40). The absence of impact craters attests to the rapid resurfacing of Io by these volcanic processes. Many large calderas, some with flows of silicate and/or sulphur lavas, were seen scattered over the entire surface. These volcanoes require an enormous source of energy; tides raised in Io by Jupiter or induction heating by Jupiter's powerful magnetic field are two possible sources.

Europa, the second large satellite out from Jupiter, is also about the size of the moon. Its surface is covered by water-ice frosts. Great lineaments crisscross its surface like fissures and pressure ridges on sea ice, as seen in the VOYAGER-2 photo-mosaic of Figure 1.41. Only a few impact craters can be identified, suggesting relatively rapid resurfacing processes. However, no volcanoes have been found on Europa; the mechanisms of resurfacing remain unknown. Perhaps the most intriguing idea for Europa is that its water-ice crust overlies a "buried"

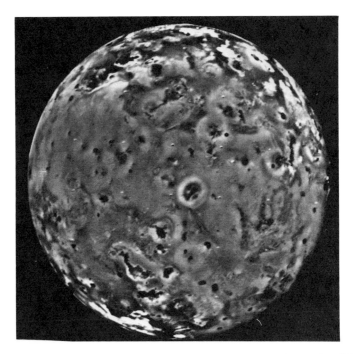

Figure 1.40 Io, the innermost Galilean satellite of Jupiter, is roughly the size of the moon. Each of the dark spots in this VOYAGER photograph is a volcanic caldera; the dark and bright ring-shape in the center shows a volcano erupting. (NASA photograph.)

Figure 1.41 Europa, the second satellite out from Jupiter, is covered by water ice. Europa is also about the size of the moon; dark and light fractures and ridges suggest a complex tectonic history. (NASA photograph.)

ocean of water, heated from within by radioactive elements distributed in the silicate body that comprises most of the satellite's mass and volume.

Ganymede is the largest of Jupiter's satellites, even larger than the planet Mercury. Figure 1.42 shows two types of terrain, either densely cratered, dark surfaces or sparsely cratered, bright surfaces. Many craters show evidence of modification by relaxation of relief. Ganymede has a low density ($\sim 1.9 \times 10^3 kg/m^3$) that suggests it has significant amounts of water-ice in its interior. There is some evidence of strike-slip faulting, an indication of plate tectonic processes similar to those found on earth.

Callisto, the outermost Galilean satellite, is also the size of Mercury and has a low density. The VOYAGER–2 mosaic of Figure 1.43 shows that its surface is a densely cratered, dark terrain. An enormous crater, with as many as 15 rings, dominates one-quarter of the satellite's surface.

So little is known about the Galilean satellites that any new data would be

Figure 1.42 Ganymede, the largest of the Galilean satellites, is larger than the planet Mercury. Its low density suggests that it is made of a mixture of silicates and ices. The large dark area is heavily cratered; the light area has ridges and grooves, indicating more recent tectonism. (NASA photograph.)

welcome. Obviously, composition, ages, and interior structure are important factors that need to be determined. These bodies are so different from the other terrestrial planets that finding common grounds for comparison will be a major challenge, yet surface forms indicate that this is not an impossible task.

1.4.2.7 The future of planetary exploration and its relevance to the earth and mankind. The terrestrial planets are, in fact, a varied group, each possessing aspects of scientific interest and some having practical uses as well. Here, some avenues of future exploration are examined and a few of these practical uses are discussed.

Planetary exploration is currently in a lull after the intense activity that characterized the first two decades of the space age. Because of decisions made in those countries most active in direct space exploration, fewer missions can be undertaken today because of their higher cost, in terms of manpower, resources, and money, than heretofore. It is expected that selective exploration, with more specific goals, will become the rule in future planetary studies.

Within the next decade, Venus is likely to be the planet of greatest interest; its geology seems remarkably like the earth's, and its atmosphere may hold

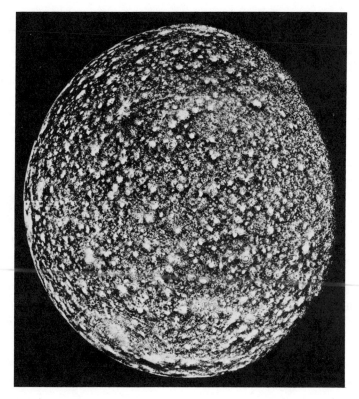

Figure 1.43 Callisto, the outermost of the four Galilean satellites of Jupiter, is as big as the planet Mercury. Callisto is entirely covered by a heavily cratered terrain. (NASA photograph.)

secrets relevant to the evolution of the earth's atmosphere. Orbiting radar systems, balloons, and short-lived landers are likely. On a longer time scale, the return to earth of samples from Venus is also possible.

Mars and the moon are extremely important, and while the United States will not return to these planets in the next decade, European and/or USSR spacecraft most probably will. The United States will concentrate its efforts on the outer planets of the solar system, namely, Jupiter, Saturn, and Uranus, and their satellites.

Study of the histories of the terrestrial planets has revealed a remarkably consistent pattern. This pattern suggests that each planet was subjected to intense bombardment by interplanetary objects early in its history, that this ceased rather abruptly, that volcanic and tectonic activity continued after the bombardment, stimulated by heat generated from internal sources, and that the subsequent evolution depended principally on the rate at which volatile elements were released from the planet's interior.

The earth, too, followed this pattern. The origin of the continents, their movement through the processes of plate tectonics, the creation of new sea floor, ice ages, and even the origin and occasional extinctions of life, all relate to this sequence. Impact cratering was important early in the earth's history, and, if recent evidence is correct, was even responsible for the extinction of the dinosaurs 65 million years ago. Climatic changes on Mars appear to be correlated with orbital changes induced by the other planets; the earth also experiences these changes, and possibly for the same reasons. The atmosphere of Venus shows what the earth's atmosphere might have been like had it been formed slightly nearer to the sun and if water could not have existed as a liquid.

Planetary exploration is, in some ways, like an insurance policy on mankind. Were man to establish self-sufficient and permanent residence on another planet, any ultimate catastrophe, either man-made or natural, to befall mankind on earth would not destroy all. Although hostilities and/or national interests might be carried out in space, it must be realized that space is a harsh environment, a common enemy. It is better for the nations of the earth to work together, to overcome its deadliness, than to work separately and to risk human lives.

Each of the planets has unique aspects that make it ripe for useful development. The moon is a close and relatively inexpensive source of building materials for large structures in space. Mars, with its rich clay soils and relatively available water, could maintain self-sufficient communities; its vulcanism indicates a wealth of minerals.

1.4.3 The Major Planets

Perhaps the single most atypical characteristic of our solar system is that the sun is a solitary star surrounded only by much smaller bodies, none of which are massive enough to sustain the thermonuclear processes which power stars. More typical "solar systems," perhaps as many as 80 percent of those in our galaxy, consist not of such a solitary star but rather contain a number of stars, anywhere from two to over ten individual stars. The fact that multiplicity seems to be typical of most stars and that typical distances between individual members of such multiple star systems are roughly of the order of the distance between the sun and the giant planets (Jupiter, Saturn, Uranus, and Neptune) leads to the speculation that the giant planets are in fact the remnants of a secondary nebular mass which was not quite massive enough to give birth to a second true star in our own solar system.

Not unexpectedly, the giant planets are extremely exotic objects in the solar system, vastly different from the terrestrial planets (Mercury, Venus, earth, and Mars). The giant planets are very large (Jupiter's diameter is about 11 times that of earth), have a much lower average density and are shrouded in thick, cold atmospheres because of their distance from the sun. Not only are these planets the remains of a star that could not quite shine, but each of them possesses a sort of "mini-solar system" of satellites ranging in number from at least two for Neptune to at least 15 for Saturn (see section discussing planetary satellites, as well as Figures 1.44 and 1.45). All except Neptune are known to be surrounded by systems of rings consisting of millions of much smaller "satellites."

Prior to exploration of these planets by unmanned spacecraft (PIONEER and VOYAGER) what we knew of these vast worlds was learned through the use of a number of terrestrially bound tools: Observations of the celestial mechanics of these planets and their satellites, along with theoretical considerations, gave us at least a notion of the nature of their interior structure; telescopic observations taught us much about the nature and composition of their thick atmospheres; later, the use of radio telescopes made possible the study of the magnetic fields of Jupiter and Saturn, and by inference provided further information on their interior structure. What was missing from this scenario were *in-situ* measurements

Figure 1.44 This montage shows Jupiter and its four largest satellites: Callisto (which appears in foreground), Ganymede, Europa, and Io. Jupiter itself is covered by light and dark bands of clouds. Superimposed on this so-called belt/zone structure are gigantic storms such as the Great Red Spot seen in this photograph. (NASA photograph.)

Figure 1.45 This montage composed from VOYAGER-1 photographs shows Saturn and its retinue of satellites. In the foreground is the crater-covered moon Dione. In the far upper right is Titan, the only moon in this solar system known to possess a substantial atmosphere. Saturn itself is surrounded by a magnificent system of rings and displays the same belt/zone cloud patterns evident in the atmosphere of Jupiter. (NASA photograph.)

of these planets' vast magnetic fields and the immense clouds of energetic, charged particles trapped within; high-resolution images of the surfaces of the numerous satellites that appeared as only fuzzy specks in earth's largest telescopes; time sequences of photographs and high-resolution temperature maps needed to understand the meteorology of these planets' vast weather systems; and closeup examination of these planets' gravity fields necessary for an understanding of their massive interiors.

1.4.3.1 Interiors of the giant planets. The giant planets are in fact protoplanets, i.e., their chemical composition is much the same as it was at the time of their formation some 4.3 billion years ago. While the terrestrial planets have evolved, losing the lighter elements (hydrogen, helium, and the noble gases) from their primeval atmospheres and forming a crust which separated from the outer core of the planets, the giant planets have probably undergone little evolution and, therefore, might have the same composition as the primitive nebula from which the present solar system has been made up. The major reason for this is the strong gravitational field of these massive planets which has enabled them to retain even the lightest elements. Their composition is mainly hydrogen and helium; they have no crust although the mantle becomes progressively liquid, eventually attaining a fluid metallic state, because of the increasing pressure toward the centers of these planets (the pressure at the center of Jupiter is roughly 10 million times that of atmospheric pressure on earth). The temperature within these planets rises with increasing pressure so that deep within their core temperatures might well range from tens of thousands of degrees centigrade to as high as 1 million degrees centigrade.

What we know, or at least think we know, about the interior of these giant planets has so far been based almost entirely on the theoretical efforts of scientists who specialize in constructing computer simulation models of these planets. Such modeling has lead to the speculation that metallic hydrogen, a substance unheard of here on earth, makes up a large fraction of the interior of Jupiter and perhaps Saturn. In the past decade the use of fly-by spacecraft has allowed us to more accurately determine the masses of Jupiter and Saturn and, more importantly, obtain some measure of the relative masses of the core and mantle of these planets. The accurate tracking of these spacecraft as they skirt by these planets allows the determination of the harmonic components of the planets' gravitational fields. Measurement of these harmonics indicates the departure of the actual distribution of mass within the planet from that of a uniform spheroid. Comparison of these gravitational harmonics with those computed for various theoretical models has played a major role in the formulation of our notions about the interiors of the giant planets. The most recent theoretical work indicates a trend among the giant planets in that each of them seems to have central rock and ice cores about 15 times more massive than the earth surrounded by hydrogen- and helium-rich mantles and atmospheres both of which rapidly diminish in mass as one goes outward among the giant planets.

1.4.3.2 Atmospheres of the giant planets.

Each of the four giant planets is shrouded in a thick and very cloudy atmosphere. The depth of Jupiter's gaseous atmosphere, though controversial among scientists, is perhaps as great as 1,000 km or about 1/70th of the radius of the planet. This is indeed very thick when compared to the 10 km deep layer of air that contains the majority of weather phenomena seen on earth. The atmospheres of the giant planets are so cloudy that all that we are able to see of these planets are patterns in the colorful cloudtops. In the case of Jupiter and Saturn (and perhaps Uranus and Neptune as well), the clouds are arranged in an alternating series of light and dark cloud bands running parallel to lines of constant latitude. The lighter bands are known as zones; the darker bands are called belts. These belts and zones, along with some storms superimposed upon them, have been observed through earth-based telescopes for the past 300 years or so (see Figure 1.46). On the basis of these at-

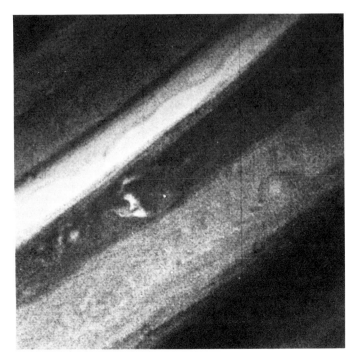

Figure 1.46 This high-resolution close-up of Saturn's banded atmosphere was taken by VOYAGER-1. The darker bands are called belts and the lighter ones are known as zones. Superimposed on this belt/zone pattern are a number of smaller-scaled disturbances. The smallest features visible in this photograph measure ~ 175 km across. (NASA photograph.)

mospheric features, it was deduced that both Jupiter and Saturn rotate quite rapidly (a day is about ten hours long). It was also observed that the equatorial regions of these planets seemed to rotate considerably faster than the rest of the planet. The PIONEER-11 fly-by of Jupiter and the VOYAGER-1 fly-by of Saturn allowed man his first closeup look at the polar regions of these two vast worlds. Interestingly, it appears that the alternating belt/zone pattern of clouds is replaced by more mottled cloud structures in the polar latitudes. The mottled appearance of these clouds in the polar regions suggests that they are due to numerous convection cells formed as a result of a large internal heat source first observed by PIONEER's infrared radiometers. The PIONEERs and later the VOYAGERs indicated in fact that as much heat is radiated from deep within Jupiter as is received from the sun. For Saturn, internal heating may be twice as powerful a source of energy input to the atmosphere as is solar heating. We do not know the exact nature of these internal heat sources, though possible explanations range from the decay of radioactive substances within these planets to perhaps the slow but steady shrinking of Jupiter and Saturn under their own immense gravitational fields. VOYAGER-2 should give us our first indication as to whether or not Uranus and Neptune also have strong internal heat sources, though the outcome will have to await the spacecraft's fly-by of these distant worlds in the late 1980s.

In the 1950s, long before the existence of the powerful internal heat sources was known, it was suggested that the rapid rotation of Jupiter and Saturn acted to convert vertical convection currents into the belt/zone cloud structure. The basic idea was that rising motions in the zones lead to the freezing of ammonia, one of the common gases in Jupiter's and Saturn's atmospheres, producing the white cirrus clouds observed in the zones. At the same time, the rapid rotation of Jupiter would deflect the diverging winds at the level of the cloudtops into strong westward winds on the equatorward side of the zone and strong eastward winds on the zone's poleward side. Much the same thing, though in reverse, was believed to occur due to sinking motions in the belts. The strong zonal winds or jet streams at each of the belt/zone interfaces would thus keep the belts and zones confined along parallels of constant latitude. This explanation, which was basically the same as that used by Hadley in the late eighteenth century to describe trade winds on earth, had much credibility in that the strongest winds were indeed observed to be situated at the belt/zone interfaces on Jupiter. Presumably, most of the energy driving the strong zonal jet streams would thus be provided by immense convection cells. The discovery of the internal heat sources by PIONEER seemed to uphold the model. Recent dynamical studies made possible by time sequences of high-resolution VOYAGER images of Jupiter seem to indicate that it is the motion of smaller-scale vortex features (eddies typically 100 to 1000 km in size) superimposed on the more regular belt/zone pattern which provides the vast amount of energy needed to drive the jet streams (see Figure 1.47). Though the two ideas are not mutually exclusive, the notion that small-scale eddies are responsible for the zonal winds is a novel one for Jupiter. The discovery that smaller-scale eddies in earth's atmosphere play a major role in driving our own global winds was one of the most important in modern meteorology.

At the time of this writing the preliminary wind measurements from VOYAGER-1 images of Saturn have just been obtained. They seem to indicate that on Saturn, though the clouds are arranged in much the same belt/zone pattern as on Jupiter, the jet streams do not necessarily lie along the interfaces between belts and zones. Thus, the long-held belief that "trade wind-like" motions lead to the belt/zone cloud patterns may be incorrect.

In addition to the possible important role played by Jovian eddies in maintaining the mean zonal flow, the eddies themselves, which range in size from the Great Red Spot (nearly 25,000 km across) to small vortices just at the limit of the VOYAGER camera resolution, are fascinating objects of study. The Great Red Spot is an immense oval-shaped vortex with counterclockwise winds of over 100 m/s around its periphery. The entire planet earth would fit several times over into this gigantic anticyclone (see Figure 1.48). This large eddy has been observed from earth for at least the past 100 years and may even have been seen at

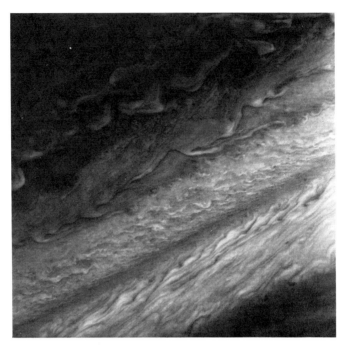

Figure 1.47 This beautiful VOYAGER photograph of Jupiter's cloudtops shows evidence of two classes of atmospheric circulation: mean zonal flow seen in the swift jet stream along the lower right of the photograph and vortex or eddy motions evident in the upper left. Current studies of jovian dynamics indicate that it is the eddy motions that feed their momentum into the zonal jet streams. (NASA photograph.)

the dawn of the telescopic age by Hooke in the late seventeenth century; it is thus not only the largest storm in the solar system but also the oldest. Infrared measurements made by the VOYAGER spacecraft indicate that it is slightly colder than its surroundings, an observation which most scientists attribute to rising motions within the spot. VOYAGER's close-up look at the Jovian cloudtops has also allowed the first detailed study of the wind fields associated with the three smaller (earth-size) White Ovals which lie south of the Great Red Spot. These eddies, though smaller and lacking the brilliant red color associated with the Red Spot, are observed to also be anticyclones with peripheral winds reaching a maximum of over 100 m/s. Most recent studies of the Great Red Spot and White Ovals have concentrated on trying to identify these storms' major energy sources. Most theories that attempt to describe these large-scale Jovian eddies can be divided into one of two schools of thought: Some believe the storms are driven by *vertical* motions intrinsic to their anatomy; others believe the large vortices are driven by the strong *horizontal* wind shears in which they sit. The most preliminary results seem to indicate that the Great Red Spot, and perhaps the

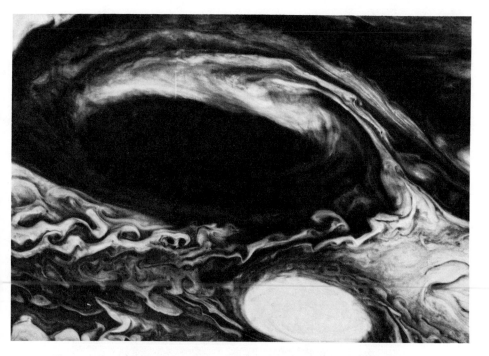

Figure 1.48 Jupiter's Great Red Spot and a smaller White Oval to the lower right are seen in this photograph. While the circulation of the Great Red Spot and White Ovals are similar, the difference in the color of their cloudtops is evident and as yet unexplained. (NASA photograph.)

White Ovals, are driven by the horizontal wind shears. The result, if further substantiated, would indicate that though small-scale eddies (on the scale of 100 to 1,000 km) drive the mean zonal winds, the largest eddies (on the scale of 10,000 km) are in turn driven by the mean zonal winds.

Information on the dynamics of Saturn's atmosphere is much harder to obtain because of obstruction by a fairly thick ammonia haze, rendering details in the underlying cloudtops indistinct. Nevertheless, several large-scale features which may be similar to those in Jupiter's atmosphere have been seen, including one feature which looks like the Great Red Spot, at least in shape and color.

Comparisons between the atmospheric circulations of Jupiter and Saturn and that of earth present an extremely important and ongoing challenge to fluid mechanics and to general circulation theory as developed in dynamical meteorology. In a sense, studies of the winds and storms on these giant planets provide a sort of large-scale experiment for the terrestrial meteorologist, who is now able to study the effects of an increased diurnal rotation rate and a strong internal heat source on a planetary atmosphere.

Two incidental but interesting notes concern the compositions of the atmospheres of Jupiter and Saturn. Since the giant planets are protoplanets, accurate measurements of the compositions of their atmospheres should lead to the knowledge of the composition of the primitive nebula from which the solar system formed and should allow us to test theories on the formation of the solar system. Determinations of the abundance ratios of helium and deuterium (heavy hydrogen) to hydrogen are especially important because of their cosmological implications. In most versions of the big-bang cosmology, deuterium and most helium are supposed to have been created in the first seconds of the universe. These gases are extremely difficult to measure in stars, galaxies, and interstellar dust clouds because of the physics of these objects. Direct measurement of hydrogen and helium in the major planets (first in Jupiter and Saturn by spacecraft infrared experiments) will allow us a test of the big-bang theory, which predicts an amount of these gases that can be compared with observations. In addition to the major components, hydrogen and helium, the atmospheres of Jupiter and

Figure 1.49 This nightside view of Jupiter obtained by Voyager-1 shows both auroral activity above the horizon in the upper right, as well as gigantic lightning storms measuring thousands of kilometres across. It may be that these lightning storms play an important role in producing some of the molecular constituents of the Jovian atmosphere. (NASA photograph.)

Saturn contain methane and ammonia and a host of even more complex organic compounds identified by VOYAGER's spectrographs. It is believed that lightning and ultraviolet radiation lead to the formation of these complex organic molecules and even possibly to the assimilation of prebiotic molecules such as amino acids (see Figure 1.49). This natural chemical laboratory is then extremely helpful in inferring the conditions under which life was created on earth.

1.4.3.4 The magnetic fields of the giant planets.   The fact that Jupiter and Saturn possess magnetic fields was one of the major discoveries of the modern science of radio astronomy. As early as 1952 it was found that Jupiter was emitting radio noise much greater than the blackbody or thermal radiation expected from its low temperature; Jupiter is, in fact, the brightest radio object in the sky. This radio noise is on two wavelengths—decimeter and decameter. The former is "synchrotron" radiation emitted by solar wind electrons which are caught by the lines of magnetic force at about 7 Jovian radii from the planet and spiral around these magnetic field lines with subsequent radio emissions. The latter come from about three sources somehow associated with certain longitudes on the planet itself. At least one of these sources is strongly triggered by the location of the satellite Io in its orbit. From these radio observations much was deduced about the Jovian magnetic field long before the magnetometers on board the PIONEERs and VOYAGERs made measurements during their fly-bys. On the basis of earth-based observations, it was found that the magnetic field of Jupiter is roughly that of a dipole inclined, like the earth's, by about 11° to the axis of rotation. The PIONEER observations indicate that this dipole is displaced from the actual center of the planet by about one-tenth of the planet's radius. From the observations of periodicity in the radio noise, the rotation period of the magnetic field can be determined; both earth-based and spacecraft observations indicate a period of 9 h 55 min 30 s for Jupiter. VOYAGER-1 has observed radio noise from Saturn with a periodicity indicating a rotation of 10 hrs 39.4 min. The strength of Jupiter's magnetic field is about 5 G or roughly 10,000 times stronger than the earth's magnetic field. The magnetic field of Saturn has a strength of about 1 G and seems to be roughly aligned with the axis of rotation.

The radio emissions are tied to the magnetic field which is in turn believed to be generated by currents of electrically conductive substances within the mantle. Hence, it is believed that the radio rotation periods are the same as the rotation period of the mantle of these planets. This electrically conductive substance may well be the metallic hydrogen previously mentioned, for at pressures of about a million atmospheres solid molecular hydrogen would be transformed into a conductive "metal"; that is, each electron would no longer be bound to a proton (the hydrogen atom's nucleus) but would be free to move through the crystal lattice of hydrogen nuclei and hence could conduct electrical currents.

Though the basic nature of Jupiter's magnetic field was known on the basis of terrestrial observation, it was the PIONEER–10 fly-by in the early 1970s which indicated that Jupiter's strong magnetic field had entrapped a system of immense radiation belts much larger and more energetic than earth's own van Allen radiation belts. This immense cloud of charged, energetic particles extends out to roughly 80 times the Jovian radius on the dayside of the planet. The ambient solar wind acts to push this cloud out to about 160 Jovian radii on the nightside. The source of most of the particles within this cloud is believed to be the solar wind, though the particle populations in the innermost portions of the magnetosphere may be enriched by ions from the upper atmosphere of Jupiter and the surfaces of the inner satellites. Some of these inner satellites, most conspicuously Io, are observed to leave observable tunnels in the particle cloud as they plow along in their orbits. The energies and flux rates associated with these trapped energetic particles are large enough to deliver lethal radiation dosages nearly instantaneously even at quite some distance from Jupiter itself. In fact, the electronic components on board the PIONEER spacecraft barely survived this unexpected rough treatment. VOYAGER components were specially designed with high radiation levels in mind, but even then, VOYAGER–1 suffered some substantial transient side-effects upon its closest approach to Jupiter.

We do not know whether or not Uranus and Neptune have similarly immense magnetic fields. Again, we may have to await VOYAGER–2's fly-by of these planets later in this decade.

1.4.3.5 Future exploration of the giant planets. Aside from VOYAGER–2's fly-bys of Saturn, Uranus, and Neptune, only one program of future direct exploration of the giant planets is currently being pursued in somewhat reluctant earnest: Project GALILEO. The GALILEO mission to the Jovian system involves the use of a Jupiter-orbiting spacecraft for studies of the planet and its satellites over a longer term than was possible during the brief PIONEER and VOYAGER encounters. In addition, a second smaller spacecraft designed as an entry probe would fly into the thick Jovian clouds providing the first *in-situ* measurements of composition, wind, and pressure in the Jovian atmosphere. Current problems with launch vehicles may well force anxious scientists to await a rendezvous during the latter years of this decade. At the same time, the placing of a large telescope in earth orbit called for by NASA's Space Telescope Project, may allow long-term, high-resolution studies of the dynamics of the atmosphere of Jupiter.

More speculative programs of exploration, such as missions involving Saturn orbiters and even Titan landers, as well as more extensive missions to Uranus and Neptune, have received some limited attention and even more limited financial support. However, the future is not entirely bleak, for at the very least, the thousands of pictures and millions of bits of other information already collected by the PIONEERs and VOYAGERs promise to keep scientists busy for years to come.

1.4.4 Chemical Composition of Members of the Solar System

The objects in the solar system, that is, the planets, their satellites, asteroids, meteorites, and comets, were formed from a well-mixed primordial nebula of chemically and isotopically uniform composition. (A particular chemical element can have various isotopic forms, with one or two more or less neutrons than usual in the nucleus of each atom, together with a certain number of protons.) At some time between the formation of the elements and the beginning of condensation of the less volatile material, this nebula must have been a mass of gas which was chemically well mixed (homogeneous), the temperature being so high that no solids could have been present. Otherwise, spatial variations in the isotopic composition of many elements would be anticipated. About 4.5 billion years ago, as the temperature decreased, the less volatile substances condensed and eventually accumulated into planetary objects.

The chemical composition of the matter from which the solar system formed is fairly well known. With the exception of the lightest elements, chemical elements are present in the sun, in the same relative proportions as they were present in the primordial gas cloud. Astronomical observations of the intensities of absorption lines in the spectrum of the sun gives the most direct information on the relative concentrations of the elements in solar matter. They show that hydrogen and helium are by far the most abundant elements in the sun. Oxygen, nitrogen, and carbon, although less abundant than hydrogen by a factor of greater than a thousand, are next in abundance. Their amounts exceed those of all the heavier elements combined.

Determinations of the abundance ratios of helium and deuterium to hydrogen are especially important because of their cosmological implications. Deuterium and most of the helium are supposed, in big-bang cosmologies, to have been created in the first seconds of the universe. The proportions of these gases are extremely difficult to measure in stars, galaxies, and interstellar dust clouds because of the difficulty in properly understanding the physics of these objects, and especially the partition of light elements in them. However, direct measurements of hydrogen and helium densities in the major planets made from the infrared observations of the VOYAGER spacecraft, will allow a test to be made of the big-bang theory. The amounts of these gases predicted theoretically can be compared with those measured experimentally.

1.4.4.1 Processes that determine chemical composition.

For the nonvolatile elements, that is, those elements and their compounds which do not, at ordinary temperatures, show any appreciable vapor pressure, a more accurate source of information on their relative abundances can be derived from the chemical composition of meteorites. The relative amounts of these elements in most stony meteorites appear to be very similar to those in the sun. In particular, this is true for the trace elements present in the so-called "carbonaceous type 1 chondrites."

The different chemical fractionation (separation by evaporation and distillation) processes that have been active during the time when the planets, their satellites, and the other objects in the solar system were formed can be divided into three types. These are fractionation processes in the gas phase, the separation of gases from condensed material, and processes of differentiation between condensed phases (separation of solids). Undoubtedly, the most important type is the second; the separation of gases from the condensed material causes the difference in the composition between the outer and the inner planets. The third type includes magmatic fractionation, which played an important role in the formation of the surface material and, presumably, of the metallic cores of the planets.

The separation of gases from condensed material during the formation of the solar system has been extremely effective. This can readily be seen from the small amount of neon in the terrestrial atmosphere; only 10^{-11} of the neon originally associated with the earth is now present in the air. However, it is not certain whether this small fraction was part of a residual thin atmosphere or was retained by the condensed material, in an either absorbed or dissolved form. The temperature at which the separation took place determined the chemical composition of the condensed material. Obviously, in the outer parts of the solar system the temperature was low enough for water and ammonia and, in part, methane, to be in their solid forms. In the inner part of the solar system the large masses of these hybrids present evaporated with the gases.

1.4.4.2 Oxidation.

By far the most abundant element in the universe as a whole, in the sun and in the interstellar medium, is hydrogen. The hydrogen to oxygen ratio in all this matter is of the order of a thousand. Yet planetary objects show considerable degrees of oxidation. The atmospheres of Mars and Venus contain carbon dioxide, whereas the earth's atmosphere also contains free oxygen. Meteorites, though considerably less oxidized than the surface rocks of the earth, frequently contain oxidized iron. The mantle of the earth appears to consist of magnesium silicates, probably containing a fraction of ferrous oxide.

The state of oxidation of the surface of the terrestrial planets is easy to explain. The retention of a small fraction of the water present in the primordial gas cloud, the subsequent photolysis of the water vapor in the atmosphere (that is, the splitting of a water molecule into its constituent hydrogen and oxygen atoms), and the loss of hydrogen so formed must have led to the presence of free oxygen in the terrestrial atmosphere and to the oxidation of carbon compounds on Mars and Venus. To explain the presence of oxidized iron in meteorites and in planets, it is necessary to assume either that hydrogen was separated from oxygen before the planetary objects accumulated or that the temperature was relatively low during a stage when the iron was in a finely divided form. The degree of oxidation prevailing in the interior of the terrestrial planets, in particular, the occurrence of oxidized iron in the silicates of their mantles, is not known. However, it is evident that the majority of meteorites contain ferromagnesium minerals, with an iron content much too high to have resulted from direct condensation out of a gas phase of solar composition. Oxidation of the iron must have occurred either at a relatively low temperature, far below that of the melting point of the minerals present in the meteorites, or from a medium that was highly depleted of hydrogen. The high degree of oxidation at the surface of the terrestrial planets must certainly be considered as a secondary phenomenon, caused by the escape of molecular hydrogen from the gravitational field of these planets.

The free oxygen present in the earth's atmosphere is generated by biological photosynthesis, that is, the process by which solar energy is used by chlorophyll-bearing plants to convert carbon dioxide and water into complex cells and oxygen. A comparison with Mars and Venus, however, shows that the surfaces of these planets are also highly oxidized; their atmospheres are of carbon dioxide. The presence of carbon dioxide can only be understood if the escape of molecular hydrogen from these planets is considered. The hydrogen is formed via a chemical reaction with reducing material such as metallic iron or silicates containing bivalent ferrous iron, or also by the photolysis of water. The study of the chemical composition of the atmospheres of Mars and Venus will undoubtedly lead to a better understanding of the history of the terrestrial atmosphere.

The fact that surface rocks of the earth contain relatively rare elements in highly enriched form is taken to be a consequence of magmatic processes that occurred when the surface temperature of the earth was substantially higher than it is today. Impurities in rocks have a lower melting point than the rocks themselves, and hence tend to accumulate on the surface, leading to the enrichment of certain materials at and near the earth's surface.

1.4.4.3 Asteroids and meteorites. The composition of the terrestrial, or inner, planets cannot be greatly different from the average composition of meteorites. This can be deduced from the observation that the mean densities of the planets, when account is taken of the increased density due to the high pressures prevailing in their interior, are similar. Indeed, it has been assumed for almost a century that meteorites give a good indication of the internal composition of the terrestrial planets. The high frequency of iron meteorites is taken as an indication that the meteorites come from a missing planet, a planet which, according to the rule of Titius and Bode, should have existed between the orbits of Mars and Jupiter. The asteroids were either formed by such an event or the strong gravitational influence of Jupiter prevented the accretion of these objects into a planet, in which case they are the original planetesimals, that is, the objects formed as the early results of condensation. There are additional indications supporting the hypothesis that the core of the earth is essentially made of iron.

The chemical history of the moon must have been different from that of the earth, and also from those of meteorites. At present, it still appears to be uncertain whether these differences can be explained by differences in the magmatic processes or by differences in the condensation history of earth and moon, or both. The two small satellites of Mars appear to be more closely related to the asteroids and the meteorites than to the terrestrial moon. It would be most interesting to learn to what extent the Martian satellites resemble meteoritic bodies.

The chemical composition of the satellites of the outer planets provides valuable information on the origin of the solar system. Exact values for the densities of the moons of Jupiter have recently become available. These indicate that the densities of the inner moons are higher than those moving at greater distances from the planet. The difference in the densities indicates that ices have been lost during the formation of the inner satellites. This can be explained only by assuming that Jupiter acted as a heat source that was able to evaporate water and ammonia from satellites in its vicinity. It will be most interesting to investigate whether there are gradients of density for the satellites of Saturn which show similar trends to those of Jupiter.

Progress has recently been made in the field of the chemical composition of asteriods through measurements of the albedo and of the absorption spectra of these small bodies in orbits between Mars and Jupiter. About 40 percent of those in inner orbits, increasing to more than 95 percent of those beyond 3 AU, are identified as of carbonaceous chondrites, especially because of their low albedo of from 3 to 5 percent. About 12 asteroids, in the orbits between 2.5 and 2.9 AU, are tentatively identified as being composed of iron, or of mixtures of iron and silicate. This identification has been made from the increase in reflected light toward the red end of the visible spectrum that is characteristic of iron.

1.4.4.4 Comets.

Volatile compounds are undoubtedly present in comets. This is natural if, as Jan Oort proposed, they represent that small proportion of the 10^8 such bodies inhabiting the outer reaches of the solar system whose orbits are perturbed by the major planets into orbits which cross that of the earth.

1.4.4.5 Conclusions.

All this information leads to a good knowledge of the chemical elements that were in the primordial nebula from which the solar system was formed. The abundances of the individual nuclear species can be computed from the isotopic compositions of the elements and from the abundances of the individual elements. It was found in 1947 that the nuclear abundances show certain regularities and appear to obey certain rules. In conjunction with these rules, it can immediately be seen that these values are determined by nuclear properties. Chemical properties of the elements are only important for secondary fractionation processes to which the mixture of primordial material was subjected during the later stages of the formation of our solar system.

Knowledge of the relative abundances of nuclear species should make it possible to recognize the individual nuclear reactions that led to the formation of the elements of which the solar system is composed. However, it was obvious many years ago that there was no single, simple, nuclear reaction, and no single set of thermodynamic data for the conditions under which the reactions could have taken place, which could possibly explain the observed abundance distribution of all nuclides.

Certainly several different nuclear reactions must have been responsible for the formation of the nuclides in their present relative amounts. In particular, a reaction must have taken place which led to the formation of unstable nuclear species that were rich in neutrons and which predominated at high mass numbers. Such reactions could conceivably have taken place in supernovae by neutron buildup on a fast time scale. Neutron buildup processes acting on a slow time scale, as well as processes leading to neutron deficient nuclear species, and those which yield reaction products in amounts corresponding to thermodynamic equilibrium conditions, may also have taken place in the interior of stars. Such considerations fit a variety of observational astrophysical facts well. The matter from which the solar system was formed was thus a mixture of material that originated from several completely independent sources.

1.4.5 Meteorites

In 1803, the Académie des sciences in Paris acknowledged that stones could indeed fall from heaven; this made their scientific study legitimate. Since then, meteorites have mostly been studied as unusual rocks, cosmic errants from another planet. During the past 20 years or so, however, emphasis has shifted away from a geological approach to the study of meteorites toward a more fundamental approach. Meteorites are now primarily valued for the record they have preserved about processes which were active during the pre-planetary stage of solar system evolution, for example, the condensation of solid matter from an apparently gaseous nebula and the subsequent accretion of solid particles into progressively larger planetary objects. Nonetheless, it is important to bear in mind that this chemical and physical evidence is contained within rocks all of which have experienced some processing within a planetary environment and whose effects must often be disentangled to reveal the primordial record. In addition to retaining an imprint of processes occurring in the early solar system and on their parent planets (probably objects substantially smaller than the earth or moon), meteorites record interactions with the recent interplanetary medium. Thus they serve as natural space probes.

The framework for the classification of meteorites has changed remarkably little over the years, although its basis has tended to shift from mineralogy to chemical composition, thereby reflecting improvements in analytical techniques. About 5 percent of those meteorites seen to fall to the earth's surface are predominantly composed of metallic iron. These are further classified by means of their contents of minor and trace elements, such as nickel, germanium, and gallium, into about a dozen groups. These compositional variations were apparently caused by chemical fractionation during the formation of asteroid-sized planetesimals in the nebula, followed by further fractionation during crystallization of their metal cores; the iron meteorites are believed to be the surviving fragments of this process.

An additional 8 percent of the meteorites broadly resemble terrestrial igneous rocks. Although compositionally diverse, these meteorites share the characteristic that they are chemically differentiated with respect to the mean composition of solar system matter. Petrographic and chemical analysis of such meteorites has clarified many details concerning igneous processes in planetary objects. In these studies, as in many other areas of petrography, the development of electron probe microanalysis has been of inestimable value in obtaining accurate compositions of individual minerals.

An important inference from the observation that iron and differentiated stone meteorites were once molten is that a source of heat existed at the time of their formation. The nature of that heat source is controversial. However, the recent discovery of evidence for the short-lived radionuclides ^{26}Al and ^{107}Pd in the early solar system makes it plausible that radioactive decay could have generated sufficient heat to melt objects with diameters as small as a few kilometers. (In the isotopic nomenclature, the superscript numeral refers to the mass of the isotope, relative to that of a hydrogen nucleus, and the letters identify the chemical element; thus ^{26}Al is that isotope of aluminum which has a mass of 26 atomic mass units (AMU).)

1.4.5.1 Chondrites.

More than 80 percent of the meteorites have not been produced by igneous processing. These are termed chondrites, from the presence within them of abundant rounded rock fragments, usually less than 1 mm in diameter, known as chondrules. The origin of chondrules is a long-standing problem in meteoritics. The majority were clearly once molten droplets, before solidification and incorporation into chondrites, but the nature and location of the heating process are unknown. The study of lunar samples has shown that impacts into planetary surfaces can produce droplets of molten rock, but insufficiently to supply the abundance of chondrules in chondrites.

Whatever the mode of formation of chondrites, their bulk composition closely resembles that of the solar system as a whole, except for depletion of the most volatile elements, such as hydrogen, oxygen, carbon, and the noble gases. Those chondrites least depleted in volatile elements are termed carbonaceous, because they contain organic matter, now known to be of nonbiological origin. Their minimally fractionated composition and other apparently primitive features make them currently the most actively investigated probes of early solar system processes. In addition, their organic molecules, of controversial origin, exhibit very large isotopic variations in hydrogen, nitrogen, and, to a lesser extent, carbon. These may be used to diagnose the conditions in which pre-biotic matter formed in the early solar system. Thus they may possibly have profound implications for theories of the origin of life on earth.

Besides serving as a basis for classification, the fact that meteorite compositions fall into a restricted number of reasonably tightly defined groups indicates that they were formerly contained within a relatively small number of parent bodies. Their residence in these planetesimals has left a record in even the most primitive meteorites. Many, perhaps most, stone meteorites achieved their final form by consolidation of a mixture of rock fragments within the surface regions of their parent bodies. Such meteorites are termed breccias; many contain exotic fragments, apparently derived from the objects whose impact fragmented, stirred, and lithified those planetesimal surfaces. It is noteworthy that, in most cases, this projectile residue closely resembles carbonaceous chondritic material. Brecciated meteorites may also contain indigenous grains which have been irradiated, while on their parent body surfaces, by the solar wind and solar flares. The presence in such grains of solar wind–noble gas atoms and trails of lattice damage ("tracks") induced by passage of charged particles, respectively, testify to this irradiation. The discovery of abundant impact-produced breccias on the lunar surface greatly increased interest in this mode of rock formation.

Even meteorites that have never been thoroughly melted frequently show textural evidence of thermal metamorphism (transformation of form, by heat). This has generally acted to bring originally unequilibrated minerals closer to compositional equilibrium. In some cases, it has resulted in isotopic equilibration enabling the data and temperature of metamorphism to be determined.

Carbonaceous chrondrites show little evidence of thermal metamorphism, but have instead experienced significant aqueous alteration. Apparently, the surface regions of their parent bodies retained liquid water for a sufficiently long time for substantial secondary mineralization to occur, although this does not appear to have perturbed their bulk chemical composition. Also, several constituents of these chondrites escaped alteration, presumably by being incorporated after the period of aqueous activity was ended; these meteorites, too, are breccias.

Despite the prevalence of secondary processes, the chondrites still retain a record, often a subtle one, of chemical processes which apparently occurred during the formation of the solar system. Three broad fractionation patterns are apparent among chondrite compositions. These involve refractory, siderophile (iron), and volatile elements. The resolution of these patterns has depended on the refinement of several analytical techniques, most notably that of neutron activation.

Each fractionation pattern involves a set of elements that show systematic trends of depletion or enrichment in bulk analyses of different chondrite groups. Thus, one pattern involves elements such as calcium, aluminum, magnesium, and titanium, which are believed to have exhibited refractory behavior in the early solar system. They formed solid compounds at relatively high temperatures, above about 1,400°K, during cooling of an initially hot gaseous nebula. These elements are depleted in ordinary chondrites, with respect to their primordial abundances, and enriched in most carbonaceous chondrites (Figure 1.50). This pattern is generally believed to have been caused by the movement of early formed condensate relative to the gas of otherwise primordial composition, leaving some regions of the nebula depleted of, and others enriched in, those elements. The presence within carbonaceous chondrites of inclusions, usually a few millimeters in diameter, and enriched in these elements, lends support to this idea. However, the observed detailed compositions of these inclusions and the predictions of condensation theory differ significantly.

The siderophile elements tend to enter the metal phase in preference to the silicates. Their fractionation pattern in chondrites (Figure 1.51) shows evidence for the systematic loss of a mainly iron metal phase with respect to the primordial composition, coupled with relatively small variations in the degree of oxidation of the remaining iron. This fractionation of metal from silicate is generally believed to have occurred during accretion of fine-grained condensate into planetesimals, but the actual mechanism involved is unclear.

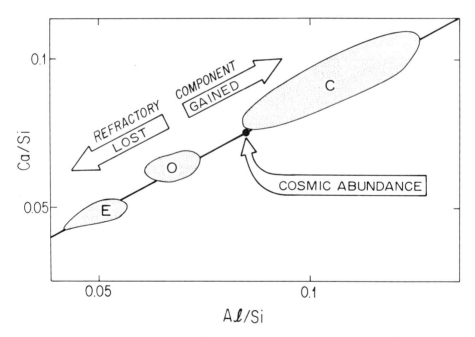

Figure 1.50 The contents of refractory elements, such as calcium (Ca) and aluminum (Al), normalized here to a common reference element, silicon (Si), vary markedly between the different groups of chondritic meteorites, known as enstatite (E), ordinary (O), and carbonaceous (C). These trends may have been caused by the gain or loss of a refractory condensate in the early solar nebula, as illustrated.

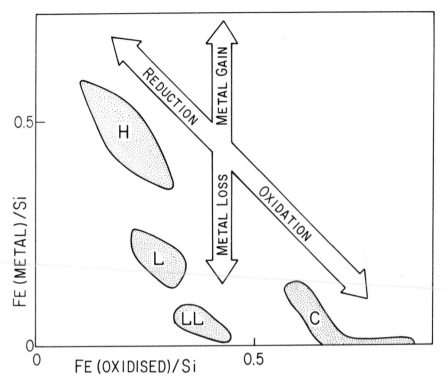

Figure 1.51 Iron varies in both abundance and state of oxidation among different chondritic meteorite groups. Here the ordinary chondrites have been divided into high iron (H), low iron (L), and very low iron (LL) groups. The original "cosmic" abundance of iron corresponds to the diagonal line marked "reduction-oxidation," which intersects the composition of the most primitive carbonaceous (C) chondrites; all other chondrites are relatively depleted of iron.

Those elements which are thought to have condensed below about 1,000°K in the nebula are considered to be more or less volatile and are depleted in chondrites relative to their primordial abundances. Depletion of the most volatile elements, such as bismuth, thallium, and carbon, frequently correlates with their degree of metamorphism. Less volatile elements, such as sodium, germanium, and sulphur, show depletions which are apparently related to their calculated nebular condensation temperatures. Whether this relationship is a continuous function of condensation temperature, reflecting progressively less efficient accretion of material as the nebula cooled, or whether chondrites contain components essentially purged of their volatiles by a reheating event is controversial.

1.4.5.2 "Cosmic" abundances.   Implicit in the discussion is the concept of an identifiable primordial abundance distribution. It happens that the most volatile-rich type of carbonaceous chondrite has a composition, for virtually all elements, equivalent to that of the sun, and hence of the solar system as a whole because the sun is by far the most dominant object therein. By taking a particular meteorite type as being compositionally equivalent to the primordial solar system, not only can the elemental abundances in the nebula be established with far greater precision than is possible for direct measurements of the sun, but also precise isotopic compositions, which are seldom accessible by solar spectroscopy, can be determined. The establishment of a reliable set of "cosmic" elemental and isotopic abundances has two important consequences. First, it provides a baseline for studies of planetary chemistry, as illustrated above for chondrites. Second, it permits rigorous tests to be made of theories of nucleosynthesis. In fact, one of the major achievements of the past 25 years has been the clarification of various modes of element formation in astronomical objects and the role that each has played in producing solar system material. For example, it is now clear that supernovae played a dominant role in the production of most solar system elements. The creation of solar systems such as that in which the earth occurs may well be causally linked to the occurrence of such supernovae.

Coupled with an improvement in understanding of how and where the elements were made, has been equally spectacular progress in determining when they were made and also when they were incorporated, first into solid grains, and later into planets. It is now known that the production of the elements began about 15 billion years ago and continued for approximately 10 billion years, at which time a fragment of interstellar cloud became separated from the rest and collapsed to form the protosolar nebula. The time scale for this process is determined from measurements on meteorites based on the decay of ^{129}I and ^{244}Pu, both now extinct, into isotopes of xenon. These give an interval of about 100 million years between the cessation of nucleosynthesis and the formation of the solar system. By contrast, similar measurements of excesses of ^{26}Mg, resulting from the decay of extinct ^{26}Al, indicate a period no longer than a few million years. The longer time suggests a relationship with the rotation of the galaxy, in which star, and thus solar system, formation could be triggered by the passage of the nebula into a galactic spiral arm.

The record of these extinct radioactivities can also be used to determine the relative ages of material produced in the early solar system. Thus, measurements of ^{129}Xe, the daughter of ^{129}I, suggests that most meteorites, and their constituents, formed within a relatively brief period of about 10 to 20 million years. Again, however, this interpretation is controversial in view of observations based on abundances of ^{87}Sr, formed by the decay of long-lived ^{87}Rb. These latter indicate a formation interval that is nearly an order of magnitude longer (Figure 1.52).

1.4.5.3 Age of the solar system.

The decay of ^{87}Rb has also been used to establish the ages of meteoritic material relative to the present day; because most meteorites were formed at the birth of the solar system, these measurements actually give the age of the solar system. The best estimate of this value is

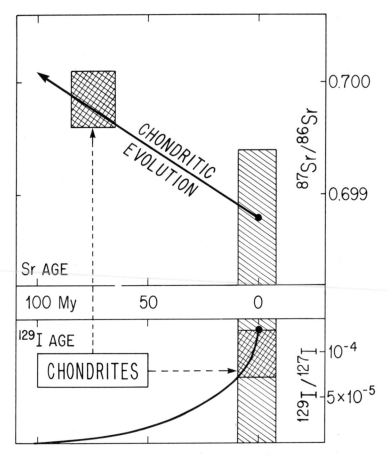

Figure 1.52 Diagram of isotopic ratios showing that the production of chondrites spanned a period of about 75 million years, based on variations in the amount of ^{87}Sr, from the decay of radioactive ^{87}Rb, which they incorporated at their formation. Analogous calculations based on ^{129}Xe contents, from the decay of ^{129}I, give a much shorter interval of 10 to 20 million years.

currently 4.57 billion years, a result confirmed by measurements based on the decay of other long-lived radionuclides, notably ^{235}U, ^{239}U, ^{40}K, and ^{147}Sm. Similar studies have also established that, although some of the differentiated meteorites were formed very early in solar system history, others have significantly younger ages, placing a severe constraint on the nature of the heat source responsible for their production.

An assumption basic to many aspects of radiochronology is the concept that the solar nebula was once isotopically homogeneous. Indeed, virtually all natural samples analyzed to date have demonstrated highly precise isotopic uniformity, if allowances are made for the predictable effects due to radio-nuclide decay, charged-particle interactions, and mass-dependent fractionations. Tremendous interest has focused recently on a few exceptions to this general uniformity. These consist of meteorites, or inclusions in meteorites, which exhibit, in one or more elements, isotopic anomalies apparently resulting from irregular mixing of the products of different nucleosyntheses. For example, many different types of meteorite show small but real, systematic variations in the proportion of ^{16}O, relative to ^{17}O and ^{18}O (Figure 1.53). Some inclusions in carbonaceous chondrites can exhibit excesses of this isotopic ratio ranging up to 5 percent. In a few cases, similar inclusions can also show a variety of other isotopic effects. Substantial mass-dependent fractionation can occur in elements, such as silicon and magnesium, whose involatile nature normally precludes such effects. Enrichment and/or depletions can occur for individual

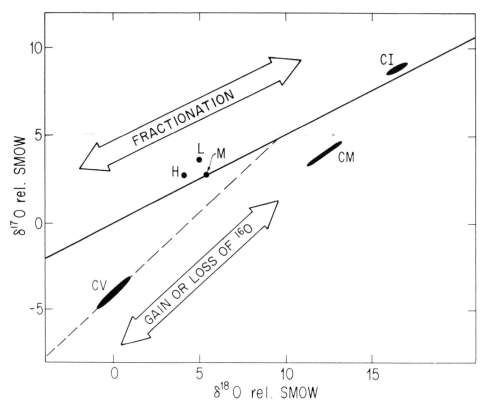

Figure 1.53 The relative proportions of the three oxygen isotopes vary among meteorite groups in a way which cannot be explained in terms of normal geochemical processing, illustrated by the arrow labeled "fractionation." All terrestrial samples, for example, lie on the solid line parallel to that arrow, as do samples from the moon (M). Compositions lying off that line, as for most meteorite groups, are probably due to variable proportions of an anomalous ^{16}O-rich component. The terms CI, CM and CV refer to subgroups of the carbonaceous (C) chondrites. Isotopic ratios for $^{17}O/^{16}O$ and $^{18}O/^{16}O$ are expressed as fractional deviations, in parts per thousand, from Standard Mean Ocean Water (SMOW).

isotopes of many elements, including calcium, titantium, samarium, and neodymium (Figure 1.54). These studies clearly show the presence of more than one nucleosynthetic component. These revolutionary discoveries were made possible by developments in mass spectrometry and associated preparatory techniques in connection with the radiochronological dating of lunar rocks. These developments are also responsible for the high precision with which the age of the solar system is known.

Isotopic anomalies of nucleosynthetic origin have also been detected in noble gases extracted from meteorites. Thus, several carbonaceous chondrites contain an apparently monoisotopic ^{22}Ne component, xenon made within a red giant star by the so-called "slow-neutron-capture" process, and another xenon component possibly resulting from the decay of an extinct superheavy element of atomic number around 114. For these, and many other anomalies, a major problem has been the identification of the host material in which the anomaly is

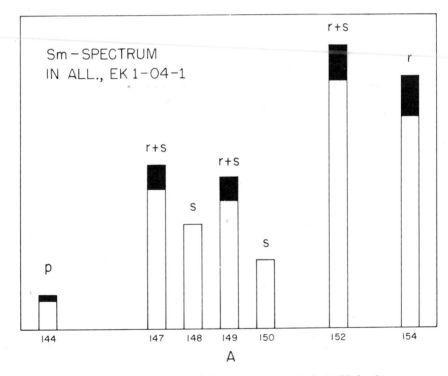

Figure 1.54 Schematic representation of the mass spectrum (A, in AMU) for the isotopes of samarium extracted from an inclusion in the allende carbonaceous chondrite. Open bars correspond to "normal" samarium; dark areas represent the anomalous excesses, exaggerated by 50 times, in five of the isotopes. Those isotopes made only in the slow-neutron-capture process (s) are normal, whereas those made in the rapid-neutron-capture process (r) or the proton-capture process (p) show excesses. Clearly, this sample reflects abnormal mixing of the products of those three modes of element formation.

preserved. This bears upon the question of whether the anomalous material was injected into the early solar system as gas or was contained within the interstellar grains.

1.4.5.4 Evidence for a gaseous nebula. The possibility that interstellar grains could have survived incorporation into the protosolar system calls into question an assumption adopted earlier, namely, that the solar nebula was once sufficiently hot to be gaseous, and hence that protoplanetary material was formed in the nebula by condensation from such gas. Isotopic homogeneity is commonly taken to imply an early mixing with an efficiency that is normally regarded as being characteristic of the gaseous state of matter. However, the only direct evidence that temperatures in the nebula were once high enough to vaporize refractory lithic (stony) material appears to be an observation that some inclusions in carbonaceous chondrites have distribution of rare-earth elements which apparently demand an origin by condensation from a vapor (Figure 1.55). The refractory element fractionation pattern is taken by some as evidence for a hot nebula, but discrepancies between theory and observation reduce the force of this argument. There seems to be no compelling astrophysical requirement that nebula temperatures were as high as the $2,000°K$ needed to volatilize refractory elements.

The concept that the nebula was initially gaseous has led to considerable interest in theoretical models capable of describing the condensation from such a gas. Most attention has focused on models based on assumptions of thermodynamic and chemical equilibrium, an approach capable of generating precise, testable predictions. The primitive nature of chondrites has resulted in the use of their observed compositions and/or mineral chemistry as tests of these models. Although equilibrium-based condensation theory has been highly successful in predicting the broad outlines of both chondritic and planetary chemistry and mineralogy, significant disagreement between theory and observation becomes apparent at a more detailed level of scrutiny. Thus, chondritic abundance patterns do not always correspond exactly to fractionation of predicted condensates, and many chondritic components, identified on one basis or another as surviving nebular condensates, have compositions that differ markedly from those predicted.

1.4.5.5 Locations of meteorite parent body formation. However particulate matter was formed in the early solar system, it subsequently accreted to form larger objects, eventually producing the present planetary system. The number, size, and original location of the meteorite parent bodies are not well known, but the present best estimates place their number at between 10 and 30, and their diameters in the range of a few kilometers to a few hundred kilometers. Current evidence appears to favor an origin for most, if not all, meteorite parent bodies in the asteroid belt. Considerably progress has recently been made in comparing spectral reflectance features of asteroids with those of actual meteorites.

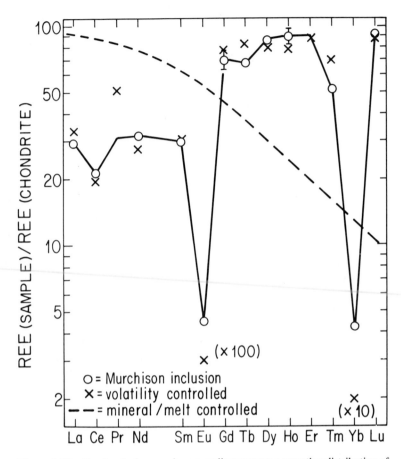

Figure 1.55 Geochemical processing normally generates a smooth redistribution of the rare-earth elements (REE) shown on the horizontal axis, as illustrated by the dashed line which corresponds to the distribution of REE between an igneous mineral and the melt from which it formed. By contrast, the distribution measured in an inclusion from the Murchison C chondrite shows irregularities which match very well some predictions based on the volatilities of the different REE. This constitutes powerful evidence that the Murchison material was once sufficiently hot to have been vaporized, presumably in the solar nebula. (Adapted from W. V. Boynton et al., *Lunar and Planetary Science,* XI (1980), 103.)

Thus, at least one asteriod appears to have a surface very similar to those of certain differentiated stone meteorites, and carbonaceous chondrite-like material appears to be more abundant in the asteroid belt than among recovered meteorites. However, plausible parent asteroids for the most abundant ordinary chondrites have not yet been identified spectroscopically. Also, despite the observational links with asteroids, dynamical problems still plague explanations of how the meteorite population could have been ejected from asteroidal parents and thence transported to the earth. For example, the way in which an asteroid-sized object can wholly or partially break up is not well understood at present, nor is the manner in which the products of such a disruption may be perturbed from an asteroidal orbit so that they can reach the earth. Whatever trajectories they took on their passage to earth, meteorites were exposed, for periods ranging typically from about a hundred thousand years to over a thousand million years, to the interplanetary environment. This can further modify a meteorite in two ways. This may occur, first, by impact with other interplanetary solids and, second, by irradiation by cosmic rays. The former can produce shock damage in minerals, extending possibly to catastrophic disruption of the meteorite, and the latter causes nuclear reactions which can measurably alter the isotopic composition of some elements. Quantitative estimates of such isotopic effects can be used to calculate the time during which a meteorite was following its interplanetary trajectory and also to study secular variations in the cosmic radiation.

1.4.5.6 Summary.   Thus, the application of refined analytical techniques to meteorites can supply clues to the formative processes of the solar system, and of the galaxy itself. It can also serve as "the poor man's space probe," revealing details of processes occurring in interplanetary space.

For reasons of brevity, this review has been kept impersonal, but one name demands inclusion in any account of recent progress in meteoritics. The name is that of the late Harold Urey, whose dedication to the peaceful application of science and whose passionate curiosity concerning the origin of the Solar System made him the founding father of cosmochemistry. To his memory this section is dedicated.

1.4.6 Comets

> There are as many comets in
> the sky as fish in the oceans.
>
> *Johann Kepler*

1.4.6.1 Historical landmarks.   The unexpected apparition of a bright comet has always stirred the fantasy and interest, but also the fear, of mankind. Even before Christ, observations of comets had been recorded, but many centuries elapsed before it was recognized that comets are celestial bodies and not transient phenomena of the terrestrial atmosphere. They are, in fact, the least massive and the most extended members of the solar system.

Scientific observations of comets started in the fifteenth and sixteenth centuries. Sir Isaac Newton fulfilled a prophecy made about 1,700 years earlier by Seneca, a Roman statesman, philospher, and scientist (he was also Nero's teacher), who had predicted that "one day a man would be born who would explain the orbits of the comets and give the reasons for their peculiar motions." Newton discovered that the laws of gravity valid on earth were also applicable to celestial bodies, and he worked out a general mathematical description of the orbits of celestial bodies. Somewhat earlier, Kepler had found his famous three laws, in which he accounted for planetary orbits. His first law is that these orbits are ellipses, with the sun at one of the two foci. But while planetary orbits are nearly circular, most comets move in highly elliptical orbits. Using these results, Halley calculated the orbit of the bright comet of 1682 and came to the conclusion that it was identical to that of the comets seen in 1531 and 1607. Thus he found the first periodic comet, one that comes close to the sun at intervals of 76 years. The regular apparition of "comet Halley," as it was named later, could then be retraced to 86 B.C.; there are two still earlier observations, in 239 and 446 B.C., the latter mentioned by Chinese historians.

While the calculation of the orbits and positions and the prediction of the reappearance of periodic comets made rapid progress, the true nature of these spectacular phenomena remained obscure until the end of the nineteenth century. By using photography and spectroscopy in astronomy, the first rough estimates of the size and chemical composition of comets could be made.

1.4.6.2 Cometary research before the space age.   Due to the increasing number and quality of telescopes, more and more comets were discovered. By 1950, their number was about ten per year, not quite half of them being periodic and returning to the inner planetary system at intervals of between 3 and about 150 years. The others appear unexpectedly; they come from regions far outside the zone occupied by the planets and, after a single visit, disappear again. Among these are most of the bright comets which occasionally are even visible to the naked eye.

In the 1950s, the Dutch astronomer Jan Oort investigated the orbits of the nonperiodic comets and developed new basic ideas about their origin and evolution. The American scientist Fred Whipple, using observations of their evolution in brightness and size while they approach and recede from the sun, high-precision positional observations, and the results of spectroscopy, gave a theory of their structure and composition. The German astrophysicist, Ludwig Biermann explained the development and kinematics of a comet's gas tail by an interaction of cometary material with the solar wind.

Before Oort's work it was not clear whether comets are members of the solar system or inhabit interstellar space and only rarely visit the vicinity of the earth. From a statistical investigation of their orbits, Oort found that the semimajor axes of the elongated ellipses of cometary orbits are at distances from the sun of up to about 1 light year. This is more than a thousand times the distance of the remotest planet, Pluto, from the sun and is almost one-quarter of the distance to the nearest star from the sun. Furthermore, Oort deduced that this large shell of comets, the "Oort Cloud," contains about 100 billion members which slowly move around the sun with periods of about a million years. No comet was found which, beyond doubt, came from interstellar space.

Comets may have been formed about 4.5 billion years ago, either together with the sun and the planets or, more probably, in a neighboring interstellar cloud. From the present star population in the vicinity of the sun and their average velocity, it can be deduced that in these last 4.5 thousand million years several thousand stars passed through the region of the Oort Cloud. There they perturbed the motions of the nearby comets rather strongly. Some of these comets are ejected from the solar system, while others are pushed into the inner planetary system and eventually become visible. Here the great planets, especially Jupiter, influence their orbits by gravitational attraction, so that either they are moved out of the planetary system or, in rare cases, they are forced into smaller orbits and become periodic comets.

During the major part of their lives, the comets are so far away from the sun that they are hardly exposed to solar radiation. Approaching the sun to within the distance of Jupiter's orbit, a comet experiences an extraordinary change; the heat from the sun causes evaporation and sublimation of the surface material and the outstreaming of gases and small dust particles to form a huge atmosphere of extremely dilute matter. This measures up to several hundred thousand kilometers in diameter around the small nucleus. It is this atmosphere, the coma, which is seen because the dust particles reflect the sunlight and the gases emit light by fluorescence. Especially the brighter comets develop tails in which the cometary matter is streaming away from the nucleus. Both coma and tail combine to create the spectacular picture which we observe as a comet (Figure 1.56), while the nucleus itself is too small to be seen.

For many years there was some controversy about whether a cometary

Figure 1.56   Comet Mrkos observed on 25 August 1957, from Mount Palomar. The upper straight filamentary gas tail is readily distinguishable from the lower curved, diffuse dust tail. (Palomar Observatory, California Institute of Technology.)

nucleus consists of many small, meteoric pieces kept together by their mutual attraction (the "sandbank model") or is one solid block of material. Whipple, about 30 years ago, succeeded in bringing together all observations in his "icy conglomerate model." According to this, the nucleus of a comet is a mixture of frozen gases enclosing dust particles and heavier atoms, a kind of "dirty snowball." Spectroscopy of the coma has revealed some of the chemical constituents of the gaseous atmosphere and their abundances. Most of the molecules and the radicals in the coma are secondary products created by chemical processes and by the dissociation of the molecules of the nucleus by the incident solar radiation.

By sublimation and dust ejection during the comet's active phase near the sun, a bright comet loses about 100 million tons of material in one orbit; this is roughly 0.1 percent of its total mass. This limits its lifetime to between 10,000 and 100,000 years. Obviously, the comets that are thus destroyed have to be replenished from the reservoir of the Oort Cloud. The values given, together with the number of visible comets, are in accord with the mechanism by which distant comets are brought into the inner solar system. Furthermore, small deviations of some comets from a strict Keplerian orbit can be explained by these continuous mass losses; these act on the comet like a rocket, with the comet gaining speed by the matter being expelled into the direction opposite to its motion.

Comets are not only destroyed by this kind of mass loss. Fragmentation into several pieces or even complete disintegrations have been observed, especially when comets come very close to the sun. Some meteor streams can be identified as the remnants of comets or as debris lost by comets along their paths. Thus the November Andromedids are the leftovers of comet Biela which had a period of six and a half years and was last observed in 1852.

Bright comets are generally accompanied by one or two tails, a straight narrow one pointing almost directly away from the sun and a shorter, diffuse tail lagging somewhat behind (Figure 1.56). The dynamics of the diffuse tails, consisting of small dust particles, are explained by the radiation pressure exerted on the dust by the photons of solar light radiation. However, the much larger accelerations shaping the long, straight gas tails could not be explained by this mechanism; the light pressure on the gas molecules is far too small. Biermann showed that the solar wind, the stream of charged particles moving away from the sun, would produce the observed properties by interaction with the ionized cometary tails. Blowing through interplanetary space, the solar wind pushes the ionized cometary molecules away just as the wind on earth blows the smoke from the top of a chimney. By measuring the small deviations of the tails from the antisolar direction and the velocities of the tail ions, it was possible to deduce various parameters of the solar wind which were later confirmed by direct measurements from space probes.

1.4.6.3 Advances made using radio and ultraviolet astronomy.

Because of low-energy transitions between their vibrational and rotational states, molecules exhibit spectral lines in the millimeter and centimeter wavelength ranges. After some failures, the search for such radiated lines was successful in 1974 when comet Kohoutek appeared. Besides the radicals CH and CN previously found in the visible spectrum, two carbon hydrates, HCN and CH_3CN, were detected. Very probably these are primary products of the nucleus. Only a few months later, the neutral water molecule H_2O was found in the radio spectrum of another comet, after a hitherto unidentified red line in the visible spectra of several comets had been attributed to the H_2O^+ ion. These findings are of great importance with respect to the chemical composition of cometary nuclei. A search for other molecules which are supposed to be abundant in comets, such as formaldehyde, ammonia, and methane, has not yet been successful. But this does not mean that these substances are absent; some observational problems make it very difficult to detect them.

In the visible spectrum of a comet, the strongest lines are those of the carbon molecule C_2 and the radical CN. Furthermore, some other compounds, like C_3, CH, NH, NH_2, and OH, as well as several metallic atoms have been observed. "Forbidden" oxygen lines were also observed (that is, lines which are too weak to be observed under normal conditions due to the selection rules of atomic theory but which can be seen in an extremely tenuous gas). The abundances of the coma gases could be deduced from some of the line intensities. However, calculations of the integrated brightness led to the conclusion that the total gas production must be about a hundred times larger than the values formerly deduced, and that some other substances must be emitted in much larger quantities. This assumption was confirmed in 1970, when the first ultraviolet spectra of two bright comets (Bennett and Tago-Sato-Kosaka) were made from two earth-orbiting satellites (OGO-5 and OAO-2). Both comets were found to be surrounded by an atmosphere of very rarefied neutral hydrogen, about 10 million km in diameter. Neutral hydrogen emits its strongest line, the Lyman α resonance line, in the far ultraviolet at a wavelength of 121.6 nm. This hydrogen coma has been observed around several comets, among them the small periodic comet Encke, and it seems to be a common property of comets.

A comparison of the production rates of hydrogen with those of the hydroxyl radical (OH), also observable in the ultraviolet spectrum, yielded the result that H and OH were produced in comparable amounts. Since they are both dissociation products of water, this confirms the assumption that frozen water is one, if not the main, constituent of a cometary nucleus. Furthermore, characteristic spectral lines of atomic carbon (C), oxygen (O), and sulphur (S) and some other species have recently been identified in the ultraviolet spectra of some comets. These have been made by rocket observations and by the International Ultraviolet Explorer (IUE) satellite.

1.4.6.4 Plans for a cometary space probe.   Plans to send a space probe to the vicinity of a comet have been discussed for the last 20 years. There are various reasons why such a mission is of great scientific value. Comets are as

old as the sun and the planets. They are the only members of the planetary system which have retained most of their original properties, being far from the deteriorating influence of solar radiation during the majority of their lives. Therefore, they contain information on the origin of the solar system. However, in order to extract this information, more must be known about the nuclei of comets and the physical processes taking place in their comae and tails. Only by means of *in-situ* measurements will it be possible to gain decisive new results.

It is evident that a space probe mission to a "new" comet, coming for the first time into the solar system, would be of the greatest value. This is, however, ruled out (so far) by the unpredictable appearance of a nonperiodic comet and the impossibility of computing the space probe trajectory to a sufficiently high precision during the limited time between the detection of the comet and the required launch date. The interest of the scientific community has, therefore, concentrated on a mission to comet Halley, the periodic comet that is brighter and still more active than any other periodic comet. It resembles those comets which appear for the first time.

Comet Halley has a period of 76 years; it will come close to the sun again in 1986 when it will be readily observable by astronomers on earth who can complement the measurements of the probe. Unfortunately, comet Halley has a retrograde orbit, which means that it moves around the sun in a sense opposite to that of the earth and the other planets. This is a serious constraint for some experiments, because the relative velocity between the comet and the space probe cannot be less than the high value of about 55 km/s. The close encounter will, therefore, be brief; it is planned that the probe will fly through the coma in about half an hour. Furthermore, the dust density around the nucleus is not well known. This constitutes a serious hazard for the probe because, if it comes too close to the nucleus, the fast-flying dust particles could damage or even destroy it. These problems call for careful considerations and preparations for the instrumentation.

The Halley probe will investigate the coma and the tail of the comet. It will lead to a better understanding of the physical processes taking place during its active phase, especially the mechanisms of dissociation and ionization of the primary molecules, the nature and behavior of the dust, and the comet's interaction with the solar wind. For a thorough investigation of a cometary nucleus, another comet whose orbit will allow a slow encounter, or even a rendezvous in which the probe will accompany the comet for months, would probably be a better target.

The United States has, during recent years, considered a mission to two comets, Halley and Tempel 2. This would have combined the possibilities of a fast fly-by and a rendezvous. Comet Temple 2 has a period of $5\frac{1}{4}$ years, and it moves in a prograde orbit which allows a rendezvous with the probe. After its flight to comet Halley where it should release a secondary probe to investigate this comet, the spacecraft should proceed to Tempel 2. It should accompany the comet for about one year. However, these plans had to be abandoned due to a budget that made it impossible to develop the ion propulsion system necessary for a rendezvous flight in due time.

The scientific community is delighted that the European Space Agency (ESA) has recently accepted plans for a fly-by mission (termed GIOTTO) to comet Halley, probably using the Ariane rocket as a launcher. Besides the preparations for the instrumentation of the probe, ground-based observations and observations from earth-orbiting satellites have to be co-ordinated, and a deep-space telemetry network has to be prepared to receive signals from the probe. The launch is scheduled for July 1985. If this chance is missed, another 76 years would elapse before this interesting comet could be studied. It is anticipated that the results of this first flight to a comet will be as surprising and spectacular as those of the recent PIONEER and VOYAGER flights to the large planets.

1.4.7 Cosmic Dust

1.4.7.1 Knowledge of the cosmic dust population.   Although it is not a very conspicuous part of this solar system, cosmic dust does provide numerous natural phenomena which have been observed by mankind since the earliest times. The term cosmic refers strictly to solid particles from the cosmos, or the more distant parts of the universe; less strictly, it encompasses any extraterrestrial dust in space. Such cosmic dust is a major feature of distant galaxies, often seen as dark bands obscuring their luminous star population. Such dust in the Milky Way galaxy, termed interstellar dust, is seen to darken it. The spiral arms contain concentrations of submicron-sized grains, termed dust lanes, close to the active regions of star formation. Sometimes such dust is illuminated by stars and reflects their light. In the Pleiades (Seven Sisters), the dust is "stratified" and aligned along the local magnetic field. Study of these reflection nebulae can thus offer information on the local magnetic fields. Another two visible manifestations of the indigenous cosmic dust population within the solar system (interplanetary dust) have been even more awesome to mankind over the years; these are meteors and fireballs.

Meteors and fireballs are now known to be the spectacular entry of solid interplanetary particles into the earth's atmosphere. Their arrival is generally seasonal, reflecting the existence of distinct streams of matter in the solar system. Observation of their trajectories across the night sky has been possible without major scientific resources; studies made over the last century have shown that many streams are associated with a specific comet orbit. The appearance of the parent comet itself has, of course, been associated with high significance. The major source of dust in the solar system and an equally prolific source of gas, a comet may produce a visible tail stretching from close to the sun out to a distance of 200 million km.

These cosmic dust phenomena have been visible to man since prehistoric times. From time to time, man has also found larger bodies, such as iron meteorites. Reporting his findings of the Challenger expedition in 1883, Sir John Murray correctly identified, in his deep-sea sediments, the nickel-iron spherules to be of cosmic origin. Most of Murray's cosmic spherules have resulted from the evaporation of meteor particles in the earth's atmosphere; the primary source of all such matter is the comet. Although comets cannot yet be viewed as a mineral source in space, the selective preservation of such components of cometary dust in these deep-sea sediments for long geological periods may prove to be significant.

Direct access to the space environment itself by space research has led to a vastly increased understanding of the wide range of phenomena involving cosmic dust. Man would certainly not have reached his present level of understanding of cosmic dust without the space age; equally, man's exploration in the space environment would have produced a few very nasty shocks without the basic knowledge of dust derived from previous terrestrial observations.

1.4.7.2 Direct measurements in space.

With the advent of space vehicles, the expectation of their possibly hazardous encounter with cosmic dust particles, or even with the larger components of this population such as meteorites, led to the assignment of a high priority to cosmic dust study. On the early SPUTNIK, EXPLORER, and VANGUARD satellites, sensors were placed to detect the impact of such space dust. The instrumentation was often crude, for example, a simple microphone sensor. Such early experiments often yielded data which now have to be partly rejected, but through the continuous improvement of such techniques, in-flight analytical techniques can now exceed the sensitivity achievable within the ground-based laboratory.

The highest reliability of early satellite measurements in the investigation of the hazards of the space environment was undoubtedly that of the EXPLORER-16 and -23. The satellites were fitted with an array of pressurized cans which were punctured by particles impacting at speeds greater than those achieved within the laboratory at that time. The puncture rate was accurately assessed over an extended period of time providing design guidelines for space vehicles and component reliability. The satellite showed that the average incidence rate of particles was quite uniform over many months. The demonstration that hypervelocity penetrations were actually taking place clearly showed the need for information on the impact process itself at these high velocities. Thus, detector technology received a massive stimulus, as did research into the high-pressure physical state of matter. For the scientific and astronomical world, EXPLORER-16 and -23 showed that the penetrating particles were sufficient in number to form part of the meteor population observed previously from the ground. In space, particles were observed at much higher fluxes and smaller masses than on the ground. Following these measurements, the three PEGASUS satellites (from 1966 onward) offered in their folding wings several hundred square meters of an electrically active layered metal sheet to assess the frequency of interception of larger masses of dust particles. PEGASUS was important in bridging the gap between space measurements and the faintest meteors that could be detected from the ground, thus co-ordinating the differing research interests.

The detection of cosmic dust by penetration of pressurized containers in space had numerous virtues, and a multitude of satellite experiments were designed to exploit this technique. The more reliable of the penetration experiments showed that some of the microphone data indicating the presence of a large belt of particles about the earth was unfounded. In the drive toward the achievement of manned lunar exploration in the late 1960s, a new trend developed, namely, the return of space material to the laboratory. More detailed scientific analysis is possible in this way, as is a more considered scientific judgment. Windows on the APOLLO vehicles used for development were recovered with clear evidence of the impact by space dust. It was evident, as the lunar landings approached, that the moon would become one of the richest sources of information on dust impacts. Such is the effect of impacts on the moon that it was uncertain whether spacecraft could land safely without being engulfed in the surface dust. The first landing of the APOLLO-11 lunar landing module showed such fears to be exaggerated; APOLLO-11 also brought back lunar rocks valuable to the study of dust and to many other fields. Such rocks, exposed to space over a much longer period than could ever be achieved in man's exploration of space, were returned to the earth for study. Following this, the USSR, using their automated lunar landing craft, drilled beneath the surface to a depth of 2 m.

The need for scientific analysis of the material from space galvanized several hundred scientific groups throughout the world into action. Many had no previous involvement with the space age. For cosmic dust studies, the impact crater size distribution was extended down to masses less than a picogram (10^{-12} g). The effects of impact crater saturation, hypervelocity ejecta splash buildup, and, for the first time ever, the modifying effects of the solar wind on solid surfaces (solar wind sputter) were noted. Later APOLLO-17 astronauts left cosmic dust sensors on the moon's surface, which showed dust particle levitation and transport to be a permanent feature of lunar sunrises and sunsets, as well as the expected meteoroid impact activity. Complementing the exploration of the earth's near environment on such lunar missions, automated instrumentation for dust studies in space was developing fast, following the discovery by W. M. Alexander and J. F. Friichtenicht, using a 2MV van der Graaff accelerator, that the plasma produced at impact could be detected electrically. Early instrumentation on the Orbiting Geophysical Observatories (OGO-1 to -4) showed that such detection was possible, but the small dimensions of the sensors precluded a good chance of interception. It was PIONEER-8 and -9 which first measured the mass, velocity, and direction of incidence of impacting dust particles by this technique. For the first time the orbits of dust particles in the solar system were found and their original directions calculated. New information was obtained from the sensors sent into interplanetary space on PIONEER-8 and -9. It was found by O. E. Berg that a population of small, very high-velocity particles originated from the direction of the sun. The velocities were so high that their retention in the solar region was precluded; they were being expelled into interstellar regions by solar radiation pressure. Their source is still a matter of debate; highest on the list of possibilities is the breakup of larger meteoroids colliding long after their expulsion from comets as they spiral into the sun under the Poynting–Robertson radiation drag effect.

Deeper into space, exploration by PIONEER-10 and -11, which were equipped with penetration arrays, showed that near Jupiter the gravitational attraction of this major planet increased the dust population some hundredfold. PIONEER-10 and -11 gave a foretaste of the more complex measurements made on the very recent VOYAGER spacecraft, now based on high-quality imaging. Particles with sizes in the 10 mm region were found in a ring about Jupiter. Dust particles are ejected from volcanoes on one of Jupiter's satellites, Io; through a combination of interacting magnetic, electric, and gravitational forces, the particles find their way into the ring.

Exploring the inner portions of the solar system, the Federal Republic of Germany/United States satellites HELIOS-1 and -2 (from 1973 onward) extended dust measurements from the earth's orbit to two-thirds of the distance to the sun. They confirmed the flux of outward moving particles discovered by PIONEER-8 and -9 and gave more extensive orbital measurements on the statistics of the distribution of dust. For the first time, a mass spectrometer measured, with modest resolution, the composition of the space dust. Developments of that technique are now being selected for the ESA GIOTTO mission to intercept comet Halley in 1986.

Imaging in space with sensitive photodetectors has largely overcome the limitations which restrict ground-based study of the sunlight reflected by the dust population in the solar system. Although in favorable circumstances it is visible, even to the naked eye, as a cone extending above the horizon, accurate measurement was limited both by natural phenomena such as airglow in the atmosphere and by artificial "pollution" such as reflected street lighting. By rockets and by satellites, accurate photometry and polarization measurements have given information on the large-scale distribution which could be only partly determined by actual spacecraft journeys. A development of this technique has been selected by the International Solar Polar Mission (NASA/ESA), and it is being considered for the GIOTTO comet Halley mission.

1.4.7.3 Terrestrial-based studies stimulated by space research.

With the extension of an observational technique to outer space, there is created a stimulus for the study and simulation of the observed phenomena within the laboratory. Space research has thereby led to the encouragement of the use of accelerators based on explosive charges, gas guns, plasma accelerators, and electrostatic and linear dust accelerators to achieve high velocities. Values as high as 70 km/s have been reached, but only for very limited size ranges.

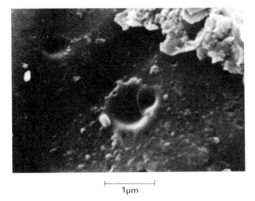

Figure 1.57 Cosmic dust impact micro-craters formed in the early solar system. Preserved on grains inside a meteorite (Murchison) dated at 4.2 million years, they give testimony to the role of cosmic dust in the formation of the solar system. (Photograph courtesy of I. Hutcheon and Pergamon Press.)

⊢————⊣
1μm

Lunar surface research must remain one of the largest stimuli to laboratory techniques. Developments such as the energy dispersive X-ray analyzer and ion beam instruments have been stimulated by the need to analyze particles of sub-micron ($< 10^{-6}$ m) dimensions (Figure 1.57). The need for computer simulation of the impact process itself was a direct consequence of the necessity to perform experiments at very high velocities. The importance of calculations as a tool for

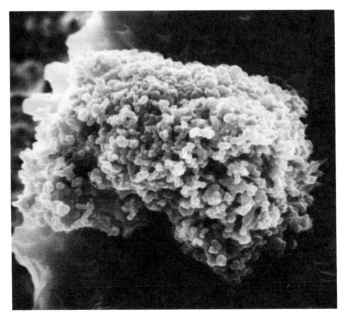

Figure 1.58 A cosmic dust particle today, as studied by modern techniques. This was collected by a U-2 aircraft in the stratosphere; very probably it was released from a primeval icy comet nucleus. (Photograph courtesy of D. E. Brownlee and John Wiley and Sons.)

understanding the role of the equation of state in high-pressure impact flow is now recognized.

Using U-2 aircraft flying over California, D. E. Brownlee has successfully recovered and analyzed pristine cosmic dust particles from space (Figure 1.58). These particles are so small that they enter the atmosphere without destructive heating. Populations of particles with compositions attributable to specific meteorite families (such as carbonaceous chondrites) have been successfully diagnosed.

1.4.7.4 Current research.   A large fraction of the mass of the universe is invested in cosmic dust. This population of small solid particles is clearly a medium through which interstellar clouds grow and accrete. In collapsing they form stars, and in some cases solar systems. Cosmic dust is also significant in the final stage of the stellar life cycle, since stars which collapse and explode into novae and supernovae emit copious fluxes of dust grains. Cosmic dust or hence interstellar grains enter the earth's atmosphere and survive entry without increasing their temperature to more than $320°K$ ($50°C$). Although the seeding of life on earth and the introduction of genetic material from such sources must remain speculative at this stage, the possibility provides ample stimulus for such studies.

1.5 APPLICATION OF SPACE TECHNOLOGY TO GEODYNAMICS AND GEODESY

1.5.1 Introduction

1.5.1.2 Problems in geodynamics.   Geodesy is the scientific study of the geometrical shape of the earth. Satellite tracking data have been used to show that the earth is pear-shaped, departing from the expected ellipsoidal shape by up to 10 or 20 m. Geodynamics describes and explains the motions of the earth. The dozen or so plates, which constitute the earth's crust, move over the mantle beneath. This concept, introduced by Alfred Wegener, is sometimes called "continental drift."

Despite the rapid progress that has been made within the last decade, there are many important questions in geodynamics that cannot yet be answered. For example, how are the plates moving at the present time? Are the movements smooth or jerky? How do the local movements caused by earthquakes at active plate boundaries contribute to the gross movement of the plates? How do the plates deform in response to the driving forces? How is strain distributed near active plate margins, and how does the strain change with time? What relationship, if any, is there between variations in polar motion and other geodynamical phenomena such as great earthquakes? Are large-scale vertical movements taking place at the present time?

What these questions have in common is that progress toward finding their answers depends at least to some extent on measuring the relative positions and movements of points on the earth's surface, over distances ranging from a few kilometers to many thousands of kilometers. In order to measure the movements of an average plate (which has a typical velocity of a few centimeters per year) within a reasonable time, and because of the necessity of detecting possible episodic changes in these motions, measurements have to be made to an accuracy of about 2 cm (20 mm) at the largest distances.

A possible observable effect of large-scale tectonic activity is the motion of the polar axis of the earth, that is, the axis about which the earth spins once per day. The classical astronomical methods of measuring the polar motion and the earth's rotational period appear to have an accuracy of somewhat better than 1 m in pole position and a few milliseconds in time. Measurements of polar motion made using the Doppler tracking of artificial satellites provide the pole position to an accuracy of 0.3 m, but do not give better results in universal time (UT). Theoretical estimates of the change of polar motion due to a large earthquake suggest that this accuracy must be improved by over an order of magnitude (that is, the measurements must be made to an accuracy of a few centimeters) to be able to study possible relationships between these global parameters and other geodynamical phenomena. At present, the Bureau international de l'heure (BIH) in Paris reports five-day averages of the position of the pole, using both astronomical and Doppler data, and the period of the earth's rotation again averaged over five days. It is necessary to have measurements at intervals of about one day (or preferably less), and to a greater accuracy than reported by BIH, in order to detect a change of the polar motion due to a large earthquake.

There are several reasons why classical ground-based geodetic surveying methods, namely, leveling and trilateration (measuring the lengths of the sides of the triangle formed by three points on the earth's surface), are impractical for making these measurements over distances greater than a few tens of kilometers. First, such measurements cannot be made at all over the oceans. The major reason is that ground surveys necessarily have to be made in a sequence of line-of-sight measurements between points up to a few tens of kilometers apart; the resulting accumulation of random (and possibly systematic) errors soon brings the uncertainty in the position to more than a few centimeters. The best trilateration is good to about three parts in 10^7, an error which exceeds 3 cm beyond about 100 km; the best leveling method accumulates random errors at the rate of a few millimeters times the square root of the length of traverse in kilometers, that is, about 3 cm in 100 km. Second, ground surveys are time-consuming, particularly for leveling, and they can be compromised in seismically active areas by the motion actually taking place during the survey. Third, the cost of monitoring vertical movements two to three times per year in even a few tectonically active areas over, say, an area of a few hundred kilometers by a thousand kilometers and with a line spacing of 100 km, is prohibitive (over ten million dollars).

1.5.1.2 Space methods. The earliest space method, namely, the photography of satellite tracks against the star background, is insufficiently accurate to be useful for studies of geodynamics. New methods have been developed to meet the needs described. There are two principal space geodetic techniques: laser ranging to or from satellites and microwave interferometry These are able to provide observations to the accuracy required for geodynamic studies; however, newer methods, such as radar altimetry, may also be able to do so, and more cheaply. The remainder of this section is devoted to a description of the laser and interferometric methods, to plans for using them for geodynamical research, and to geodetic operations.

In these space-related systems, the concepts are simple but the practical realization is very difficult. To measure the position of a point to an accuracy of 2 cm over distances of hundreds of kilometers or more requires a high level of technical sophistication both in the hardware to make the measurements and in computer modeling to interpret the results. The hardware required includes accurate time and frequency standards, short-pulse duration lasers, and wideband recording systems. Factors which must be modeled with extreme care include the earth's gravity field, earth tides, ocean loading, and the effects of earth's precession, nutation and diurnal rotation, as well as both special and general relativity. Development of the required technology is not cheap, but the cost can be justified on grounds of the scientific importance of the information obtainable by space methods and the inability of other methods to provide these measurements. In contrast to the development costs, the operational costs of the new equipment are moderate. Particularly in the case of frequently repeated precise vertical position measurements involving traverse lengths greater than a few kilometers, the costs are considerably less than those incurred using conventional methods.

In addition to the positioning systems described above, space technology is making other significant contributions to geophysics. For example, detailed mapping of the earth's magnetic field has been obtained from a satellite (MAGSAT), launched in October 1979, into an orbit at an altitude of almost 1,000 km. The geomagnetic field has been observed using both a scalar magnetometer measuring the total field to an accuracy of about 3 nT or gamma (10^5 gamma $= 1$ G) and a vector magnetometer which was accurate to about 6 nT or gamma in each component. Also, knowledge of the earth's gravity field has improved enormously since the advent of satellite geodesy nearly two decades ago. Satellite laser ranging, and radar altimetry from GEOS–3 and SEASAT–1, have enabled a representation of the earth's gravity field to be constructed at high resolution, complete to harmonic degree and order 36. NASA, ESA, and INTERCOSMOS are now studying dedicated gravity field mapping satellites to determine the gravity field with an accuracy of several milligals and with about 100 km resolution. This is roughly equivalent to obtaining the geoid to an accuracy, at this resolution, of a few decimeters (tenths of meters).

1.5.2 Laser Ranging

1.5.2.1 Principle of method. In laser ranging, a short-duration pulse is fired from a laser on the ground and reflected back from a cube corner retroreflector mounted on an earth-orbiting satellite. The time of flight is measured accurately; since the orbits of the satellites used for this purpose are accurately known, the laser position can be determined (in geocentric coordinates) by measurements made when the satellite is in different parts of the sky.

1.5.2.2 Lunar laser ranging. The principal use of lunar laser ranging for terrestrial geodynamics is to provide information on long-term changes in the polar motion and rotation rate of the earth. Lunar laser ranging began with the APOLLO Lunar Laser Retroreflector Experiment. Retroreflectors were placed on the moon by the astronauts of APOLLO-11, -14, and -15, and by the unmanned USSR LUNA-17 and -21. The objectives of the lunar laser ranging experiments were to study relativity, the lunar orbit, lunar dynamics and structure, and terrestrial geodesy. The first lunar laser ranging was accomplished shortly after the landing of APOLLO-11, and ranging has been done routinely at the McDonald Observatory of the University of Texas at Austin from 1969 to the present. The mean uncertainty of the range, and hence the accuracy to which the earth–moon distance has been measured, using the McDonald Observatory system over the last ten years is about 12 cm (0.12 m). For lunar laser ranging and for the Transportable Laser Ranging System, the accuracy given is the standard error of a "normal point," which is the average of observations in a single satellite observing pass (about ten returns for lunar work and about 100 for LAGEOS). For all other stations mentioned here, the accuracy given is the standard deviation of a single observation about the mean. Range accuracy translates into station position accuracy in a complicated way, but the uncertainty in each component of position is of the same order as, or slightly larger than, the systematic error contribution to the range uncertainty. The McDonald Observatory is the only continuously operating lunar ranging station in the world at present, although several other stations have ranged to the moon successfully. The stations at Crimea, USSR, and at Orroral Valley (near Canberra, Australia) obtain returns from time to time. A laser station at Wettzell, Federal Republic of Germany, and another at Grasse, France, are expected to be operational for lunar ranging in 1981. Lunar laser ranging stations at Dodaira, Japan, and Haleakala, Hawaii, United States, are under development.

1.5.2.3 Satellite laser ranging. The geodynamical objectives of satellite laser ranging are to provide information on polar motion and earth rotation, to establish baselines for the determination of plate movements and internal plate deformations, to measure crustal deformation in tectonically active areas, to study global deformations of the earth due in particular to tidal forces, to study the earth's gravity field, and to determine precise orbits for certain satellites such as SEASAT-1.

Laser ranging to man-made satellites has been performed routinely since the middle 1960s using about a dozen satellites equipped with cube corner retroreflectors. Global deformations of the earth as revealed by periodic variations of the earth's gravitational field are detectable by satellites at altitudes of up to 1,000 km, but these are masked somewhat by the effects of atmospheric drag. In 1975, the French Satellite, STARLETTE, a dense sphere covered with retroreflectors, provided information on this topic from an altitude of 700 km. Research on the earth's gravity field and on crustal movements was done with such relatively low-altitude satellites until the launch of LAGEOS in 1976. LAGEOS is now the main satellite used for geodynamics. It is built on the same principle as STARLETTE, but it is much larger and heavier and should remain in its orbit at 5,900 km altitude for at least 8 million years. A Bulgarian satellite covered by 84 retroreflectors will be launched into a circular polar orbit at a height of 1,000 km; the geodynamics program is being co-ordinated by the Central Laboratory of Superior Geodesy of the Bulgarian Academy of Sciences.

Today there are about two dozen laser ranging observatories around the world, with more being constructed or planned. The range accuracy of these stations varies from about 3 cm, at best, to about 1 m. NASA operates fixed observatories at Goddard Space Flight Center, Maryland, United States, and, in cooperation with the United States Air Force, at Patrick Air Force Base, Orlando, Florida. Their range accuracy is about 3 cm for the Goddard laser and 10 cm to low-altitude satellites for the Patrick laser. The Smithsonian Astrophysical Observatory (SAO) operates observatories at Natal, Brazil, Orroral Valley, Australia, and Arequipa, Peru, with a range accuracy to LAGEOS of from 10 to 20 cm. The Institut für Angewandte Geodäsie operates a laser ranging station at Wettzell, Federal Republic of Germany, whose accuracy is about 4 to 5 cm. The Delft University of Technology operates a laser station at Kootwijk, Netherlands, with a current range accuracy of about 10 cm. Another high-accuracy laser station now in operation, and one which is also transportable, is that of the Groupe de recherches de géodesie spatiale at Grasse, France, with a range accuracy of about 20 cm that will soon be improved to 10 cm. Similar precisions are obtained at the Finnish station of Metsähovi. There are also stations operating, to 1 m accuracy, at San Fernando, Spain, Dionysos, Greece, and Zimmerwald, Switzerland. INTERCOSMOS operates 12 laser stations with 0.6 to 1.5 m accuracies. New laser systems with accuracies in the centimeter and decimeter (0.1 m) region are being developed in the German Democratic Republic, Czechoslovakia, and Poland. NASA also operates eight mobile laser stations (MOBLAS). These are located at Goddard Space Flight Center, Owens Valley Radio Observatory in California, the Goldstone Deep Space Network Station in California, Kwajalein Island in American Samoa, Geraldton in Western Australia, the Haystack Observatory in Massachusetts, and at Fort Davis in

Texas. These stations were originally designed for high-precision orbit determination of SEASAT-1 as well as for geodynamic purposes. In their present locations they are now participating in several experiments. These include a validation and intercomparison experiment between laser ranging and very long baseline interferometry (VLBI) methods, plate motion measurements between North America, Australia, and sites on the Pacific plate, and plate stability measurements for these plates. The range accuracy of the first three MOBLAS units is about 5 cm, and about 10 cm for the others; plans are being made for upgrading their performance.

Several weeks are required to move and set up the MOBLAS stations at new sites. A much more highly mobile transportable laser ranging station (TLRS) has been constructed by the Department of Astronomy of the University of Texas at Austin. This station is mounted in one self-contained truck that requires only two hours to be dismantled or to be set up at another site. The designed range accuracy of the TLRS is 3 cm or better. Similar systems are planned in the Federal Republic of Germany and in the Netherlands.

In order to achieve the accuracy goal of 2 to 3 cm, it is necessary to use the highly precise laser stations in pairs to measure relative position. The TLRS and the MOBLAS will be used with fixed base stations.

1.5.2.4 Space-borne laser ranging. From 1975 to 1979, NASA studied the possibility of reversing the positions of lasers and retroreflectors, that is, of placing the laser in earth orbit and the passive retroreflectors on the ground. This system might be useful for rapid surveying of the relative positions of sites within a few tens of kilometers of each other but spread over fairly large areas such as the tectonically active region of southern California. The studies indicate that accuracies of the order of 2 cm in all position components can be attained with existing equipment and that the system may be cost-effective if the relative positions of a sufficiently large number of ground targets are required. These studies were discontinued in 1980, pending the outcome of studies on and prototype development of the Global Positioning System (GPS) described below. A decision will be made in 1982 or 1983 on which system, if any, should be further developed for geodetic use. Studies for the development of such a system are also under way in the Federal Republic of Germany.

1.5.3 Very Long Baseline Interferometry

1.5.3.1 Principle. The second basic method of space geodynamics involves microwave radio interferometry. The idea is straightforward: Two radio antennae simultaneously receive radio noise from a distant extraterrestrial radio source. By cross-correlating the output of the two antennae, the time difference between the arrivals of the signals at the two stations and the rate of change of that difference can be determined. By observing a number of sources repeatedly throughout a day, and by combining the resulting measurements in a least-squares adjustment, the vector baseline between the observing sites can be recovered. The measurements are also sensitive to the positions of the radio sources, the orientation of the earth in an inertial reference system, and the movements of the earth's crust caused by earth tides and ocean loading. Model parameters for these effects are recovered in the least-squares adjustment, and so are the baseline parameters. The observations are also affected by the earth's ionosphere and troposphere, which slightly change the speed of propagation of the signals through those parts of the atmosphere. These effects can be removed by calibrations based on other instruments at the observing sites. To date, all radio sources used in geodetic microwave interferometry have been extragalactic sources, usually quasi-stellar objects, which have negligible proper motions. Except for very small, but known, source structure effects, these sources can be considered as fixed points in inertial space.

Historically, radio interferometry was developed by radio astronomers using antennae separated by distances of a few hundred meters to a few tens of kilometers. The signals received at one antenna could be sent to the other via a relatively simple communications link, and the cross-correlation information could be generated in real time with no recording required. In geodetic applications, however, the observing sites may be separated by hundreds to thousands of kilometers and to date there has been no economically feasible way of transmitting the signals received at one site to the other. In the technique called very long baseline interferometry (VLBI), the signals are recorded on magnetic tape at each station and correlated later at a central analysis facility. To accomplish this, the signals (which are recorded at very high frequencies, typically 3 to 8 GHz) must be converted to some lower frequency via an extremely stable frequency standard. The most successful VLBI programs have used hydrogen maser clocks as frequency standards. These are expensive and sophisticated devices which are not yet available commercially. The tape recorder requirements are stringent; it is necessary to record information at a rate in excess of 100 megabits per second (10^8 bits/s), the equivalent of recording 20 television channels simultaneously. Offsetting these technological difficulties in VLBI are two overriding advantages; VLBI stations can be operated at large distances apart and they can be made small and highly mobile.

1.5.3.2 Space methods.

Radio signals from earth-orbiting satellites can also be used for VLBI measurements. Lunar dynamics has been studied using a combination of quasar noise and radio signals from the Apollo instrumentation packages on the moon. A promising recent development is the use of radio signals from the Global Positioning System (GPS) satellites. Although their main purpose is for navigation, they offer an opportunity for specialized use in geodynamics research.

The geodynamical objectives of VLBI measurements are the same as those for satellite laser ranging (except that the VLBI results are obtained in an inertial reference frame). The technological improvements of VLBI observatories required to achieve the determination of the baseline length to an accuracy of 2 to 3 cm will also bring benefits in radio astronomy.

1.5.3.3 Attainable accuracy. Radio astronomers first attempted to make a precise determination of the vector separation between VLBI observatories in the late 1960s; later, many determinations were made between the Haystack Observatory in Massachusetts and various observatories in California. The accuracy of these measurements has improved from about 16 cm in 1974 to about 4 cm recently. Measurements have been made of the baseline between the Haystack Observatory, the Owens Valley Radio Observatory in California, the National Radio Astronomy Observatory in Greenbank, West Virginia, the Onsala Space Observatory of the Chalmers University of Technology in Sweden, and the 100-m Effelsberg radio telescope of the Max Planck Institute for Radio Astronomy in Bonn, Federal Republic of Germany.

Occasional geodetic VLBI measurements have been made at stations of the NASA Deep-Space Network (DSN), which operates tracking facilities at Goldstone in California, Madrid in Spain, and Tidbinbilla in Australia. VLBI capability at the 50 cm accuracy level is required for the control of the navigation of interplanetary spacecraft; thus, the necessary equipment is in place at all the DSN stations. However, the operational determination of polar motion and earth rotation requires measurements to be made every one to five days. This almost continuous use of antennae is not possible at large radio observatories whose primary responsibility is unrelated to geodynamics programs.

This problem will be resolved when the Polaris network of three VLBI observatories becomes fully operational in 1983. This network is being constructed by the United States National Geodetic Survey (NGS) in co-operation with NASA; it will consist of stations at Fort Davis in Texas, Westford in Massachusetts, and Richmond in Florida. It will provide daily estimates of polar motion and earth rotation, with expected errors of about 5 to 10 cm in pole position and only 100 μs or 0.1 ms in time. These stations will also serve as base stations for crustal deformation measurements made using mobile VLBI stations. NASA is also building two transportable VLBI data systems to be used primarily for establishing temporary base stations for plate motion studies and to support mobile VLBI measurements of crustal deformation.

Two mobile VLBI stations have been built, one of which has been used since 1974 for measurements at sites in southern California. Demonstrations of the mobile VLBI system have shown a 6 cm accuracy when compared with NGS measurements over 400 km. A second, much more mobile station will be in operation in 1980, and both will be used for crustal deformation measurements.

The European Space Agency (ESA) is considering establishing a VLBI network of up to ten existing radiotelescopes and a geostationary satellite. This would provide real-time operation which is particularly important for the detection and study of earthquakes and other sudden geodynamical events.

1.5.4 Local Surveys

1.5.4.1 Global Positioning System satellites. The Global Positioning System (GPS) will consist of at least 18 satellites in high orbits (at 20,000 km altitude) placed so that at least four satellites will always be visible from any point on the earth's surface. The possibility of using the radio signals continuously broadcast by these satellites for precise geodetic surveying has been recognized. The advantages over astronomically based systems are that the ground receivers can be made very small and easily transportable (because of the high signal strength) and that the relative positions of two such receivers can probably be determined to 3 cm, or better, in a few minutes. The limitation is that the receivers must be sufficiently close together (probably only 100 to 200 km apart) so that the uncertainty of a few meters in the satellite position does not contribute significantly to the baseline length error. Systems of this type may make it possible to make frequent measurements of crustal deformation and strain changes over large tectonic areas both rapidly and economically.

1.5.4.2 Doppler measurements. Another technique for precise relative positioning, to an accuracy of a few centimeters over distances of up to 1,000 km, involves a satellite making sequential measurements of the distance to, and the Doppler shift from, a network on earth-based responders. Further studies using this principle are under way in Europe.

1.5.5 Space Programs in Geodynamics

1.5.5.1 Organizational aspects. Substantial efforts have been made in many countries to develop space technology for applications to geodesy and geodynamics. In the United States, a consortium of five federal agencies has been formed to coordinate such efforts, with working groups on specific problems such as the selection of sites for mobile station operations, local supporting surveys, and data management. ESA has formed an Earth-Oriented Working Group to study earth science applications; the Geodynamics Working Party of this research group is formulating a European program that is similar to that in the United States. Several countries in other parts of the world are also considering geodynamics programs within the framework of INTERCOSMOS.

1.5.5.2 Aims. Accomplishment of the geodynamics objectives involves repeated and systematic measurements of position and position changes at global, continental, regional, and local scales. International efforts are under way to establish a worldwide network of laser ranging and VLBI observatories which will measure polar motion, earth rotation, and tectonic plate movements, and which will also serve as base stations for mobile stations measuring crustal deformation over regional distance scales.

1.5.5.3 Comparative studies. It is important that both the laser ranging and VLBI systems be used in parallel, with a considerable overlap of the network of observation sites for the two types of stations. The reason for this is the
necessity to check the validity of the observations which, as pointed out above,
cannot be done with ground surveys. The sources of error and bias in laser ranging and VLBI are almost certainly independent, so that cross-checking of the two
methods can be done by continued comparison of the measurements. Astrometric observations of the stars using astrolabes and other instruments provide
useful supplementary information, particularly on the relationships between the
different coordinate systems used and the inertial coordinate system.

The observing strategy at the fixed observatories is simple: to make as
many VLBI observations and to perform laser ranging on as many satellite
passes as budgetary and other constraints allow. The strategy for the measurement of crustal deformation in tectonically active areas, however, is very much
more complex. The number of mobile stations is small, and the number of
geophysically important problems that might be studied with such equipment is
large. The number of areas where crustal deformation is known, or inferred, to
be taking place, at rates that are large enough to be measured at the present levels
of inaccuracy (a few centimeters), is also large. After consultation with advisory
groups, a NASA plan for sites to be occupied by mobile VLBI and laser ranging
stations has been agreed upon. The first priority is to observe crustal movements
in the western United States because of the economic and social importance of
the earthquake hazards there, particularly along the San Andreas fault system.
Less frequent observations will be made in the rest of North America, primarily
to determine the stability of the North American plate, to establish initial baselines in the seismic region of Missouri, to monitor uplift in the eastern United
States and Canada due to post-glacial recovery, and to measure crustal deformation in Alaska.

The first operations outside the United States will involve observations of
the relative movement of the Pacific, North American, Australian, and South
American plates. These will be followed by observations of the movement of the
Nazca plate and of crustal movements in Andean South America, Central
America, and the Caribbean and the north coast of South America. Later, observations will probably be made in Australia, New Zealand, and elsewhere in the
Pacific. When the European and other space geodynamics programs begin, it is
expected that close cooperation with NASA will evolve naturally through existing international activities such as the International Geodynamics Project and
the Commission on International Co-ordination of Space Techniques for
Geodesy and Geodynamics (CSTG), jointly established by COSPAR and the International Association of Geodesy. The exchange of laser ranging and VLBI
data is already being accomplished under international agreements such as project MERIT (monitoring of earth rotation and intercomparison of techniques) of
the International Astronomical Union (IAU) and the International Union of
Geodesy and Geophysics (IUGG), and the joint IUGG/COSPAR project

EROLD (earth rotation from lunar distances). MERIT is a cooperative international effort to compare satellite laser ranging, lunar laser ranging, VLBI, Doppler satellite tracking, and astronomical observation techniques in order to determine better the polar motion and earth rotation which are essential parameters in geodynamics. EROLD is another international campaign; its program aims to gather lunar ranging observations and to reduce them to produce accurate ephemerides and to compute geodynamic parameters as speedily as possible.

1.6 MATERIALS SCIENCES IN SPACE

1.6.1 Introduction

1.6.1.1 Melting ice—a simple example.   Let us begin with a simple experiment: the melting of ice both in a terrestrial laboratory and in a space laboratory (Figure 1.59). Under similar conditions of atmospheric composition and temperature, but under different conditions of gravity, that is, "1 g" (unit value of the acceleration due to gravity, 9.81 m/s^2) on earth and approximately one-ten-thousandth of 1 g (10^{-4} g), or "micro-gravity," in space, several differences may be discerned:

1. The overall shapes of the melting ice block develop differently.
2. The interface between the solid and the liquid state (or phase) assumes different appearances.
3. Spatial separation into a solid and a liquid phase only occurs on earth because gravity pulls the liquid downward.
4. The time for complete melting at 1 g is appreciably longer in space. Both on earth and in space heat is transferred to the melting ice by conduction and by radiation; on earth heat is also transferred by convection currents within the liquid and within the surrounding atmosphere. The formation of drops under 1 g conditions increases the total surface area, and thus increases the heat transfer to the ice.

These observations demonstrate, either directly or indirectly, all areas of "materials science and processing" under microgravity conditions that are of interest. They are:

1. Interfacial and fluid physics [(1), (3), and (4) above].
2. Interfacial reactions including phase changes (2) and (3) above; these include solidification, or melting, and physicochemical reaction kinetics.

While (1) is of a more fundamental character, (2) is oriented more strongly toward applied materials science, that is, materials processing. Solidification

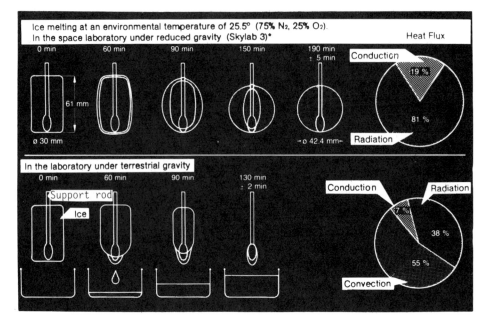

Ice melting at an environmental temperature of 25.5° (75% N₂, 25% O₂).
In the space laboratory under reduced gravity (Skylab 3)*

Heat Flux

0 min   60 min   90 min   150 min   190 min ± 5 min

Conduction

19 %

81 %

61 mm

ø 30 mm

≈ø 42.4 mm

Radiation

In the laboratory under terrestrial gravity

0 min   60 min   90 min   130 min ± 2 min

Support rod

Ice

Conduction   Radiation

7 %

38 %

55 %

Convection

Figure 1.59   Diagram showing the behavior of melting ice in space (upper part) and on the earth (lower part). (Composed from G. H. Otto and L. L. Lacy, AIAA Paper 74–1243, by ASN (P. Kleber), 1974.)

and physicochemical reaction kinetics need to be understood in connection with technological processes, such as the casting of metallic alloys and the combustion of gaseous fuel in chemical reactions, respectively.

1.6.1.2 Processes occurring under microgravity conditions. From a scientific standpoint, the entire field of materials science and technology may be linked to the gravity field by the factors shown in Figure 1.60, which illustrates the areas receiving attention within the various materials science activities in space research. The fact that the common denominator of "reduced gravity" applies to all experiments carried out in space and from a wide variety of subjects has proved to be extremely beneficial to the broad subject of materials science in space. Crystal growers, fluid physicists, biologists, metallurgists, physicians, and thermodynamicists join to exploit the microgravitational environment to study "g-related" phenomena. This has led to numerous contacts being established between scientists from these diverse fields; highly valuable cross fertilization of ideas has taken place. As an example, the understanding of fluid flow phenomena in crystal growth process control and product quality (defects, chemical inhomogeneity, and interface stability) is much more detailed now than it was a few years ago. Similarly, metallurgical processes, such as

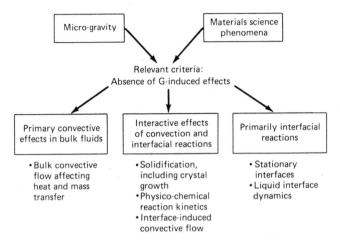

Figure 1.60 Summary of micro-gravity related materials science topics.

casting, welding, alloy formation, and directional solidification, are being studied in great detail; much is to be expected from systematic investigations carried out under the unusual experimental conditions of a microgravity environment.

1.6.2 Interfacial and Convection Phenomena

1.6.2.1 Interfaces between solids and liquids or between liquids and vapors.   Under microgravity, it is difficult to obtain planar interfaces on liquids; Figure 1.61 (a) (right) shows the surface of water in a cup as a curved surface at all positions, not just at the edge as it is at the earth's surface (left). The liquid surface makes an angle of contact with the wall, α. Liquids assume a spherical shape because no gravitational pull is present to "deform" the surface (a hydrostatic deformation). The forces of surface tension (or considerations of the surface energy at the liquid–vapor interface) cause a liquid sample to assume the smallest possible size, that is, that its surface takes up a spherical shape. Depending on the contact angle, a measure for interfacial (or surface) tension, α, the liquid–container wall interaction is termed either wetting (for α between 0° and 90°) or nonwetting (for α between 90° and 180°). These shapes are shown in the upper and lower parts of Figure 1.61 (b), at the left under conditions of 1 g and in the center under microgravity conditions. Wetting signifies a strong adhesive force between the liquid and solid surfaces (that is, the solid–liquid interface). Nonwetting indicates a weak, or even zero, adhesive force; equivalently, high repulsive forces occur at the solid–liquid interface. In order to obtain a planar interface under microgravity conditions, the shapes of the required containers are

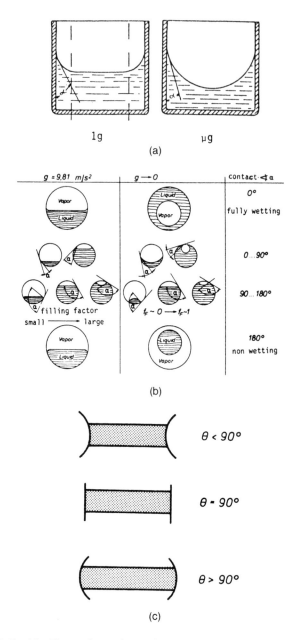

Figure 1.61 (a) The meniscus of water in a cup at 1 g and at μg (microgravity). (b) Surface energy and contact angle of liquids are responsible for differences in interface shapes at 1 g and at μg, for example. (c) Container wall shapes required for planar surfaces of a liquid at various contact angles, θ. (Adopted from: H. Oertel, Report BMFT 7907, 1980; B. W. Wahl, Raumfahrtforschung Nr. 6, p. 273, 1969; and H. U. Walter, Haus d. Technik-Veröff. Nr. 391, p. 41, 1977; respectively.)

very different from those used at the earth's surface. Figure 1.61(c) shows planar liquid surfaces for three values of the angle of contact between the liquid and the wall, ⊖ , and the corresponding wall shapes that are needed. The shapes of liquid surfaces, that is, the shapes of liquid–vapor interfaces, become research topics. These are even more important when kinetic or dynamic effects, which result from physical and/or chemical reactions occurring at such interfaces, are considered. For example, the mixing of two liquids in contact with each other may induce convective movements; Figure 1.62 shows that these are more easily visible under microgravity conditions than on the earth, because the earth's gravitational field tends to damp out the convective flow. In this respect, liquid bridging between bubbles, as shown in Figure 1.63, and liquid filaments, illustrated in Figure 1.64, represent other fundamental areas of research worthy of investigation in space.

1.6.2.2 Convection.   In bulk fluids, that is, gases or liquids, convection in its most elementary form may be started by "heating from below." Hot parts of a fluid are less dense than the cold parts. The hotter and lighter fluid rises while the colder and heavier fluid sinks; the conditions for characteristic convection cells have thus been established. Depending on the geometry and size of the container, and on the specific properties of the fluid, that is, whether it is a gas or a liquid, its density, viscosity, and heat content, the appearance of convection cells changes. Figure 1.65 shows these so-called Benard cells on the surface of a liquid heated from below. In this case, each cell exhibits a characteristic hexagonal appearance. Silicon oil, with finely dispersed aluminum particles, is used for better visualization of the phenomenon.

Such convective movements determine the transfer of energy (that is, heat, measured by the temperature) and mass (that is, the bulk of the fluid) within the fluid. Apart from convection, heat and mass are also transferred from one volume element in the fluid to another by diffusion (that is, movements on the

Figure 1.62   Instabilities at the interface between two liquid phases may be caused by temperature, concentration, and/or surface energy differences. They result in convective movements in the liquids. (Adopted from H. Oertel, Jr., Report BMFT 7907, 1980.)

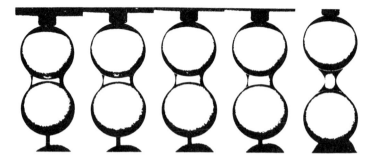

Figure 1.63 The shape of the liquid bridge between two solid spheres varies as a function of its surface energy and the contact angle; here it is shown for different separations between the spheres. (From W. Leiner, B. Schindler, P. J. Sell, Report BMFT–FBW 11–19, 1977.)

molecular, or atomic, scale). Excluding forces applied externally to the fluid, such as magnetic or acoustic fields, the forces which drive convection are caused by density differences. These, in turn, are due to concentration and temperature differences, that is, gradients, within the fluid; these are gravity-dependent. Thus, important research topics in the microgravity environment are the transport of energy and mass in a fluid by convection.

Thermal convection can be studied by optical methods, for example, using an interferometer; contour lines of equal density within the liquid are displayed. Figure 1.66 illustrates lines of equal horizontal and vertical density differences between two plates heated from one side. Large vertical density differences at the upper horizontal plate indicate an area of upward motion. An area of downward motion is accordingly visualized by a large number of fringes at the lower horizontal boundary.

Clearly, thermal convection due to density differences, observed in a 1 g environment, will hide many other smaller effects. This fact is illustrated by considering a surface energy determined effect, the so-called Marangoni convection

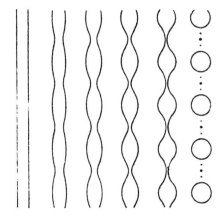

Figure 1.64 The surfaces of flowing liquid filaments exhibit ripples; these may become so marked that the stream disrupts into large and small bubbles. (Adopted from H. Oertel, Jr., Report BMFT 7907, 1980.)

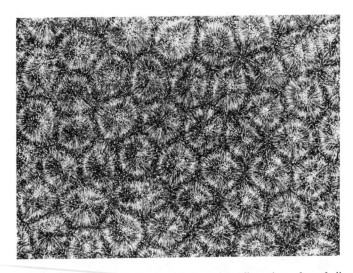

Figure 1.65 The appearance of hexagonal convection cells on the surface of a liquid heated from below; the liquid consists of fine aluminium particles suspended in oil and is in constant motion. (From H. Oertel, Jr., Habilitationsschrift, University Karlsruhe, 1979.)

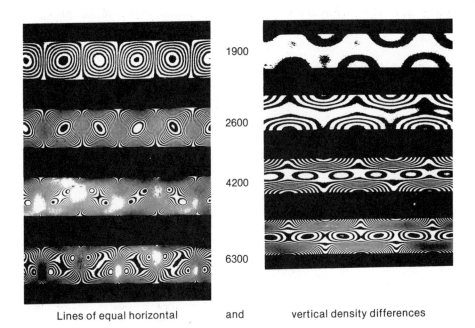

1900

2600

4200

6300

Lines of equal horizontal   and   vertical density differences

Figure 1.66 Lines of equal horizontal and vertical density differences within a fluid exhibiting convection. (From H. Oertel, Jr., Int'l Symp. Flow Visualiz., Bochum (Sept. 9–12, 1980), 1980.)

(Figure 1.67) resulting from Benard cells such as those shown in Figures 1.65 and 1.68(a). This type of convection is caused by unbalanced gradients of the surface tension at the free (upper) surface of liquids. The surface tension gradients can be due either to temperature or to concentration differences along the surface. These give rise to a surface flow from the region of low to high surface tension, for example, from hot to cold, as shown in Figure 1.67. If the gradient is maintained, by installing a source of heating and cooling, Marangoni convection will develop as a continuous flow which penetrates into the bulk of the liquid. In a microgravity environment, Marangoni convection, which is damped out to a large extent in an earth-based laboratory, will predominate; it can thus be verified and measured in space much more accurately than on earth.

1.6.2.3 Applications of interfacial and convection phenomena in space. The potential for the applications of interfacial and convection phenomena is substantial. Materials science and processing depend to a great extent on reactions involving the fluid phase. Chemical reactions and phase changes take place mostly at interfaces or surfaces; these are driven by temperature and concentration gradients which, in turn, are gravity-dependent. Particular examples are solidification, including crystal growth (Figure 1.68(b)), and physicochemical reactions. Besides their application to materials sciences, these fluid physics research topics are fundamental and are applicable to astrophysics, geophysics, meteorology, and other disciplines.

1.6.3 Solidification

The process of ingot casting introduces the subject of solidification. The melt within a container has to freeze (or solidify). This requires that heat be extracted from the melt through the walls of the container. In Figure 1.69, this is demonstrated, with more heat being taken out through the bottom and side walls and less through the top surface. The solidification, or freezing, front progresses into the interior of the melt in an irregular way, because of the uneven temperature distribution which induces thermal convection currents. The solidifying material reveals a host of different morphologies, depending on the material and parameters such as temperature, concentration of alloying

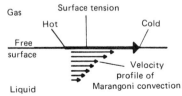

Figure 1.67 A temperature gradient at the surface of a liquid causes Marangoni convection on the free fluid surface. (From D. Schwabe, A. Scharmann and F. Preisser, *Materials Sciences in Space,* ESA SP-142, p. 327, 1979.)

Figure 1.68 (a) Diagram of Benard convection cells in a fluid heatd from below. (b) Convection patterns established in a crystal growth experiment. (From D. Schwabe, A. Scharmann and F. Preisser, *Materials Sciences in Space*, ESA SP-142, p. 327, 1979).

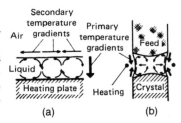

elements, and state of convection, which are gravity-dependent. The growth front morphology determines the microstructure, that is, the shape and size of the grains, and the distribution and segregation of alloying elements.

The solidification process determines the characteristics of most metallic and many nonmetallic materials, be they crystalline (metals, alloys, or ceramics) or amorphous (glasses of all types). Because of its significance in materials science and engineering, the field of solidification processes is usually divided into two categories: the solidification kinetics of metals, alloys, composites, and glasses that are normally polycrystalline or glassy in nature and the growth of single crystals.

1.6.3.1 Solidification of polycrystals.
For any liquid material to crystallize, nuclei have to be provided, that is, centers have to be present from which solidification may progress. A homogeneous, that is, a direct and unaided, nucleation from an absolutely clean, non-agitated melt does not easily oc-

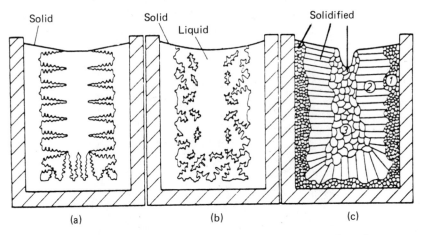

Figure 1.69 The solidified microstructure of an ingot depends strongly on the state of convection within the melt. (a) undisturbed (or dendritic) growth leads to microstructure of type 2 (in c); (b) convection will break off dendrite peaks and yield type 3 microstructure (in c). (Adopted from M. C. Flemings and R. Mehrabian, Symp. Solidif. Metals, MIT Cambridge, Mass. (Jan. 28, 1970), 1970.)

cur, and then only when supercooling below the usual freezing temperature is considerable. The walls of the container often initiate the process of nucleation, and thus solidification of the alloy. A particularly good example of this effect is shown in Figure 1.70.

Microgravity processing promises interesting new possibilities here. Since containerless work with melts will be feasible, such nucleation effects will be absent. In fact, microgravity simulation experiments with very fine melt droplets in oil suspensions, and also free-fall experiments in drop towers, have shown considerable supercooling to occur. Solidification occurring under such supercooled conditions results in an amorphous, or glassy, structure; in other words, nucleation is suppressed. Metallic glasses which could be formed in space are of great theoretical and practical interest.

Since the purpose of studying solidification is to achieve improved properties by controlling process variables in conjunction with certain parameters of the materials, it is essential to understand how convection in the molten phase affects the overall solidification process. The intrinsic heat and mass transfer occurs at and around the solid–liquid interface, which is the internal phase boundary that, ultimately, must pass through every solidified or crystallized object. Intensive study of the solid–liquid interface reveals interesting and complicated morphologies. An example of the influence of gravity on solidification involves the case of dendritic freezing. Dendrites (branched, tree-like crystals) are formed when castings and metallic ingots solidify. Specifically, dendrites are now

(a)   (b)

Figure 1.70   The microstructure of metallic alloys may be influenced by nucleating agents applied to the inner wall of the mould. (a) shows the appearance without nucleants and (b) shows the appearance with 6 percent nucleant. (From P. R. Sahm and Fl. Schubert, Sol. Casting Metals, Sheffield (July 18–21, 1977), p. 381, 1977.)

known to be an "unstable" mode of crystallization, in the sense that the geometry of a dendritic solid–liquid interface departs from the simple smooth form of a plane which would be the expected geometry from macroscopic heat flow considerations alone. Indeed, the dendritic form is generally a microscopic feature that demonstrates the complexity of heat and mass transfer over certain scales of size, despite the attempt to control the flow of heat from a casting on a larger scale.

Figure 1.72 shows a cluster of dendritic crystals growing in a molten material which is slightly supercooled below its freezing point. The material is a transparent organic substance (succinonitrile) chosen because it reproduces many key features of metallic solidification. Despite the fact that the molten material was originally uniformly cooled, a simple smooth, spherical, crystal does not form; instead, a complex, microscopically branching crystal grows in various directions. It is clear that the rates of growth of the various dendrites, shown as a, b, c, and d in Figure 1.71, are different, depending on the angle between the dendrite growth axis and the gravity vector. Dendrite c, oriented closely to the gravity vector, clearly grows faster than any of the others; these respectively grow more slowly as their orientation departs farther from the gravity direction. This example can only be explained by the presence of microconvective currents. These effects also occur in a great variety of casting and crystal growth processes of technological importance.

Macroconvective flow, that is, currents with dimensions comparable to the container size, may be much more directly visualized and measured than those shown by the dendrite experiments of Figure 1.71. They are responsible for inhomogeneities on macroscopic dimensions. Space experiments have shown that large ingot solidification may yield more homogeneous materials. Figure 1.72(a) shows manganese-bismuth being separated from bismuth at 1 g, because

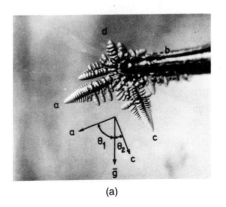
(a)

(b)

Figure 1.71   (a) Cluster of dendritic crystals growing in a slightly supercooled melt. The direction of gravity, g, is indicated, along with the orientation angles of the dendritic growth axes. (b) Same cluster of crystals one minute later. (M.E. Glicksman, 1980.)

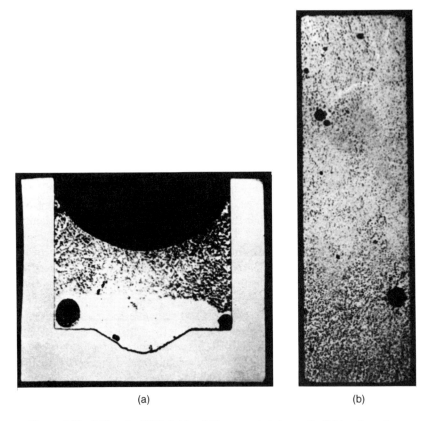

(a)   (b)

Figure 1.72   (a) Ingot solidified at 1 g yields segregated phases; the lighter phase rises
to top. (b) Under microgravity conditions a homogeneous distribution is reached.
(From P. Pant, TEXAS II Final Report, p. 48, 1978.)

of density differences. Figure 1.72(b) shows a homogeneous distribution of
manganese and bismuth reached under rocket-borne, microgravity conditions.

1.6.3.2 Applications of solidification phenomena in space.   The
few examples mentioned reflect only a small sample of the space potential of applications of polycrystalline solidification. The microstructure of a material
determines its properties which have engineering applications; new magnet,
superconductor, and bearing materials may, for example, be obtained. In certain cases, larger ingot volumes, with homogeneous microstructure and properties, may also be expected. Composite materials, obtained by combining complete material sections having different properties, are imaginable. Also entirely
new concepts of materials technology, drawing on the wetting and nonwetting
behavior of melts with respect to solid surfaces, appear to become possible.

1.6.4 Crystal Growth

1.6.4.1 Growth of single crystals.

The main task for a crystal grower is to make reproducible a well-determined material. Because there is a lack of basic scientific knowledge, this is not always possible. For all its importance, the physics of crystal growth remains poorly developed. No single crystallization theory is available; problems facing crystal growers, for example, the relation between structure, composition, and properties, are too often only being approached empirically. In the terrestrial laboratory, gravity plays an important part in crystal growth, even though its precise effect can only be estimated. The opportunity to produce single crystals in a microgravity environment is expected to contribute to knowledge in this field.

As in the solidification of polycrystalline materials, gravity-related phenomena determining single crystal growth are thermally and concentration-induced convection, as applied to the dynamics of the solid–liquid growth front. This statement applies to crystal growth both from the liquid and from the gaseous phase. An example of each illustrates the point. Figure 1.73 shows silicon crystals with striations on their surfaces. These surface striations on silicon crystals are caused, according to present understanding, by an inhomogeneous distribution of the minor (dopant) elements added to the silicon. This, in turn, is probably caused by convective instabilities ahead of the advancing solid–liquid interface. It is expected that completely homogeneous microstructures, with no striations, will be obtained under microgravity conditions where regions of stability may be extended over a broader set of experimental parameters. Interesting new insights have been gained from experiments in which the role of Marangoni convection (Figure 1.69(b)) on zone refining and

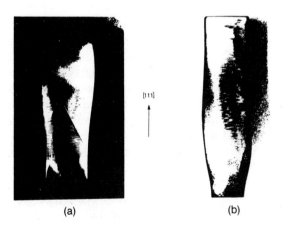

[111]

(a)   (b)

Figure 1.73 Single silicon crystals show striations after growth from the melt, irrespective of whether the melt was (a) rotated or (b) fixed. (From A. Eyer, R. Nitsche and H. Walcher, Status Seminar BMFT, DGLR–Bericht 79-01, p. 105, 1978.)

crystal growth was studied through direct visualization of the macroconvective currents.
The growth of crystals from the vapor phase occurs in three stages:

1. Evaporation of the source species.
2. Transfer of the gaseous components from the source to the crystal.
3. Deposition of the crystalline material.

Natural convection increases the mass transfer, but it also gives rise to imperfect crystals and to uncontrolled nucleation. This should be avoided, for high-quality crystals, in favor of forced convection due to the evaporation and deposition reactions. A microgravity environment appears to be particularly suitable for achieving better controlled convection of the vapor phase.

1.6.4.2 Applications of crystal growth in space.   Single crystals are a prerequisite for studying the crystallographic, as well as the microstructural and hence electrical, magnetic, dielectric, optical, and mechanical properties, of solid materials. In numerous applications they are required for the production of optical and electronic devices, such as optical lenses, integrated circuits, detectors, and power devices. Microgravity processing promises better quality materials, with fewer defects and with a better controlled structure. It should be stressed, however, that the capability of producing better crystals in space does not automatically imply less complicated devices; new technological schemes to realize this potential will have to be developed.

1.6.5 Physicochemical Reactions

1.6.5.1 A simple example–burning.   Physicochemical reactions are an extremely important realm of materials science and processing. A particularly convincing case is made by considering a burning flame (Figure 1.74), whose appearance is very different at 1 g and under microgravity conditions. There is an intimate interaction between a physical phenomenon, here gravity-driven convection due to thermal gradients, and the chemical reaction delivering the necessary heat. At 1 g the combustion is faster because the required ambient oxygen is conveyed to the reacting interface by convection currents; under microgravity every molecule required for the reaction is being slowly transported to the interface by diffusion.

1.6.5.2 Critical point phenomena.   The critical point is important when considering the transition of a fluid from two-phase to one-phase regimes, that is, liquid–vapor transitions. As shown in Figure 1.75(a), the isotherms have a broad flat appearance in the density-pressure diagram. There, the isothermal compressibility of the phase in question becomes very large; this peculiar property is sensitively affected by density variations. In a finite container, a critical

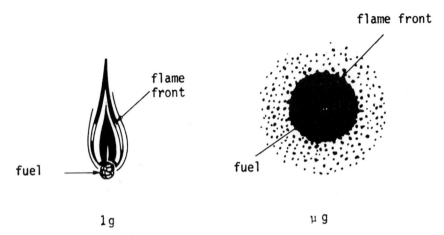

flame front

flame
front

fuel

fuel

1 g

μ g

Figure 1.74 A flame's appearance at 1 g (left) and at micro-gravity (right) shows striking differences. (Adopted from H. Wenzl et al., Kristallzucht in Weltraum, KFA-Bericht Nr. Jül–1182, 1975.)

point fluid will exhibit density variations, as a function of height (H) in the container, due to its own weight. Figure 1.75(b) shows specific heat variations with temperature at two values of H. Such density variations may be visualized by optical means (interferometry). For a homogeneous critical point fluid of finite geometry, microgravity conditions promise critical point measurements that are much more accurate than those obtainable in an earth-based laboratory.

1.6.5.3 Space applications. An understanding of the thermophysical properties of fluids is an essential basis for numerous technical processes. The necessary equations of state very often have to be provided by costly empirical methods of measuring physicochemical data. Important technological processes are found in gaseous and liquid phase reaction kinetics, for example chemical vapor deposition for crystal growth and coatings, electrophoresis and combustion kinetics.

1.6.6 Concluding Remarks

It was once said "in the material lies the message." Materials are at the beginning of everything that is "tangible"; their processing and treatment can only be improved through scientific research. With the advent of microgravity, an important parameter pervading the terrestrial laboratory is eliminated. In the very short history of materials research in space, there have already been a few surprises. New insights have also been transferred to "terrestrial research thinking." It is not unreasonable to predict that many new discoveries will be made in materials science through investigations carried out under microgravity conditions in space.

169

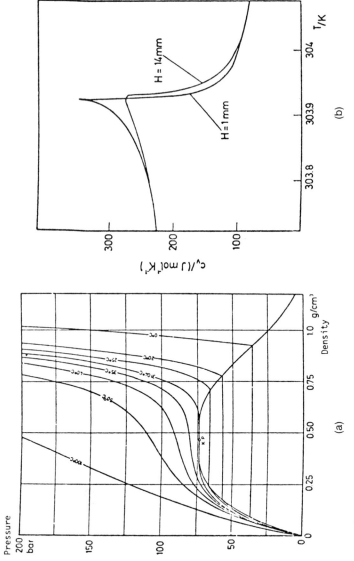

Figure 1.75 The properties of critical point fluids are sensitively influenced by density varia-tions. (From J. Straub and W. Wagner, Status Seminar BMFT, DGLR Bericht 79-01, p. 185, 1978.)

1.7 SPACE BIOLOGY AND MEDICINE

1.7.1 Man in Space

On 12 April 1961, Yuri Gagarin was sent into space orbit and returned to the earth safely after 1 hour and 48 minutes. This first spaceflight became the beginning of a new adventure for human beings that would capture worldwide attention. It was followed by many other journeys into outer space, the most distant of which were the APOLLO flights to the moon, in which astronauts left their spaceship, explored the lunar surface and returned to the earth. Flights into space multiplied in number and became progressively longer, and in 1980 a record duration for a flight, of 185 days, was achieved by cosmonauts Leonid Popov and Valery Ryumin on board the SALYUT-6 orbital station.

The success of manned space flights to date results from long technical and scientific preparations. Space presents surroundings that are hostile to man's existence, with extreme living conditions. Inside the cabin of a spaceship, human beings are protected against the particularly intense ultraviolet radiation, vacuum, and major temperature changes of space. They are, however, inevitably under the influence of weightlessness or, more precisely, microgravity, as well as that of cosmic radiation. The first objective of space medicine was to anticipate the effects of these influences on humans and to find ways of protecting them from these effects. It seems that this objective has been attained; nonetheless, survival during long space flights, on orbital stations, or for interplanetary flights, still poses problems.

Space biology and medicine have another objective. Cosmic radiation and gravity are two environmental factors that affect all living organisms; life was created and living species developed in their presence. Although it is possible to achieve increased gravity by using a centrifuge, the effects on humans of prolonged weightlessness cannot be studied on earth. It is only in parabolic flights of aircraft that the effects of gravity can be minimized, and then for only a few tens of seconds. Prolonged immobilization or immersion in water can simulate some aspects of weightlessness. Although human responses, notably those of the heart and blood vessels, in these situations are similar to those observed in the space environment, they are often of weaker amplitude. Thus they do not allow predictions to be made of the reaction of the human body to prolonged periods under weightless conditions. On the other hand, clinostats, that is, rotating devices that can be spun at varying speeds, and used largely for studying geotropism (gravity determined growth) in plant physiology, can only counteract the effects of gravity; they are, therefore, insufficient for simulating weightlessness.

Studies of the biological effects of cosmic radiation conducted at ground level are limited in scope. This is because the only devices that can be used to modify its intensity and, consequently, its influence on living organisms are shielding devices. In addition, such ground simulations of the space environment are far from perfect, in that the primary cosmic radiation has a complex nature covering a wide range of energies. By contrast, laboratory accelerators produce radiation of a well-defined nature and energy, and then only for limited periods of time. The conclusions drawn from such ground-based experiments may not be appropriate in a practical space context since combined effects of weightlessness and cosmic radiation occur in space flight. Finally, space studies should allow us to determine the possibility of existence of extraterrestrial life. It may be possible to show, using automatic probes and radio astronomy, that life is not unique to our planet.

1.7.1.1 The effects of weightlessness. The effects of weightlessness on human beings are better known than those on animals, thanks to manned space flights. These effects are seen in the cardiovascular, vestibular, and locomotor systems, that is, in the performance of those organs whose functions are known to be related, under normal conditions, to the presence of gravity.

Considering first the cardiovascular system, the performance of the heart depends on the distribution of the blood mass within the different parts of the body considered and on the position of the subject. In space, the absence of gravity is accompanied by a redistribution of the blood mass; from one to two liters of blood leave the lower part of the body and are concentrated in the head, neck, and thorax. This explains the facial swelling of astronauts and the shrinkage of the lower limbs (known as "chicken legs.") It is anticipated that these changes in blood distribution may cause important endocrinal (hormonal) and metabolic (chemical) modifications; dilation (expansion) connected with this accumulation of the blood is bound to have repercussions on pituitary and adrenal secretions. The result is the loss of water and minerals and an increase of urinary excretion and a decrease of the volume of plasma, the colorless liquid component of blood. These reactions, in turn, stimulate new responses of each organism to permit it to adapt to new conditions of reduced gravity. After a few days, the water loss ceases and the weight becomes stable. The heart itself seems to be affected very little: Recent studies on SALYUT-6 showed only a slight decrease of blood pressure and an increase in cardiac output.

It is evident that the cardiovascular system of an astronaut works under abnormal conditions during the entire flight. In fact, body position changes cannot bring about the reflex phenomena that allow the body, when on the ground, to avoid the accumulation of blood in the lower limbs when the subject is in a vertical position. This explains the occurrence of minor problems, after return to earth, when an astronaut moves from a horizontal to a vertical position; these are dizziness, nausea, and fainting and all are characteristics of a cardiovascular deconditioning. Fortunately, these effects are short-lived, and their importance can be minimized through the use of low-pressure apparatus that stimulates a movement of the blood to the lower limbs.

The vestibular system (the balance-sensing system) is modified by reduced gravity. Man's equilibrium, posture, and precision of movements depend on reflex phenomena that originate in muscles, tendons, articulations (joints), the skin, the eyes, and, especially, in the vestibular system of the inner ear. The wall of the latter is covered by a layer of cells (epithelial cells) against which rest small calcium-rich crystals (otoliths). Due to their weight, these exert pressure on the cilia (small hair-like projections) that cover the upper surface of the epithelial cells; cilium orientation can be modified by inclination or movements of the head, and this change stimulates these cells and the vestibular nerve ends in contact with them. Messages are then sent to the central nervous ssytem, and, in particular, to the cerebellum (brain); these, acting on the muscles, reestablish the equilibrium of the organism. This vestibular activity is obviously upset in space, where the otoliths exert no pressure on the epithelial cells. Moreover, since the absence of weight also affects the nerve endings in the skin, tendons, and joints, spaceflight often results in vestibular disorder. This is the classic "motion sickness," with nausea and dizziness, which appears in the first days of flight, and which disappears later. Nevertheless, these disorders are important; they are unpredictable and vary greatly from one subject to another, upsetting the astronauts' planned activities. This is a particular problem for short-duration flights, such as the forthcoming SPACELAB missions.

The locomotor system, namely, the muscles and the skeleton, is also influenced by the action of gravity. In space, the lack of weight of the body and of the objects handled by the astronaut results in a partial resting of the skeleton and muscles. The intervertebral discs stretch, thus explaining the customary, though temporary, increase in height of the astronauts; this can be as much as several centimeters. Phenomena of atrophy (wasting away) occur in the skeleton and muscles. Demineralization of the bones takes place and is accompanied by a loss of calcium through the urine, representing about 0.5 percent of the total calcium content of the body per month. These changes in the locomotor system are reduced by daily physical exercises, but they do not disappear entirely until return to earth.

The effects of weightlessness can be seen, in differing degrees, in all the organs and functions of the human body: haemotological, digestive, and immunological modifications have also been described. In the same way, the composition of microflora (microorganisms in the bowel) seems modified in flight and contains more bacteria resistant to antibiotics than before the flight.

All these facts explain the interest in continuing space medicine studies, not only for practical purposes but also to obtain a better understanding of human physiology. It is probable that future research will lead to more knowledge of the regulatory systems of blood circulation and of plasma volume and to a better understanding of the relation between the vestibular apparatus and other systems which determine the equilibrium (position of the body). Also, studies of calcium modification in conditions of reduced gravity represent a new way of investigating mechanisms of osteoporosis, a condition very commonly affecting the elderly in which bone substance is lost.

Studies of reduced gravity effects cannot be limited to effects on human beings: gravity has an effect on the whole biosphere. Indeed, since the flight of the dog, Laïka, on 3 November 1957, many species have been sent into space: these include humans, animal and plant cells, bacteria, yeast, mushrooms, protozoa, amphibians, fish, and mammals, as well as seeds and plants. Despite being of an exploratory nature, this research holds great scientific interest and cannot be considered separately from space medicine studies. Such investigations, carried out with simple biological material, such as animal or human cell tissue, should improve knowledge of the influence of gravity on basic vital processes. They should also lead to a better understanding of the complex responses of higher-order organisms and their various mechanisms and regulatory systems.

Of course, some space experiments have resulted in negative findings; these include an absence of any effect on human cell cultures, as shown in experiments carried out on SKYLAB and SALYUT, and a lack of influence on cell infrastructure or on certain embryonic developments. At the cell level, the effects of gravity can seem negligible compared with other factors, such as the forces of molecular attraction. Nevertheless, reduced gravity has been shown to have a stimulating effect on the proliferation of *Paramecium,* a single-cell organism, and on some bacteria, as reported by United States, USSR, and French scientists. Also noted were probable modifications in membrane permeability and in the cell chromosomal apparatus.

USSR and United States experiments have also been carried out on rats aboard a COSMOS spacecraft, using a centrifuge to bring gravity near to that at the earth's surface. The skeletal changes noted in human beings were again observed and studied more precisely: These included a decreased growth rate, an inhibition of the periosteal osteogenesis (formation of bone structure), and the appearance of osteoporotic-type structure, observed after approximately three weeks of weightlessness. The same experiments have also shown that antigravity muscles are those most affected by conditions of reduced gravity. Biochemical and histological investigations, simultaneously carried out on endocrine organs, have given evidence on the role of stress in observed responses during the first days of spaceflight.

With regard to plants, the existence of geotropism has been known for many years; the orientation of roots and stems depends on gravity, which acts in a certain direction. Here the gravity perception system is well understood. It is made up of cells or statocysts (sensors) having small starch granules (amyloplasts), heavier than the other parts of the cell. The latter move under the influence of gravity and exert pressure on vesicle membranes. This leads to a synthesis or to a release of compounds such as indolylacetic acid, responsible for cell growth and, therefore, the bending of the roots and stems. In space, the amyloplasts float inside the cells and the indolylacetic acid is spread uniformly. Multidirectional growth results, and it can be shown that the general form of the plant depends on the position of the seeds in relation to the surface of the support at the beginning of the space experiment. Again, these observations are preliminary and a deeper knowledge of plant growth mechanisms will only be obtained through new studies carried out on earth as well as in space.

1.7.1.2 The effects of cosmic radiation. Primary cosmic radiation is made up of two parts: galactic cosmic radiation, resulting from the explosion of supernovae, and radiation of solar origin, related to solar flares. Primary cosmic radiation is composed essentially of atomic nuclei, protons (or hydrogen nuclei), alpha particles (or helium nuclei), and nuclei of highly charged atoms or heavy ions. These particles, of great energy and arriving isotropically (uniformly from all directions), interact with nitrogen and oxygen atoms in the atmosphere; the consequence of this is a secondary corpuscular radiation of very complex nature irradiating the whole surface of the earth.

Its effects on the biosphere are still little known, even though some physiological changes in shielded animals or plants have been reported. The dose received on earth is very weak, about 40 mrad/year at sea level. The maximum permissible dose per year, according to present standards, is about 5 rad in humans, that is, 125 times higher. In space, the daily dose inside the cabin is higher, about 15 to 30 mrad/day but this dose still seems negligible, at least for short-duration flights. However, irradiation could be a hazard for manned spaceflights, especially in the Van Allen belts. Although beneath these belts the absorbed doses are weak, within the belts the dose is much higher and incompatible with the survival of human beings. Outside the belts, the calculations of USSR scientists show that the annual dose of the galactic cosmic radiation alone can reach 50 to 100 rem, which would obviously constitute a real danger for humans during an interplanetary mission. In addition, during a space mission, it is possible that an unpredictable solar flare may produce intense solar cosmic radiation up to a thousand times more intense than galactic radiation. In such a case, an astronaut could receive a lethal dose of about several hundred rads.

Finally, part of the danger of cosmic irradiation comes from the presence of heavy ions of galactic origin; these particles, often of very high energy, can easily enter the space station cabin, penetrate into human organisms, and deposit energy along their paths which could correspond to several hundred or even several thousand rads. Therefore, it can be feared that, although the total mean doses seem acceptable, any part of an organism directly hit by heavy cosmic ray particles could be seriously damaged. In human beings these particles are responsible for temporary visual phenomena, the light flashes described by astronauts since the APOLLO flights. Various investigations have been carried out based on a correlative method for detecting the paths of heavy charged particles through biological objects. These experiments have shown that a single heavy ion can decrease the developmental capacities of seeds or encysted eggs (eggs within a cyst) and can be responsible for genetic changes. It can be assumed that the modifications in mutation and recombination rates, already described in some species, such as the fruitfly *Drosophila,* were related to these particles.

Taking into account all these data, it appears essential to define, on an international level, standards of radiation safety, and to limit the permissible exposure of an astronaut to less than 1 percent of the maximum allowed dose. In the absence of specific shielding devices now under study, cosmic rays constitute a limiting factor in long-duration spaceflight and, especially, in interplanetary flight.

1.7.2 Exobiology

1.7.2.1 Planetary biology. Exobiology, or the search for extraterrestrial life, is a further objective of space research. It is also the most interesting from a philosophical point of view. Considering only the solar system, and assuming that life can only develop in physical conditions similar to those of earth, it now seems very unlikely that living organisms can be detected anywhere other than on earth. Nevertheless, life exists on earth at all latitudes and in climatic conditions as extreme as those of the polar regions and the deserts.

On Mars, where the probability of finding life seemed, in the 1960s, to be highest anywhere in the solar system, neither of the two American VIKING missions of 1976 were able to discover the existence of life, or even any trace of organic material. Some early results, such as the release of oxygen and carbon dioxide from Martian soil samples, led to preliminary indications of life on Mars. It now seems that these responses were due only to chemosynthetic processes related to the presence of highly reactive peroxides used in the experiments. However, the Martian soil was only investigated at two sites, both practically without water, whereas the Martian polar regions certainly contain water. It is, therefore, not possible to reject completely the idea of detecting life on Mars. Thus, while Mars continues to be of prime interest in exobiological research, such investigations may be extended to other celestial bodies, such as Jupiter and Titan (a satellite of Saturn), as well as meteorites and comets.

1.7.2.2 Planetary quarantine. The possible existence of extraterrestrial life led to quarantine measures being undergone by the astronauts following the first lunar missions. It also resulted in precautions being planned when samples from those planets which may contain living organisms are brought back to earth. Conversely, it is essential that probes launched from earth do not contaminate planets that could support life; before launch, they must be decontaminated or, better still, sterilized. It is possible that the vacuum, ultraviolet radiation or heat encountered during the spaceflight could sterilize their surface. In fact, the combined action of all these factors can destroy bacteria or even bacterial spores located on the outside of the spacecraft. However, it can happen that parts of the spacecraft are not exposed to solar radiation, and it has been shown that bacterial spores can, in such a case, resist the other space dangers. All equipment inside the spacecraft, intended for external use upon landing on another planet, must also be sterilized; this is not an easy task.

The possibility of contaminating a planet is related to the surface characteristics and the atmospheric conditions of that planet. For example, the surface temperature of Venus is too high to permit the existence of life. The estimation of contamination risk depends, therefore, on studies of the capacity for survival of microorganisms cultivated in environmental conditions simulating the atmosphere or surface of the planet concerned.

1.7.3 Benefits and Practical Applications

The results already obtained, the perspective of longer and longer spaceflights as well as various theoretical considerations, explain the importance of biology and medicine to space programs and justify the continuation and extension of the present research. New investigations in the coming years will be carried out in orbital stations and SPACELAB. Among the desired objectives are:

1. Investigations of the capacity of the human body to adapt to long periods of reduced gravity.
2. The development of basic biological studies closely associated with research carried out on human subjects.
3. The use of primates whose physiological and metabolic responses are similar to those of human beings.
4. The development of new equipment which, in flight, would permit various tests and multiple measurements of physiological parameters. Much of the current data are relative only to measurements made before and after the spaceflight.

It seems probable that future research programs will lead to a better knowledge of the physiology of living organisms. The effects of gravity and cosmic rays on these organisms will be studied in order to understand better the mechanisms of the development of organisms and the emergence of species. Life began in the terrestrial gravitational field, and the form and structure of the organisms were determined by the effects of gravity. It is known that gravity can determine the position of the bilateral embryonic plane of symmetry in some species; this is the plane in which the left and right sides of a species appear mirrored. Thus it would be of great interest to study, from well-chosen models, the influence of prolonged periods of weightlessness on the processes of development and evolution.

A practical aspect of biological research and space medicine is the application of the technology developed for space to everyday medicine. The contribution of automatic survey techniques developed for astronauts to the monitoring of cardiac or casualty patients, for example, is well known. Progress in the design of artificial limbs is another application of space technology. Also, techniques in progress for nontraumatic recording (that is, using external sensors) of the blood circulation in the brain of astronauts can be used in medicine. Finally, with regard to therapeutics (the use of drugs to cure diseases), it is possible that the use of space may allow the fabrication of better quality drugs. Electrophoresis (the separation of particles by an electric field), the efficiency of which is improved in weightless conditions, could lead to more active and less toxic vaccines.

In conclusion, it is considered that the continuation and extension of present research programs in biology and medicine, with increasing international cooperation, will permit long-duration spaceflights to be carried out safely and will increase man's knowledge and improve man's living conditions.

1.8 THE SEARCH FOR EXTRATERRESTRIAL INTELLIGENCE (SETI)

Heaven and earth are large, yet in the whole of space they are but as a small grain of rice. . . . It is as if the whole of empty space were a tree, and heaven and earth were one of its fruits. Empty space is like a kingdom, and heaven and earth no more than a single individual person in that kingdom. Upon one tree there are many fruits, and in one kingdom many people. How unreasonable it would be to suppose that besides the heaven and earth which we can see there are no other heavens and no other earths.

Teng Mu, thirteenth-century philospher
(translated by Joseph Needham)

For many hundreds of years, philosophers and scientists have occasionally thought about the possibility of the existence of other worlds, and of life on them. In early times there was no real astronomical or scientific evidence on which such suppositions could be based. They stemmed more from a fundamental question. "Why should life be unique to the planet earth?"

After the Copernican revolution, which established the true nature of the solar system, speculations on the existence of life elsewhere, including intelligent life, began to increase in number. In the nineteenth century there was much excitement about the possibility of intelligent life on the moon and on Mars. Early in the twentieth century the USSR biochemist Aleksandr Ivanovich Oparin and the British biologist J. B. S. Haldane first formulated their ideas of the origin of life. Their concepts, which remain valid today, visualized the emergence of primitive life forms from a mixture of organic compounds existing in the waters of the ancient earth.

As biologists, chemists, and physicists continued to unravel the basic mechanisms of life, it became clear that these concepts about the origin of life on earth were indeed reasonable. As the understanding of biological evolution improved, a picture emerged of a sequence of events, occurring over an immense period of time, which led ultimately to the extraordinary diversity of life that is seen on earth today. The sequence is as follows:

1. Formation of sun and earth.
2. Evolution of organic compounds.
3. Origin of life.
4. Emergence of simple life forms.
5. Evolution of complex life.
6. Intelligence.
7. Cultural evolution.
8. Civilization.

It was natural, as understanding developed, for another important question to be asked: Could such a sequence of events take place on planets other than the earth? In 1959, Joshua Lederberg, in the United States, proposed the term "exobiology" to embrace the search for and the study of extraterrestrial life. It did not seem likely at the time that intelligent life existed elsewhere in the solar system. Nonetheless, it seemed possible that microbial life forms could exist elsewhere in the solar system, and particularly on the planet Mars. The two VIKING spacecraft sent to Mars in 1976 carried experiments to search for such life. The results were essentially negative; however, it is conceivable that microbial life could have existed on Mars in the past, or could indeed exist at locations on Mars other than those where the VIKING spacecraft landed.

It is likely that life has emerged on some planets orbiting some stars elsewhere in the galaxy and the universe, just as it has emerged on the earth. It is also likely that it would, in some cases, have passed through various stages of biological evolution to a state of intelligence.

1.8.1 Rationale for the Existence of Extraterrestrial Intelligence

The central arguments for the existence of extraterrestrial civilizations are as follows:

1. It is now believed that the birth of stars is normally accompanied by the birth of an orbiting retinue of planets, just as happened in the solar system. While such planetary systems have yet to be identified, it is likely that they will be detected by the end of this century.
2. There are hundreds of thousands of millions of stars in the galaxy. Thus, there could be tens of thousands of millions of planets. Occasionally, one of these planets is likely to have suitable surface and atmospheric conditions for the origin of life. Suppose that these "other earths" are rare, that

they occur only once in a million solar systems. Simple arithmetic then shows that there would be tens of thousands of sites where life could begin in the Milky Way galaxy. Because there are many tens of thousands of millions of galaxies in the Universe, the number of possible life sites becomes very large.

3. Given a suitable planetary environment and hundreds of millions of years, most exobiologists believe that simple life forms will emerge.

4. Given many more hundreds of billions of years, the principles of biological evolution indicate that life is likely to become more complex, and in some cases to proceed to the development of intelligence, cultural evolution, and civilization.

The long spans of time required for the evolution of complex life are available. The sun and earth are 4.5 billion years old. Some stars in the galaxy and in other galaxies may be 10 to 20 billion years old. Therefore life, even in an advanced form, could have been present on the planets of stars in the galaxy before the sun and earth existed, and there could be extraterrestrial civilizations many millions of years older than civilization on earth. The largest element of uncertainty in estimating the number of other civilizations is the lifetime of a civilization. More precisely, the question is as follows: How long does a civilization remain in a technological and communicative phase? Estimates of the answer range from 50 to a billion years. They are also likely to be life-bearing planets following the path of terrestrial development, but well behind in biological evolution.

The rationale for the existence of extraterrestrial intelligence leads inexorably to an important and fundamental question. Is there a way in which we can detect the existence of extraterrestrial intelligence?

1.8.2 SETI Methods

1.8.2.1 Direct investigation.   There are many possible ways of detecting extraterrestrial intelligent life. One way might be to send manned or unmanned spacecraft on missions to nearby stars to examine their planets. The problems here are formidable. Depending on the design, size, and propulsion system of the vehicle, the trip could take thousands or millions of years, the energy consumption would be prodigious, and the cost would be extremely high. Thus there are no serious proposals for interstellar missions to be a reasonable candidate for initial attempts to detect intelligent life elsewhere.

1.8.2.2 Microwave observations.   The situation becomes dramatically easier if attempts are made to detect the radiation emitted by other civilizations. Such signals could be detected comparatively easily against the natural background noise; they would be expected to have characteristics which would distinguish them from other radiations. Such signals would use a type of radiation which requires an acceptable power at the source, which is not absorbed by the interstellar medium, which can be generated and detected comparatively easily, and which could be transmitted by a technologically based civilization.

A close examination of all possible forms of radiation leads to the electromagnetic spectrum. In turn, examination of the huge range of this spectrum leads to a region of the radio spectrum between 1 GHz (10^9 Hz) and 100 GHz (10^{11} Hz), which is known as the microwave window. This is the quietest region of the spectrum; it is the region where it is easiest to detect a faint signal emanating from another civilization against the noise of the natural background. To illustrate this principle, a whisper cannot be heard in a room filled with loud voices in conversation, but it can easily be heard when the room is very quiet.

The idea of discovering other intelligent species by detecting their radio signals forms the core of most of the SETI concepts being seriously pursued today. It was first advocated in 1959 by Giuseppi Cocconi and Philip Morrison. They suggested that a search be made at frequencies near the hydrogen line (at the low-frequency end of the microwave window at a wavelength of 21 cm) for signals emitted by advanced extraterrestrial civilizations attempting to establish contact with man. In 1961 the radio astronomer Frank Drake carried out the first search using a radiotelescope in the United States. He pointed his telescope at two nearby stars, Tau Ceti and Epsilon Eridani, and listened for two months; he did not detect a signal. Since then, two dozen other searches have been carried out, and numerous scientific meetings and papers have included discussions on all aspects of extraterrestrial intelligent life.

In 1977 a thorough study of SETI was conducted in the United States. Among the conclusions of this NASA report are the following:

1. It is both timely and feasible to begin a serious search for extraterrestrial intelligence.
2. A significant SETI program, with substantial potential secondary benefits, can be undertaken with only modest resources.
3. Large systems of great capability can be built if needed.

The report also recommends the initiation of a SETI program, and it points out that the quiet microwave window is rapidly being filled with sources of radio frequency interference, high-powered signals from vast numbers of transmitters on the surface of the earth and in earth orbit. It also underlines the fact that SETI is intrinsically an international endeavor because of its significance for mankind as a whole. Nikolai Kardashev, in the Soviet Union, also argues for the importance for mankind of this search.

1.8.2.3 Other methods.

Searches in the infrared region of the spectrum could become a possibility in the future. The detection of X-rays, ultraviolet radiation, or visible light emitted by laser beams is less likely, and the detection of heavy charged particle beams is only a remote possibility. Ronald

Bracewell has suggested searching for space probes from other civilizations within the solar system. It is also possible that messages transmitted from a powerful beacon on earth might be picked up by another civilization which might then reply. The problem with this approach is that it could take hundreds of years, or even millenia, to detect the reply.

1.8.2.4 Optimum tactics for SETI.

Listening in the microwave region of the spectrum is proposed by most SETI investigators as being the best method. Some propose that a concerted SETI program should not be carried out, but that an accidental discovery of artificial signals might be made in the normal course of radio astronomical observations of the sky. However, most scientists feel that the chances of making the detection of extraterrestrial intelligence accidentally are remote, because the usual radio astronomical observing procedures are not optimized for detecting signals of intelligent origin. Success in SETI is likely to come only from a concerted effort by mankind to listen with several of the most sensitive and sophisticated instruments possible, searching many regions of the sky in many different frequency bands.

The central strategy for SETI is to listen and not to transmit. There are, in fact, few serious proposals to transmit signals from a beacon on the earth. It should be noted, however, that man has been transmitting a huge variety of radio signals into space over the past five decades. In recent years, some of these transmissions have become comparatively strong, for example, those used for planetary radar exploration.

The signals sought are either those from beacons transmitting deliberately for the purpose of establishing communications or those emitted by other civilizations for their own purposes. The detection problem is formidable. Man does not know where to look in the sky, nor with what sensitivity, nor at what frequency, nor at what time. Much effort has, therefore, gone into the development of search strategies which are likely to succeed in SETI. In some of these strategies the whole sky is searched by sweeping the beam of a radiotelescope back and forth in the hope of detecting either very strong transmissions, which might be generated by very advanced civilizations far away, or lower-power transmitters that are fortuitously close. Other strategies envisage the use of large radiotelescopes to observe suitable stars (those like the sun) for minutes or even hours, and hence to achieve a very high sensitivity.

Tackling the problem of frequency band selection, Bernard Oliver and John Billingham first proposed that spectrum analyzers should be constructed to listen to many different channels at once. In this way, an elusive signal may be detected more rapidly than would be the case if the channels were explored one at a time. Signal processing systems to identify the signal against the background noise are being designed.

The problem of allocating frequency bands to transmitters and for listeners is the responsibility of the International Telecommunication Union (ITU). In 1979 the World Administrative Radio Conference discussed the subject of SETI. It was decided that no portion of the spectrum should be reserved for listening for signals of intelligent origin, but simply that member nations should be informed that such searches will be taking place. SETI observers will use the rather narrow frequency bands set aside for radio astronomy, in which there is no allocation for transmissions; however, they really need access to much broader protected frequency bands for comprehensive exploration of the spectrum.

If the microwave window becomes increasingly congested with radio frequency interference over the next two decades, it will not be possible to carry out SETI from the earth. It will be necessary to construct radiotelescopes in space or on the moon to escape from the interference. In the early stages of SETI, this may be a costly way to proceed, although in the future it could be preferable to observe from space for other reasons (as pointed out by M. Subotowicz in Poland and Nikolai Kardashev in the Soviet Union). In the next two decades, however, it is important to attempt to set aside at least some portions of the microwave region for SETI listening purposes. All nations should seriously consider the problem and communicate their views through their appropriate national organizations to ITU.

No signal of extraterrestrial intelligent origin has yet been detected. This is not surprising; SETI is still in its infancy. The searches to date have been carried out in the United States, the USSR, and Canada. Interest is spreading, and observers in other nations will soon undertake searches of their own.

In SETI both informal and formal channels of international exchange and cooperation are needed. A broader understanding of SETI by the international community is of great importance in fostering a favorable climate for the evolution of more sophisticated search activities in the future. The people of planet earth have a common interest in learning about other people on other planets far out in the galaxy.

1.8.3 The impact of SETI on mankind.

SETI will make an impact on the peoples of the earth; psychological, sociological, cultural, national, and international effects of any discovery of extraterrestrial intelligent life are anticipated.

A long and dedicated search that fails to find an extraterrestrial intelligence will not be a waste of time, however. Important developments in technology, with applications to many other aspects of civilization, will be made. Knowledge of the physical universe will increase. The global organization of a search for interstellar radio messages should provide a cohesive and constructive influence on mankind. It should strengthen belief in the uniqueness of *Homo sapiens,* his civilization and planet. Man should learn better how precious is human culture, how unique his biological patrimony, painstakingly evolved over three or four thousand million years of tortuous evolutionary history.

If a single extraterrestrial signal were to be detected, it would be known immediately that it is possible for a civilization to maintain an advanced technological state and not to destroy itself. It might even be learnt that life and intelligence pervade the universe. The sharpness of the impact of simple detection will depend on the circumstances of discovery. If genuine signals were to be found after only a few years of a modest search, there is little doubt that the news would be sensational. If, however, signals were detected only after a protracted effort over generations with a large search system, the result might be less conspicuous.

It is likely that the early announcements of the detection of deliberate signals may turn out to be mistaken, and not be verified by further study and observation. They may be natural phenomena of a new kind, or some terrestrial signal, or even a hoax. It is important to stress the need for verification, because even a single genuine detection would have enormous importance.

Of course, it is very difficult to foresee the content of a signal except in the most general way. A signal could be a beacon, a deliberate transmission specifically for the purpose of attracting the attention of an emerging civilization like the earth's. Alternatively, it could be a "leakage" signal, similar to television broadcasts or radars, not intended for detection by other civilizations. Whatever the signal, it will only be a one-way transmission. Also, any message in such a transmission would be a message between cultures, not between individual persons. Human analogies are evident here; there is a long-continued interest in great books from the past. The Greek philosophers are studied afresh by each generation, without any hope of interrogating Socrates or agruing with Aristotle.

The information content of any signal from another civilization could be rich. Its study would continue for decades, even for generations. Books and universities would be more suited for the news than the daily bulletins. If the signal were deliberate, decoding it would be relatively easy, because the signal should be anticryptographic, that is, made to reveal its own language coding. If the message comes by radio, both transmitting and receiving civilizations will have in common at least the details of radiophysics.

Some have worried that a message from an advanced society might make man lose faith in his civilization, might deprive him of the initiative to make new discoveries if it seems that there are others who have made those discoveries already, or might have other negative consequences. But man is free to ignore an offensive extraterrestrial message. Man is under no obligation to reply. If man chooses not to respond, there is no way that the transmitting civilization can determine that its message was received and understood on the distant planet earth. Thus the receipt and translation of a radio message from the depths of space seem to pose few dangers to mankind. Instead, it holds promise of philosophical, and perhaps even practical, benefits for all of humanity.

By holding a conversation with another civilization, succeeding generations of mankind may gain a wealth of new knowledge; this could range from an understanding of the past and the future of the universe to physical theories of the fundamental particles of which the universe is made and to new biologies.

Man might be able to converse with distant and venerable thinkers on the deepest values of conscious beings and their societies. Man may then become linked with a vast galactic network.

APPENDIX 1.1: UNITS AND PHYSICAL QUANTITIES

A.1.1.1 SI Units

The Système international (SI) units system is used in this book as much as possible. This is based upon the system of units using the meters (m), kilogram (kg), and second (s).

Preferred multiples of these units are thousands (10^3 times, or three orders of magnitude) or thousandths (10^{-3} times); for example, for distances, kilometers km (10^3 m) are used, as are millimeters mm (10^{-3} m) and micrometers or microns, μm (10^{-6} m).

A1.1.2 Standard Prefixes

The complete list of such prefixes is as follows:

1. Tera-, T (10^{12} times)
2. Giga-, G (10^9 times)
3. Mega-, or million, M (10^6 times)
4. kilo-, thousand, k (10^3 times)
5. milli-, or thousandth, m (10^{-3} times)
6. micro-, or millionth, μ (10^{-6} times)
7. nano-, n (10^{-9} times)
8. pico-, p (10^{-12} times)

Other prefixes sometimes used are as follows:

1. myria- (10^4 times)
2. hecto- (10^2 times)
3. deca- (10 times)
4. deci- (10^{-1} times)
5. centi- (10^{-2} times)

A1.1.3 Distance

For distance measurements, in addition to multiples of meters defined above, the Ångstrom unit, Å (10^{-10} m, or 0.1 nm) is sometimes used for wavelengths. Another measurement of distance used in astronomy is the astronomical unit, the distance between the sun and the earth. Another is the radius of the earth R_E, the radius of Jupiter, R_J, of Mercury, R_φ, of Mars, R_σ, of Venus, R_V, and of Saturn, R_S. Another measure is the light year, the distance that light travels in one year at a speed of 3×10^8 m/s. It is almost 10^{16} m.

A1.1.4 Mass

For measurements of mass, the kilogram (kg) is the unit, though the unit from which it is derived, the gram (g), is sometimes used. The mass of the sun, $M_\odot$, is used in astronomy.

A1.1.5 Time

For measurements of time, minutes (min), hours (h), days (d), weeks, months, and years are also used.

A1.1.6 Derived Units

For derived units, the same metric principle is used. For example, for velocity, or speed, the unit is meters per second, m/s or ms^{-1}. Larger velocities are measured in kilometers per second, km/s or kms^{-1}.

Another example is volume, for which the unit is a cubic meter (m^3). The density, or mass per unit volume, is measured in kg/m^3 or kgm^{-3}. The number density, the number of particles per unit volume, is measured in m^{-3} or $/m^3$.

A1.1.7 Temperature

For temperature, measurements are made in degrees Kelvin (K). Absolute zero (0°K) occurs at -273 degrees centigrade, or Celsius (°C); alternatively, water freezes at 273°K, or 0°C.

A1.1.8 Energy and Power

The unit of energy is the joule (J). For example, a mass of 1 kg moving at a speed of 1 m/s has an energy (kinetic energy) of 1 J; 1 J is equivalent to 10^7 ergs or to 0.24 calories. The energy of a solar flare is between 10^{22} and 10^{25} J.

The rate at which energy is produced (or expended) is the power produced (or consumed). A rate of 1 J/s is termed 1 watt (W). Multiples of this unit of power are kilowatts (kW) or megawatts (MW). The power flux is the energy crossing unit area in unit time; for example, the energy from the sun at the top of the atmosphere, passing through 1 m^2 per second, that is, the solar constant, is 1.37 kW/m^2.

One electronvolt (eV) is the energy (1.6×10^{-19} J) gained by an electron when accelerated through a potential difference of one volt (V). Energetic charged particles have energies of kiloelectronvolts (keV), or megaelectronvolts (MeV); cosmic rays have energies of gigaelectronvolts (GeV).

Electrical power is also measured in watts; 1 watt is produced in an electrical conductor when a current of 1 ampere (A) flows, driven by an applied potential difference of 1 volt (V). For example, 1 kW is produced by a steady current of 5 A drawn from a 200-V supply. Currents flowing in the ionosphere have a current density conveniently measured in ampere per square kilometer. An electric field is measured in volts per meter or volts per kilometer.

A1.1.9 Magnetic Field

A magnetic field is measured in Teslas (T) or in gauss (G) (1 T is equivalent to 10^4 G). In the magnetosphere, the geomagnetic field strength is a few nanoteslas, or gammas (1 gamma is 10^{-5} G).

A1.1.10 Electromagnetic Spectrum

The frequency of an electromagnetic wave, the number of cycles per second (c/s), is now measured in Hertz (Hz). Radio waves in the electromagnetic spectrum range from:

1. Very low frequencies (VLF, 3 to 30 kHz) through
2. The low-frequency (LF) band (30 to 300 kHz)
3. The medium-frequency (MF) band (0.3 to 3 MHz)
4. The high-frequency (HF) band (3 to 30 MHz) to
5. The very high-frequency (VHF) band (30 to 300 MHz) and beyond to the gigahertz region.

A1.1.11 Gravity

The acceleration due to gravity at the earth's surface is 9.81 m/s^2. Small departures from this value are measured in milligals; 1 milligal is 10^{-5} m/s^2 (or 10^{-3} cm/s^2).

A1.1.12 Pressure

The atmospheric pressure at the earth's surface is approximately 1,000 millibars (mb). One millibar is equivalent to a force of 10^2 Newtons (N) per square meter (m^2).

A1.1.13 Angles

Angles are measured in degrees (°). There are 60 minutes of arc in 1°, and 60 seconds of arc (″) in 1 minute (′). The diameters of distant astronomical objects are measured in seconds of arc to an observer on the earth. Angular resolution is similarly measured.

A1.1.14 Nuclear Radiation

An astronaut is exposed to radiation from trapped energetic charged particles, galactic cosmic rays, and solar flares. The radiation dose is defined in terms of energy absorbed per unit mass of material (for example, body tissue). One rad is equal to the absorption of 100 ergs (10^{-5} J) of energy from any type of ionizing radiation by 1 g (10^{-3} kg) of any material; 1 mrad is one-thousandth of this.

APPENDIX 1.2: CHEMICAL SYMBOLS USED IN THIS BOOK

α	alpha particle, doubly charged helium
Al	aluminum
Ar	argon
C	carbon
C_2	molecular carbon
C_3	molecular carbon
CH_3	methyl-
CN	-cyanide
CH_3CN	methyl cyanide
CO	carbon monoxide
CO_2	carbon dioxide
Ca	calcium
D	deuterium, heavy hydrogen
e	electron
Fe	iron
H	hydrogen (1 AMU)
HCN	hydrogen cyanide
H_2O	water
He	helium
I	iodine
K	potassium

Mg	magnesium
N	atomic nitrogen
N_2	molecular nitrogen
NH	-imide
NH_2	-amide
NO	nitric oxide
Ne	neon
O	atomic oxygen
O_2	molecular oxygen
OH	hydroxyl-
p	proton, nucleus of hydrogen atom (unit charge, Q = 1)
Pd	palladium
Pu	plutonium
Rb	rubidium
S	sulphur
SO_2	sulphur dioxide
Si	silicon
Sm	samarium
Sr	strontium
U	uranium
Xe	xenon
−	(superscript) negatively charged particle
+	(superscript) positively charged particle
number	(superscript) mass of atom in atomic mass units (AMU)

APPENDIX 1.3: ACRONYMS AND ABBREVIATIONS USED IN THIS BOOK

AKR	Auroral kilometric radiation
AMU	Atomic mass unit
ANS	Astronomical Netherlands Satellite
ATS	Applications Technology Satellite
AU	Astronomical unit
AXTF	Advanced X-ray Telescope Facility
BIH	Bureau international de l'heure
COBE	Cosmic Background Explorer
COSPAR	Committee on Space Research
CRAND	Cosmic ray albedo neutron decay

CSTG	Commission on International Co-ordination of Space Techniques for Geodesy and Geodynamics
DC	Direct current
DSN	Deep-Space Network
ECHW	Electron-cyclotron-harmonic waves
ECW	Electron-cyclotron waves
EROLD	Earth rotation from lunar distances
ESA	European Space Agency
EUV	Extreme ultraviolet
EXOS	A Japanese magnetospheric satellite
GEOS	Geostationary scientific satellite of European Space Agency
GPS	Global Positioning System
GRO	Gamma Ray Observatory
HEAO	High-Energy Astronomical Observatory
H–R	Hertzprung–Russell diagram
IAA	International Academy of Astronautics
IAU	International Astronomical Union
ICHW	Ion-cyclotron-harmonic waves
ICW	Ion-cyclotron waves
IMP	International Monitoring Platform
IR	Infrared
IRAS	Infrared Astronomy Satellite
ISEE	International Sun-Earth Explorer
ISO	Infrared Space Observatory
IUE	International Ultraviolet Explorer
IUGG	International Union of Geodesy and Geophysics
ITU	International Telecommunication Union
LAMAR	Large area modular array
LHR	Lower hybrid resonance
MERIT	Monitoring of earth rotation and intercomparison of techniques
MHD	Magnetohydrodynamic (or hydromagnetic)
MOBLAS	Mobile laser stations
NASA	National Aeronautics and Space Administration of the United States of America
NGC	New general catalog (of galaxies, primarily)
NGS	National Geodetic Survey of the United States of America
OAO	Orbiting Astronomical Observatory
OGO	Orbiting Geophysical Observatory

OPEN   Origin of plasmas in the earth's neighborhood
OSO   Orbiting Solar Observatory
PEDAS   Potentially environmentally detrimental
            activities in space
SAO   Smithsonian Astrophysical Observatory
SAS   Small Astronomy Satellite
SETI   Search for extraterrestrial intelligence
SIRTF   Shuttle infrared telescope facility
SMM   Solar Maximum Mission
SOLRAD   Solar radiation monitoring satellite
SOT   Solar optical telescope
TKR   Terrestrial kilometric radiation
TLRS   Transportable laser ranging station
TMR   Terrestrial myriametric radiation
UHRW   Upper-hybrid-resonance waves
URSI   International Union of Radio Science
UT   Universal time
UV   Ultraviolet
VLBI   Very long baseline interferometry

ACKNOWLEDGEMENTS

Dr. Michael J. Rycroft served as general editor for the COSPAR contribution, coordinating the activities of individual section teams and ensuring the uniformity of the text.

Contributions to Section 1.1, compiled by Prof. L. G. Smith (United States of America), were provided by Dr. A. D. Danilov (USSR), Prof. K. Hirao (Japan), Prof. M. Ya. Marov (USSR), Dr. H. Rishbeth (United Kingdom), Prof. Satya Prakash (India), and Dr. E. Thrane (Norway). Comments received from Prof. R. Knuth (German Democratic Republic) and Prof. K. Rawer (Federal Republic of Germany) have been incorporated.

Contributions to Section 1.2, compiled by Dr. J. Lemaire (Belgium), were provided by Prof. W. I. Axford (Federal Republic of Germany), Dr. J. B. Blake (United States), Dr. L. F. Burlaga (United States), Dr. C. R. Chappell (United States), Prof. A. J. Dessler (United States), Dr. V. Domingo (Spain), Dr. T. E. Eastman (United States), Prof. C. G. Fälthammar (Sweden), Dr. L. A. Frank (United States), Dr. R. Gendrin (France), Prof. K. I. Gringauz (USSR), Dr. D. Jones (United Kingdom), Dr. W. P. Olson (United States), Dr. G. A. Paulikas (United States), Dr. H. Rosenbauer (Federal Republic of Germany), Dr. Y. C. Whang (United States), and Dr. D. J. Williams (United States).

Contributions to Section 1.3 were provided by Prof. H. S. Hudson (United States), Dr. Z. Svestka (Netherlands) and compiled by Prof. L. E. Peterson (United States). Information received from Dr. R. M. Bonnet (France), Dr. R. Harms (United States), Prof. S. Hayakawa (Japan), and Prof. S. L. Mandelshtam (USSR) was also used in the preparation of this Section 1.3.

Contributions to Section 1.4, compiled by Prof. S. K. Runcorn (United Kingdom), were provided by Dr. D. Gautier (France), Dr. J. F. Kerridge (United States), Dr. Rhea Lüst (Federal Republic of Germany), Dr. M. C. Malin (United States), Dr. J. A. M. McDonnell (United Kingdom), Mr. J. L. Mitchell (United States), and Prof. H. E. Suess (United States). Observational information in this section was based on work in observatories and laboratories in all parts of the world and on information brought back by planetary missions of the United States and USSR.

Section 1.5 was prepared by Dr. E. A. Flinn (United States) and reviewed by Prof. I. I. Mueller (United States); comments received from Prof. J. Kovalevsky (France) and Prof. A. G. Massevitch (USSR) have been incorporated.

Contributions to Section 1.6, compiled by Prof. P. R. Sahm (Federal Republic of Germany), were provided by Dr. M. E. Glicksman (United States), Dr. Y. Malméjac (France), and Prof. A. S. Okhotin (USSR).

Contributions to Section 1.7, compiled by Prof. H. Planel (France) and edited by Dr. J. B. Anderson (United Kingdom) and by Prof. T. H. Jukes (United States), were provided by Dr. E. A. Ilyin (USSR), Dr. H. P. Klein (United States), Dr. E. Kovalev (USSR) and Dr. R. S. Young (United States).

Contributions to Section 1.8, compiled by Dr. J. Billingham (United States) and Prof. R. Pešek (Czechoslovakia), were provided by Dr. H. Djojodihardjo (Indonesia), Dr. F. Drake (United States), Dr. R. Edelson (United States), Dr. A. T. Lawton (United Kingdom), Prof. G. Marx (Hungary), Prof. M. Morimoto (Japan), Prof. Yash Pal (India), Prof. C. A. Ponnamperuma (United States), Dr. M. Rees (United Kingdom), Dr. J. C. Ribes (France), and Prof. M. Subotowicz (Poland).

Chapter 2

CURRENT AND FUTURE

STATE OF SPACE TECHNOLOGY

ABSTRACT

This chapter is the second of four general review chapters. It was prepared with the assistance of an international team of experts organized by the International Astronautical Federation (IAF). The purpose of this chapter is to review the current situation and likely future developments in space technology. The applications of space technology are reviewed in several other chapters: scientific applications in Chapter 1; remote sensing and meteorology in Chapter 3; and educational satellite broadcasting in Chapter 6. This chapter therefore concentrates on the technology of the satellites themselves. Furthermore, since some aspects of current technology are reviewed in Chapter 5 (remote sensing and meteorology) and Chapters 6 and 7 (communications), this chapter emphasizes technologies which are now being studied or developed and which could become operational in the next two decades. It is hoped that this chapter might be useful to Member States in developing long-term policies for space technology.

2.1 SPACE APPLICATIONS TECHNOLOGY

2.1.1 Communications Satellite Technology

Communications via satellites is increasingly becoming a stimulus to general telecommunication development. In addition to long-distance communications, new services are being planned: electronic mail, video conferencing, and packet-switched data networks. Satellites for television broadcasting directly to homes are also currently under development. In total, 95 satellite communication systems are operational or planned through 1985, including three international systems (INTELSAT, INTERSPUTNIK, INMARSAT), five regional systems, 70 national systems amongst which 36 use INTELSAT satellites, and eight military or experimental systems.

Communications are surely the main applications field of space in the near future both for developed and developing nations. The tremendous increase in demand requires rapid improvements of the technology to get larger capacities, higher reliability, and reduced cost. For a more detailed discussion of the technology relating to communication capacity, see Chapter 7.

2.1.1.1 Operational systems

2.1.1.1.1 International Telecommunications Satellites. Communications satellites found their first operational application in transoceanic international telecommunications. This service has been provided since 1965 by the International Telecommunications Satellite Organization (INTELSAT), an intergovernmental organization with a membership of more than 100 countries. The INTERSPUTNIK organization offers similar service using the Union of Soviet Socialist Republic's STATSIONAR satellites. INTELSAT is now using its fourth series of satellites, and is introducing the INTELSAT-V series. It provides telephone trunk service and television transmission to those countries with high-volume international traffic and single-channel-per-carrier demand assigned service, called SPADE, to countries around the world whose need is not yet sufficient to justify a dedicated domestic system. This domestic service includes telephone, television distribution, service to off-shore oil drilling rigs, and other special services. Demand has continued to grow at a rate exceeding 20 percent per year.

INTELSAT-V will provide communications in two of the up and down link frequency bands allocated to the fixed-satellite service. It uses the 6 and 4 gigahertz (GHz)[1] band used by all preceding INTELSAT satellites, achieving frequency re-use through spatial and polarization isolation. In addition, the satellite has 14/11 GHz spot beams to provide coverage to high-traffic areas, using beams which can be steered by command from the ground.

Two basic types of earth stations are in use in the international INTELSAT

---

[1] One gigahertz is 1 billion cycles per second.

194

service at 6/4 GHz, the so-called standard-A and standard-B stations. The standard-A stations are large, are of very high performance, and offer very high capacities. The standard-B stations are smaller, of lower capacity, and lower cost. However, they do not make as efficient use of the space segment, and consequently higher space segment charges are incurred. The choice of station depends on the economic trade-off among traffic demand, earth stations costs, and space segment charges.

2.1.1.1.2 National Systems.   Following the initial application of geostationary satellites for transoceanic communications came the development of systems dedicated to national needs. In some cases, the needs are to cover long distances, where satellite links are more cost effective than terrestrial systems. In others, the needs are to provide reliable communications to remote areas. For long distances, the applications are heavy route telephony, both wide and narrow band data, and network television distribution. For remote areas, the applications are light route telephony (down to as little as one telephone circuit) and data and television delivery for terrestrial distribution.

Domestic systems are characterized by antenna beam coverage restricted as nearly as possible to the national territory being served. For systems using the 6/4 GHz bands, generally one beam covers the national territory. In the higher frequency bands, if large areas must be covered, several beams are used. Examples are the ANIK-C, SBS, and proposed Australian DOMSAT systems at 12 GHz, each of which uses four spot beams. For single-beam systems, the available frequency band generally is used twice by using opposite polarizations. For multibeam systems, both polarization and spatial beam separation are used to permit frequency band re-use.

Earth stations also vary more widely than in the INTELSAT sysem, ranging from the very large 30-m antennas down to 1- or 2-m antennas. Most common, however, are 6- to 9-m antennas for 6/4 GHz systems, 3- to 8-m antennas for 14/12 GHz systems, and 2- to 4-m antennas for 30/20 GHz systems.

The main operational domestic systems are: ANIK-A and-B (Canada), PALAPA (Indonesia), STATSIONAR (USSR), COMSTAR (United States), RCA SATCOM (United States), WESTAR (United States); and SBS (United States). Those soon to be placed in operation are: ANIK-C (Canada), TELECOM I (France), INSAT (India), and CS-2 (Japan).

2.1.1.1.3 Maritime Mobile Systems.   After meeting long-distance, fixed telecommunications needs, satellite technology was applied to meeting mobile communications needs of ships at sea. This technology offers the reliability which is unavailable from shortwave radio, the principal medium used when ships are more than a few kilometers distant from the VHF shore radio stations. Voice, teletype, and facsimile are the main services provided by satellite, with data service being available as well. The advantages to the shipping industry of having reliable communications for business, weather, and safety information are obvious, with resulting operational efficiencies making maritime mobile satellite communications cost effective. The MARISAT (United States) satellites currently provide operational communications for the Atlantic, Pacific, and Indian oceans.

The satellite–ship links of maritime satellite systems operate at 1.6 and 1.5 GHz. At these frequencies, earth station antennas have wide beams, thus easing the problem of antenna pointing from a moving ship. Antenna systems at the present time are 1 to 1.5 m in diameter and are protected from the elements by a radome. A steerable mount is required, and total above-deck weight is in the vicinity of 265 kg. For this reason, as well as for cost reasons, installations have been restricted to fairly large ocean-going vessels. Satellite–shore links operate at 6/4 GHz. For the MARISAT system, six strategically located shore stations currently provide connection to national telephone systems.

In 1978, INMARSAT was formed by the agreement of 40 countries to participate in the establishment of an internationally owned maritime mobile satellite communications system. This will be established through the launch of two MARECS satellites with dedicated payloads and the inclusion of 1.6/1.5 GHz transponders on three INTELSAT-V satellites. The first MARECS was launched in 1981, ensuring continuity of service as the MARISAT satellites reach the end of their useful lives.

Several national maritime mobile systems are at present in planning. The USSR has notified the International Telecommunication Union (ITU) of plans to implement its VOLNA systems, and Japan is planning an aeronautical–maritime experimental satellite (AMES) for communications with much smaller ship terminals, which would be suitable for small fishing vessels.

2.1.1.1.4 Direct Broadcasting Satellites. The most recent development of satellite communications technology has been for direct broadcasting of television to community or individual home receivers. Such a service finds application for a variety of reasons. For some countries, it is to extend basic service to areas difficult to cover with terrestrial broadcasting because of distance or terrain; for others, it is to provide additional channels of television more economically than can be done terrestrially; for some, where the radio-frequency spectrum for conventional television is already congested, it provides additional spectrum for broadcasting. Individual geographic, demographic, or technological circumstances have prompted a number of countries to make firm plans to implement direct broadcasting satellite systems.

Direct broadcasting satellite systems are generally characterized by high-power satellite signal transmissions capable of being received with a small antenna, 1 m in diameter or smaller, and a simple receiver. The high signal power is achieved with a combination of spot beam satellite antennas providing high-gain and high-power traveling wave tube amplifiers. One spot beam is usually sufficient to provide coverage of an entire country, except for those countries with large land masses. For these cases, a multiplicity of spot beams are used rather than one large beam to obtain the advantage of antenna gain.

Since there is neither an unlimited amount of radio-frequency spectrum nor unlimited capacity for satellites in the geostationary orbit, it was deemed necessary in ITU regions 1 and 3 to plan spectrum-orbit use by allocating channels in specific orbit locations to all countries. This was done at the 1977 World Administrative Radio Conference (WARC) of ITU dealing wih broadcasting satellites. In general, five television program channels were allocated per spot beam or service area. Decisions on the allocation of the spectrum-orbit resource in region 2 were deferred until 1983 when a Regional Administrative Radio Conference will be held.

Receiver terminals suitable for individual home reception of direct broadcasting satellite transmissions range from as small as 0.6 m in diameter to as large as 1.8 m, with the smaller sizes being more suited to urban applications and the larger sizes being acceptable in rural areas.

Firm implementation plans for direct broadcasting satellites have been developed by several countries. France and the Federal Republic of Germany are cooperating on pre-operational satellites for their respective countries— TV-SAT for the Federal Republic of Germany and TDF-1 for France. The satellites are expected to be launched in early 1984 using Ariane launch vehicles.

Japan also has firm plans to establish an operational system, beginning in 1984, with the BS-2 program. The satellites will be similar to the experimental BSE satellite which operated from April 1978 to mid-1980. The antenna beam is shaped to provide coverage of outlying islands at some distance, as well as the main islands of Japan. Japan's N2 launch vehicle will be used.

Other countries also planning direct broadcasting satellite systems are Canada, the Scandinavian countries, Luxembourg, the United States, Australia, the Arab countries, the United Kingdom of Great Britain and Northern Ireland, Italy, and Switzerland. The list is expected to grow.

2.1.1.2 Experimental systems.

Although some new technology can be tested adequately on the ground prior to operational deployment, a major portion of the necessary new technology requires research and development (R and D) orbital test data on performance and lifetime to ensure acceptably low risk prior to initiation of an operational program. Further, new technologies require overall system experiments in association with the appropriate terrestrial systems. Participation by typical users in these experiments is essential in order to develop user acceptance and demand and to demonstrate mission cost effectiveness.

From 1960 to 1973, the National Aeronautics and Space Administration (NASA) took the dominant role in communications satellite R and D with the Applications Technology Satellite (ATS) series, especially the ATS-6, launched in May 1974. ATS-6 demonstrated a large deployable multibeam antenna, three-axis stabilization and pointing capability by an on-board processor, an L-band direct broadcasting experiment, propagation experiments at 20 and 30 GHz, a data relay experiment between ground and low orbit via geostationary orbit, an L-band experiment for maritime and aeronautical use, and other new space technology.

At the same time, ambitious experimental communications satellite programs emerged from other countries, especially European countries, Canada, and Japan. These programs were to establish operational domestic or regional communications satellite systems and to stimulate industry in a high-technology area.

In Europe, satellites for the first French–German program, SYM-PHONIE, were launched in December 1974 and in August 1975. The Italian satellite SIRIO was launched in August 1977. Some of the technology proven by SIRIO was transferred to SIRIO-II, to be launched by the European Space Agency (ESA) in 1982. In May 1978, the Orbital Test Satellite OTS-2 was successfully launched, and extensive performance tests and telecommunication experiments have been carried out by ESA. The performance of OTS during its first $2\frac{1}{2}$ years in orbit has verified the European Communications Satellite (ECS) design concepts. ECS is a European regional communications satellite and will begin service in 1982. Technologies of the Orbital Test Satellite (OTS) bus have also been transferred to MARECS, the ESA maritime satellite launched by Ariane in December 1981. The Communications Technology Satellite (CTS) HERMES was launched by NASA in January 1976, as part of a joint United States–Canadian program in which Canada designed, assembled, and operated the satellite. CTS was used for $3\frac{1}{2}$ years by the Canadian Department of Communications (DOC) and NASA for experimental applications of communications technology at 14/12 GHz. As a pilot project of ANIK-B, Canadian DOC has performed direct broadcasting reception experiments using the ANIK-B 14/12 GHz transponder. In Japan, the communications satellite for experimental purposes (CS) SAKURA was launched in December 1977, and the broadcasting satellite for experimental purposes (BSE) YURI in April 1978. More than two years of successful experiments on CS and BSE led to confidence that the design concepts of the two experimental satellites can be satisfactorily transferred to the operational ones: CS-2, to be launched in 1982, and BS-2, to be launched in 1983.

In addition to the above-described systems, many other countries are planning to have their own domestic or regional communications satellite systems during the next 10 years.

2.1.1.3 Trends in satellite technology.

The technical trends in the design of communications satellites and their subsystems are motivated by the requirements (a) to improve the performance and enlarge the capacities of the systems by using higher transmission frequencies, multibeam satellite antennas, frequency re-use, etc., (b) to enlarge the applications field by developing small earth terminals, and (c) to reduce the cost by developing large platforms, as discussed later.

2.1.1.3.1 Frequency Range.   In the early phases of fixed satellite communication, most satellite communications experiments were carried out using frequencies of 6 GHz (up link) and 4 GHz (down link) to have good compatibility with terrestrial microwave relay systems. After that, the 6/4 GHz frequency band was chosen for operational communications satellites, and this band became heavily crowded as the number of user terminals increased.

Since 1974, various experiments have been carried out using higher frequency bands, above 6/4 GHz, as follows:

Propagation experiments	
ATS-6 (United States, 1974)	20,30 GHz
ETS-II (Japan, 1977)	12,37 GHz
Communication experiments	
CTS (Canada, 1976)	14/12 GHz
SIRIO (Italy, 1977)	18/12 GHz
CS (Japan, 1977)	30/20 GHz
ANIK–B (Canada, 1978)	14/12 GHz
OTS–2 (ESA, 1978)	14/12 GHz

These frequency bands will become operational in the very near future. On the other hand, the ITU Radio Regulations assigned new bands for communications satellites, including broadcasting satellite and satellite-to-satellite communication, in the frequency range from 30 GHz to 275 GHz.

2.1.1.3.2 Beam Shaping and Multibeam Satellite Antennas.   On-board antenna systems in modern communications satellites are required to accomplish such functions as:

— Improving equivalent isotropically radiated power (EIRP) over prescribed areas through pattern shaping.

— Reducing power outside desired coverage areas.

— Allowing frequency re-use by both spatial and polarization isolation.

Shaped-beam antennas and multibeam antennas are necessary for communications satellites to meet these needs. Three distinct types of shaped-beam or multibeam antennas have evolved: reflectors, lenses, and arrays.

Reflector-type antennas are attractive because of their design simplicity, inherent bandwidth, low mass, and low cost. Two types of reflector antennas are used to produce shaped beams. One type is composed of a parabolic reflector and a cluster of feed horns. This type is also applicable for multibeam use. However, in this case, feed structure causes excessive blockage and consequent high sidelobe levels for all configurations except offset-fed types. The most typical multibeam antennas in use are INTELSAT-V communications antennas.

The second type of shaped-beam reflector antenna is composed of a shaped reflector and a feed horn. The features of this type are simple structure and capability of operating over several frequency bands. The shaped-beam horn-reflector antenna, developed for the Japanese medium-capacity CS, operates in the 6/4 GHz and 30/20 GHz bands. The reflector is shaped to produce the desired beam shape, which efficiently covers the main Japanese islands.

Lens antennas are attractive for multibeam use because there is no blockage due to feed clusters. In the classic solid dielectric lens, the random fluctuations in refractive index of dielectric material result in excessive sidelobe levels. A constrained lens is composed of RF transmission lines, interconnecting small pickups, and radiating elements. Several experimental models with constrained lenses have been constructed and tested.

Array antennas are able to produce not only ordinary multibeam, but also variable-shaped beams. On the NASA Tracking and Data Relay Satellite (TDRS), the multiple access (MA) service uses an array of 30 S-band helix antennas to form individual return link antenna beams pointing at each user. These beams are formed at the ground segment and 20 simultaneous beams are provided. The MA forward link uses 12 of the same helix to generate a simple forward link beam. Various satellite systems employing phased arrays have been studied, including a communications satellite system for mobile users and a rapid-scan area coverage communications satellite system.

2.1.1.3.3 Frequency Re-use.

The capacity of a communications satellite is significantly increased if the same carrier frequency is used twice or more in the satellite. Two means of frequency re-use are foreseen.

Frequency re-use by means of narrow-beam satellite antenna. The frequency re-use can be achieved by the use of multiple satellite antennas with narrow beamwidths serving different areas on the surface of the earth. The re-use of this kind was well demonstrated by the INTELSAT-IV–A satellite, in which the same frequency bands are used in the west hemispheric beam and in the east hemispheric beam, thereby doubling the capacity in those bands.

Frequency re-use by means of orthogonal polarizations. The twofold use of the same carrier frequency is made possible when two carriers are transmitted on orthogonal polarizations. Frequency re-use of this kind is being used in domestic or regional satellites which usually serve a rather small area. PALAPA satellites, which serve Asian countries, and United States domestic satellites such as COMSTAR and RCA SATCOM are in this category. A technical problem associated with the re-use of this kind is depolarization due to precipitation. A number of studies have been made to estimate the degree of degradation of polarization purity and development effort has been made to compensate the depolarized signal at the earth station.

Future satellites will pursue frequency re-use by combining a large number of narrow beams and orthogonal polarizations, as foreseen on the INTELSAT-VI satellite.

2.1.1.3.4 Satellite Transmitters.   Since the inception of active communications satellites, traveling wave tubes (TWTs) have played a fundamental role in on-board transmitter development. Hundreds of TWTs are now in space and only TWTs have a good history of operation in space.

The major competition to TWTs comes from the longer lifetime and the reduced mass of solid-state amplifiers. Recent advances in gallium arsenide field-effect transistors (FETs) have made possible the design of solid-state satellite transmitters. FETs can now challenge TWTs at C- and X-band with power outputs of several watts.

The low-noise FET preamplifier (4 GHz) used in the Japanese CS and the high-gain driver amplifier (11 GHz) used in CTS are providing the base of FET space operation verification. An experiment with a 1-watt FET amplifier (4 GHz) in space is being planned in the Japanese ETS–IV in 1981.

TWT improvements are also under way, especially in the field of high-power outputs (above 100 watts) for use in broadcasting satellites. High-power and high-efficiency TWTs utilizing tapered helix, dispenser-type cathodes, and multicollector technologies are being developed.

Future satellites are gradually expected to have FET transmitters with output power around 10 watts. Power levels from 10 to 100 watts would become the competitive region for FETs and TWTs.

2.1.1.3.5 Satellite Switching System.   In a multiple-beam satellite system, a switching circuit is required on board to route signals from one beam to another. When the communications satellite network is based on the time division multiple-access (TDMA) system, such an on-board switching circuit could be realized by the dynamic switching matrix that is operated to synchronize the TDMA system.

This highly advanced multiple-access technique is called the satellite switched (SS) TDMA system and can be considered the first step toward a satellite with sophisticated signal-processing capability.

Studies on the SS/TDMA system have been conducted over a number of years by various organizations, including INTELSAT, COMSAT, Bell Laboratories, and KDD, and several types of experimental packages have been developed that have proved the technological feasibility of implementing on-board switching systems.

The first communications satellite system that is expected to use SS/TDMA is Advanced WESTAR, one of the domestic satellite systems of the United States, which is scheduled to come into operation in 1982.

2.1.1.3.6 Small Earth Terminals.   Present and future communications satellites are tending to use higher power, wider bandwidths, and more sophisticated technology, thus permitting the use of smaller and less expensive earth terminals. A small earth terminal network can provide not only conventional voice channels for international, regional, and domestic communications but also data channels. Such small earth terminals are also foreseen for communications during disasters and other situations where easier transportation is required.

The INTELSAT system, operating initially with earth stations using large 30-m antennas, has been developing smaller terminals. Satellite Business Systems (SBS) in the United States will provide services using all-digital systems to make the most of small earth terminals installed on or near the customers' premises. This system will use the 14/12 GHz bands and 5- to 7-m diameter antennas.

Japan has developed various kinds of communications satellite systems using small earth terminals operating in the 30/20 GHz bands and 6/4 GHz bands. For example, a satellite link is formed between a very small earth terminal and a master earth station which is connected to the terrestrial telecommunications networks. This system uses the 6/4 GHz bands. A very small earth terminal with an offset parabolic antenna (aperture size 2.4 m X 1.5 m) can transmit up to three channels.

Ship–earth terminals which are now being used worldwide in the IN-MARSAT system consist of above-deck (1.2 m antenna and 1.8 m radome) and below-deck equipment. These terminals are designed to be used by rather large vessels such as tankers, cargo ships, container ships, and passenger ships. Development of a new earth terminal for small ships such as fishing ships is now under way in Japan. This terminal will reduce the size, weight, and cost of the above- and below-deck equipment, but the performance (with 0.85 m antenna and 1.2 m radome) will remain the same as the existing terminals.

2.1.1.3.7 Large Structures for Communications.   The development of large structures in space and of the necessary transportation systems, discussed in Section 2.2, will offer an attractive way to increase space communications capabilities at minimum cost.

2.1.1.4 Trends in system technology

2.1.1.4.1 Multiple Access.   Communications satellites provide a cost-effective means for communicating across wide geographic areas. Multiple-access techniques permit communications satellites to use the capacity of the satellite transponders efficiently when a large number of earth stations must have access to the satellite simultaneously. There are three basic multiple-access techniques: frequency division multiple access (FDMA), TDMA, and code division multiple access (CDMA).

FDMA is a technique in which carriers from different earth stations access a satellite transponder with different frequencies. The single channel per carrier (SCPC) system in which the earth station carrier contains only one voice or data channel is an FDMA technique used in the INTELSAT system.

TDMA is a technique in which multiple earth stations share a common transponder by transmitting bursts of carrier signals on a sequential and non-overlapping basis. The first commercial TDMA system was the TELESAT

TDMA system (Canada), which went operational in 1975. Many pre-assignment TDMA systems are planned for international, regional, and domestic satellite communications systems. When demand assignment is required, the TDMA network can be more efficient and flexible. TDMA systems with demand assignments are being tested with present communications satellites in the United States and Japan. SS–TDMA systems will be also planned as the next promising technique for future INTELSAT and domestic satellite systems. SS–TDMA is an effective technique for maintaining interconnectivity among satellite transponders when a satellite has multiple spot beams.

The application of spread spectrum techniques to communications satellites systems provides several remarkable capabilities, such as anti-jamming, message privacy, and multiple access. CDMA is usually implemented as a spread spectrum multiple-access (SSMA) technique, and experiments will be carried out in the near future.

2.1.1.4.2 Intersatellite Communications.   Studies have been made on the intersatellite communications between two geostationary satellites. Direct satellite-to-satellite relay can improve global communications over an area larger than the maximum coverage of a single geostationary satellite by avoiding double hops. There are still some problems to be solved before implementation: accuracy of antenna tracking, wide-band modulation and demodulation equipment, and selection of frequency bands.

Nevertheless, NASA is now developing the Tracking and Data Relay Satellite System (TDRSS) for global satellite tracking and control and for data relay to low-earth orbit satellites and the Space Shuttle. Two geostationary satellites (TDRS–East and TDRS–West) are able to maintain 85 percent and 100 percent of continuity of coverage for satellites orbiting at altitudes of 200 km and 1,200 km, respectively.

2.1.1.4.3 Site Diversity.   To meet the growing demand for both international and domestic communications, frequency re-use techniques and technology for the frequency bands above 10 GHz are being developed.

But at frequencies above 10 GHz, signal attenuation due to rainfall is severe; for example, 10 to 20 dB or more attenuation occurs in heavy rain. Therefore, technological development to overcome rainfall attenuation is necessary to ensure continuity of communication.

A site diversity system is recognized as the most effective technique to overcome rainfall attenuation. As heavy rainfall is generally limited to small cells, it is possible in the site diversity system to select a satellite path with low attenuation if two earth stations are installed 10 to 30 km apart and combined by a terrestrial transmission line. By using the site diversity system, link availability is remarkably improved compared to conventional systems because outage time depends on the joint probability of rain attenuation in two different paths.

Experiments in site diversity systems are divided into two categories. One is propagation research in the 14/12 GHz and 30/20 GHz bands, experiments which are carried out using COMSTAR (United States), ETS-2, CS (Japan), and other satellites. The other is site diversity system development, especially with TDMA. These developments are carried out or planned in several countries and experiments using communication satellites in orbit are planned.

2.1.1.5 Economics of future systems.

Communications via satellites already represent a significant economic activity, of the order of $2 billion per year, and are growing at a very rapid rate, with current projections reaching $10 billion by the beginning of the 1990s.

In this expanding market, larger and larger satellites are being used or planned, with a significant increase in cost effectiveness resulting from economy of scale. While many domestic or regional systems use spacecraft 500 to 600 kg in mass, INTELSAT-V, the first of which was launched in 1980, has a mass of about 1,000 kg, and INTELSAT-VI will be 1,800 kg. This growth will improve spacecraft cost effectiveness, as characterized by the cost per unit of communications payload mass, by a factor of about three.

Still larger platforms of 5,000 kg or even much more are envisaged to further improve cost effectiveness. They would also offer a number of extra advantages:

1. Efficient utilization of the already crowded geostationary orbit.
2. Interconnection between on-board payloads, potentially resulting in better utilization of the frequency spectrum and savings in operational expenditures.

They are, however, likely to suffer from a number of drawbacks:

1. To fill their large capacity, several payloads must be installed on a single platform (perhaps a dozen or more payloads), resulting in a reliability problem until repair and maintenance in space become economically feasible.
2. Since they would be launched infrequently, the payload would have to be large enough to cope with traffic growth and would be poorly used initially.
3. Very large platforms may be penalized because of the complexity of their designs, resulting, for instance, in a less favorable structural index than today's compact satellite designs.
4. They cannot be launched until transportation systems more capable than Ariane IV (1,800 kg) or Space Shuttle plus the Inertial Upper Stage (IUS) (2,300 kg) become available.

An attractive alternative to very large multimission platforms is a cluster of satellites co-located at the same orbital position through close station keeping;

this would permit gradual building up of the constellation, reasonably cheap replacement of the failed satellites, and use of available launchers. It could also provide relatively easy interconnectivity via satellite-to-satellite links and could reduce the need for in-orbit spares to the extent that interchangeability could be achieved. Future service requirements are expected to justify constellations of rather efficient satellites of the INTELSAT-V or-VI category.

Finally, another way to improve cost effectiveness of communications satellites is to increase spacecraft lifetime. Life of 7 to 8 years is already considered normal; the current goal is 10 years, requiring improvement in the reliability of critical equipment, particularly power amplifiers and batteries.

2.1.1.6 Developments in the USSR

2.1.1.6.1 International Systems.   To promote satellite communications on a wide basis, in 1973 the USSR together with other socialist countries established the INTERSPUTNIK System providing broad coverage including the northern areas of the globe. The system uses the MOLNIYA–3 and STATSIONAR (RADUGA, GORIZONT) satellites operating at 6/4 GHz with 12-m antennae.

INTERSPUTNIK serves the countries of the socialist community and some developing nations. It is open to any other interested countries. The system can be used for the exchange of almost all major types of communication (telephone, TV, digital, etc.). The system played an important role in relaying overseas TV coverage and radio programs during the XXII Olympic Games held in Moscow in 1980. Now work is under way for further improving the system, including an increase in capacity through use of the 14/11 GHz band and a reduction in ground station cost through the use of smaller antennae.

2.1.1.6.2 National Trunking Communications Systems.   The USSR domestic satellite system is the largest in the world. It has been based on the MOLNYA and STATSIONAR satellites and on a wide network of transponder and receiving stations with antennae ranging from 12.5 to 2.5 m in diameter (Orbita, Mars, Moskva). The system also incorporates the central digital stations with antennae of 32, 25, and 15 m (Dubna, Vladimir, Lvov).

The Orbita and Moska stations, the most numerous, now number nearly 100 and use single-beam antennae of 12 and 2.5 m operating at 6/4 GHz.

The USSR domestic systems, besides being used for a great volume of television, radio, and facsimile have also been employed efficiently for newspaper photocopy distribution from Moscow to Khabarovsk, Irkutsk, and other distant cities. The inhabitants of these towns, situated thousands of kilometers from Moscow, now receive the national newspapers almost simultaneously with Moscovites. This system is also used for links with the scientific research ships of the USSR Academy of Sciences as well as for manned spacecraft and space station control when these vehicles are out of range of the USSR ground control stations.

2.1.1.6.3 Maritime Mobile System. For ocean navigation, the Soviet Union is planning to use both the INMARSAT international system and the VOLNA domestic system. The latter will permit service to ships either in temperate or arctic waters. This system can be used on a wide basis for navigation in northern areas through lease of communication channels by interested countries. Preliminary experiments toward this goal were conducted by the Sibir atomic powered ice-breaker in the polar basin in 1978.

2.1.1.6.4 Telecommunications System. In 1976, the EKRAN telecommunications system was established in the USSR for relaying central TV programs to a broad network of simple and low-cost receiving terminals. The transmission of TV pictures from the satellite uses frequencies within the range of routine terrestrial television broadcasting (700 MHz) with an on-board transmitter power output of 200 W and a gain coefficient for the transmitting antenna of 33.5 and 26 dB (at the center and on the edge of the radiation zone, respectively). Two types of receiving terminals are in use: the so-called standard-A for regional terrestrial TV transmitters of high and medium power output and standard-B for low-power TV retranslators or cable networks. There are now about 1000 such terminals.

The EKRAN system is simple and effective, but further expansion of its service zone is a problem because of the International Radio Regulations restraints on power flux density to areas of neighboring countries. This problem has been resolved by the new Moskva TV system designed to serve any area of the country, including the Far East and its European part. The system operates via STATSIONAR (GORIZONT) satellites at 6/4 GHz with a higher output (40 W at the antenna) and with a highly directional radiation pattern (the transmitting antenna gain coefficient is 30 dB). The Moskva ground station is simple in its design, is small, and doesn't need continuous skilled maintenance. The directional pattern of the receiving antenna is accurate to $\pm 1°$.

The satellite systems provide telecasts for 75 million people out of a total of 200 million Soviet regular television viewers.

Further development of the telecasting and broadcasting systems to simple and low-cost ground receiving terminals in the 12 GHz frequency band is being planned.

2.1.2 Earth Observation Technology

In the technologically advanced countries, remote sensing from space is a cost-effective method of acquiring new data to update existing information. In several developing countries, however, remote sensing has provided an opportunity to obtain the first generation of information on some of their earth resources.

2.1.2.1 Current technology. The technology and equipment that were successfully tested during the early stages of space exploration are now finding wide applications in the global exploration of the resources of the earth. These include, among others, the use of electromagnetic sensors placed on board various space platforms to detect, identify, evaluate, and monitor earth resources. Since 1972, experimental satellites such as LANDSAT-1, 2, and 3 and SOYUZ 5 and 6 and experimental space stations such as SKYLAB and SALYUT 4, 5, and 6 have collected data over most parts of the globe in the course of their respective missions. These data have found applications in agriculture, forestry and range management, hydrology, geology, oceanography, energy resources, environment, flood and disaster warning, land use, and mapping and charting.

In addition to these systems are the meteorological satellites which comprise the Global Observing System (GOS) of the World Weather Watch. This consists of geostationary meteorological satellites, launched by the United States (GOES), Japan (GMS) and ESA (METEOSAT), and a system of polar-orbiting meteorological satellites operated by the USSR (the METEOR satellite series) and the United States (the NIMBUS and TIROS-N series). These polar-orbiting satellites operate in sun-synchronous orbits at altitudes of 800 to 1,500 km. Five satellites have been launched in the GOES series and two each in the GMS and METEOSAT series.

Data received from these satellites can be used to determine profiles of atmospheric temperature and moisture content, more accurate sea-surface temperatures, cloud cover and height data, snow-melt forecasting, surface water boundaries, water vapor sensing with greater atmospheric depth and atmospheric ozone data important to environmental studies. Such information now constitutes a major aspect of the international program initiated in 1978 called the Global Atmospheric Research Program (GARP). The goal of this program is to increase mankind's understanding of atmospheric processes and help improve his forecasting ability. These satellites are currently being used as part of the World Weather Watch Program in monitoring and tracking typhoons and hurricanes in the seas and oceans of the world and in keeping a close watch on the desertification effects of droughts, particularly in the Sahel region of Africa.

From August 1959 to date, at least 30 earth observation satellites have been launched into space, mostly from the United States and the USSR. The most recent ones include METEOSAT-2 (1981) (ESA); LANDSAT-3 (1978); Heat Capacity Mapping Mission (1978) and SEASAT (1978) (United States); SOYUZ-22 (1976), SALYUT-6 (1977), and SOYUZ-30 (1978) (USSR), and BHASKARA-2 (1981) (India).

A new Soviet experimental satellite launched in July 1980 has three different sensors aboard. A multispectral scanner MSU-SA with four LANDSAT-type spectral bands, 600 km swath-width, and 170 m instantaneous field of view (IFOV), an eight-channel multispectral scanner FRAGMENT-2 with 80 m IFOV in the visible, near and medium infrared, and a three channel electronic scanning device MSU-VA in the visible and near infrared with 30 m IFOV.

2.1.2.2 Future systems.   The future development of space technology will result in new generations of more versatile and compact space platforms and sensors. This trend is very apparent in the field of remote sensing from space. A thermal infrared band (10.4–12.6 μm) on LANDSAT–3 makes a global analysis of the earth's surface temperature possible. The thematic mapper on the proposed LANDSAT–D will have six and possibly seven spectral bands that are narrower and of higher radiometric and spatial resolution (from 80 m at present to 30 m) than the multispectral scanner (MSS) channels on board LANDSAT–1, 2, and 3. India plans to launch a multipurpose satellite, INSAT, into a geostationary orbit in 1982. INSAT will have a meteorological sensor on board, furnishing data both in the invisible and infrared.

The French satellite SPOT, scheduled for launch in 1984, will have an instantaneous field of view (IFOV) of 10 to 30 m. High resolution photographic space imagery taken from SOYUZ–22 and SALYUT–6 by means of the six channel multispectral camera MKF–6 (GDR–USSR) can be used for basic maps at scales as large as 1:50,000. Such imagery could therefore replace or supplement standard aerial photography and also be of use in areas with small agricultural fields (down to 0.2 hectare).

A major drawback in the use of sensors operating in the visible band (0.5–0.8 μm) of the electromagnetic spectrum for earth resources survey is their inability to penetrate cloud cover. This is of major concern, particularly in the wet tropics where there is persistent cloud cover most of the year. To overcome this handicap, a number of ideas have been proposed, including the use in the immediate future of all-weather microwave radar sensors. The attractive characteristics of the radar system include its multiple look-angles, increased discrimination, ability to penetrate clouds and rain, and sensitivity to textural differences. ESA plans to employ active and passive microwave sensors on an earth application satellite. The SEASAT satellite, launched in 1978, also carried a single frequency (L-band) 25-m resolution horizontally polarized imaging radar. A multifrequency imaging radar system was on board the Space Shuttle flight of November 1981. Canada is also considering a synthetic aperture radar (SAR) satellite to assist in navigation and petroleum exploration in its northern ice-infested waters.

2.1.2.3 Application objectives.   Other future technological prospects in earth survey from space include the use of low-latitude satellites which can collect data over the equatorial zone with increased frequency. Along this line, Indonesia, in co-operation with the Netherlands, is considering a tropical earth resources satellite (TERS) in an equatorial orbit to provide optimum coverage of the tropical regions. Single-purpose satellites for application in such areas as hydrology and agriculture may also become a regular feature of earth resources sensing in the immediate future. Furthermore, the United States, the USSR, Japan, India, Canada, France, the Federal Republic of Germany, the German

Democratic Republic, and ESA are all currently developing various sensor systems for specialized remote sensing missions.

2.1.3 Navigation and Position-Determination Technology

Position determination is one of the many terms used to describe methods which are common to navigation and to applied geodesy and geophysical research. But, whereas the navigator needs to know his position in relation to his point of departure and his destination in real time, the geodesist or the scientist engaged in solid earth research requires this information with respect to known points of a reference system with generally better accuracy but is willing to accept some delay between the moment the measurement is made and the time the position coordinates are determined.

Also, since both disciplines have extended their activities from local operations to global activities, they now need techniques which work day and night independently of weather and other interfering factors.

Terrestrial position-determination systems have been developed to provide continuously available services which are accurate, reliable, and simple. Radio-determination techniques have overcome the deficiencies of optical observations, and satellite methods finally have demonstrated their unique capability of doing all this on a worldwide basis. The main advantages offered by satellites are the synoptic view and repetitive coverage of the earth's surface with relatively few spacecraft.

2.1.3.1 Applications. Position determination of geodesy and geophysics spreads over a wide range of scientific and commercial activities. In the field of earth sciences, high-precision position information in terms of absolute and relative values is required to:

1. Establish a permanent reference system for triangulation at the earth's surface.
2. Observe the earth's rotation (pole movement) with an accuracy and, above all, a resolution so far not reached.
3. Observe variations in the earth's crust (tides, tectonic movements).

Requirements for a similar degree of accuracy arise in connection with the study of ice flow rates (glaciology) and in the interest of other related scientific disciplines, such as climatology and hydrology.

Commercial applications of geodesic measurements include topographical surveys for a multitude of civil engineering projects (construction of dams, nuclear power plants, roads, etc.). Cartography can be improved through more exact reference points for photogrammetry. Perhaps the greatest need for improved position-determination services will emerge in the field of exploration and management of our limited natural resources and our environment. Mineral, gas, and oil prospecting and exploitation, both at sea and on land, involve a number of problems related to the initial determination of position of the prospecting team, redetermination (often after long time intervals) of a certain location, and drawing of a basic topographic grid. There are also many potential requirements for a position-determination function in systems for data collection purposes (localization of drifting balloons or buoys, wildlife migration monitoring, etc.).

2.1.3.2 Current technology. The use of satellite techniques for position determination in geodesy and geophysics is almost as old as the history of artificial earth satellites itself (TRANSIT-1B, ANNA-1B, SECOR, etc.). Satellites dedicated to geodesy and geodynamics have been launched (EXPLORER, GEOS, DIAPASON, DIADEME, STARLETTE, LAGEOS) and significant progress has been made with their utilization.

These systems have employed a variety of methods of position determination, including radio interferometry, satellite laser ranging, Doppler techniques, and the use of precision time standards to allow one-way range measurements. In the following paragraphs some examples are given with the intention of characterizing the possibilities and trends that can be seen in this rapidly progressing discipline.

At present five navy navigation satellites (NNSS, formerly denoted TRANSIT) are maintained in circular polar orbit at altitudes of about 1,000 km. The user determines his position on the surface of the earth by establishing his coordinates with respect to the known ephemerides of a satellite. He utilizes the Doppler shift technique of observing the apparent frequency changes of a radio signal received from a moving object. It became evident that more highly processed data could be used for geodetic applications, and since its inception in 1960, NNSS has been used by individual land surveyors and by international teams of geodesists all over the world. Parallel efforts in improving the computational treatment of the Doppler data reduced the error in position fixing below the 1-m mark.

The United States Department of Defense currently has an advanced navigation satellite system, the Global Positioning System (GPS) (or NAVSTAR) under development. The program plans call for the deployment of 18 satellites by 1986 in circular 20,000-km orbits as an operational system, with the possibility of providing extremely accurate three-dimensional position fixes. Undoubtedly, this system will find extensive utilization in the geodetic community if civilian access to GPS is approved.

The principle of position determination is based on the simultaneous reception of four navigation signals which allow independent range and Doppler calculations to be performed. Again, further processing of the measurement data is expected to provide the geodetic community with global absolute positions in the 1-m accuracy range. Considerable advances toward the adopted goal of positioning by Doppler satellite to an accuracy of 10 cm or better have been made, thus allowing scientists to study the geodynamic phenomena.

In addition, the satellites of GPS offer an important new opportunity for geodetic position determination. A United States concept called satellite emission interferometric earth surveying (SERIES) makes use of GPS radio transmissions by employing the techniques of very long baseline interferometry (VLBI). The phases of the electromagnetic signals transmitted from the satellite yield an extremely precise determination of the baseline vectors, i.e., the distances between the observing antennas. A number of groups have performed experiments and have demonstrated subdecimeter three-dimensional accuracy on baselines of several hundreds of kilometers. These extraordinary experimental results have encouraged the development of transportable VLBI stations for exclusive use in geodetic and geodynamic projects; much of the Geodynamics Program Plan of NASA, for example, is devoted to discussions of possible operations with mobile VLBI stations in tectonically active regions in order to monitor the small relative motions of crustal plates (1–10 cm/year) and thereby to study dynamic processes related to earthquakes.

Another technique, laser ranging to satellites, which was originally developed to improve the precision of satellite orbit determination, has proved itself an invaluable geodetic tool to determine station position. The principle of position determination by laser ranging is based on the determination of the distance from a point on earth to satellites of known position. In contrast to the systems described above, the space segment is passive, consisting of optical reflectors mounted on the surface of the satellite. In the past 15 years, about a dozen satellites equipped with retroreflectors have been launched. In parallel, technological progress with laser telescope technology has improved laser ranging precision rapidly from the meter level to centimeter accuracy. Ranging precisions of 1 to 2 cm are now being achieved by several fixed stations, and design goals for mobile laser stations are also aimed at this precision, which is required for regional crustal deformation investigations. The main thrust of this evolution is expected to achieve better mobility, to equip more satellites with high-quality reflectors, and to reduce system unit costs.

2.1.3.3 Future outlook.

It should be noted that many methods are still in an experimental stage and that their potentially most rewarding application, earthquake prediction, is untried. The technology will change rapidly, as is indicated by the number of new concepts under discussion.

One interesting new technology is a method that inverts the principle of satellite laser tracking. It has been introduced under the name spaceborne laser ranging and consists of a satellite that carries the laser system, with the retroreflectors on the ground. Such a system could be used to monitor crustal movement over most of the world's seismic areas and to detect possible motions prior to large earthquakes.

Space techniques provide an efficient way to improve the precision of geodetic measurements for position determination. The practical value of the extraordinary technological progress may not be immediately obvious, but what may be of interest today to geodesists alone may well tomorrow end up serving the needs of other disciplines, in particular those that are concerned with the management of the planet's resources and its environment.

2.1.3.4 Developments in the USSR. The USSR has established a worldwide high-precision operational space navigation system using the TSIKADA low-altitude satellites to serve a great variety of users; passenger and cargo seagoing ships, floating bases, off-shore oil drilling rigs, and the USSR Academy of Sciences scientific research vessels.

The adoption of this system has enhanced the economic efficiency of the maritime and fishing fleets by increasing the operational speed, extending the fishing time, reducing accidents, reducing the loss of fishing tackle, and improving the operational efficiency of fleet dispatcher control. At present, several hundred ships are making use of these services.

Future domestic navigation systems will offer higher accuracy of observations and reductions in cost of the users' equipment, thus allowing more ships to use the system.

2.1.4 Scientific Mission Technologies[2]

2.1.4.1 Study of the earth and its atmosphere. Much work has been done in the past 20 years on both national and international scales, partly United Nations supported. Programs such as the International Geophysical Year (IGY), the International Quiet Sun Year (IQSY), the Global Atmospheric Research Program (GARP), and the Monsoon Experiment (MONEX) have gathered quantities of significant data for scientific interpretation. The current situation is characterized by the following:

1. Basic understanding of the near-earth environment has been reached, and projections to planetary atmospheres and/or ionospheres and magnetospheres are possible.

2. The compatibility of data and measurements must be improved to secure global coverage, easy interpretation, and convenience of data exchange.

3. For future missions, long-term stability of the measuring equipment, high realiability, and wide dynamic ranges of the parameters to be measured will be necessary.

4. Planning of complex experiments embracing simultaneous rocket, satellite, and ground measurements should be emphasized to obtain data sets having high significance.

[2] A more complete account of the scientific nature of these subjects may be found in Chapter 1, "Current and Future State of Space Science."

5. Experiments using refined experimental techniques will be needed to study interactions between processes in the middle atmosphere, the magneto-sphere-ionosphere, and the solar wind.

In near-earth scientific research we will see a period of evolutionary development consisting of continuous refining of the measurement methods and techniques for complex interpretation of the data and the formulation of generalized theoretical relationships describing the processes involved in this environment. Extension to other planets is necessary. New experiments will require both scientific originality and higher precision.

2.1.4.2 Astronomy and astrophysics.

The possibility of avoiding atmospheric disturbances by placing telescopes in orbit is a great step forward in the astronomical and astrophysical research which has been pursued through the ages. Such space telescopes will operate in different spectral regions, ranging from infrared through the visible and ultraviolet spectra to X-ray frequencies. Observations over great distances and close examination by specialized deep space probes will complement each other. This is a truly rich field for an internationally planned long-term research program. Some very successful international missions have already been performed, e.g., the small astronomical satellite UHURU, IRAS, and the submillimeter telescope BST–IM on SALYUT 6. Much should be done in the future, necessitating strong support by the scientific community. This research is philosophically motivated, in that it concerns the basic question of the origin of the universe and gives rise to new, extremely sensitive equipment both to perform long-range searches and to facilitate *in-situ* inspection utilizing telemetry.

This field is closely connected with the study of the planets and the interplanetary matter discussed below, and such interdependence should not be neglected in setting goals for the research. Implementation should be in the frame of a long-range research program, which needs to be funded in view of both national goals and international co-operation or competition. Significant findings in this discipline which have contributed to the solution of man's problems on earth include stability of the environment, similarity of evolutionary conditions, testing of physical theories in a much broader context than that permitted by terrestrial limitations, etc.

2.1.4.3 Study of the planets and the interplanetary medium.

The past years have returned significant data on all planets of the solar system except the very outer ones—Uranus, Neptune, and Pluto. Planetary probes have explored the satellites of Mars, Jupiter, and Saturn as well. Most of these results were accomplished by the USSR and the United States, but the Federal Republic of Germany and France have also contributed; for example, the experiment "Stereo" (France–USSR) and the HELIOS probe (Federal Republic of Germany–United States). Matter has been brought back from the moon by United

States astronauts and by the robot sampler LUNA, launched by the USSR. The robot United States VIKING spacecraft have tested the soil of Mars for signs of life. Scientists are eagerly looking forward to sample returns of planetary matter from Mars or Mercury, but the cost of such projects has prohibited their implementation up to now.

The methodology of exploration generally consists of the following steps:

1. Earth-based observation.
2. Fly-by with deep-space probes using video equipment and physical sensors of different kinds; for example, to sense plasma parameters, particles, radiation, magnetic fields, etc.
3. Orbiters, especially for atmospheric and magnetospheric research, long-term observation of processes, and video mapping.
4. Landers and rovers to make measurements on the surface and inspect planetary soils for possible life and to do chemical and physical analysis—including microanalysis—of the planetary matter.

Analysis of the upper mantle and the crust of the planet and seismological investigations are valuable goals for detailed planetary studies of the future. The extension of these carefully planned experiments to comets (like Halley) and to asteroids seems obvious and should be undertaken, bearing in mind the great scientific interest of such interplanetary bodies.

Deep-space research highlights the limitations of presently available technologies. Energy for on-board power of deep-space probes cannot be delivered by solar cells beyond Jupiter; radioactive thermoelectric generators or other autonomous energy sources are required. Precision spacecraft orientation, long-range communication, high data rates, ultra-high reliability, and on-board artificial intelligence require developments on the frontiers of modern technology. Lifetimes of 2 to 15 years are required for such probes, demanding thorough testing, new concepts in hardware designs, and new solutions in electronics and sensors.

A balance between inter and outer planetary missions and the study of interplanetary matter should be established, taking into account the fact that preparation and implementation of planetary exploration spacecraft require 10 to 15 years. The importance of sustained funding in such projects is obvious.

One example of good long-term planning is the Japanese effort to fly a planetary probe as part of their national program by 1984. The Soviet Union will fly a Venus probe in collaboration with France around the same time. The United States also plans to launch an atmospheric probe to Jupiter (GALILEO) in 1985.

We conclude that a vigorously executed solar system research program will help to foster disciplinary science by extending our knowledge of the ranges and scales of phenomena of significance to man, enlarging our basic knowledge of nature and giving rise to new breakthroughs in advanced technology. Further, the "experimental art"—the technology of scientific experimentation—will receive great impetus from the pursuit of these far-reaching goals. Space research also constitutes a peaceful form of co-operation or competition among the developed nations and an avenue for co-operation with the developing nations, via multilateral or bilateral agreements for mutual benefit such as INTERCOSMOS and NASA co-operative programs in the framework of principles developed by the United Nations Committee on the Peaceful Uses of Outer Space.

2.1.5 Technology for Experiments in the Space Environment

2.1.5.1 Motivation.[3]   The common scientific goal of space research in the life sciences and materials science is to use those features of the space environment which cannot be reproduced on earth: the virtual absence of gravity, access to that part of the cosmic ray spectrum which cannot be reproduced on earth, and the availability of a virtually infinite source of near vacuum.

2.1.5.1.1 Materials Science and Space Processing.   Low-gravity experiments provide unique opportunities to expand knowledge in materials science. On earth there are well-known gravity-induced phenomena that occur on a macroscopic scale within liquids and gases. The influence of gravity in solids is of less significance; however, solids are produced from liquid or gaseous phases whose behavior can be strongly influenced by gravitational forces. The study of fluid physics in a microgravitational environment is therefore of prime importance as a part of materials science, in addition to its relevance as a study in its own right.

There are many well-known effects due to gravity in fluid systems, including buoyancy, convection, sedimentation, and segregation. One important effect is gravity-driven convection in fluids. The main problem is that convection currents are, in most cases, oscillatory or turbulent. In a microgravity environment, the relative importance of convection, diffusion, and surface forces may be changed drastically. In space, crucible-free zone melting of single crystalline materials or metals is possible without restriction by density effects or surface tension.

Studies of these aspects of materials science in a space environment, coupled with information obtained from ground-based experiments, should lead to a new understanding of the underlying mechanisms. Fluid physics, chemistry, metallurgy (including composite materials and glasses), and single crystals represent specific areas of activity.

2.1.5.1.2 Biology and Medicine in Space.   The practical goals of research on manned space missions has been the study of space biology and medicine. However, there is a need for biological and medical research in space

---

[3] For a discussion of the applications of this technology, see Chapter 1, "Current and Future State of Space Science."

which is primarily prompted by intrinsic scientific interest. In this area, the research objectives are to make use of the unique properties of the space environment to study the normal nature and properties of living organisms.

Neither the microgravity environment nor the complex spectrum of space radiation can be produced or effectively simulated in ground-based laboratories. Consequently, studies of the influence of these factors on living matter will only be possible to the extent that experimental facilities in space become available. Both factors have at least two characteristics of exceptional biological interest: They have not been encountered by living organisms throughout the entire history of their terrestrial existence and evolution, and living organisms display varying degrees of tolerance of each factor, permitting varied and systematic quantitative experiments to determine the nature and extent of their actions. For biology and medicine, the space environment therefore represents a new and powerful research tool: It makes possible experimental investigations into problem areas in which theory is in no position to make trustworthy predictions.

Our current state of knowledge derives from experiments carried out in space over the last two decades, notably those conducted during recent years in the American SKYLAB and the Soviet COSMOS Biosatellite and SALYUT series. Of the problems that have so far aroused interest, and continue to do so, many involve fundamental issues in the biosciences. Cardinal questions concern the biological significance of earth's gravity. Gravity strongly affects the behavior of the liquid and semisolid phases which comprise living matter. It always existed as it does today, and the evolution of all organisms has therefore taken place under its constant and pervasive influence.

2.1.5.2 Tools to provide a microgravity environment.

Decisive factors in performing research under near weightlessness are the duration and the quality of low-gravity conditions. Capabilities extend from drop towers and aircraft to orbiting space platforms. Typical microgravity durations are as follows:

Drop towers: up to 4 seconds.

Aircraft: 10 to 50 seconds.

Sounding rockets: 4 to 6 minutes.

Space facilities: several days to several months.

The quality of the low-gravity environment is influenced by both natural laws and physical characteristics of the system chosen. Low-g experiments have therefore to be planned and executed in view of the possible perturbing accelerations. Figure 2.1 gives a typical example of calculated levels for a low-orbit manned space system (Shuttle/SPACELAB).

2.1.5.3 Experimental facilities.

This section presents typical examples of experimental facilities that are reusable and are capable of simultaneous use for several purposes. Further technical characteristics are de-

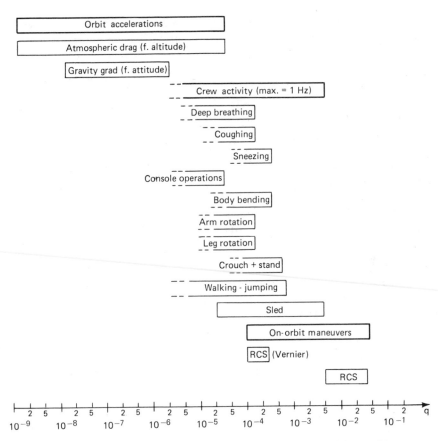

Figure 2.1 Calculated gravitational acceleration levels for low-orbit manned space systems.

termined by the fact that most of these facilities are operated in manned orbital systems, thus implying, for example, safety precautions. Other driving technological requirements in the field of materials science derive from specific test requirements; for example, furnaces which operate at temperatures up to 2,000°C demand a considerable amount of energy compared to traditional space experiments. The introduction of new measuring and observation techniques to observe and quantify phenomena under microgravity conditions also requires advanced concepts. In the field of life sciences, besides the requirement for sophisticated measuring equipment, the payload technology is influencd by the nature of the experimental subjects themselves, e.g., cells, tissues, low vertebrates, small animals, and man. Therefore, although many of the technical problems represent no major problem on earth, the application of these techniques in space requires a high level of technological efforts.

2.1.5.3.1 Human Physiology—Vestibular Research (an ESA Project). The SPACELAB-mounted SLED facility (Figure 2.2) is designed for the examination of the equilibrium system of humans and animals. It consists of a test stand which can be rectilinearly accelerated and also oscillated along the longitudinal axis of SPACELAB. Mechanical stimulations are provided in three preselected axes of the head at various programmed gravity levels up to 0.2 g. The experimental package is mounted in the head area to investigate the effects of linear accelerations and optokinetic and calorie stimulation on visuovestibular coordination and sensory integration in weightlessness. The design calls for monocular motion stimulation and visual target setting, with simultaneous recording of induced movement of the other eye.

The experiments are expected to cast light on the understanding of basic physiological phenomena with respect to the three main sensory systems (vestibular, somatosensory, and visual) which subserve static and dynamic spatial orientation as well as control posture and locomotion.

2.1.5.3.2 Human Physiology—Cardiovascular Research. Previous physiological observations in the cardiovascular system during exposure to microgravity conditions indicate that hydrostatic pressure differences are immediately eliminated. This causes a rapid displacement of body liquids (mainly blood) in the human body. This massive redistribution interferes directly with cardiac activity, reflexly influences the control of circulation, and changes the pattern of salt and water excretion by the kidneys. The mechanisms underlying most of the observed events are, however, poorly understood. Although ground-based simulation models (e.g., long-term bed rest, water immersion, and lower body negative pressure exposure) must be fully exploited, essential aspects require sequential measurements with sophisticated techniques in the space environment.

The NASA lower body negative pressure device (LBNB) (Figure 2.3) has been used primarily for studying these physiological events. It was also used at a later stage for conditioning the cardiovascular system during prolonged exposure to weightlessness, leading to the development of mobile devices such as the Soviet "CHABLIS" suit.

The special requirements for in-space monitoring and recording have already prompted rapid progress in this area, but because of the relative expense of investigations in space, the importance of well-conceived experimentation using new and developed technological aids cannot be overemphasized.

2.1.5.3.3 Cell and Developmental Biology. The role of gravity in individual cells, on lower forms of life, and in fertilization and embryonic development has great biological significance. The modern view of adaptation is that the environmental forces impose certain "problems" that organisms have to "solve," and that evolution by means of natural selection is the mechanism for solving these problems. In this context, it is poorly understood how cells and organisms sense gravitational force fields and what effect gravity has on growth

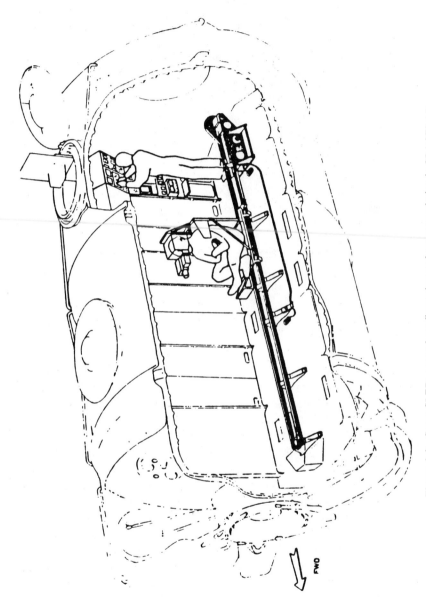

Figure 2.2 Space SLED: general configuration and accommodation in SPACELAB.

FWD

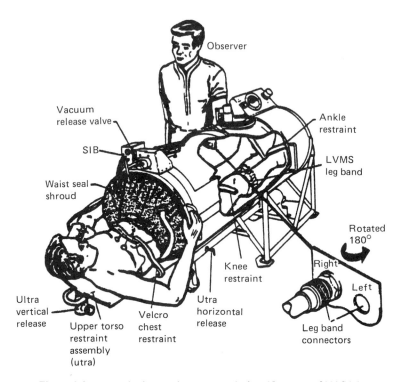

Figure 2.3 Lower body negative pressure device. (Courtesy of NASA.)

and development. The use of the weightless state for unmasking the factors that operate at normal gravity is of decisive importance in these studies.

In order to facilitate these fundamental investigations, concerning mainly the field of cell and molecular biology, plant physiology, and radio-biological aspects, a multi-user experimental facility called BIORACK (Figure 2.4) has been proposed for development by ESA. This facility includes life support, environmental support, sample preservation, storage and examination facilities, and 1-g reference centrifuges for on-board control experiments.

The results are expected to demonstrate that new scientific knowledge can be gained by establishing a correlation between the removal of gravity and the observed effects in order to identify the actual point in the biological system where gravity acts and to quantify the effects.

The BIORACK facility is designed for, and expected to fly on, several missions in low earth orbit on SPACELAB.

A typical example of space flight hardware in this field is the Cytos experiment, developed by the Centre national d'études spatiales (CNES, France), which was exposed to space environmental conditions on the SALYUT–6 Soviet space station (see Figure 2.5).

The purpose of the Cytos experiment was to investigate the possible effect

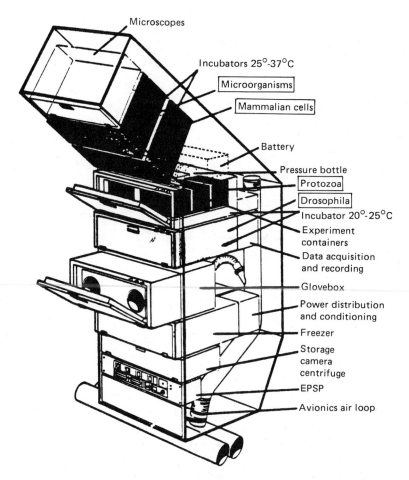

Figure 2.4 BIORACK (European Space Agency), a multi-user experiment for use on SPACELAB.

of microgravity on cell proliferation kinetics. In order to obtain these results, the experiment required a complex facility comprising incubators with highly constant temperatures ($25° \pm 0.1°C$), culture media bags containing fixative devices, automatic activation and de-activation mechanisms, and continuous temperature recording and transmitting to ground.

2.1.5.3.4 Materials Science—Fluid Physics. The science of fluids in a low or microgravitational environment is an extremely broad field, of which only a few areas have been investigated and many important areas are still being discovered. Although some progress has been made in this direction in recent years, much effort still needs to be devoted to the systematic identification of all areas of possible activity in the field of microgravitational fluid dynamics and to

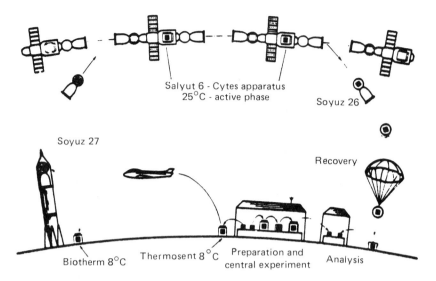

Figure 2.5 The Cytos experiment (CNES, France). After joining with SALYUT-6, the culture box was transferred into the Cytos apparatus and maintained at a constant temperature.

the subsequent critical assessment of the nature and type of their relevance to basic knowledge.

There are three fundamental motivations for conducting fluid science studies in space:

1. Interest in the subject *per se*. Many fluid phenomena of basic scientific interest could presumably be reproduced, in the absence of gravity, either more neatly or under conditions where the values of many controlling parameters are far removed from those achievable on earth, thereby greatly improving our knowledge. Interface science is particularly relevant, having many potential ways of taking advantage of the gravity-free condition.

2. Fluid physics is relevant to materials science because of its controlling effect on the quality of the final product. This is probably the source of strongest motivation.

3. Long-duration exposure to microgravitational environments permits some phenomena, especially in surface science, to be conducted on a larger and hence more visible scale. The application of fluid physics to fluid systems already used in space is also important. Heat pipes, phase-change thermal capacitors, and superfluid helium coolers are only three examples arising from the need for spacecraft thermal control. Additional examples are flammability under microgravity and the behavior of biological fluids. Such studies will undoubtedly lead to better designs and improved understanding.

It must be emphasized that it is not possible to foresee the results of many space experiments in which some characteristic parameter differs by several orders of magnitude from those involved in the terrestrial laboratory, and that the first aim would be to gain more knowledge of what will happen to fluid systems in space. In order to obtain these results, special multipurpose facilities are being developed. Two examples are the fluid physics module (FPM), developed in Italy and to be flown on the first SPACELAB mission, and the fluid experiment system (FES), under development by NASA and to be flown during future SPACELAB flights.

Fluid physics module. A precision apparatus is under development for the investigation of fundamental fluid physics problems, such as natural and enhanced column stability, convection, drop coalescence, liquid/vapor interfacial forces, etc.

The FPM permits the establishment of a liquid bridge between two parallel coaxial discs and the application of various types of perturbations (rotation vibration, misalignment, heating, electrical field) to the liquid being studied. Results of the experiment are recorded photographically.

The FPM (Figure 2.6) consists of a cylindrical structure fitted with two piston discs. The fluid to be studied can be injected through one of the discs. This disc can be moved axially, thereby enabling the length of the floating zone to vary. The discs can be rotated separately, at the same or at different speeds, and in either direction. One disc can be vibrated axially at different frequencies and with different amplitudes. The form and diameter of the end plates can be modified according to the experiment objectives. Special containers can be mounted and rotated with the help of the end plates. Temperature gradients and a difference in electric potential can be established between the two discs. The test chamber is airtight and will accept different fluids, with and without tracers. An air circulation and liquid recovery system is provided to clean out the test chamber in case the floating zone is broken and to control temperature and moisture inside the test chamber.

Phenomena developing in the test chamber (shape and size of the floating zone, motions inside the floating zone) can be observed and filmed at right angles to the lighted meridian plane for recording shape and speed in the meridian plane along the axis of the rotation and for recording motions at right angles to the longitudinal axis. Operation of the FPM requires the attention of the payload specialist. The main characteristics of the equipment are as follows:

Liquid volume: from 0 to 1,300 cc (step: 0.8 cc).

End-plate diameter: up to 100 mm (different shapes, sizes, and materials).

Rotation speed of the plates: from ±5 to ±99.9 rpm (step: 0.1 rpm); separate operation possible.

Oscillations: 0.1 to 1 Hz (step: 0.01 Hz); 0 to 0.5 mm amplitude.

Liquid injection speed: from 0.01 to 20 cc/s (step: 0.1 cc/s).

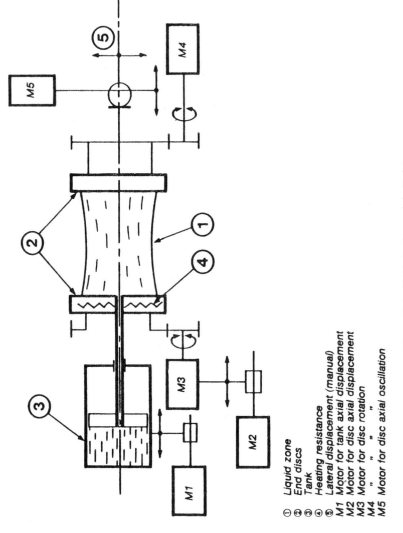

Figure 2.6 Fluid physics module (Italy).

① Liquid zone
② End discs
③ Tank
④ Heating resistance
⑤ Lateral displacement (manual)
M1 Motor for tank axial displacement
M2 Motor for disc axial displacement
M3 Motor for disc axial rotation
M4 " " " "
M5 Motor for disc axial oscillation

Thermal capability: ambient to 60°C (feeding end plate).

Electrical potential: ± 100 V to DC at one end plate.

Photographic recording: Two 16-mm cine cameras.

Electric power required: 360 W maximum.

Fluid experiment system. FES provides the capability for conducting fluid phenomena investigations using optical observation techniques in the micro-gravity environment of orbital flight. Its data capability consists of holographic photographs, from which flow patterns can be visualized along with full in-terferometric information, a separate Schlieren (flow visualization) video system, several small temperature probes which can be located within the fluid, triaxial linear accelerometers, and triaxial rate-gyros. Its holographic com-ponents (see Figure 2.7) have the following performance characteristics:

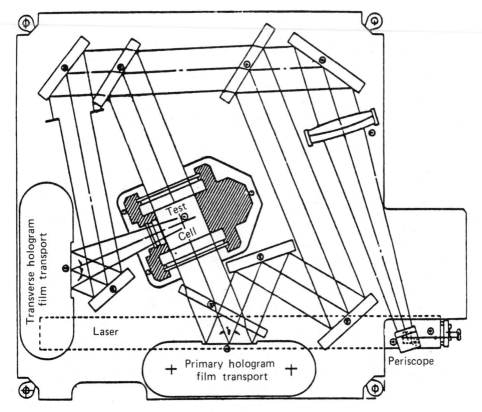

Figure 2.7   Holographic components of the fluid experiment system. (Courtesy of NASA.)

Optics for laser holography.

Predetermined time–temperature profiles generated with control rates up to 1°C/min from ambient to 200°C.

Gradient plates within the cell controlled to 0.01°C within a 50°C range.

Thermal gradients throughout the test cell not exceeding 0.02°C.

Uniform illumination (2 percent) of image field.

Resolution of optics better than 100 lines/mm at film plane.

Transverse microscopy having 25X to 50X magnification.

Typical experiments might be as follows:

Fluid convection.

Phase transition.

Solution growth of crystals.

Surface behavior.

Bubble behavior.

Chemical reaction and precipitation kinetics.

Electrochemical deposition studies.

2.1.5.3.5 Materials Science—Crystal Growth. Single crystal growth investigations under near-zero gravity conditions require specialized equipment (furnaces). The main aim in designing the apparatus was to facilitate experiments with a wide range of materials used in manufacturing semiconductor devices. Furthermore, requirements had to be met for precision temperature measurement and maintenance, as well as particular requirements for on-board apparatus construction such as minimum mass, dimensions, and power consumption; high reliability and easy maintenance; and safety conditions for carrying out the experiments.

Semiconductor materials can be produced using the "Crystal" apparatus on board the USSR space station SALYUT 6 by four techniques: directional crystallization, sublimation, gas-carrier transfer, traveling melt solvent.

The electrically heated furnace allows temperature conditions to be maintained as follows: hot zone temperature as high as 200° to 1,200°C, temperature gradient of 150° to 200°C/cm, steady-state gradient when the furnace temperature changes, and furnace body temperature during the process less than 53°C.

A computer sets the operating temperature, the thermal soak time, and the zone cooling parameters, (rate and duration) to maintain the pre-set temperature with an accuracy of 1° to 3°C; performs the operation time program; controls the motion motor; measures the temperature of the melting zone, the cold thermojunction, and the furnace body; indexes these temperatures, processing stages, and parameters specifying the apparatus efficiency; and generates signals for remote control of the process parameters.

The mirror heating facility (Figure 2.8) is particularly suitable for investigating crystal growth using the zone melt or traveling solvent methods. It can be used, for example, to perform the following experiments:

Zone crystallization of silicon.

Traveling solvent growth of cadmium telluride.

Traveling heater method of III-V compounds (e.g., indium-antimony).

Crystallization of silicon spheres.

Unidirectional solidification of cast iron.

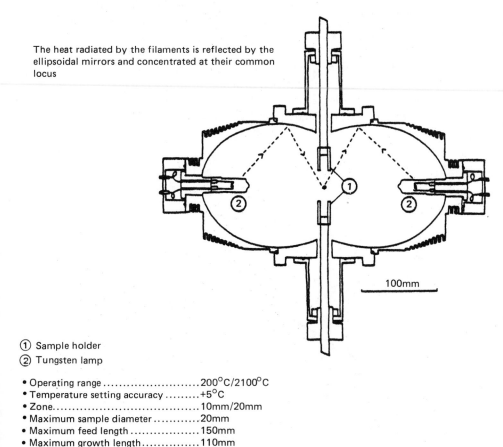

The heat radiated by the filaments is reflected by the ellipsoidal mirrors and concentrated at their common locus

100mm

① Sample holder
② Tungsten lamp

• Operating range .... 200°C/2100°C
• Temperature setting accuracy .... +5°C
• Zone 10mm/20mm
• Maximum sample diameter 20mm
• Maximum feed length 150mm
• Maximum growth length 110mm
• Turn velocity .... 0.1 to 10 rpm
• Feed velocity .... 10 to 10mm/min

Figure 2.8   The mirror heating facility for crystal growth in space.

The facility consists of an optically heated zone furnace and ancillary devices, e.g., a pulling and turning device. The key elements of the zone furnace are two ellipsoidal mirrors with coincident optical axes and a common focus (see Figure 2.8). Halogen lamps acting as heat sources are located in the other two foci. The actual space for samples (the melt zone) is located at the common focus. Samples can be inserted here by means of two holders perpendicular to the optical axis. Two viewing ports are provided for optical monitoring and pyrometric temperature measurement or control.

2.1.5.3.6 Materials Science—Metallurgy. The heating facility for low-temperature gradients (Figure 2.9) is a multipurpose facility for different types of experiments such as crystal growth, unidirectional solidification of eutectics, etc. It can be used, for example, for experiments on the unidirectional solidification of aluminum-zinc (Al-Zn), aluminum-aluminum/copper (Al-Al$_2$Cu), silver-germanium (Ag-Ge), and indium-antimony/nickel-antimony (InSb-NiSb); on the growth of lead telluride; and on thermodiffusion in tin alloys.

The furnace allows for parallel injection of three cartridges that can be heated with three independently controllable heating elements, so that a variety of temperature profiles (also isothermal) can be achieved. Thermal insulation is provided by axial heat shields, a low-conductivity radiation shield, multifoil insulation, and an outer protective shield. Vacuum and noble-gas supply provi-

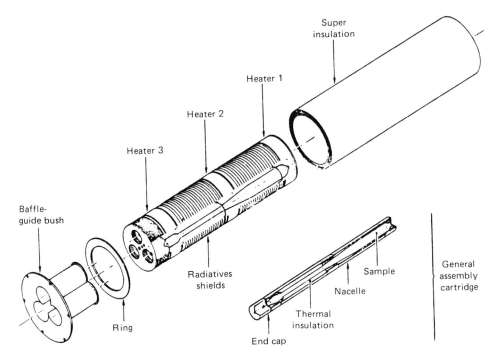

Figure 2.9 Internal view of the multipurpose low-temperature gradient furnace.

sions are part of the facility. Quenching can be obtained by purging the furnace with helium gas. Its technical characteristics are as follows:

Volume: sufficient for 3 cartridges of 275-mm length, 25-mm diameter.

Maximum temperature: 1,200°C on the cartridge.

Gradient: up to 100°C/cm.

Maximum heating rate: 1.5 hours to reach 1,200°C.

Cooling rate: from 1,200° to 50°C achievable in 4 hours (fastest rate) or up to 20 hours (slowest rate).

2.1.5.4 Outlook.   Experimental research in the microgravity environment is a rapidly growing space discipline which will lead to advanced scientific knowledge and new applications. The research will develop through utilization of existing and planned manned space systems. There is also a clear trend to utilize fully automated orbital platforms which will provide microgravity conditions, energy, power, control mechanisms, etc., and will guarantee the recovery of samples and test objects. Examples of fully automated systems are the biosatellites of the COSMOS series, the Space Shuttle long-duration experiment facility, and the projected materials experiment carrier (NASA) and MINOS (CNES).

2.1.6 Emerging Technologies

2.1.6.1 Space solar power systems

2.1.6.1.1 Background.   This section deals mainly with the solar power satellite (SPS) concept. An SPS would be a primary electrical power source generating electrical power from solar energy in geosynchronous orbit, transmitting the power to earth via focused microwave beams (in the most commonly studied configuration), and collecting and converting the beams into useful electricity on the earth's surface. The concept was suggested in 1968 by Dr. Peter Glaser and preliminary feasibility studies were conducted from 1973 to 1977. In 1977, the United States launched a three-year Satellite Power System Concept Development and Evaluation Program managed by the Department of Energy, with major contributions from NASA, to generate information on all aspects of the program: systems definition; health, safety and environmental factors; economics; international and societal issues. As of this writing, no decision has been made by the United States Administration on the continuation of the program.

ESA has sponsored a study on the usefulness of this concept for Europe; other studies done in France, the Federal Republic of Germany, and Czechoslovakia have been presented to Congresses of the International Astronautical Federation (IAF) in 1978, 1979, and 1980, some of which include discussion of the possibilities for utilizing nonterrestrial materials in SPS fabrication.

A number of other energy system concepts have been suggested that involve the use of satellites for terrestrial power:

1. Orbiting nuclear reactor power systems, with microwave transmission of power to earth. This concept has been evaluated by the United States to a limited extent. It is quite complex, and the nuclear safety improvement resulting from space waste disposal is counterbalanced by the high environmental risk at launch.

2. Solar reflecting satellites (mirrors) in earth orbit. Solar energy would be reflected to earth to augment ground-based solar power plants, allowing night operation. These satellites have also been suggested for street illumination, an application that has yet to be assessed.

3. The power relay satellite. Power systems on the earth's surface or in low orbit would transmit power by microwave to geostationary satellites, which then relay (or reflect) the microwave energy to ground stations. This system would be competitive with cryogenic power transmission, whose potential also has yet to be assessed. As part of a space system, it represents a different trade-off between launch cost and system complexity.

4. Laser power transmission. This concept utilizes lasers instead of microwaves to transmit power from the satellite. It is attractive because it allows small-diameter energy beams, thereby using less ground area and allowing lower power levels per unit. However, this concept requires considerable technological development and also suffers from lower efficiency than is estimated for microwave power transmission. Atmospheric attenuation may constitute a further problem. In a multiple-satellite system (some in low orbit, some in geostationary), laser beams could be used for power transmission from low to geostationary orbit, provided the necessary technological development can be accomplished.

5. The use of materials derived from the moon and perhaps subsequently from asteroids. Nonterrestrial materials have also been suggested for construction of SPS in order to reduce both the cost and the terrestrial environmental impact of space transportation. However, this approach would require development of moon-based mining, manufacturing, and launch facilities, as well as permanent moon occupation. The cost of the research, development, and implementation of such an approach would be very large, and it should be undertaken only if a decision is made to produce many SPSs.

Although some of these other energy system concepts offer interesting possibilities, their application appears more remote than those of the basic concept studied by the United States, which is substantially identical to the original Glaser concept.

2.1.6.1.2 Description of the Solar Power Satellite Concept.   For the purpose of the United States DOE–NASA study, a "reference" SPS was designed, so that the various studies could be conducted on a common basis. Although this system may be subject to major changes, it represents an integrated system design fulfilling the SPS mission and requiring no technological "breakthroughs."

SPS is intended to be a baseload electrical generating plant, providing utilities with the same service as nuclear or fossil-fuel-fired power plants. The reference system would generate 5-gigawatt (5 GW)[4] power for a conventional power grid. The satellite is expected to have a mass between 35,000 and 50,000 tons. It consists of the following elements:

1. A rectangular planar structure built from a graphite-composite material. This structure supports the solar array. The total area of the satellite platform is about 55 km^2 (10.5 km north to south and 5.2 km east to west). Because of the reduced loads in the microgravity space environment, the structure can consist of very thin frames.

2. The solar array, which converts sunlight into direct-current electricity. Two conversion options are considered in the reference system: silicon cells in a flat array with no concentration, and gallium-aluminum-arsenide thin-film cells with rows of V-shaped reflectors providing a concentration ratio of 2. Efficiency of conversion is expected to be 17 percent for silicon and 20 percent for gallium-aluminum-arsenide. A power distribution system collects the DC electricity and controls the power fed to the transmitting system.

3. An orbit and attitude control system, using small rockets as thrusters, orients, and platform so that its longest axis stays in the north–south direction, i.e., perpendicular to the orbital plane, and keeps the solar array facing the sun. It also adjusts the orbit to keep the satellite in its geostationary orbit location.

4. A microwave power transmitting system. This system is mounted on a plane structure 1 km in diameter suspended in a yoke, attached to the main structure by a mechanical turntable incorporating a rotary joint for power transmission so that the antenna can be continuously pointed toward the earth. The antenna is composed of about 7,000 subarrays, each fed by a microwave klystron transmitter and radiating through a few hundred waveguide slots. The subarrays are phased together to form a single coherent beam focused at the center of the receiving antenna. Ninety percent of the beam energy reaches earth within a circle of 5-km radius perpendicular to the boresight. The microwave frequency is 2.45 GHz, in a frequency band reserved for industrial applications.

---

[4] A gigawatt is 1 billion watts, or 1 million kilowatts.

The ground segment of the system consists of the ground receiving and rectifying antenna (rectenna) and the utility interface equipment (DC to AC inverters, breakers, etc.). The rectenna is composed of a very large array of dipoles, each feeding a rectifier diode, wired together. The rectenna has an elliptical shape (typically 13 km N–S by 10 km E–W). In the reference design, the microwave power flux at the center is 23 mW/cm² (a quarter of sunlight flux) and less than 1 mW/cm² at the rectenna edge. An exclusion area around the rectenna can be provided; in the DOE/NASA reference design the maximum power flux at the edge of this exclusion area in 0.1 mW/cm².

The satellite would be constructed in geostationary orbit in about six months. A crew of about 600 would be needed: 500 in geostationary orbit, 100 in low orbit. The proposed transportation system consists of four major subsystems:

1. The heavy lift launch vehicle (HLLV) will carry heavy loads from ground to low orbit. It is a two-stage, vertical-launch, winged, horizontal landing, reusable vehicle with 424,000 kg payload to low orbit.

2. The cargo orbit transfer vehicle is an independent reusable electric-engine-powered vehicle which transports cargo from the HLLV delivery site in low orbit to the assembly point in geostationary orbit.

3. The personnel launch vehicle is a modified version of the present Space Shuttle, able to carry 75 passengers, which will be used to support construction and to transport crew from ground to low orbit.

4. The personnel orbit transfer vehicle is a two-stage reusable chemical-propellant vehicle providing fast transfer of personnel and critical supplies between low and geostationary orbits.

2.1.6.1.3 SPS Technology Required.   Although scientifically feasible, the technologies are not yet fully developed, and a large R and D effort will be required. The main technological requirements are the following:

1. Low-cost, high-efficiency, space-qualified photocells. It is possible that some of the effort devoted to development of terrestrial photocells will benefit SPS applications, but the requirements are sufficiently different that a substantial development of space cells will be needed.

2. Power transmission by radiation (microwave, laser, etc.). This is a specific requirement of SPS but could benefit (especially in laser technology) from both military and nuclear fusion technology.

3. Large space structures. These may be developed before the end of the century for other uses in space, e.g., large antennas.

4. Earth orbital stations. Permanent occupancy of space is also possible before the year 2000.

5. Heavy-load transportation system. The heavy-launch vehicles need specific development for SPS, since other planned missions in the next 20 years seem compatible with the Space Shuttles or other vehicles. At present, the highest priority in this field is to attain reliable and economic utilization of the Shuttle before developing larger reusable launch vehicles.

It is evident that general development of space techology and of ground solar energy technology will help to meet some of the technological requirements of the SPS. However, power transmission and space transportation of heavy loads require specific development starting in the next five years if the first power satellite prototype is to be deployed by the year 2000.

2.1.6.1.4 Impact on Earth's Environment. *Microwave biological effects.* The power flux at the rectenna edge is to be less than 1 mW/cm², which is one-tenth of the value allowed in the United States for ocupational exposure, and is recommended at 0.1 mW/cm² at the edge of the exclusion area surrounding the rectenna. However, there is considerable controversy about what value is safe; for example, the recommended USSR limit is only 0.01 mW/cm², ten times lower. A larger exclusion zone, increasing required land area, may become necessary. Flying animals (birds, bees, etc.) cannot be excluded from the beam, but preliminary experiments on bees in the United States suggest that there is no discernible effect on such creatures. Nevertheless, studies of microwave biological effects have to be continued, especially on vertebrates, to estimate the long-term hazard for man. Accidental mispointing of the microwave beam cannot occur, because the beam focus would then be lost (it depends on a "pilot" beam from the ground). If this were to happen, the transmitting elements in the satellite would each point in a different direction and the total energy reaching the earth would be so low as to be harmless.

Biological effects on space workers. Experience in working in space is limited to a few missions of four to six months, all in low earth orbit. Biological effects are severe but fortunately disappear after return to earth. Further studies are necessary to evaluate not only the consequences of space work on assembly and maintenance crew health, but also on space mission duration and crew productivity, which may have a big impact on satellite construction cost, and on radiation exposure of workers in geostationary orbit, well above the protection from cosmic radiation and the solar wind offered by the earth's van Allen belts.

Launch vehicle pollution. Hydrogen-burning rockets produce large quantities of steam and some nitrogen oxides during their flight. Effects on climate, ozone level, and on the ionosphere have been extensively studied in connection with the United States Space Shuttle, but they are not well-known for the much larger HLLV exhaust.

Microwave effects on the ionosphere. Earth is surrounded by a high-altitude layer of gas ionized by solar radiation. This layer, the ionosphere, plays an important role in communications. There is concern that too powerful a microwave beam could disturb the ionosphere and thereby communications, not only locally but also on a global basis, due to the stimulation of free-electron mobility in the ionosphere. For this reason, the maximum flux density of the proposed beam has been limited to 23 mW/cm², which is assumed to be safe. (Recent experiments with big radiotelescopes have indicated that the "safe" level may be considerably higher.) Another aspect of communications impact is interference with radar and communications equipment. The official bandwidth assigned to SPS is 100 MHz, but it is so powerful a transmitter that it would produce radio-frequency noise all over the spectrum, as would also nonlinear interaction of the beam with the ionosphere. The consequences of these interferences may be severe, but they could be mitigated by the evolution of communications equipment in the next century.

2.1.6.1.5 Economic Aspects.   An SPS development project could be one of the costliest ever planned in the history of mankind. The most recent figures published by the United States Department of Energy,[5] in billions of 1977 United States dollars, are as follows:

Early studies	$ 8.5
SPS development, including demonstrations	23.
Investment needed to build the first "production" satellite	57.5
First production satellite and rectenna	13.5
Average cost per system (5 GW)	11.5

The investment includes the cost of developing the space transportation system (half the total investment cost), which may have a more general use, and other space technologies (e.g., large structural design and deployment), which are likely to be developed for other purposes (e.g., communications antennas).

2.1.6.1.6 SPS Potential Applications.   The demand for electricity will continue to grow at a fast pace during the next 50 years. The requirements will be met mainly by large electrical power plants supplying electricity to major utility grids. Many reasons (such as minimal technical or economic size, problems of transmission and pollution control) suggest a push toward increasing amount of power produced in these "energy parks": 3 to 5 GW may be typical. These large central stations could be nuclear breeder reactors, advanced coal-fired thermal plants, or fusion reactors. All have their disadvantages:

[5] R. Piland, "Reference System Characterization and Cost Overview," Proceedings of SPS Programme Review, USDOE, July 1980.

1. The environmental questions raised by the development of nuclear energy and particularly by the breeder reactor are well-known (risk of major accident, radioactive wastes, diversion of plutonium, etc.).

2. Fusion reactor feasibility is not yet established, costs have yet to be estimated, and fusion power plants will not be free from radioactive risks.

3. The fossil-fuel thermal plants produce large amounts of carbon dioxide, which may lead to a dangerous greenhouse effect, as well as more conventional air pollution, acid rain, and major cost and environmental impacts due to transportation requirements for coal.

4. Fusion, fission, and conventional thermal plants reject about 150 percent more heat to the environment than they produce as useful electricity.

Ground-based solar power plants, although admirably suited for decentralized applications, are not suitable for centralized production, primarily due to their need for energy storage. The SPS may be a promising way to produce electricity from the sun in populated and industrialized countries where powerful utility networks have to be fed.

According to the report of the World Energy Conference (Istanbul, 1976), the world electricity demand in the year 2000 will reach about 22,000 terawatt-hours and in the year 2025 about 43,000 terawatt-hours. The additional demand from 2000 to 2025 amounts to 21,000 terawatt-hours, corresponding to 3,000 to 4,000 GW of additional generating capacity.

Accounting for the demand for both peak-load power plants and decentralized power plants, the number of large (3 GW) power plants to be built between 2000 and 2025 to meet additional electricity demand is close to 500. Replacement of old power plants, built before 1980, would increase this figure by 25 percent.

However, SPS could not be used in all cases because of the limitations; for example, at higher latitudes the rectenna area becomes too large, and it is often difficult to find suitable sites for rectennas in densely populated or mountainous countries. These constraints may be mitigated by the development of new concepts allowing smaller power receiving areas and by siting of rectennas in offshore locations adjoining large power demand centers such as seaport cities.

2.1.6.1.7 SPS Military and Political Issues.   SPS cannot be used as a weapon. The microwave beam is not lethal and cannot be focused outside the rectennas. Because these huge satellites are very sensitive to military aggression, implementation of such a system implies hope and confidence that continuous peace in the world will be achieved.

In summary, SPS could produce renewable electricity from the sun for heavy electric-power load centers. A major development program would be required as well as studies of environment impact. Due to the worldwide interest in such systems, their global environmental impact, and their military vulnerability, their implementation seems more suited to a framework based on international co-operation, on a worldwide basis, rather than as a national project.

2.1.6.2 Extraction and use of nonterrestrial materials

2.1.6.2.1 Background. The prospects for utilization of nonterrestrial materials were first seriously considered by Gerard O'Neill in his studies of space colonization. A more important and nearer-term rationale for considering nonterrestrial materials is their potential use in fabricating satellite power systems.

The necessary extraction and processing technology for lunar or asteroidal materials has been discussed at conferences in the United States since 1974. Extensive studies were performed at the Lunar and Planetary Institute in Houston, Texas (United States of America), and elsewhere, and have been published in some detail. Current efforts, mostly in the United States, are aimed at developing laboratory-scale models of the extraction and chemical processing methods suggested by these studies.

The bulk of this effort to date has been devoted to lunar materials, to which the extensive microgravity materials processing experiments being carried out in the SALYUT space laboratory do not apply because of the finite gravity of the moon. There is, however, considerable interest in the prospects for using asteroidal materials and also in earth-orbital processing of both lunar and asteroidal materials, activities that would benefit considerably from results of the SALYUT experiments.

2.1.6.2.2 Lunar Materials. *Resources.* Lunar soil and rock samples retrieved by the United States APOLLO astronauts revealed a range of elements. Major elements, each constituting more than 1 percent of the lunar surface, are oxygen, silicon, aluminum, calcium, iron, magnesium, and titanium. Minor elements (0.1 to 1 percent) include chromium, manganese, sodium, potassium, sulfur, and phosphorus. Small but significant "trace" quantities of hydrogen, helium, carbon, and nitrogen, key elements for space manufacturing, are also present. The useful products which could be derived from these sources are structural materials (metals, reinforced metals, glasses, ceramics, and cements), thermal materials (refractories, heat-transfer fluids, insulation), abrasives and cutting tools, electrical materials (conductors, electrodes, magnetic materials, and electrical insulation), fibrous materials, plastics and elastomers, sealants, adhesives, coatings, lubricants, industrial chemicals (detergents, cleansers, solvents, acids, and bases), and biosupport materials. Key industrial elements *not* available on the moon in adequate quantities are carbon, nickel, molybdenum, tungsten, vanadium, and niobium, all useful for various steel alloys, and zinc, needed for certain aluminum alloys. Titanium alloys also require molybdenum, and any large requirements for water will necessitate bringing hydrogen (fortunately the lightest element) from earth.

Extraction and processing. The principal processes identified for extracting key minerals from the lunar soil are direct electrolysis of molten silicates, magnetic recovery of iron, anhydrous thermic reduction, carbochlorination, and acidic or basic leaching. The governing principles for selection of any of these processes is that the final cost of the processed material at the site of use (e.g., in geostationary orbit) be less than the cost of earth processing and delivery of the same material to orbit from the earth's surface. Thus, the fraction of terrestrial materials (e.g., catalysts, precision equipment, etc.) must be minimized. The energy source would most probably be solar, but nuclear power remains a potential option. The cost of mining the basic ores on the moon, including the costs of electric power and all personnel and equipment hauled there from earth, has been estimated at 12 to 37 United States dollars (1978) per ton.

Delivery to space-based factory. Because there is no lunar atmosphere, materials can best be launched from the moon using an electromagnetic accelerator (basically a linear electric motor, such as has been used on experimental railroads in Japan, France, the Federal Republic of Germany, and the United States of America). This device, called a "mass driver" when adapted for vacuum environments, is now being evaluated and developed in the United States. In the mass driver, a payload (e.g., processed lunar material) is placed in a container equipped with a superconducting coil in which a high electric current flows. A series of magnetic-field impulses is then applied by a series of electromagnets placed along a long linear track, powered by either a solar or a nuclear electric generator. Each impulse is synchronized to match the container's trajectory, just as is done in the particle accelerators used for high-energy physics research. When the container is slowed and recirculated by the same electromagnetic process, the payload flies off into space. Projected guidance and control techniques appear to be adequate to deliver the payload to any specified orbital target within at most a few tens of meters. Experimental mass-driver models have operated successfully, and it is likely that this technique for delivering lunar materials to space-based factories could be developed in the same time frame as the satellite power system, should it be decided to proceed.

2.1.6.2.3 Asteroidal Materials.

Carbon is one of the key materials not present in sufficient quantities on the moon. Based on analyses of meteorites that have fallen to earth, it appears that many earth-approaching asteroids (called the Apollo-Amor class of asteroids) contain carbonaceous chondritic materials. Further, the asteroids also contain iron, nickel, germanium, gallium, precious metals, and other valuable minerals. Access to these materials requires extremely small transportation energies because of the negligible gravity of asteroids. It has been proposed that Apollo-Amor asteroids be sought by astronomers and that they be retrieved to earth orbit using mass drivers as propulsion devices, the propellant for the mass driver consisting simply of chunks of the asteroid itself. Asteroids represent an almost limitless source of raw materials in the future, although the time frame for their development as such is probably at least several decades behind that of satellite power systems and lunar material utilization.

2.1.6.3 Space settlements

2.1.6.3.1 Background. The concept of settlements in space, as contrasted with planetary or lunar colonies, was first suggested by Konstantin Tsiolkovsky in the early 1900s. Aside from several conceptual suggestions, little was done with the idea until about ten years ago when Professor Gerard O'Neill began to develop a detailed technical scenario for such settlements. A series of publications by O'Neill and others, both in the technical community and in the popular press, brought the concept to the point of serious consideration by several governments, and a number of small research studies and "workshop" courses were funded to further develop the technical, economic, and social aspects of space settlements.

O'Neill first suggested space habitats as a means both for indulging the pioneering spirit of mankind and as a potential long-term solution to terrestrial limits to growth. As the SPS effort began to grow, however, it became obvious that the construction of power satellites in orbit, using nonterrestrial materials, could provide employment for the inhabitants of a space settlement, and since 1975 much (although not all) of the rationale for the concept of space settlements has been based on their prospective role in solar-power satellite construction.

2.1.6.3.2 Description. A variety of space habitat configurations have been suggested, ranging from a small 200-person "construction shack" to enormous hollow cylinders 25 km long and 7 km in diameter housing several million people. The basic concept is that of a rotating structural shell (e.g., a cylinder or sphere) using nonterrestrial mass (such as lunar rocks and soil) as "earth" on the inner surface of the shell. The entire shell rotates about its axis, aligned in the sunward direction, to provide simulated gravity at the "earth" surface; on the axis of rotation the gravity would be negligible. Hence, living quarters would be located on the inside of the high-gravity shell, workshops and assembly factories near the axis of rotation. External solar power plants would provide power, and hinged sun-facing mirrors would open and close slowly to simulate earth-normal daylight and nightfall. Agricultural shells outside the main shell would grow crops in fulltime sunlight.

Settlements could be located in any stable orbit devoid of regular solar eclipses. O'Neill's original suggestion was to utilize orbits around the stable Lagrangian libration points in the moon's orbit around the earth, called L-4 and L-5. Considerations of launch trajectories for lunar materials, however, suggest other stable orbits in the earth–moon system. Each settlement would be essentially self-supporting, using lunar and eventually asteroid materials for internal construction, for manufacture of power satellites, and for replicating new settlements. Key materials, complex instruments and equipment, and people would still need to be transported from earth, but this would constitute only a tiny fraction of settlement requirements.

2.1.6.3.3 Applicable Technologies.   Construction of space settlements depends directly on the development of the lunar material extraction, processing, and transport capabilities described earlier. The mass driver required for lunar material transport and later for asteroid retrieval is perhaps the key "pacing technology" item, with demonstration of lunar material processing systems a close second. Other key factors are those associated with the launching, construction, and maintenance of any large structure in space, for example, the relative roles of people and teleoperator equipment (robots), the logistics of material supply and assembly, protection against cosmic and van Allen belt radiation, dynamic stability and thermal control of large structures subject to frequent eclipsing, etc. In the more distant future, the psychological, physiological, and sociological requirements involved with long-term habitation of space settlements will need to be established.

2.1.6.3.4 Future Prospects.   Current studies on mass-driver and lunar material processing technology are funded at very low levels and only in the United States. Many unfunded independent research studies are also being undertaken, primarily by universities and individuals. It is not expected that substantial efforts in this direction would be financed by any government or industry unless and until some evidence has appeared to indicate the prospects for economic return, for example, a decision to begin production of solar power satellites.

2.2 SUPPORT TECHNOLOGIES

2.2.1 Spacecraft and Space Platform Technologies

2.2.1.1 Structures

2.2.1.1.1 Introduction.   As we progress to larger and larger systems in space, the demand for economy becomes paramount. If large space systems using large space structures are to be economically competitive with terrestrial systems, the costs of the systems and the costs of transporting them to low-earth or geosynchronous orbits must be minimized. Specified performance at minimum cost will dictate the way in which large space systems are designed and made.

The cost of large space structures is influenced by several factors. Transportation costs, for example, can be minimized by reducing the total weight of the structure as well as by achieving high packaging density for the trip into orbit. Fabrication costs are reduced if the structure is made of simple elements and joints having a high degree of commonality; the ease with which the structure can be assembled is a major factor in determining assembly cost. It follows that the key to controlling both the performance and the cost of large space structures, and hence of large space systems, is the selection of the structural material to be used.

The advent of the United States Space Shuttle as an operational system will provide a unique new capability to deliver materials and support the construction of large systems in space. Future space antennas, platforms, and radiometers will no longer be limited in size to deployable configurations that can be packaged to fit unmanned launch vehicles. New space construction techniques, such as deployment and assembly of terrestrially made structural elements and in-orbit fabrication and assembly, and alternative materials for construction in space are being developed. A discussion of the capabilities and status of space construction techniques and materials is presented below.

2.2.1.1.2 Structural Concepts.   Structures required for the satellites of the future are expected to fall into two broad categories: advanced manned space systems that can support somewhat larger crews than current manned vehicles, and space platforms, antennas, and solar power systems that are more complex, and in some instances considerably larger, than any spacecraft launched up to this time. The next generation of manned spacecraft is typified by the NASA concept of a space operations center (Figure 2.10). Such a manned space system would use structural approaches similar to those used in recent manned vehicles such as the SKYLAB and SALYUT, but the spacecraft will be somewhat greater in size to accommodate larger crews that will stay in space for longer times and conduct more complex operations than possible with current spacecraft.

The second category of future space systems, large platforms, antennas, and solar arrays, will require the maturing of structural technologies currently being developed for the purpose of placing large unmanned systems into space using components which are deployable or erectable in orbit. The structural technologies being evolved, while not entirely new, are being given greater emphasis because of the benefits that can accrue from larger platforms and antennas.

2.2.1.1.3 Deployable Structures.   Deployable structures are those that are fabricated on earth, folded into compact packages for transport to orbit, and then unfolded in space to perform their required functions. Deployable components have been used in the past in such spacecraft as EXPLORER-49, SKYLAB, and SOYUZ. EXPLORER-49, an unmanned radio astronomy satellite, had 230-m long deployable antenna booms, while the manned SKYLAB and SOYUZ spacecraft used deployable solar panels to provide electrical power.

Deployable platforms upon which scientific experiments and earth applications instruments may be mounted are currently being studied for use with an advanced power system (Figures 2.11). The platform, consisting of three truss-type arms and a central core, would be transported to low-earth orbit by the United States Space Shuttle and deployed there for subsequent mating with an

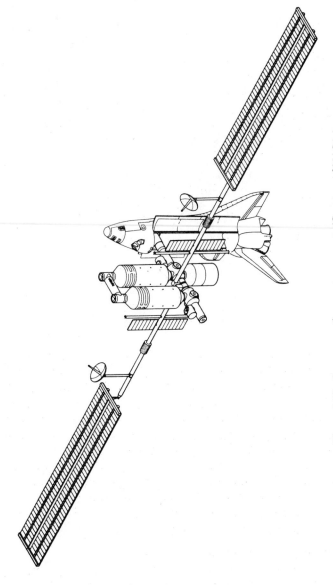

Figure 2.10 Space operations center concept. (Courtesy of NASA.)

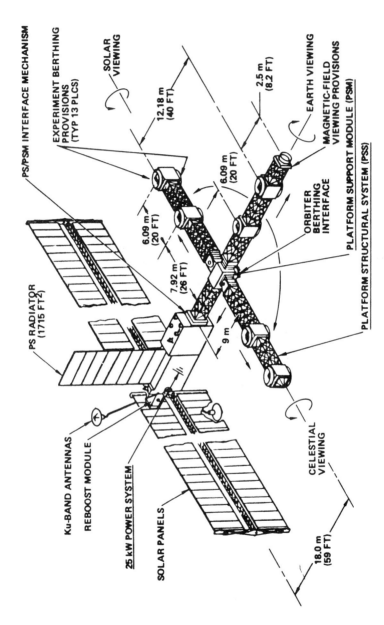

Figure 2.11 Science and applications space platform mated with a 25 kw power system (United States).

PS/PSM INTERFACE MECHANISM

EXPERIMENT BERTHING PROVISIONS (TYP 13 PLCS)

SOLAR VIEWING

MAGNETIC-FIELD VIEWING PROVISIONS

EARTH VIEWING

PLATFORM SUPPORT MODULE (PSM)

ORBITER BERTHING INTERFACE

PLATFORM STRUCTURAL SYSTEM (PSS)

CELESTIAL VIEWING

PS RADIATOR (1715 FT²)

Ku-BAND ANTENNAS

REBOOST MODULE

25 kW POWER SYSTEM

SOLAR PANELS

12.18 m (40 FT)

6.09 m (20 FT)

6.09 m (20 FT)

7.92 m (26 FT)

2.5 m (8.2 FT)

9 m

18.0 m (59 FT)

orbiting solar power system. Deployment of the platform is accomplished by electrical or hydraulic motors that actuate foldable and extendable struts. The deployed platform would have an approximately 46-m span of its unfolded arms. Another deployable system that uses foldable struts to deploy into a repeating-pattern open truss platform has undergone preliminary testing by NASA. This configuration consists of repeating tetrahedral elements.

Deployable antennas of large size have also received considerable attention in recent studies because of their increased gain and thus improved efficiency. Several concepts for deployable antennas are being developed to provide antennas with diameters up to approximately 100 m. One of these concepts, the wrapped rib antenna, has been used in a 9-m diameter antenna on the Applications Technology Satellite (ATS-6). In this concept, the flexible meridional ribs are wrapped around the hub during transport to orbit and are then released in a controlled manner to achieve the full diameter. Another deployable antenna concept is the hoop-column design, a low-mass configuration that positions an outer hoop around a central column with a system of tension cables (Figure 2.12). For antennas having diameters larger than 100 m, deployable truss modules have been studied for use as building blocks that can be joined to form a large antenna.

In addition to deployable platforms and antennas, deployable structures for power generation are still receiving considerable attention because of the ever-increasing demand for electrical power on future spacecraft. This drive for larger solar arrays is currently manifested in the solar electric propulsion system (SEPS). The initial mission conceived for SEPS was that of providing a low-thrust, long-duration electric propulsion system for extended space flight.

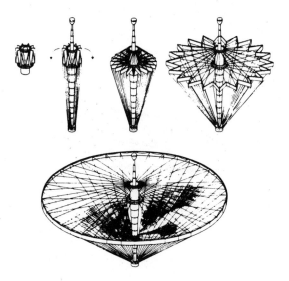

Figure 2.12 Deployable hoop-column concept.

However, in recent studies the SEPS solar array configuration has been looked upon as the basic structure that could provide an additional power source for the Shuttle. The power extension package (PEP), shown in Figure 2.13, would use a 25-kW, 73-m deployable solar array system based on the SEPS to provide additional on-board power and longer mission times for the Shuttle. Another space power system, the 25-kW power system (Figure 2.11), would also use a deployable solar array developed from the SEPS concept and could be mated with a number of satellites that may need an external power supply. One such spacecraft is the science and applications space platform (SASP) shown in Figure 2.11.

The current technology for deployable structures has been demonstrated with antennas and solar arrays having major dimensions in the 10 m to 30 m range, and sizes of these deployable systems have been projected to be feasible up to 100 m in diameter or length. However, key technology and cost considerations that must be addressed in developing larger deployable structures are (a) reliability, (b) dynamic response of the structure during deployment and operation, and (c) the cost of such systems with increasing size.

Reliability of deployable systems will become more difficult to achieve with increasing size as more joints, members, and actuators are required. It is interesting to note that even the modest sizes of deployable structures used up to this time have not been completely trouble-free in their operation. Dynamic behavior of deployable structures will also require extensive analysis and testing to assure adequate performance of the structure. Preliminary studies of cost as a function of size for deployable, erectable, and in-orbit fabricated structures have shown deployable systems to be less efficient in packaging and thus to incur larger transportation costs as structural size increases. This effect is shown in Figure 2.14. Size limits for deployable structures have also been investigated as a function of stiffness and dynamic behavior with the conclusion that a platform

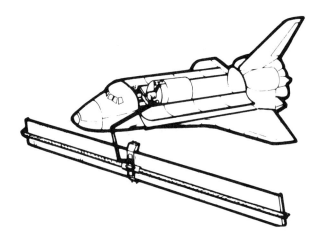

Figure 2.13   Power extension package used with the Space Shuttle.

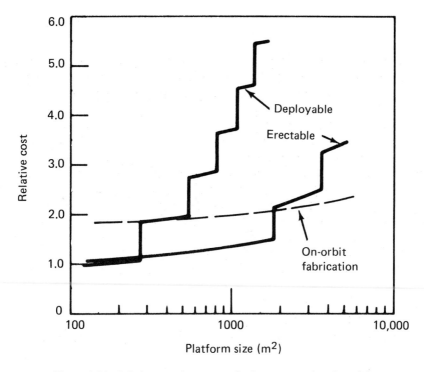

Figure 2.14   Relative costs for structural subsystem as a function of size.

span of approximately 250 m may be the upper limit in achieving satisfactory mass and stiffness characteristics.

2.2.1.1.4 Erectable Structures.   Erectable structures are generally considered to be those in which all of the elements and structural members are fabricated on earth, transported to orbit, and assembled there either automatically or with manned support. This type of structure in general lends itself to denser packaging than is possible with deployable structures. Thus, for the same payload volume, a larger structure can be produced with the erectable approach than with the deployable approach. However, studies to date have shown that auxiliary equipment must be transported to orbit to assist in assembling the erectable elements, and the higher packaging density of the erectable structures is partially offset by the required additional equipment.

Despite auxiliary equipment requirements, analyses indicate that the high-packaging density of erectable structural members permits achievement of mass-limited payloads in the Shuttle, rather than volume-limited payloads as with many deployable concepts. Figure 2.15 shows a typical erectable platform being assembled from the Shuttle payload bay. In this concept, the structural members are tapered hollow half-columns which allow each member to be nested within the adjoining member. In this manner high-packaging density can be achieved.

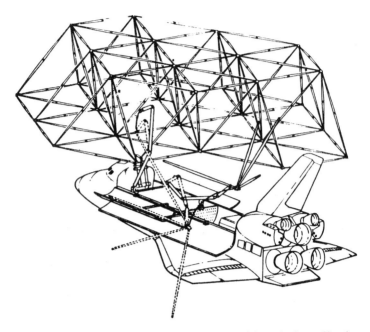

Figure 2.15 An erectable structure being assembled from the Space Shuttle.

Assembly of near-term erectable structures is likely to require extravehicular activity by crew members. For that reason, experiments have been conducted underwater in neutral buoyancy tanks to determine the time required to assemble the structural elements of erectable systems. These studies have provided limited data for assembly of small numbers of structural members, but additional tests with somewhat larger structures and with different member sizes are considered necessary to more accurately assess the required construction time in orbit.

Automated assembly is also being studied for the construction of erectable platform structures. It may be possible to assemble platforms with diameters up to 500 m from the Shuttle using auxiliary equipment attached to the Shuttle. However, larger platforms may require self-contained automatic assembly equipment that does not rely on the Shuttle. Design and development testing of such equipment is considered a necessary step to assembly of large erectable structures in orbit.

The technology required to assemble large structures is currently being developed through the design and testing of candidate structural elements. Also, manned tests in simulated space conditions are being conducted to determine the time required to assemble such structures in space. Additional work is required to develop automated assembly equipment and to gain experience in simulating assembly of large numbers of structural elements. As noted for deployable structures, dynamic behavior and costs of placing large erectable structures in orbit

246

must be resolved before full operational use of these systems can be achieved. The costs shown in Figure 2.14 show that erectable structures will be less costly than deployable systems for platform sizes greater than approximately 300 m.

2.2.1.1.5 Fabricated Structures.

Structures fabricated and assembled in orbit hold the greatest promise for achieving low-cost structures for large systems. In recognition of the potential of in-orbit fabrication techniques, NASA has sponsored work to design and develop automatic beam builder concepts. Two structural concepts have received considerable attention in these efforts: a geodetic beam configuration and a triangular beam configuration. The development of automatic beam builder machines for use in orbital fabrication operations is currently under way and their full development will provide the capability to build structural systems with dimensions as large as 1 km or greater.

The geodetic beam fabrication machine being developed will automatically fabricate either straight or curved longitudinal cylindrical beams in space from earth-fabricated rods and will attach earth-prefabricated beam end fittings. The machine being developed, when fully loaded with material, is capable of producing 200 geodetic beams as large as 130 m long by 1.70 m in diameter with graphite-thermoplastic rods having 2-mm square cross sections. By prefabricating the rods, beam closeouts, and end fittings on each, high-energy operations can be performed on earth rather than in space. The only space processing required is joining of the rods at the node points of the lattice and shear cutting-to-length of the beam. The geodetic beam machine weighs 3,960 kg empty and it can be loaded with up to 19,710 kg of beam rod stock material on an initial Shuttle launch. The peak power required for the geodetic beam machine is proportional to the beam fabrication rate. The major power usage is the node joining subsystem. At 4,500 W, a fabrication rate of about 3.0 m/min is projected. On a first flight carrying the geodetic beam machine, approximately 26 km of beams could be produced. The length of geodetic beam structure that could be fabricated with the amount of material that could be transported into orbit in one shuttle resupply flight is 31 km.

The development of automated structural fabrication capabilities in space will undoubtedly proceed from ground and flight testing of automated beam builders to in-orbit construction tests conducted from the Shuttle in low earth orbit (LEO). Eventually, very large structural systems will be fabricated from a space operations center.

Initially, an orbital flight test program using the Shuttle is required to validate the capabilities of the beam builder and provide needed data on the construction process and on the structural elements fabricated in space. These structural objectives should be accomplished in the process of constructing a useful space-fabricated platform. This platform might be used, for example, for integration of scientific experiments relating to microwave power transmission, communications, earth surveys, and other technologies requiring long-duration orbital activity.

While the size of the Shuttle's cargo bay might seem restrictive in placing large structural systems in LEO, studies have shown that a geodetic beam builder can produce a sizable structural platform from just one shuttle flight. This result comes from the geodetic beam builder's ability to convert bulk materials such as advanced composites, which can be packaged in a high-density form (300 to 800 kg/m^3) into very low-density (0.0006 to 0.08 kg/m^3) truss structures in space. For example, the beams required for a platform approximately 1 km^2 in size can be fabricated with a beam builder and material supplied by three Shuttle launches. For many structures fabricated in space from high-density stock, the primary factors limiting the size of structures are the Shuttle's power limits, mission duration, and possibly crew size rather than cargo bay capacity. Also, the launch intervals, determined by Shuttle fleet size and ground support operations, will affect the time necessary to construct a large space structure and therefore will indirectly affect the feasible size of structures fabricated with Shuttle support.

Development of a triangular beam builder is also under way under the sponsorship of NASA. A ground demonstration machine has been built and is currently undergoing ground tests. In its initial configuration the triangular beam machine fabricates aluminum beams; however, conversion of the machine to produce composite beams is currently being pursued. While aluminum has been frequently used in spacecraft structures and can be used in automatic beam builders, its thermal expansion characteristics make it undesirable for some large space structures that require close figure control.

2.2.1.2 Materials for space structures

2.2.1.2.1 Requirements. Materials being considered for space structures must satisfy the basic requirements of high stiffness, low density, and long life. Temperature extremes and temperature gradients must be accounted for and in many cases thermal distortion must be low. Most structures will be required to have high damping, and the materials used must exhibit low outgassing for some systems (e.g., optical systems). In many cases in which very large structures are being considered, materials must be amenable to fabrication in space.

2.2.1.2.2 Candidate Materials. Because of the extensive experience and manufacturing capability that has been gained in the aerospace industry with aluminum, there is a strong desire to use this material whenever possible. However, it is clear that aluminum will not satisfy many performance requirements. Therefore, alternate materials must be considered. Other candidate materials include titanium, steel, magnesium, beryllium, and graphite-resin composites. The composite materials utilize the unique graphite fiber properties of high stiffness and strength along with a negative coefficient of thermal expansion.

2.2.1.2.3 Comparison of Material Characteristics.   Structural stiffness is proportional to Young's modulus (E) for the material. Therefore, maximum structural stiffness can be provided by maximizing (E). Minimum mass is required to minimize transportation costs, and mass is proportional to the material density (ϱ). Hence, an important parameter for evaluating materials for space applications is the ratio E/ϱ. The higher the E/ϱ ratio, the more efficient the material.

Long slender columns will be used to construct many large space structures. Low-mass columns designed to carry orbital transfer or other mechanical loads while undergoing minimum deformation must be made of materials with high modulus-to-density ratios (E/ϱ), i.e., high specific stiffness. A comparison (Figure 2.16) of the modulus-to-density ratios for various materials indicates that graphite-resin composites provide significantly greater specific stiffness than do metals.

Aircraft applications indicate that, for equivalent performance, structures made of composite materials may well cost less than structures of aluminum. Placing a given area of aluminum structure in orbit will require more Shuttle flights than would be required for the same area of graphite-epoxy structure, assuming mass-critical payloads. At or above current Shuttle transportation costs, it is cost effective to use graphite epoxy rather than aluminum because the lower cost of aluminum structure is more than offset by the higher cost of transporting it to orbit. The cost of graphite preimpregnated material has been steadily decreasing over the past several years. With increasing use, the cost will

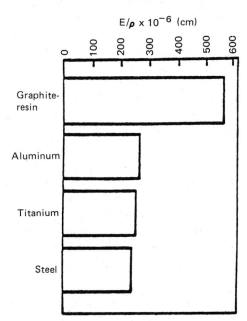

Figure 2.16   Comparison of specific stiffness.

continue to decrease. In addition, fewer parts and less raw material waste are associated with composite structures, which tends to lower the fabrication and assembly costs.

In evaluating a material for long life, the material's resistance to fatigue and to ultraviolet (UV) and particulate radiation must be considered. Structural temperature variations during the orbital period are caused by full exposure to the sun and earth shadowing, and these variations cause cyclic thermal stresses and distortions. Typical variations are shown in Figure 2.17.

Most large space structures must have a minimum life of 30 years to be economically viable. A space structure in LEO for 30 years will be subjected to 164,000 major thermal stress cycles; a structure in geosynchronous earth orbit (GEO) will be subjected to 11,000. Materials which provide high fatigue resistance and low thermal stress offer the maximum potential for long life structures.

Thermal distortions and stresses are proportional to the material's coefficient of thermal expansion (α). The unique negative α of graphite fibers can be balanced against the positive α of the matrix material, resulting in a net α of zero for the composite. Thermal distortions and stresses are also proportional to the structural thermal gradients. Thermal gradients can be minimized by maximizing the material's thermal conductivity (K_t). Hence, an important parameter for evaluating a material's thermal distortion characteristics is the ratio K_t/α.

In addition to the material conductivity, material surface condition is important in determining the amount of heat a material will absorb. A surface painted white, for example, will reflect a large amount of heat.

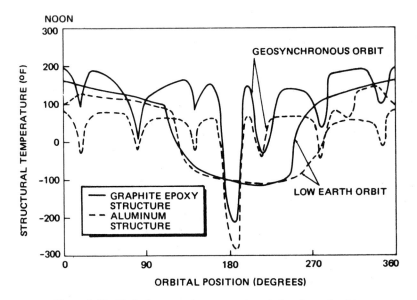

Figure 2.17   Typical structural temperature variations in earth orbit.

Space structures will be subjected to various dynamic inputs that will cause them to vibrate. The time that is required for the vibration to decay to a negligible level is inversely proportional to the damping characteristics of the material. Since it is important that decay times be low, materials with high damping characteristics will be required.

Advanced composite materials possess many attractive characteristics and have the potential to satisfy the performance requirements of large space structures. Graphite epoxy has been used in the aerospace industry for approximately two decades and although the manufacturing and engineering technology data base is not as large as for aluminum, it is rapidly increasing.

Large space structures must perform in the unique environment of space. Although some elements of this environment affect all materials in more or less the same way, others have significantly different effects on composites than on metals. Before the potential of advanced composite materials can be fully realized, certain key behavioral characteristics must be determined or more fully investigated. These characteristics include the response of composite materials to long-duration, high- and low-temperature thermal aging and cyclic thermal environments, the response to ionizing (charging) radiation, and the corrosion characteristics of composite/metallic joints in space.

2.2.1.3 Power generation.
On-board sources of electric power for spacecraft consist of chemical fuels, as used in fuel cells or batteries, nuclear reactors, radioisotopes, or photovoltaic (solar-cell) converters.

2.2.1.3.1 Photovoltaic Converters.
By far, the most widely used source of spacecraft power is the sun. All solar cells used in space to date are manufactured from silicon, with operational efficiencies currently in the 12 to 16 percent range. Most spacecraft have employed body-mounted solar cells on a spin-stabilized vehicle, developing 25 W/m² of array surface with specific mass of the order of 150 kg/kW. Rigid arrays arranged in "wings" on three-axis stabilized spacecraft typically develop 65 W/m² at a specific mass of 65 kg/kW for up to seven years.

More recent developments utilizing low-density honeycomb cores and low-mass Kapton or carbon-fiber face sheets have demonstrated 80 W/m² and 30 to 40 kg/kW. The 25-kW array being developed by NASA could have a specific mass as low as 18 kg/kW.

Further improvements now in the research and development phase include the use of gallium arsenide or gallium-aluminum-arsenide materials, which show promise of attaining efficiencies above 20 percent, and multibandgap cells capable of using a larger fraction of the solar spectrum and therefore capable of efficiencies of around 30 percent.

2.2.1.3.2 Batteries.
Batteries are a necessary element in solar space power systems because they can provide power during eclipses or other occultations of the sun. Since conventional nickel-cadmium batteries can represent up

to 20 percent of total spacecraft mass, there is considerable motivation to develop higher-performance devices. The most promising of these is the nickel-hydrogen battery, which can double the energy density of current nickel-cadmium cells (30 to 44 W-h/kg, as compared with nickel-cadmium energy densities of 10 to 15 W-h/kg).

2.2.1.3.3 Fuel Cells.   The fuel cell is suitable for high levels of on-board power over relatively short periods, e.g., a week or two. The hydrogen-oxygen fuel cells used on the APOLLO flights developed 1.4. kW, weighed 100 kg, and had a design life of 400 hours. The significant progress achieved in this area is reflected in the fuel cells currently used in the Space Shuttle: 12 kW, 90 kg, and a design life of over 5,000 hours. Further minor improvements are expected to bring projected lifetimes up to 50,000 hours for these units.

2.2.1.3.4 Radioisotope Generators.   The United States has utilized radioisotope thermoelectric generators (RTGs) on planetary lander spacecraft (APOLLO's ALSEP and the Mars VIKING) and on outer-planet probes (PIONEER-Jupiter/Saturn and VOYAGER). Such systems are necessary when solar occultation periods are lengthy or where distances from the sun are very large.

Capable of power levels up to several hundred watts, these units can attain specific mass levels of about 100 kg per kilowatt of electric power generated (kWe), but unlike photovoltaic power sources they do not require batteries during eclipse periods. Extension of radioisotope power sources to higher power levels requires replacing the low-efficiency thermoelectric converters (typically 5 to 8 percent) by higher-efficiency dynamic thermal-electric power conversion systems (vapor or gas-cycle engines). Such systems could deliver power levels up to several kilowatts at efficiencies up to 25 percent and specific mass in the same range as RTGs.

2.2.1.3.5 Nuclear Reactors.   The United States flew a test flight of a nuclear reactor (SNAP-10A) on a spacecraft in the 1960s, and the USSR uses them regularly on surveillance satellites (Topaz reactors). These reactors develop somewhat greater power levels than the RTG (5 to 10 kWe). The United States has considered preliminary designs for much more advanced power reactors utilizing heat pipes as power sources for electric propulsion systems, with power levels ranging up to 400 kWe, but has not yet developed such reactors.

2.2.2 Spaceflight and Operations Technologies

2.2.2.1 Astrodynamics and flight dynamics.   The motion of a space vehicle can be divided into two components: orbit and attitude. Orbital motion is the motion of the center of mass and may be thought of as an extension of classical celestial mechanics. The motion of a spacecraft about its center of mass, often referred to as libration, belongs to the field of attitude dynamics. In

general, for a mission to be successful, both orbital and attitude motions demand precise control.

There are numerous situations of practical importance where it is desirable to maintain a satellite in a fixed attitude with respect to the earth. For example, proper functioning of communications satellites with directional antennas, weather satellites scanning cloud cover, or earth resources satellites monitoring crops, water, etc., requires varying degrees of orbit and attitude control. Unfortunately, even though a spacecraft may be positioned precisely at launch, it tends to deviate, in time, from this preferred orientation under the influence of external disturbances in the form of micrometeorite impacts, solar radiation pressure, gravitational and magnetic field interactions, and, if the spacecraft is below 1,000 km altitude, atmospheric drag. This leads to undesirable orbital perturbations and librations that must be controlled for the successful operation of the spacecraft.

2.2.2.1.1 Present State of the Art.

Broadly speaking, methods evolved over the years for attitude and trajectory control may be classified as active and passive techniques.

Active control involves expenditure of energy, which is a very expensive commodity aboard an instrument-packed spacecraft. The main advantage of this procedure is its ability to provide control to almost any specified degree of accuracy. Using mass expulsion (thrusters), reaction, or momentum wheel systems, it is now possible to control the satellite attitude to within $0.1°$, and orbital error can be reduced to a few meters for most near-earth and synchronous operations. Even re-entry trajectories can now be controlled with an accuracy that permits predictions of splashdown or ground landing within a few square kilometers. With an on-board computer capable of monitoring the spacecraft position and orientation using high-precision sensors, our ability to control orbital and librational motions is governed entirely by the amount of fuel made available to the spacecraft. This in turn also governs its lifetime. Once the fuel is expended, there is no mechanism for control left and the spacecraft is no longer useful. This is equivalent to discarding an expensive automobile every time its fuel tank is empty.

Passive control procedures, which use no power and are more relevant for attitude stability, can provide necessary pointing accuracy if the demands are not too severe.

Perhaps the simplest form of essentially passive control of satellite librations is illustrated by turning the whole spacecraft into a gyroscope. A rigid spinning body tends to maintain its orientation in inertial space in the absence of any external disturbing moment. Normally, for stable operation a spacecraft is spun about its maximum moment-of-inertia axis. Almost all spin-stabilized satellites have an axis of symmetry. Any undesirable wobble about the two transverse axes is usually minimized through the use of a nutation damper. Of course, the main advantage of any passive control system is a relatively longer life at a cost of limited, coarse control. Our experience with spin stabilization of spacecraft is extensive and appropriate technology is well developed, as substantiated by literally hundreds of such spacecraft launched to date.

There are several other procedures for passive control of attitude which employ non-uniformities of environment in conjunction with the physical properties of the satellite. The significant forces available for passive stabilization of a spacecraft arise from gravitational, solar, magnetic, and aerodynamic effects. A number of primarily scientific satellites have been launched in the past, and there are some in operation today that are controlled by such passive procedures. It must be pointed out, however, that among all the above-mentioned passive control procedures, spin and gravity-gradient stabilization have proved more successful, with others showing limited effectiveness in isolated situations. Furthermore, it must be emphasized that passive orbit control, particularly using solar radiation pressure, is limited to analytical studies. Although several proposals to this end (solar sail) are being explored at the moment, feasibility experiments have not yet been planned.

2.2.2.1.2 Future Trends.   In the early stages of space exploration when spacecraft tended to be small, mechanically simple, and essentially inflexible, elastic deformations were relatively insignificant. Numerous investigations involving active and passive stabilization procedures and accounting for internal as well as external forces have been carried out assuming satellites to be rigid. However, in a modern space vehicle carrying low-mass deployable members, which are inherently flexible, this is no longer true. This aspect can be emphasized through several illustrations:

1. Ever-increasing demand on power for operation of the board instrumentation, scientific experiments, communication systems, etc., has been reflected in the size of the solar panels. The Canada/United States Communications Technology Satellite (CTS, Hermes) launched in January 1976 carried two solar panels, 1.14 m by 7.32 m each, to generate 1.2 kW.
2. Use of large members may be essential in some missions. For example, the Radio Astronomy Explorer (RAE) satellite used four 228.8-m long antennas to detect low-frequency signals.
3. For identifying extraterrestrial radio sources, the Applied Physics Laboratory of Johns Hopkins University once proposed a gravitationally stabilized tethered orbiting interferometer (TOI) consisting of two spacecraft connected by a tether 2 to 6 km long.

With this as background, space dynamicists recognize that our future efforts are aimed at construction of gigantic space stations that cannot be launched in final form but have to be constructed in space through integration of modular subassemblies. The concept of solar power satellites in geostationary orbit and designs for space settlements are being studied in depth from dynamical considerations, allowing for structural flexibility. A major area of concern is the conjugate character of the problem where flexibility affects the librational motion while the librational motion, in turn, excites flexible structures. Dynamics and control of such systems are receiving particular attention and will continue to do so for a long time.

In the near future attention will be focused primarily on the dynamics of Space Shuttle-supported experiments. Although the variety of configurations that may be encountered in the future is limitless, they do offer a degree of similarity at a fundamental level: They all represent multibody systems interconnected through rigid or flexible links. In particular, considerable effort will be directed toward understanding the dynamics and control of tether-supported payloads deployed from or retrieved toward the Space Shuttle. It is generally agreed among dynamicists that the focus of attention should be the control of large flexible structures.

As far as the future course in orbit control is concerned, analytical and technological tools and their capabilities are well established. More attention will be directed toward optimization of time, effort, and expenditure. Cost effectiveness will be the key. High-performance cryogenic engines and similar low-thrust propulsion systems for orbit transfer are likely to receive more attention. Ion propulsion systems for orbit control of broadcasting television satellites will be investigated in depth.

2.2.2.2 Space transportation systems.

Technological problems of space transportation systems arise from the increasing demands of future missions, which fall under three main categories:

1. Transportation of automated payloads of ever-increasing size for earth-oriented applications (telecommunications, observation, navigation) and space science research (planetary missions and orbital astronomy). Prospective nuclear waste transportation is a special case in this category of space transportation.
2. High-reliability spaceflight for manned orbital operations: materials processing, SPACELAB operations, spacecraft repair and refurbishment, and space construction.
3. Construction and transportation of large space structures: space stations, telecommunications platforms, power plants, and antenna dishes.

2.2.2.2.1 Present Status.   Up to now, space transportation systems have all consisted of expendable launch vehicles. There are three general types:

1. A single-stage or multistage booster that propels the payload from earth to high but still suborbital velocity.
2. An upper stage that places the payload into a low orbit.

3. An orbital transfer stage, which can be integrated with the payload, to transfer the payload to high (e.g., geosynchronous) orbit or to an escape trajectory.

A very wide range of launch vehicles exists in the United States, the USSR, Japan, India, and Europe. Figure 2.18 shows the expendable launch vehicles used over the past years. The evolution and diversification of these vehicles were not the result of a change of concept, but a response to the successive demands of missions, mainly in terms of mass in orbit, and a consequence of technology improvements, mainly in terms of performance.

Requirements now emphasize cost reduction and higher reliability. To understand the reason for these trends, one should note that material for a single solar power satellite would need 500 flights of a vehicle capable of transporting 400-ton payloads to low earth orbit.

2.2.2.2.2 Future Trends.   The primary future demand will be to reduce specific transportation cost. This can be achieved through vehicle reusability. The expendable launch vehicles used up to now were relatively expensive. Specific transportation costs into LEO were in the range of 7,000 to 10,000 United States dollars per kilogram of payload. The largest launch vehicle used to date was the United States Saturn V, which could carry 130 tons of payload into LEO.

Future launch vehicles can provide substantial cost reductions by utilizing three main concepts:

1. Reusability of the transportation system.
2. On-board system controls for launch procedures.
3. Payload recovering capability.

The United States Space Shuttle (Figure 2.19) is the first of the future reusable vehicles. Its launch mass is 2,000 tons and the payload it can carry to LEO is about 30 tons. It is partially reusable: The winged "orbiter" vehicle will be flown into orbit and back to earth more than 100 times, but the large propellant tank will be expended each time. The solid-propellant boosters will be recovered and reused for ten or more flights. The reduction in launch cost is remarkable: It is currently estimated at $1,500 per kilogram payload. The vehicle return capability has another important advantage: Expensive payloads such as the Space Telescope can be returned from space back to earth, refurbished, re-equipped, and launched again, thereby offering considerable potential cost savings.

Future fully reusable vehicles will allow additional reductions in launch cost to about $500 per kilogram, and large launch vehicles (100 tons payload) will eventually reduce costs to below $100 per kilogram. These vehicles are techni-

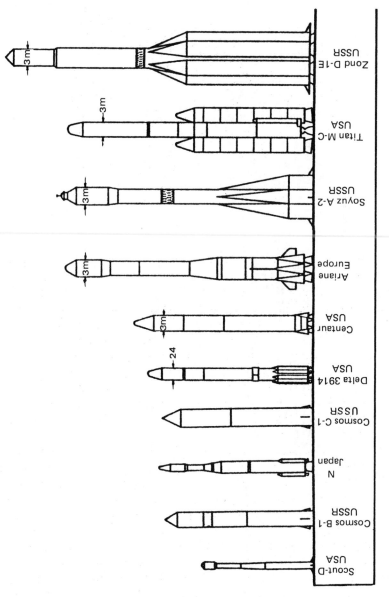

Figure 2.18 Expendable launch vehicles.

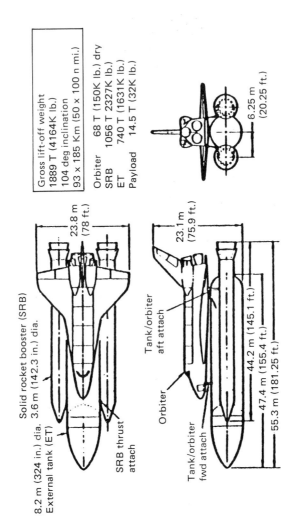

Figure 2.19 A partly reusable launch vehicle: The United States Space Shuttle.

Gross lift-off weight
1889 T (4164K lb.)
104 deg inclination
93 × 185 Km (50 × 100 n mi.)

Orbiter   68 T (150K lb.) dry
SRB   1056 T 2327K lb.)
ET   740 T (1631K lb.)
Payload   14.5 T (32K lb.)

6.25 m
(20.25 ft.)

23.8 m
(78 ft.)

Solid rocket booster (SRB)
8.2 m (324 in.) dia.   3.6 m (142.3 in.) dia.
External tank (ET)

SRB thrust
attach

23.1 m
(75.9 ft.)

Tank/orbiter
aft attach

Orbiter

Tank/orbiter
fwd attach

44.2 m (145.1 ft.)
47.4 m (155.4 ft.)
55.3 m (181.25 ft.)

cally feasible with present technology, but they require a large development investment. Such vehicles could be available in the mid-1990s if development is initiated within the next few years.

Systems reliability and the consequent vehicle complexity require considerations of simpler vehicles with fewer stages, which also reduce operational costs. Hence single-stage-to-orbit vehicles (SSTO) have gained increased interest in recent years. Thus future launch vehicles will be in marked contrast to the multistage expendable launchers widely used today.

2.2.2.2.3 Reusable Vehicle Concepts.

It is not likely that there will be only a single future space transportation system, but rather a range of systems to accommodate different needs, just as today we have short-, medium-, and long-range aircraft. Reusability can be achieved with two different re-entry technologies: winged horizontal landing or ballistic vertical landing.

Winged vehicles offer the advantage of operational versatility and flyback capability to the launch site: They can reach their home base from a larger variety of orbits. They are manned vehicles and are preferred in current United States studies. Disadvantages are the limited cargo volume and their high development cost, because they approach the theoretical limits on chemical rocket propulsion, particularly when winged SSTO vehicles are considered.

Ballistic systems do not require pilots and are more effective for heavy cargo transportation. They have the advantage of lower structural mass, perhaps a factor of two lower. They do not require any breakthroughs in propulsion technology and could make use of present-day propellant technology. They are preferred in European studies. The disadvantage of two-stage vehicles is the tow-back requirement for the first stage after landing on water several hundred kilometers from the launch site. This is avoided by going to single-stage vehicles which can return directly to the launch place, but they have the disadvantage of larger size or lower payload.

2.2.2.2.4 Orbital Transfer Stages.

For transportation from LEO to GEO or interplanetary trajectory, additional rocket systems are required. In the past, solid fuel rockets have generally been used for this purpose. The Space Shuttle currently relies on solid propulsion: the Spinning Solid Upper Stage (SSUS) or Perigee Assist Module (PAM) systems and the Inertial Upper Stage (IUS). These systems can place satellites of 1,000 to 2,000 kg mass in geostationary orbit. Cryogenic upper stages using liquid hydrogen and liquid oxygen will increase geostationary payload capability to 4 to 8 tons, thus allowing the launch of multipurpose communication platforms. The United States is now considering adapting its cryogenic Centaur stage to the Space Shuttle, and a reusable orbital transfer vehicle (OTV) has been under study for many years to transfer payloads from low earth orbit to the geosynchronous orbit. It could be operational by the mid-1990s, improving the economical utilization of space considerably.

2.2.2.3 Propulsion technology. As with space transportation systems, rocket propulsion can be considered in three categories:

1. Earth-to-orbit propulsion (launch vehicles).
2. Orbit-to-orbit propulsion (orbital transfer vehicles).
3. Attitude and orbit control propulsion.

Chemical propulsion is used in all categories. In general, liquid propellant engines are used in launch vehicles, solid propellant engines in orbital transfer vehicles, and hydrazine (monopropellant) engines for orbit and attitude control.

2.2.2.3.1 Earth-to-Orbit Propulsion. Several advanced rocket engine technologies are presently being pursued in the United States to fulfill the requirements of reusability and to make single-stage-to-orbit feasible:

1. High-chamber pressure engines increase specific performance by increasing available expansion ratios. Based on present Space Shuttle main engine technology (200 atm), chamber pressures of up to 600 atm have been suggested. Development of such engines would require major advances in engine cooling and turbopump technology.
2. Mixed-mode propulsion uses different propellant combinations in the same vehicle to optimize the trade-off between low-density/high-specific impulse and high-density/low-specific impulse propellants. Advanced dual-fuel engines, capable of sequentially burning hydrocarbons and hydrogen, would also need advanced cooling techniques (e.g., using liquid oxygen as coolant). In addition, starting an SSTO requires adaptable nozzles. One solution, called the dual-expander engine, uses concentric combustion chambers for different propellant combinations.

Another technology that has attracted widespread interest in the field of heavy launch vehicles (about 100 tons payload to low earth orbit with SSTO transportation) is the air-augmentation rocket engine, which, in its limit, could become an air-breathing booster. Although studies of winged air-breathing vehicles have shown that the gain resulting from using outside air is offset by a corresponding increase in drag and mass, recent studies of ballistic vehicles indicate payload mass as high as 10 percent of gross liftoff mass for $1\frac{1}{2}$ stage vehicles and 6 percent for SSTO vehicles. These payload ratios are substantially higher than those of pure rockets and appear to warrant further investigation.

The staged ballistic vehicles that are being considered for payloads of 400 tons and more to LEO (corresponding to about 100 tons to GEO) do not require propulsion technology breakthroughs. These reusable heavy lift launch vehicles do pose financial and project management problems, but these very large vehicles are the key to massive development of space industrialization and large space structures, as noted previously. With respect to propulsion technology, since the ballistic landing of these vehicles requires the development of engine clustering techniques, large numbers of rocket motors will have to be arranged around the vehicle base, which will have to serve as a re-entry shield.

2.2.2.3.2 Orbit-to-Orbit Propulsion.   Transportation from low earth orbit to geostationary orbit also requires high-performance reliable propulsion, especially if the vehicle is to be reused. Such an application calls for high-energy propellants (hydrogen and oxygen) rather than today's solid or storable liquid propellants.

For long-duration flights to the planets, very high-energy propellants offer distinct advantages. United States developments have included investigations of fluorine/hydrazine, for example, a propellant combination that yields a high specific impulse accompanied by excellent storability.

Electric propulsion offers a promising alternative to chemical propulsion in both types of missions. The very low propellant consumption of ion engines yields higher payloads at higher mission velocities than chemical engines. However, electric engines are capable of only very low thrust-to-mass ratios. Consequently, thrust duration has to be very long and the geostationary orbit, for example, is attained from LEO in many revolutions along spiral trajectories. This poses problems with solar electrical power systems, since the van Allen belt radiation inflicts serious damage to their solar cells. Nuclear electric propulsion has been considered by the United States, as noted earlier, as a possible answer to this problem.

2.2.2.3.3 Attitude and Orbit Control Propulsion.   The majority of attitude control and orbit correction requirements to date have been satisfied by chemical monopropellant engines. Electrical propulsion has developed to the point that it is being considered for applications in orbit trimming. Very small bipropellant rocket engines are also beginning to be used for these applications because they provide the opportunity for using main propulsion propellants for secondary applications. This "integrated propulsion" technique, used in the Symphonie satellite, leads to reduced structural mass.

2.2.2.3.4 Advanced Propulsion.   For classical missions, propulsion systems have reached a mature state of technology. Very advanced missions would need propulsion systems with performance characteristics superior to those available at present.

Large-scale transportation of lunar or asteroid material could be achieved with mass drivers using electromagnetic acceleration rather than jet propulsion. Present designs are in their infancy.

The main goal of advanced propulsion systems is to achieve high thrust at high specific impulse, a combination requiring very high power consumption. Plasma engines, a special type of electrical engine, are such devices. They achieve their high specific impulse through the high temperatures achievable in electric discharges, in combination with plasmadynamic acceleration. When electrical power is no problem, as would be the case in solar power stations, they are ideally suited for large-scale attitude and orbit control.

Another way of achieving high temperatures is to use external energy. Laser beams could transmit the energy from ground or space-based power plants to spacecraft equipped with laser engines. In small systems, this is an effective way of decreasing on-board propellant consumption and leads to increased operational life. Large systems, like launch vehicles, could be reduced to the size of present-day aircraft at comparable payloads.

2.2.2.4. Unmanned space operations. Automatic and remote-controlled operations have been developed in the past and will gain increasing importance in the future as new activities begin to develop: the assembly of large structures, manufacture of materials, nonterrestrial material exploitation, etc.

Very high long-term operational reliability is needed, a difficult requirement in the extreme conditions of space, lunar, or planetary environments. It will be a major challenge to human engineering skills.

Unmanned operations consist of the following activities:

1. Orbital operations (rendezvous, docking, propellant storage and transfer, vehicle and payload mating).
2. Orbital fabrication and assembly (materials processing, fabrication of construction elements, assembly of units, manufacture of propellants).
3. Landing and ground operations on lunar and planetary surfaces.

The introduction of the United States Space Shuttle will start a new era of space operations in LEO. Payload deployment and assembly, as well as payload retrieval for transportation back to earth, will become standard procedures. After refurbishment, these spacecraft can be brought back into space and serve for another period of years. This could reduce the cost of such spacecraft as astronomical observatories considerably.

Another new era of docking and retrieval operations can be accomplished by orbiting factories for fabrication of advanced materials, super crystals, extremely clean pharmaceutical products, and new alloys. Experiments have already begun on the automatic fabrication of large beams made from thin aluminum foils or carbon-fiber materials, as described in detail earlier.

Inspection, servicing, and repair of satellites, especially of larger platforms in geosynchronous orbit, will require the development of automatic manipulators, or teleoperators. Automatic landing on the moon and planets has been successfully demonstrated and will be another area for future activity. Operation of a small rover for exploration of the planetary surface is one of the next goals, particularly for Mars. Because of the communications time lag due to the finite speed of radio impulses, this vehicle requires a form of artificial intelligence and cannot be operated from earth. Instead, it has to choose its own way through the obstacles of the surface.

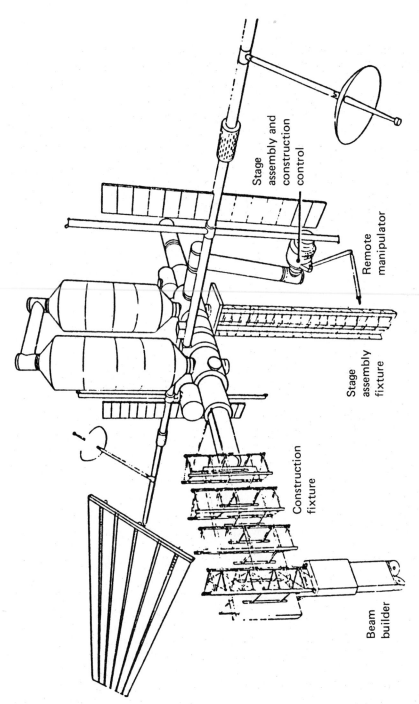

Stage
assembly and
construction
control

Remote
manipulator

Stage
assembly
fixture

Construction
fixture

Beam
builder

Figure 2.20   Space Operations Center facilities. (Courtesy of NASA.)

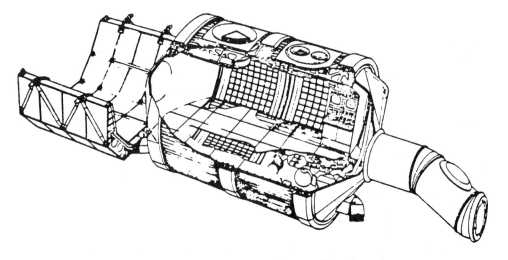

Figure 2.21 SPACELAB (European Space Agency).

2.2.2.5 Manned space systems. The role of people in space has been studied first in terms of survival and subsequently in terms of the advantages and disadvantages of manned versus unmanned operations.

The first manned spaceflights including the manned lunar landings, were demonstrations of capability. The next steps were experimental orbital laboratories for in-space research and observations (SALYUT and SKYLAB). The third step will be the use of man's skills for construction and assembly work in a permanent space station or space operations center (Figure 2.20).

The SPACELAB (Figure 2.21) is the first module developed as a step toward such a space facility. In the 1980s SPACELAB will provide the possibility for manned investigations in space, both for biomedical research and mantended experiments.

The Space Shuttle itself provides additional operational capabilities for people in space besides transportation capability into space and back to earth. Full utilization of the cargo bay would allow for accommodation of 74 people in an aircraft-type environment.

If it is decided to proceed with plans for solar power satellites, people and support facilities would be required in geosynchronous orbit. This could lead to space settlements and orbital space cities to service future space industrialization. The environmental conditions of space can be more favorable than on earth, and space offers a location for industry that will remove pollution from the earth's surface. Further, lunar and asteroidal resources might become important when rare metals are exhausted on earth. Hence it is possible that mankind's presence in space will increase dramatically in the not too distant future.

ACKNOWLEDGEMENTS

Chapter 2 was prepared with the assistance of an international team of experts organized by the International Astronautical Federation (IAF) and consisting of the following members: Mr. J. J. Dordain (France) (team Chairman), Mr. J. Billingham (United States of America), Mr. M. Claverie (France), Mr. H. J. Fischer (German Democratic Republic), Mr. J. Garibotti (United States), Mr. J. Grey (United States), Mr. S. Hieber (Federal Republic of Germany), Mr. U. Huth (Federal Republic of Germany), Mr. D. Koelle (Federal Republic of Germany), Mr. R. Lo (Federal Republic of Germany), Mr. V. J. Modi (Canada), Mr. E. Peytremann (Switzerland), Mr. O. Roscoe (Canada), Mr. H. Uda (Japan), and Mr. J. Vandenkerckhove (Belgium). Contributions were also received from many other people from various countries.

Chapter 3

RELEVANCE OF SPACE

ACTIVITIES TO MONITORING

OF EARTH RESOURCES

AND THE ENVIRONMENT

ABSTRACT

This chapter is the third of four general review chapters. Its purpose is to review the contributions that space activities are making to man's ability to manage the earth's natural resources and to live in harmony with his environment. It emphasizes the demonstrated uses of currently operating systems, but more advanced systems now in the planning stages are also briefly mentioned. The section on land observations presents examples from developing countries and the sections on the oceans, the atmosphere, and disaster relief concentrate on international projects. For a discussion of the technology that makes these applications possible, the reader may wish to consult Chapters 2 and 5; for a discussion of the implications of the adoption of space technology by developing countries, see Chapter 8.

INTRODUCTION

Earth, my Likeness!
Though you look imposing
Ample and spheric there
I now suspect that is not all.

Walt Whitman

The first views of earth from space, obtained about 20 years ago, of its active atmosphere, and of its variegated land and seascapes, were exciting, provocative, and difficult to use. There was more, much more, than the ampleness and the sphericity praised by the poet. The wealth of potential information was suspected, but it took man nearly a decade to appreciate the amount of information available and to learn how to use it.

Of course, some of the early scenes showed obviously useful patterns, of clouds, for example. However, much patience and time and comparison of satellite views with ground views were required before the seemingly obvious cloud patterns could be turned into useful information. The decade of the 1970s saw the quantification of satellite observations of the atmosphere and the oceans and the ripening of advanced interpretations of land surface patterns. The 1970s have also been years of technical advances which hold much promise for the development of vastly improved satellite systems for the coming decades.

Perhaps not entirely coincidentally, the decade of the 1970s also witnessed a growing awareness that the resources of the earth are not limitless, that if even the basic needs of the people of the earth were to be met on a continuing basis, the earth's natural resources would have to be managed and developed with care and foresight. Food production and distribution systems need to be improved, clean water supplies must be extended, clean air must be assured, and raw materials and energy must be developed.

The air, water, and land which normally support man can also turn against him at times. Violent storms, floods, earthquakes, and volcanic eruptions can cause great destruction, especially when they come unexpectedly. It seems unlikely that man will soon acquire the power to tame nature's violence, but he can learn to understand and to prepare himself for such attacks. Man can adapt his activities to the risk of these disasters and can plan emergency measures to minimize damage when disaster strikes.

If man is to live in harmony with the earth and with himself, he will have to learn to understand the natural world around him, and using that understanding, he will have to adapt himself to the earth and, to the extent possible, to adapt the earth to himself. In the past, when man did not have the power to exhaust the resources of nature, and when most production and consumption were local activities, man's intuitive understanding of his immediate environment seemed sufficient to sustain him. Now, with man's power over the environment increasing and with production and consumption based on international exchange, the traditional knowledge that shaped man's attitude toward his planet may no longer be adequate.

If man is to acquire a new understanding of his planet, he needs new tools. If the understanding is to be dynamic and global, the tools must be capable of dynamic and global application. Space technology is inherently dynamic and global in that a satellite in an orbit that passes over a particular area will continue to pass over that area regularly until the orbit is changed, and all orbits provide visibility of a large part of the earth's surface.

The full benefits of earth observation satellite technology could be realized if the earth's resources were managed on a global and dynamic basis. Traditionally, however, man's activities are planned and coordinated on a local or national basis and the information on the state of local or national resources is updated at intervals of five or ten years at best. For a single national survey, traditional survey methods, using aircraft or ground observations, are probably more cost effective than a satellite mission. The effective use of satellite data, therefore, both permits and requires new approaches to both international and national resource management activities.

New technology is generally developed and adopted most quickly when the technology can be adapted to existing activities. The earliest uses of earth observation satellites have therefore been in national programs and for static mapping applications. As the technology becomes operational, international and dynamic applications will begin to develop. In meteorology, in which operational satellite systems have existed since 1966, cooperative international programs have been in existence for some time, first for exchange of data—the World Weather Watch—and subsequently for international programs such as the Global Atmospheric Research Programme (GARP). In remote sensing, which is still in the pre-operational phase, international networks and procedures for distribution of data are still not established, and use of the data for dynamic monitoring is still largely limited to national experimental projects.

Technology should serve human needs; human needs should not be sacrificed to the needs of technology. The development of earth observation satellite systems and their applications will therefore depend on the decisions that are made as to how the earth's resources are to be managed and what will be the extent of international cooperation. If careful and foresighted management in a context of international cooperation is a priority, then the use of satellites for earth observation should develop rapidly and should make an important contribution to man's well-being.

It is of course not possible in this chapter to examine all of the applications of earth observation satellites or to adequately represent the activities in all countries. This chapter, therefore, cites specific examples in some of the proven applications in an effort to give the reader a feeling for the types of applications being developed and the benefits being sought. By citing specific cases rather than discussing theory, it is hoped that the reader will develop not only an appreciation for the potential benefits of satellite observations but also the difficulties in realizing that potential. A large number of projects in addition to those listed in this chapter have been undertaken by many countries.

3.1 LAND OBSERVATIONS

3.1.1. Data Acquisition and Interpretation

Observations of the earth's land surfaces began with the first manned missions. The astronauts and cosmonauts used hand-held cameras to take snapshots of the surface from various altitudes, at various angles, and at various times of day. These photographs, with their variable characteristics and sporadic acquisition, whetted the appetite of resource scientists but did not provide a basis for assessing the potential value of satellite remote sensing.

With the launch of the first United States LANDSAT satellite in 1972, satellite imagery became available for the first time in a consistent format with worldwide coverage. The launch of LANDSAT 2 in 1975 and LANDSAT 3 in 1978 have assured repetitive coverage for most of the land surfaces of the world. Remote sensing imagery has also been acquired from special systems on board the orbiting laboratories SKYLAB (United States) and SALYUT (USSR) and by the Indian satellites for earth observation BHASKARA. In addition, high-resolution images from meteorological satellites (NOAA and METEOR) have been used to study some surface phenomena, particularly in the field of hydrology.

A number of new systems are now in planning or under construction, ensuring that data will continue to be acquired for the rest of the decade. Two more LANDSAT satellites are under construction for launch in 1982 and 1983, the second carrying the thematic mapper, an advanced multispectral scanner with improved spatial and spectral resolution. France is developing SPOT (Satellite probatoire pour l'observation de la terre) which will have a pointable detector with ground resolution as fine as 10 m. SPOT is expected to be launched in 1984. Other future land observation satellites are being planned by Japan and the European Space Agency (ESA).

The land observation studies discussed in this chapter are based largely on data from the Multispectral Scanner (MSS) on the LANDSAT satellites since this system has provided the greatest quantity of data, the data are readily available, and the data have been used throughout the world. Catalogs and data are distributed by the EROS Data Center, Sioux Falls, South Dakota. Prices of photographic images range from $8 for a 1:1,000,000 scale black and white print to $50 for a 1:250,000 scale color composite image. Computer tapes of digital data cost $200 per image.

A LANDSAT image covers an area 185 km by 185 km, corresponding to an image 18.5 cm square at a scale of 1:1,000,000 and 74 cm square at a scale of 1:250,000. The MSS has four spectral bands in the visible and near-infrared regions of the spectrum; data are available either as black and white images of a single spectral band or a color composite image of three spectral bands. The photographic data products can be interpreted using techniques of air photo interpretation; alternatively, the data can be analyzed more precisely and quantitatively using digital data in the form of computer compatible tapes.

In those areas covered by LANDSAT ground receiving stations, images can be received once every 18 days, although some images may be less useful due to cloud cover. LANDSAT ground stations are now operating in the United States, Canada, Brazil, Italy, Sweden, Japan, India, Australia, Argentina, and South Africa and are under construction in Thailand, Bangladesh, China, and France. These ground stations can cover the surrounding area within a range of about 3,000 km. Areas outside the range of a ground station can be covered by means of on-board tape recorders, but these have limited capacity and short lifetimes, so such areas generally have only a fraction of the coverage of areas with ground stations. In a few areas of persistent cloud cover, no cloud-free coverage has yet been obtained because operations of the satellites over those areas have been limited.

Countries outside the range of a receiving station not only have less coverage but also experience longer delays in getting imagery. Under these conditions, it is generally not possible to conduct experiments requiring regular data delivered soon after acquisition by the satellite. This problem is being overcome primarily by means of the growing network of ground stations. For areas that remain outside the range of ground stations, more reliable but still limited service will be provided in the future by the Tracking and Data Relay Satellite System (TDRSS) which will obviate the need for on-board tape recorders by using two geostationary satellites to relay data from the LANDSATs to a central ground station. The limiting factor then will be the capacity of the central processing system.

Data from the USSR SOYUZ-SALYUT missions can be obtained internationally on the basis of a bilateral agreement between the USSR and the country concerned. On the basis of such agreements imagery has been acquired by USSR spacecraft over a number of socialist countries and developing countries, including Angola and Morocco.

Since photo-interpretation techniques have long been used operationally while computer-assisted interpretation is still largely experimental, most of the proven applications of satellite data are based on photo interpretation. The interpretation of satellite imagery is similar to that of aerial photography but more abstract since objects that may be recognizable on aerial photography such as trees, rock outcrops, roads, and buildings are not visible on satellite imagery due to its lower resolution. Since the satellite sensors view the land on a scale vastly different from the human scale, intuitive recognition of features can be unreliable, and more rigorous interpretation techniques are required. When aerial photographs of the area in question exist, they may greatly aid the process of establishing correlations between image features and terrain characteristics.

Computers can assist the interpretation process in various ways. The simplest and least expensive method is to use the computer to optimize or enhance the imagery for visual interpretation. The standard photographic products are often far from ideal in lightness, contrast, resolution, and spectral balance. Special processing can be used to produce an image that is optimized for a specific application. Computers can also be used for analyzing images on the basis of the image brightness in each of the spectral bands and assigning each element of the image to one of a number of classes. The correlation of these image classes with terrain features requires human intervention.

Whether image analysis is done visually or by computer, the final interpretation is done by the interpreter on the basis of his knowledge of the terrain. An interpreter will normally use whatever other information is available—maps, aerial photographs, ground survey data—to check the accuracy of his interpretations. In general, for large areas, ground surveys are most detailed and most expensive, followed by air surveys, with satellite surveys being least detailed and most economical. An interpreter may therefore use ground or air surveys extensively in the early phases of a survey project to ensure the necessary precision and later rely as much as possible on the satellite data to minimize cost.

3.1.2 Cartography and Land-Use Mapping

As the world population grows, so does the demand for food, wood products, housing, transportation, and other products and services. Each of these requires land, and the amount of new land that is both readily available and usable for particular purposes is steadily decreasing. Land-use patterns are constantly changing, commonly with forest land being converted to agriculture and agricultural land to urban use. If land is to be allocated to its most appropriate use, planners must have information on current land-use patterns and information on potential land capability. In the first case, satellite observations, verified by ground surveys, can provide up-to-date information on regional land-use patterns. In the second case, satellite images can provide some information on the distribution of soil types, vegetation classes, and other parameters which can be used in conjunction with information from other sources to determine suitability of the land for a particular use.

Bolivia began a major multidisciplinary mapping program using LANDSAT imagery in 1972. Prior to that time aerial photography had been acquired for 80 percent of the country, some of it dating back to the 1920s, and about 45 percent of the country was covered by topographic maps. The satellite remote sensing program was established to provide accurate up-to-date information rapidly and economically for areas where maps did not exist or were outdated. To cover the entire country, an area of 1,100,00 sq km, 65 images were required. Complete coverage was acquired by 1976 through the use of the on-board tape recorder; subsequently, new images have been obtained through the Brazilian receiving station.

Using 28 LANDSAT images acquired between 1972 and 1975, a series of

1:250,000 scale planimetric maps were produced covering 605,000 sq km. These maps were used in 1976 to plan the national population and housing census and served as base maps for delineating census zones. In areas where topographic maps existed the LANDSAT images were used to identify areas where either natural or man-made forces had caused changes.

Land-cover and land-use maps were produced in Bolivia for the first time in 1975 using LANDSAT imagery. In a period of two and a half years, a 1:1,000,000 scale map was produced covering the whole country. More detailed studies were done in the Oriente Boliviano, an area of 620,000 sq km in the eastern lowlands. These integrated surveys were carried out by multidisciplinary teams including geologists, soil scientists, geomorphologists, and geographers working in consultation with climatologists, botanists, and hydrologists. The results of these studies were published as maps at 1:250,000 scale including soil maps, terrain-type maps, and vegetation and land-use maps.

More recently, an integrated survey was made of the Department of Oruro to provide basic information for the planning and execution of development programs and to identify zones of high potential for agriculture or mining. Six LANDSAT images were mosaicked to cover the 53,000 sq km area and were interpreted with respect to structural geology and stratigraphy, mineralogy, hydrology, geomorphology, soils, and land cover and land use. Aerial photography was used where available and information from ground surveys was used to check the interpretation and to provide information that could not be interpreted from satellite or aerial imagery. The conclusions of the study identified areas of potential interest for mining and agriculture and identified the factors that posed problems to the development of other areas.

A study of the use of satellite data for monitoring urban expansion was carried out with LANDSAT data of San José, Costa Rica. To obtain the high resolution necessary for urban studies, computer analysis was used to classify each element of the image as urban or non-urban. Comparison of the results of the study with a map of urbanization derived from aerial photography indicated that the satellite data provided accurate information more rapidly and economically than did aerial photography.

3.1.3 Renewable Resources

3.1.3.1 Agricultural information.   In general, the accurate identification of plant species and vegetation condition requires ground observation. When large areas are to be surveyed, however, ground observations become prohibitively time-consuming and expensive. The use of aircraft or satellite surveys in conjunction with selective ground sampling can provide both accuracy and economy. Some information, such as the extent and location of cultivated land, can in many cases be interpreted accurately from satellite images with little need for ground data. In other cases, such as crop yield estimations, extensive data from other sources may be required to ensure accuracy.

The identification of crops using satellite imagery is partially successful. Since all crops contain chlorophyll, they have similar spectral characteristics, so additional information is generally needed. Since agriculture is generally seasonal, knowledge of the cycle of planting, growth, and harvesting combined with imagery acquired at different times in the growing season can be used to provide accurate identification of fields. Crop identification techniques are most accurate for areas where field size is large. In the Large Area Crop Inventory Experiment (LACIE) carried out in the United States, the area, growth stress, and state of maturity of wheat were observed, giving yield estimates with an error of about 5 percent. These techniques require frequent acquisition and rapid delivery and interpretation of satellite data.

Alternatively, satellite imagery can be used for planning the statistical sampling that is required in conventional agricultural information systems. In Costa Rica, research has indicated that LANDSAT data can be used economically for determining broad land-use classes. Aerial photography can then be used to subdivide these classes into sample units. Within randomly selected sample units, ground surveys are then conducted to identify crops and measure field areas. The delineation of strata and sampling units should be valid for about ten years before changes in land-use patterns would require restratification. It is estimated that the national estimates for major agricultural crops would have errors of 4 to 12 percent, with enumerated samples accounting for less than 1 percent of the total land area.

Infestations by plant parasites cause large losses of crops every year. If these infestations can be predicted, countermeasures can be taken effectively and economically. The prediction of infestations requires knowledge of the life cycle of the parasite, observations of weather and environmental parameters which hinder or assist the parasite's development, and the assessment of vegetal biomass which can serve as food for the parasite.

In 1976, a pilot study was undertaken by FAO to evaluate the use of data from remote sensing and environmental satellites to improve the existing methods of desert locust survey and control. The pilot study started in Algeria and was subsequently extended to Morocco, Tunisia, and Libya, the region of the Northwest Africa Desert Locust Control Commission.

The migratory desert locust (*Schizstocarea Gregaria*) is dependent on rainfall to supply soil moisture for egg hatching and for the development of green vegetation on which the locust feeds during its nymphal, nonmigratory and migratory states. Twice daily satellite meteorological data were combined with the sparse ground meteorological observations to locate and outline areas of precipitation. LANDSAT multispectral scanner data were used for an assessment of green vegetal growth resulting from the precipitation which was assessed and located using the environmental satellite data. It was found that areas with green vegetation of 5 to 8 hectares (ha) and with vegetal coverage as low as 25 to 30 percent can be positively identified by visual interpretation of the LANDSAT imagery. By using computer-assisted analysis, the limits were reduced to 2 to 8 ha and 20 to 25 percent, respectively. This information was correlated with ground observation of locust infestations with promising results.

One of the most important physical quantities needed for optimal agricultural management is knowledge of the amount of moisture in the soil. This quantity is very difficult to observe either on the ground or by remote sensing. On LANDSAT imagery, irrigated land is one of the most easily detected features. In Libya scientists are studying the usefulness of remote sensing data to identify the extent of irrigated regions in the Gefara Plain where water has been consistently overused for the last 20 years. The use of these data, along with others from the ground, is expected to help quantify the rate of recharge and withdrawal from the underground acquifers.

Microwave observations provide a good index of soil moisture, but only for the topsoil and when the vegetation is not too dense. Other indirect methods, however, can be used to infer the total reservoir of moisture in the soil. Since wet soil holds much more heat per unit volume than does dry soil, observations of day and night soil temperatures from satellites allow the amount of moisture present in the soil to be inferred. Extensive ground experiments for typical regions are necessary to calibrate this technique. Observations of wilting (water stress) in plants, which is seen in particular spectral bands, also show when the moisture has been depleted beyond certain levels, but this information is often too late to be useful in planning. Some experiments have been planned using radar systems to penetrate the soil to depths beyond the surface layer.

In India, LANDSAT data have been used for the delineation and mapping of agriculturally unproductive land affected by soil limitations such as salinity or alkalinity, water-logging, soil erosion, or rock outcrops. As the demand for food grows, it will be necessary to reclaim as much of this land as possible for agriculture. Maps showing the extent and severity of soil limitations will be required. The study carried out by the Indian National Remote Sensing Agency indicated that LANDSAT data enabled the affected areas to be mapped more rapidly and economically than with conventional surveys. Five categories of unproductive soils were distinguished on the basis of reflectivity; in order of decreasing average reflectivity these were: highly saline or alkaline soils, saline or alkaline soils, low-lying areas with high water tables, active sand dunes and barren sandy areas, and sandy soils.

3.1.3.2 Rangeland monitoring.

In general, rangelands are characterized by extensive grasslands, relatively low and highly variable productivity, and environmental limitations, notably poor soils and low and variable rainfall. Due to the extent and the relatively low economic productivity of these areas, survey techniques must be able to cover large areas at low cost. A primary consideration in range monitoring is therefore to balance the low cost of extensive data collection techniques with the high quality of the information of intensive data collection techniques. LANDSAT imagery rapidly produces general information at a cost of about $0.01 per square kilometer, while a ground survey can produce detailed vegetation data at a cost of about $100 per square kilometer. Aerial surveys will generally provide data that are intermediate in both cost and quality. A combination of satellite, aircraft, and ground surveys allows correlations between intensive and extensive data which can ultimately result in the phasing out of the more expensive intensive techniques. Ideally, management could then rely routinely on satellite data, supported as necessary by aircraft and ground sampling. Figure 3.1 shows a satellite image of an area of West Africa that is extensively used for cattle raising.

The Kenya Rangeland and Ecological Monitoring Unit (KREMU) uses LANDSAT imagery in conjunction with aircraft and ground surveys to monitor the country's rangelands, which are important for both tourism and domestic cattle raising. The LANDSAT data have been used to define ecological zones which provide the framework for aircraft and ground sampling. LANDSAT data can also be used to identify areas of relatively high productivity and to monitor the appearance of ephemeral vegetation. While it is not possible to determine animal numbers directly from the satellite imagery, it is possible, after correlation of satellite data with simultaneous aircraft observation of animal density and distribution, to estimate these parameters from observations of vegetation density and distribution. The Kenyan experience has demonstrated that inexpensive visual analysis of LANDSAT color images can yield good estimates of animal density. Similarly, vegetation patterns visible on the satellite imagery, after correlation with ground data, can be used to infer soil humidity.

Fires are often used to control rangeland vegetation, and a preliminary study in Upper Volta has demonstrated the potential of satellite imagery for monitoring the timing and extent of burning. Burned areas appear very dark on the satellite imagery and can usually be reliably distinguished from other dark features such as water bodies or laterite outcrops by their shape and relationship to the topography. In case of doubt, comparison of imagery from different years

Figure 3.1 Satellite images have depicted several places where sharp boundaries appear in otherwise rather uniform semi-arid regions. These boundaries are often the line of demarcation between places where grazing or land use is controlled and places where it is not. Such images can assist managers of grazing in semi-arid regions to optimize the use of the land and minimize environmental deterioration. (Courtesy of NASA.)

is usually conclusive since burn patterns vary. The burning generally begins about a month after the end of the rainy season when the grass is dry enough to burn but not so dry that the fire will kill the roots or eliminate the shrubs and trees. Within about a month about 80 percent of the open rangelands are burnt. While it was not possible to accurately measure the persistence of the burn pattern due to insufficiently frequent coverage, it appeared that burns would fade and disappear in about a month in the southern more humid part of the country but might persist for more than a year in the northern drier regions.

In Brazil, LANDSAT imagery has been used to study the evolution of pastureland created by clearing the tropical rain forest and planting pasture in Para State. The soil fertility diminishes progressively causing a thinning of the grass species, a gradual invasion of species unfavorable to cattle raising, and eventually a reversion to rain-forest vegetation. The LANDSAT images clearly distinguished rain forest from cleared land except in areas where reversion was advanced. Within the cleared areas, four classes were distinguished: pastureland predominantly covered with grasses (46 percent of the cleared area), degraded pastureland with areas of bare soil (15 percent of the cleared area), areas predominantly covered by bare soil (15 percent of the cleared area), and areas reverting to arboreal and herbaceous rain-forest vegetation (24 percent of the cleared area). The cleared area covered a total of 55,000 ha. Ground surveys indicated that the areas of degraded pasture and bare soil could be reclaimed to pasture by better management techniques and replanting of grasses. The areas of reversion, some of which had been abandoned, could not be easily reclaimed. Degradation and reversion were most frequent in areas that had been cleared five or more years ago.

3.1.3.3 Forest mapping.

Forest lands are often found in areas of low population density and on land unsuited to intensive agriculture by reason of climate, soil, or topography. Like rangelands, they are often areas of relatively low economic productivity where expensive detailed surveys cannot be justified. Unlike agricultural or range areas, however, forest lands change slowly; hence mapping at a frequency of once every five to ten years may be sufficient for management and planning purposes. In areas of rapid deforestation more frequent surveys may be required.

For mapping extensive forest areas, the small scale of satellite imagery and the integration due to the limited resolution are useful characteristics. For mapping at scales of 1:250,000 or less, the similarity in scale between the images and the map simplifies the compilation process, in contrast to the use of aerial photographs in which 800 images at a scale of 1:50,000 would be required to cover the same area. However, not all the desired information can be readily interpreted from satellite imagery. Commonly, cleared areas can be accurately delineated, and boundaries between different tree communities can often be observed, but the determination of economic parameters such as the volume of wood or the identification of economically valuable species must be either inferred from correlations derived from air or ground surveys or must be sampled on the ground and extrapolated on the imagery.

In the context of the Global Evironment Monitoring System (GEMS), the Pilot Project on Tropical Forest Cover Monitoring was undertaken in Benin, Cameroon, and Togo by the Food and Agriculture Organization (FAO) and the United Nations Environment Programme (UNEP). Maps of forest and other vegetative cover were prepared for the whole of Benin and Togo at 1:500,000, and for the southern two-thirds of Cameroon at 1:1,000,000 through the use of LANDSAT data to extrapolate from detailed data acquired over limited areas by ground and air surveys. The LANDSAT data were interpreted by visual analysis of standard black and white and color images. Each country was divided into ecofloristic zones based on altitude and climate, and within each zone from two to twelve vegetation classes were distinguished. These are the first such maps of these areas produced at these scales. The satellite imagery permitted the relatively rapid and accurate delimitation of the major vegetation classes of interest. A comparison of these maps with a vegetation map of the Ivory Coast derived from aerial photographic interpretation shows that the detail and the range of vegetation types mapped are very similar.

Detailed monitoring of changes in forest cover by satellite data was limited both by the resolution of the data and the infrequent coverage due to the lack of a LANDSAT ground station covering the area. Nonetheless, LANDSAT imagery is particularly suitable for identifying areas of change because it permits the rapid examination of very large areas. The experience of the project indicated that forest could be readily distinguished from degraded forest or nonforest areas and that clearings of 5 to 10 ha were usually visible. When images of two different years, preferably from the same time of year, were available, a comparison of the images gave an accurate indication of clearing. If only one LAND-SAT image was available, changes could still be detected by comparison of the image with aerial photographs or maps.

In Peru, LANDSAT imagery has been used to study the humid tropical forests of the Amazon region. The forest vegetation could be distinguished on the imagery from nonforested areas such as treeless swamps, recently cleared areas, and many, but not all, agricultural and urban areas. Broad stratification of forest vegetation appears to be possible on the basis of landform, drainage conditions, and tree density. For example, areas of terrace and hill forest are differentiable from lowland forest by their rough texture on the images and by their association with dissected rolling terrain. In the lowlands, areas of permanent inundation could be distinguished from areas of seasional flooding. Areas of the economically valuable aguaje palm, occurring as pure stands, are frequently discernible within the lowland forests.

In the Philippines, a national forest inventory based on LANDSAT data was completed in 1977 using imagery acquired from 1972 to 1976. The total forest area was mapped and classified into five categories: full canopy dipterocarps, partial canopy dipterocarps, mangrove, high-elevation mossy forest, and nonforest wetlands. The survey showed that 30 percent of the country was still covered by forest, with two-thirds of that being full canopy forest. For individual islands, the forest cover ranged from 75 percent to 24 percent. This national inventory is serving as a base for national management and planning, for example, for programming levels of exploitation on log-excess and log-deficient areas. A comparison of a 1976 image with a 1972 image of Mindoro Island indicated a decrease in the forested area of 42,700 ha, or about 9 percent. A study of mangrove forest on part of Mindoro Island indicated a decrease of about 2,400 ha between 1973 and 1976, a decrease of about 3.8 percent per year. LANDSAT data have been used in conjunction with air surveys and fisheries data to designate areas of mangrove forest, totaling about 78,000 ha, for conservation.

3.1.3.4 Water resources.

Management of water resources requires models for water demand and for water availability, which generally do not correspond in either time or space. Monitoring of water availability may involve data on rainfall, runoff, storage, and evaporation. Satellite data can contribute to the measurement of all of these parameters.

Surface water is generally readily distinguishable on satellite imagery, and maps showing the location and surface area of natural or man-made water bodies can be compiled. In temperate zones or at high altitudes, snow can represent important quantities of stored water. In humid zones an overabundance of rainfall and runoff can cause flooding. In large river basins there may be some weeks delay between heavy rainfall in the upper parts of the basin and flooding in the lower parts. If flooding is to be predicted, the response of the basin to rainfall must be known. Since the flooding can extend over a large area and since it changes rapidly, repetitive satellite coverage offers an ideal technique for observing the flooding patterns. A thorough analysis would require frequent data over a period of years, but some features of the flooding patterns can be inferred from even a single image. Such a study was made of the Mekong River basin, substantially changing hydrologists' understanding of the hydrological dynamics of the basin.

In July 1978, following heavy rains, the Gezira agricultural development project in Central Sudan was heavily damaged. Information on the extent of the damage was unavailable for some time after the flood because of the inaccessability of the whole area and the lack of communication facilities. The United States Office of Foreign Disaster Assistance obtained LANDSAT data on the flooded area on 4 and 5 August and again on 13 and 14 August. Using visual interpretation techniques to map surface water and saturated soil, the flooded area was divided into three classes: highly damaged, moderately damaged, and slightly damaged. Then the area of each class was measured. From maps, the villages in the flooded area were identified and the number of people affected was estimated. A comparison of the images from 13 August and 4 August showed a slight diminishing of the flooded area.

In Egypt, satellite data have been used to monitor the surface area, siltation patterns, and aquatic vegetation distribution for the reservoir created by the Aswan High Dam. On LANDSAT images from four different dates, the surface area of water was measured and was correlated with water level data from the ground. Between the lowest water level, on the image of 13 June 1973, and the highest, on the image of 17 September 1977, the water level rose by 15.3 m and the surface area increased from 2,717 sq km to 5,681 sq km. As the water level and the surface area change, the patterns of deposition of silt by the annual flood also change. The movement of the flood front as indicated by turbidity can be clearly monitored on the images. When water levels are high and flood volume is low, silt deposition is limited to the southernmost portion of the reservoir; as water levels decrease and as the flood volume increases, silt deposition extends northward. In general, however, both satellite imagery and hydrographic surveys indicate that siltation occurs primarily in the region from 280 to 400 km upstream of the dam.

In Bangladesh, a study has been done of the changes in the Karnafuli reservoir due to siltation. The reservoir is filled during the annual rainy season to an elevation of 33 m (above sea level), and the water surface occupies 750 sq km. At the peak of the dry season the water level has typically declined to about 22 m, corresponding to a surface area of 272 sq km. Images taken in March of 1975 and 1976 at almost identical water levels indicated a reduction in the water surface from 619 to 580 sq km between the two years. This was interpreted to indicate that siltation had produced 39 sq km of new land at that water level and reduced the capacity of the reservoir correspondingly. This heavy siltation is apparently due to shifting cultivation on the steep slopes surrounding the reservoir. On an image of March 1975, shifting cultivation appeared to occupy 1,165 sq km of a total watershed of 5,980 sq km.

3.1.4 Nonrenewable Resources

3.1.4.1 Tectonics.   Satellite imagery has created new possibilities for investigating the earth's composition and structure, for research as well as for exploration for minerals, energy, and other earth resources. Today, images from a number of satellites are successfully being applied in geological exploration. Data for relatively large areas acquired under the same illumination conditions and in several spectral bands make it possible to locate and identify various surface structures, both local and regional in character.

One of the valuable characteristics of satellite images is that they not only provide an overview of regional geological structure but also indicate surface morphology and, indirectly, topographic relief. Experience has shown that even in well-explored areas, new and useful information, such as new information on major faults, can be obtained in this way. Other structural forms, such as ring structures several kilometers to several hundred kilometers in diameter, can also be recognized. Tectonic maps containing such information make possible a better insight into the structure of our planet. Figures 3.2 and 3.3 show linear and curved features extracted from satellite data.

LANDSAT data have been used in the Philippines to assist in the evaluation of earthquake-prone areas. The data were specially processed to enhance the lineaments of the Central Luzon alluvium covered basin. This information was combined with existing lithologic and earthquake recurrence maps of the area to produce an earthquake risk map at a scale of 1:500,000 showing areas that are least susceptible, slightly susceptible, moderately susceptible, and highly susceptible to earthquakes. This map is intended for land-use planning and construction planning.

3.1.4.2 Mineral exploration.

There is growing belief that most near-surface metallic deposits have now been found, at least those detectable by direct observation and conventional techniques. Nevertheless, expectations were high that LANDSAT could spot diagnostic indicators of potential mineralization, particularly in remote or still poorly explored regions. To some extent, this has happened but the list of new discoveries remains small.

The chief indicators open to satellite detection include fractures and faults that are related to mineral emplacement, curvilinear fracture patterns and landforms that indicate underlying intrusions, and alteration products associated with primary and secondary ore bodies. The latter are typified by "gossans," which are assemblages of altered minerals (dominated by hydrated iron oxides) developed from weathering processes. Under favorable conditions, multispectral sensors can pick out iron-rich rock surfaces in vegetation-poor, well-exposed bedrock and derived soils. However, iron staining is ubiquitous because rocks of many kinds, even if not mineralized, contain iron compounds that alter easily. It is a drawback of existing systems that iron-poor, clay-rich alteration products and other mineral guides to ore are not detectable or separable solely from reflectances in the 0.5 to 1.1 μm spectral range.

A planned sensor called the thematic mapper promises to correct for this. Spectral bands centered at 1.6 and 2.2 μm are sensitive to characteristic radiation from a wider range of mineral and rock types that can be identified from existing data. The inclusion of the shortwave infrared bands in the thematic mapper and the potential indicated by the thermal bands—broadly supported by results from the Heat Capacity Mapping Mission—augur well for metals exploration.

In Brazil, satellite imagery has been used to assist in the mapping of tin deposits. The tin mineralization is associated primarily with biotite granite domes intruded into metamorphic gneisses, schists, and granites. The tin has been weathered and concentrated into rich alluvial and eluvial placer deposits. The deposits in the present drainage system are now being worked extensively, but large unworked reserves also exist in the paleovalleys and channels dating to other climatic periods and completely unrelated to the present drainage system. The tin-bearing domes are not distinguishable on either black and white aerial photography or side-looking airborne radar (SLAR) imagery.

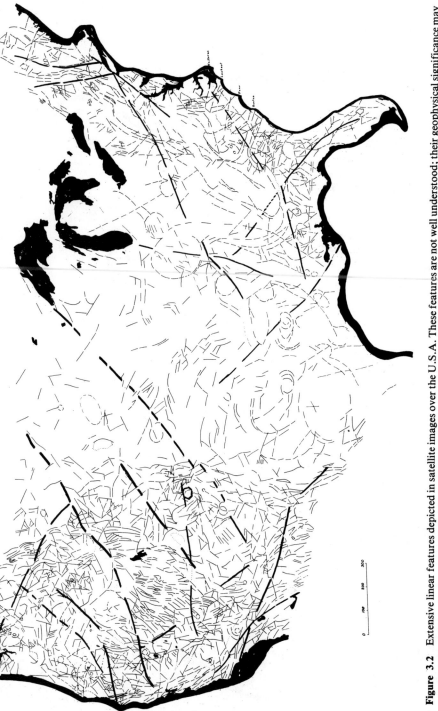

Figure 3.2 Extensive linear features depicted in satellite images over the U.S.A. These features are not well understood; their geophysical significance may relate to tectonic movements, mountain-building activity, or mineralization zones. They often are useful clues to likely places for mineral prospecting. Almost all large land masses show such linear features. (Courtesy of W.D. Carter, EROS Program, U.S. Geological Survey, and the American Association of Petroleum Geologists.)

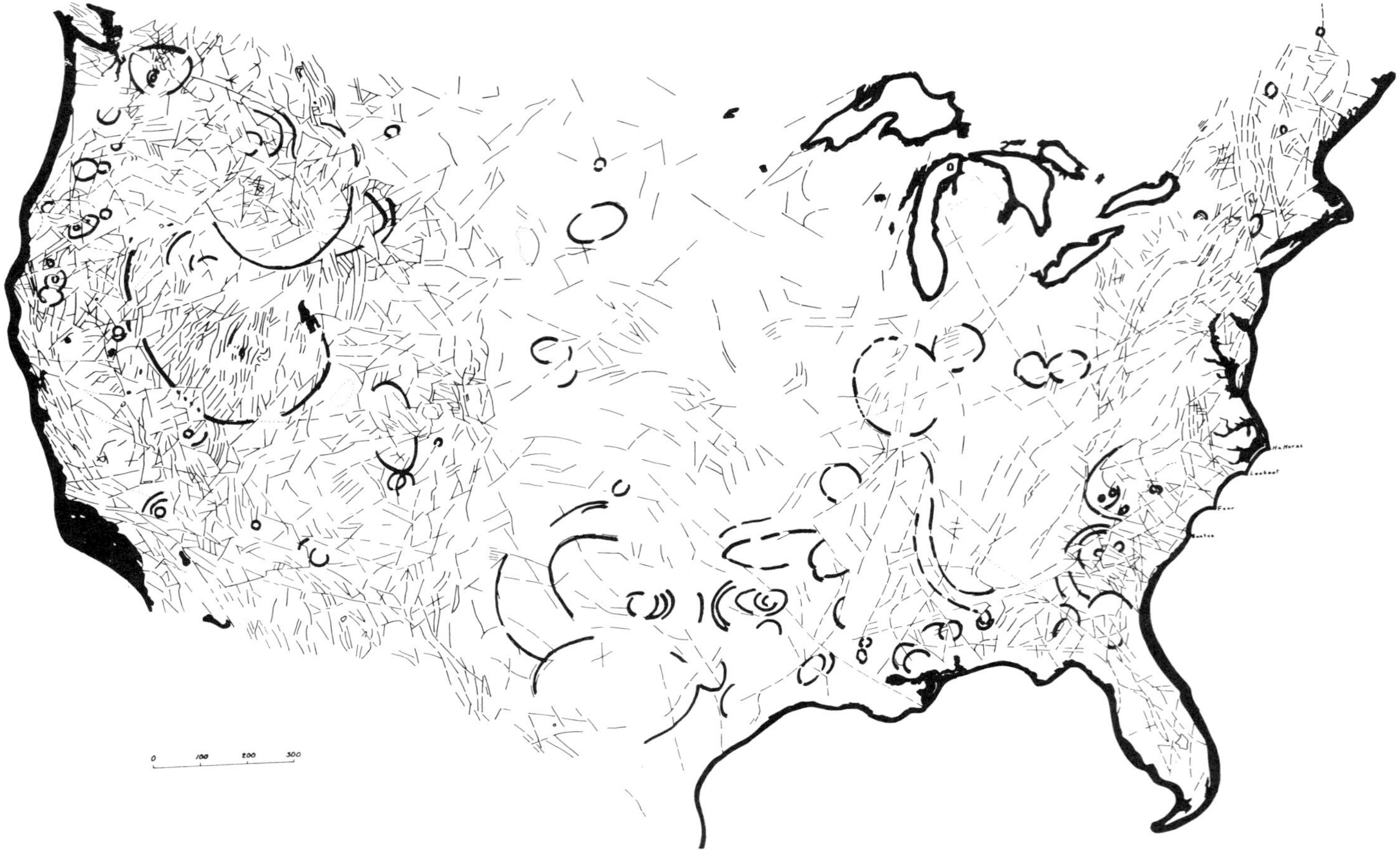

Figure 3.3 Curved features recently revealed from extensive analysis of satellite images. Some of these have been dated and found to be very ancient, some 2,000 million years old. It is believed that some of these are remnants of pre-Cambrian impact craters of the young Earth. Again, some of these features are associated with mineralization. (Courtesy of W. D. Carter, EROS Program, U.S. Geological Survey, and the American Association of Petroleum Geologists.)

On LANDSAT infrared or false color imagery, the tin-bearing granites have been differentiated from non-tin-bearing granites. Several extensive and economically important tin belts, both north and south of the Amazon Basin, are evident on the images. One tin belt can readily be followed from eastern Bolivia, through Rondonia, Mato Grosso, Amazonas, and Para States to beyond the Rio Xingu, a distance of about 2,000 km. The tin-bearing granites, interspersed with barren intrusives, outcrop along the entire belt.

In Botswana, satellite imagery has been used to help locate superficial deposits of materials for low-cost road construction. Due to the general lack of outcropping bedrock throughout the Kalahari Desert region, a calcareous material known as calcrete is being studied for use in road bases. Field investigations have shown that LANDSAT imagery can be effective for rapid initial location of calcrete occurences over very large areas within a route corridor. Color composite imagery shows changes in sand color caused by near-surface calcrete deposits. This technique was used when a materials survey had to be undertaken at short notice for a section of a road under construction. Of 16 potential sites identified on satellite imagery, calcrete was found within an area of 1 sq km or less at 14 sites.

3.2 OCEANOGRAPHY

Remote sensing of the oceans from space has advanced only slowly, and with good reason. First of all, only the surface can be observed; electromagnetic radiation cannot penetrate very far into water. Yet, as will be shown, some surprising inferences can be made about conditions well below the surface. Moreover, the atmosphere, in particular clouds, haze, dust particles, and water vapor affect observations of the sea surface made from space. Such complications can be dealt with by using active (radar-like) radio systems to "see" through atmospheric hindrances.

Since the oceans cover about 75 percent of the earth's surface, they have a large impact on man and his environment. The seas dominate our weather system, they provide resources such as food and other materials, they are essential to world commerce, and they provide recreation and many other benefits to man. On the other hand, they pose hazards such as tidal waves, they cause coastal erosion, and they are the birthplace of hurricanes and other damaging tropical storms.

It is evident that only orbiting spacecraft can effectively perform the formidable surveying task required by oceanography. The monitoring requirements include surface temperature, ocean currents and circulation, salinity, coastal dynamics, sediment transport erosion, shoaling, surface winds, waves and wave diffraction, and sea ice and its dynamics. Obviously, ship and aircraft operations would be too limited and prohibitively costly to do all this on a continuous basis.

The data from remote sensing and meteorological satellites provide a large amount of information on ocean processes. Remote sensing data are particularly useful in the coastal zone and in ice-infested waters while meteorological satellites have been used extensively to measure sea surface temperatures. Most observations of the oceans have thus been made with sensors that were really designed for observing land or atmospheric properties. This has obviously limited the development of applications of satellite observations to oceanography.

The first satellite dedicated to oceanography was the United States SEASAT launched in 1978. This spacecraft carried five ocean sensing instruments as well as laser corner cube reflectors for accurate orbit determination. The ocean sensors were: an altimeter to determine the ocean topography and the significant wave heights; a scatterometer to determine surface wind speeds and directions; a scanning multichannel microwave radiometer for measuring the sea surface temperature in all weather conditions; a visible and infrared radiometer to determine ocean color pattern and temperature; and a synthetic aperture radar to provide images of the oceans in order to determine wave and current patterns, ice fields, and coastal/ocean interaction.

3.2.1 Sea Surface Temperature

The early infrared observations from satellites showed that patterns of temperature variations across the oceans could be detected readily, especially near the great current systems where there are strong temperature gradients. Figure 3.4 is an example showing many complicated temperature structures of the Gulf Stream, off North America. First of all, the meanders of the Gulf Stream can easily be seen. These meanders change daily.

Figure 3.5 also illustrates other phenomena of interest. The large circular feature is a gyre, or eddy, which has been spawned by the meanders of the Gulf Stream and the interaction of the boundaries of the fast-moving stream with the slower-moving surrounding waters. Both cold and warm eddies can form. They move slowly away from their sources and sometimes exist for months. When they move into other waters, they form cold or warm temperature anomalies which can influence the ocean itself and also the overlying atmosphere. Most eddies of this kind are of the order of a few hundred kilometers in size. Others that are formed under different situations may be a few thousand kilometers in diameter.

Oceanographers and meteorologists are paying much attention to these eddies, which were unknown a few decades ago. Such structures probably play an important role in the transport of heat from the equator to the poles, and also horizontally across the ocean basins. In order to understand the close coupling between the atmosphere and the oceans, which is one of the most important physical processes influencing climate, the behavior of these eddies must be studied and understood. Ship surveys are the best way to study a particular eddy.

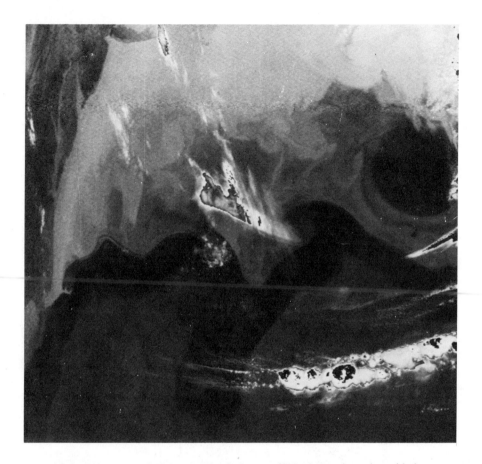

Figure 3.4 Image of the North Atlantic Ocean, off North America, taken with the infrared detector on a U.S.A. meteorological satellite. The Gulf Stream is shown after it has left the continent and heads for Europe. The Stream has meanders in it, as many rivers have, and the interaction of the Stream with the stationary boundary water produces cut-off loops that form ocean eddies or rings that move away, and which may last months. Such a ring or eddy is seen to the north of the Stream near the eastern margin of the image. (NOAA-6 photographs.)

Satellites offer the best way to determine, over an entire ocean basin, how many there are and their distribution of sizes, lifetimes, and motions.

3.2.2 Ocean Circulation

Ocean currents, transporting large amounts of energy, generally from lower to higher latitudes, have a profound influence on weather and climate. The rather mild climate of western Europe as compared to the more harsh interior is well known to be caused by the warm water masses of part of the Gulf Stream.

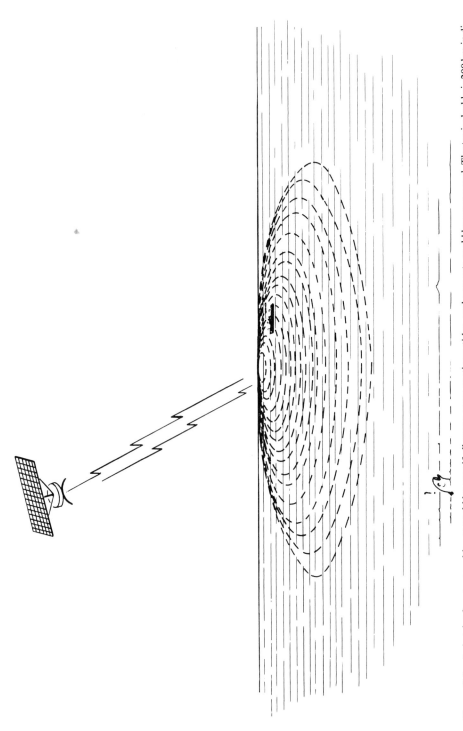

Figure 3.5 A sketch of an ocean eddy, as could be seen if all ocean wave motions, tides, and currents could be suppressed. The typical eddy is 200 km in diameter, has a 0.5 to 1 m rise (or depression) in the center, and may differ in temperature from surrounding waters by 2 to 4°C. This sketch is not to scale, the

The enormous quantities of energy transferred by ocean currents are dramatically illustrated by hurricanes and typhoons, whose primary energy source is the vast amount of heat stored in the oceans. Thus, predictions of weather and climate cannot realistically be undertaken without considering the ocean circulation. Further, knowledge of ocean currents is important for the maritime and fishing industries as well as for navigation. Knowledge of the speed and direction of these currents is also needed for studies of the dispersal of pollutants of sewage and of other waste products discharged into the sea.

Sealed bottles, with notes inside indicating the time and location of release into the waters, were once a way of tracing large-scale ocean motions. Now small buoys, with a few instruments to measure the surface pressure and air and water temperature and carrying low-power transmitters, can be followed by satellite.

Some 300 of such buoys were set out in the southern oceans during the 1979 Global Weather Experiment. The motion of the satellite with respect to a relatively slow-moving buoy produces a Doppler shift in the frequency of the received signal (similar to the shift in tone of an ambulance siren as it approaches an observer, passes, and then recedes). Analysis of these signals provides the position of a buoy at the time that the satellite passes overhead (or when the buoy–satellite distance is minimum). Figures 3.6 and 3.7 show the tracks of some of the buoys deployed in 1979. The paths traced out by these buoys provide a new source of information on the circulation of the southern oceans. Plans are being made for additional buoy experiments to study oceanic motions elsewhere, to obtain information on the year-to-year variability, and to improve buoy lifetime and the reliability of the sensors. Figure 3.8 shows the track of one buoy in the South Atlantic. This information is useful in tracing the origin of lobster larvae found on a seamount in the East South Atlantic, a valuable fishery resource. Satellite-tracked drifting buoys may aid in research on the origin and movements of fish stocks.

In addition to the ocean motions, the 1979 buoy experiment shows that the atmospheric circulation in the south is much stronger than expected. Storm depressions were measured more reliably than ever before and in more detail, giving useful data on the statistics of storms. The information is so useful that consideration is being given to operating a buoy–satellite system routinely.

Ocean eddies can be discerned not only by their temperature differences with surrounding waters but also by their dynamic structures. They rotate, and water piles up in their centers. The poet, rhapsodizing on the sphericity of the earth, would be astonished at how rough the surface really is over the oceans. Even if all waves and tides and the effects of the earth's rotation were absent, the sea surface would not be perfectly smooth. It would have hills and valleys, as well as the mounds of the ubiquitous eddies, as shown in Figure 3.5.

Two recent satellites, GEOS–3 and SEASAT, carrying very accurate radar altimeters, have provided some remarkable new information on how the ''average'' sea surface is affected by dynamical forces of moving water and by ir-

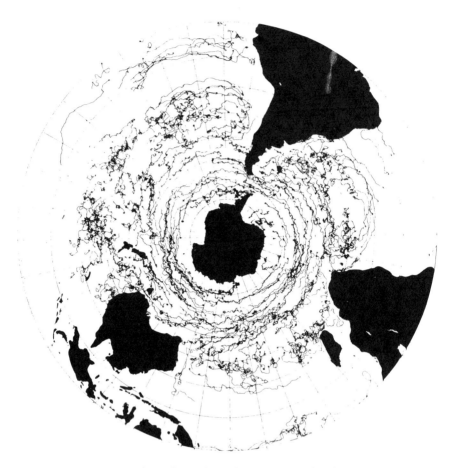

Figure 3.6 Tracks of satellite-tracked drifting buoys deployed in the southern hemisphere for the first GARP Global Experiment. (From IOS, Canada.)

regularities of the earth's gravitational field. These altimeters can measure the distance between the satellite and the sea surface to an accuracy of about 7 cm (70 mm) from heights of about 1,000 km. In order to use this information for oceanography, the real shape of the earth under the satellite has to be known, as does the exact distance of the satellite from the center of the earth. Figure 3.9 shows observations for eight satellite passes at three-day intervals over a portion of the North Atlantic at 70° W. Depressions of the sea level, by 1.3 to 1.7 m, below the expected surface are seen; these correspond to the Gulf Stream. Because the Gulf Stream is in strong motion, a depression is formed on one side of the Stream and a wall on the other side; the situation is somewhat analogous to the system of ridges and troughs (high- and low-pressure zones) in the atmosphere. Away from the Gulf Stream, it seems likely that a depression arising

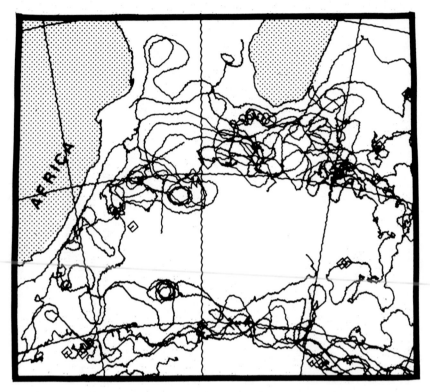

Figure 3.7 Drifting buoy tracks southeast of Africa show a closed pattern, indicating that fish larvae from the African coast could reach the waters south of Madagascar. (From IOS, Canada.)

from a cold eddy was also seen. Figure 3.10 shows a satellite track over an area of rough ocean bottom topography. Because wherever on earth there are mountains and valleys the gravity field is uneven, water acts as a natural level, and thus the sea surface reacts to the gravity irregularities. A remarkable correspondence is seen in Figure 3.10 between sea mounts and ridges and the ocean surface.

The success of these studies is prompting an intensive study of how very precise altimetry experiments should be conducted in future years. Studies are under way in a number of countries on the precision needed for the satellite orbits, and on measurements of the earth's gravitational field with sufficiently high accuracy and in fine enough detail, that the satellite altimeter observations can be used to their full potential. It is estimated that an overall accuracy of 10 cm, taking all factors into account, will give useful information on the great oceanic current systems and on the more vigorous eddies. However, an accuracy of 2 cm would be needed for making a complete study of the ocean currents. Future experiments are being planned that would achieve this goal.

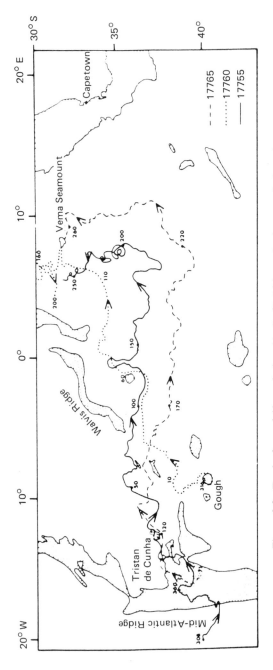

Figure 3.8 Track of a special buoy deployed by South African scientists to study the trajectory of lobster larvae, which terminated their sea voyage at Vema Seamount and were responsible for the maintenance of an important fishery. (J.R.E. Lutjeharms, in *Advances in Space Research*, Official COSPAR Journal, vol. 1, no. 4, Proceedings from the Symposium on FGGE, 1981.)

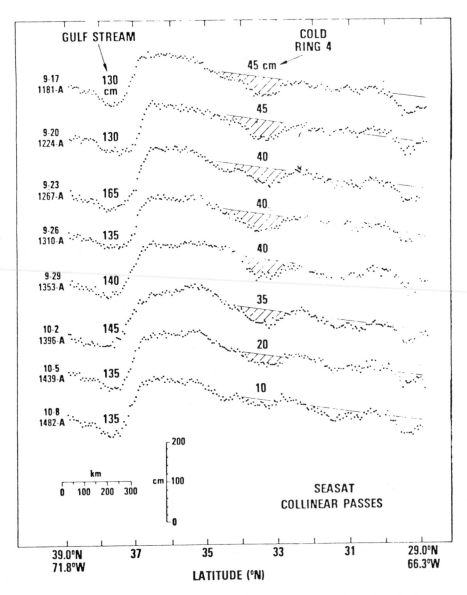

Figure 3.9 Radar altimeter observations from SEASAT over the western North Atlantic. Observations from many identical orbits were combined here to eliminate random signals arising from waves and nonsteady currents; the effects of tides were calculated and removed. The residual clearly shows the depression in the ocean surface arising from the dynamic effects of the Gulf Stream; a smaller depression is noted that is a cold eddy or ring, as confirmed by ship observations. (T. D. Allen, in *Advances in Space Research,* Official COSPAR Journal, vol. 1, no. 4, Proceedings from the Symposium on FGGE, 1981.)

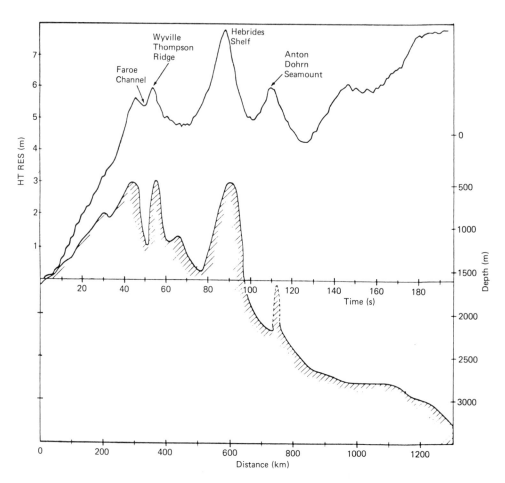

Figure 3.10 Radar altimeter observations from SEASAT over the eastern North Atlantic. Removal of wave, current, and tidal signals reveals that the ocean surface is a very faithful reproduction of the ocean bottom topography. This effect arises because the ocean surface responds to gravity anomalies that are produced by uneven topography. (T.D. Allen, in *Advances in Space Research,* Official COSPAR Journal, vol. 1, no. 4, Proceedings from the Symposium on FGGE, 1981.)

3.2.3 Sea Waves

Special radars on satellites can observe the patterns of reflections from small wind-induced waves. From these patterns, the winds near the surface can be deduced. This information is vital to marine forecasts for shipping and to the study of atmosphere–ocean interactions. Shipping and marine operations also require reliable information on ocean wave heights. The radar altimeter discussed above can provide this information. The radar pulses reflected back by each particular part of the surface are averaged to obtain the distance between the satellite and the surface. The spread of the radar pulse signals returned to the satellite are a good measure of the sea wave heights. Comparisons have shown that the satellite signal is more reliable and accurate than conventional measurements made from special buoys or than estimates made from ships.

3.2.4 Coastal Zone Color Monitoring

To the human eye, the ocean is blue (in regions where there is little microscopic life), green (where life is abundant), or murky (where there are sediments from continental runoff). The Coastal Zone Colour Scanner on NIMBUS-7 provides a quantitative way of determining ocean color, and, through the different amounts of blue, green, and red reflected by a given patch, the amount of chlorophyll can be estimated. For this new technique, the amount of data needed from the satellite is very large, and extensive computer processing is required. Once such results are assimilated and understood, the method will be valuable for studying marine productivity.

The distinctive spectral reflectance of crude oil permits satellites, under certain circumstances, to estimate the extent of an oil spill and to track and forecast the probable direction of its spread. LANDSAT imagery has been used to monitor the massive oil spill from the oil platform blowout in the Gulf of Mexico.

3.2.5 Ice

As a large component of the system that determines weather and climate, the ever-changing ice and snow masses of the earth are strong influences on the dynamic conditions of the atmosphere and oceans. Sea ice is particularly important because of its large seasonal and interannual variations. Its extent affects the solar radiation absorbed in a hemisphere and hence the earth's heat balance. Sea ice extent alters atmosphere–ocean exchanges of heat, moisture, and momentum. The ice sheets of Greenland and Antarctica are also important because they contain a large fraction of the earth's fresh water and because the fluctuations observed in the ice sheet volume serve as precursors of sea level change. An understanding of the earth's ice cover will eventually contribute to better extended range (seasonal) weather forecasts and to better estimates of the climate changes in the next few decades.

Ice is a major obstacle to the development of resources at high latitudes. Icebergs are navigation hazards and pack ice restricts shipping on northern routes. Unfortunately, polar regions are often cloudy or hazy in just those seasons when ice must be monitored. High-resolution observations using visible light are, therefore, limited at such times. Fortunately, microwaves penetrate clouds. Satellites carrying large microwave antennae image the oceans, and the ice can thus be "seen," but only with coarse resolution. New satellites are being planned with very large antennae that will improve the ability to observe ice and to determine the presence of leads (open sea tracks in pack ice) and icebergs. Microwave observations of the circumpolar ice field around Antarctica, averaged over a week or so, have shown changes in the ice pack from year to year. There are considerable interannual variations in the way that the ice forms in winter, breaks up in spring, is transported away, and melts. In some years, a large ice-free region forms in the Weddell Sea, probably indicating the presence of a strong ocean current. From these preliminary studies, it has been concluded that the oceanic and atmospheric circulations in the Antarctic are much more complex than had been known before. These circulations must be studied to find out the causes of the variability; they must be monitored to provide better information for weather and climate studies in the southern polar region since strong weather there appears to have an influence throughout much of the southern hemisphere.

Over the past decade, advances in remote sensing have made spacecraft data indispensible to scientific and commercial applications. For example, discrimination of snow and ice regions from land or ocean background and classification of ice type have been studied with both passive and active microwave remote sensing techniques. Daily and yearly variations of sea ice cover have been monitored by passive microwave images, and ice sheet elevations have been mapped by satellite radar altimeters.

3.2.6 Fisheries

Surface temperature data can be used in support of fishing activities since some species of fish prefer a particular temperature range and avoid waters that are either too warm or too cold. Some fish of great commercial importance, for example, the tuna family, feed on smaller fish which feed in turn on small life which is abundant only in regions of strong upwelling. These vertical ocean currents bring rich nutrients from the depths to the surface regions where microscopic animal and plant life form the beginning of the marine food chain. Off the west coast of the United States, satellite monitoring of cold water upwelling is used to identify regions of potentially good fishing. Data on ocean color can also be used to provide information on the distribution of nutrients and microscopic life, and therefore of fish.

3.2.7 Future Outlook

Several countries are planning to launch oceanographic satellites in the next decade to provide operational data for forecasts of marine weather and ocean conditions and for continued research on the dynamics of the oceans. Some marine-oriented satellites will carry emergency search and rescue systems (location devices, with automatic communications capability), and some will relay information to those involved in marine operations such as fishing. Both oceanic and atmospheric monitoring and research satellites will be needed for the continuing effort to understand the physical bases of climate, with its short-term and long-term variabilities, and of air–sea interactions. They will also provide support for economically important marine operations, in particular the estimation of the potential food resources of the oceans.

3.3 OBSERVATIONS OF THE ATMOSPHERE

The earth's atmosphere is a very complex geophysical system. Its basic motions are governed by forces set up by the unequal loss of heat to space between the equator and the poles; heat is transported from equatorial to polar regions. The basic motion is also strongly influenced by the earth's rotation and, to a lesser but still considerable extent, by the roughness of the continental terrain, especially the great mountain ranges.

The sun's energy that reaches the earth's surface is mostly absorbed there, heating the surface. Gases expand when heated, and when a body of gas is heated from below, buoyant forces are created. These forces cause vertical motions, giving rise, for example, to thunderstorms and other severe convective disturbances.

To a good approximation, the atmosphere neither loses nor gains gas; any motion at one place must be compensated for by flows elsewhere to maintain a mass balance. However, the atmosphere is so large that imbalances, such as high- and low-pressure weather systems, can exist for some time.

Water in its various forms is an important source and sink of energy in the atmosphere and on the earth's surface. The solar energy that is absorbed in the oceans ends up as (latent) heat energy of the evaporated water. When the water vapor condenses into clouds, the heat energy is released; this affects the energy balance and the motions in the vicinity. Water evaporated from the oceans and from other water bodies often travels great distances before it condenses. Thus, there is a wide distribution of energy within the atmospheric system.

All of the above physical ingredients contribute to the complexity of the behavior of the atmosphere, its motions, and its state, termed the weather. The temperature, amount of moisture, winds, amount of clouds, and the pressure at the surface, which is a measure of the mass of the atmosphere above any place, are the basic variables that describe weather.

3.3.1 Weather Forecasting

Satellite data were used successfully in meteorological operations as soon as the first satellite cloud pictures appeared two decades ago. It rapidly became evident that satellite cloud pictures show peculiarities that cannot be followed by means of the network of meteorological stations on the earth, where the observer has a limited field of view. Meteorologists use cloud patterns to estimate the strength and direction of the winds and to correct the locations of fronts and storms on weather maps. Satellite data are used qualitatively for synoptic analysis and forecasting; their quantitative use, through statistical and numerical methods, will lead to an improved understanding of atmospheric processes and to an improvement of weather prediction.

The possibilities opened up by the use of sensors aboard satellites have provided the background for such important international programs as the World Weather Watch (WWW) organized through the World Meteorological Organization (WMO) and the Global Atmospheric Research Programme (GARP) organized jointly by WMO and the nongovernmental International Council of Scientific Unions (ICSU). Both polar orbiting and geostationary meteorological satellites are vital in closing the gaps in the standard meteorological observing system. They are also being used for the collection of meteorological data from automatic stations and from ships, buoys, balloons, and aircraft and, further, for the dissemination of meteorological information.

Progress will undoubtedly be made in the coming years, not only by obtaining new types of information by means of satellites, the data being readily available to all countries, but also by improving the quality of information currently derived from satellite data, as well as by improving the uses made of such information, especially in the developing countries. The extended use of geostationary satellites will have a major impact on weather services in the tropics and midlatitudes, especially for observing severe storms and providing the information necessary for timely warnings.

3.3.2 Clouds

As seen from the earth's surface, clouds display great local variability in shape, motion, and even brightness, which all depend on the motions, temperature structure, and amount of moisture available. When large storm systems are present, or when clouds form at the ground to create fogs, the structure of cloudiness is not apparent. Seen from aircraft, however, cloud patterns show very large structures; these may be lines or avenues of small clouds or solid decks of clouds with slightly rippled upper surfaces, punctuated here and there by huge towers of rising cloud forms, indicating a great amount of local surface heating and buoyancy. Seen from a satellite, especially from the vantage point of a geostationary satellite at a height of about 36,000 km above the earth's surface, many very large cloud patterns can be interpreted to give important information on planetary-scale atmospheric motions, sources of moisture and its transportation to other regions, and weather. Among the earliest space observations of cloud patterns were indications of such weather phenomena as fronts, jet streams, and large storms. Figure 3.11 shows a storm system over land and Figure 3.12 shows a large ocean storm, a hurricane in the Atlantic.

Satellites that revolve around the earth above the equator at the same angular speed as that of the earth's rotation hover over a particular point and are

Figure 3.11 Image of extensive and intensive storm over the central U.S., from geostationary altitude. The image was processed by computer to preserve the detail in the brightest portions of the cloud image. (NOAA/NESS photograph.)

thus called geostationary satellites. The meteorological geostationary satellites spin at about 100 revolutions per minute for stability. Because of their spin, they act as gyroscopes, keeping very accurate timing and a very accurate pointing direction in space. Thus, sequential views from such satellites (about every 20 or 25 mins) give quite accurate indications of the movements of clouds and cloud systems. From the latter, the motions of the large storms may easily be determined and a good idea gained of the probable trajectory of the storm. Thus, timely warnings of their approach to populated areas are possible. Clouds also modify the gain and loss of heat of the atmosphere and of the earth's surface by radiation. The rapid cooling of surface air in winter under clear sky conditions is but one example of how the relationship of cloudiness to temperature is important to human activities.

The system of meteorological satellites circling the earth every 90 or 100 mins at altitudes of about 1,000 km in orbits reaching the polar regions, and those circling above the equator and remaining geostationary, provides a nearly continuous coverage of the globe. This coverage was complete during the Global Weather Experiment of 1979. Now some of the satellites have failed or have been moved, but there are definite plans to renew the satellite system to provide once

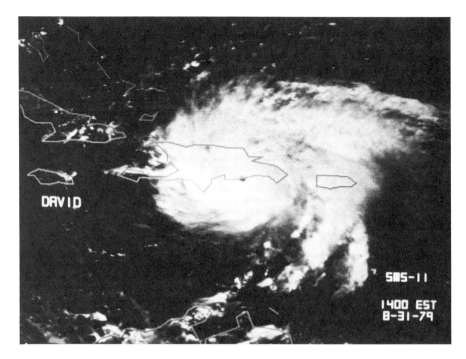

Figure 3.12 Image of Hurricane David, one of the most severe Atlantic storms of this century, as the center of the storm was about to strike Cuba. Such images, taken every half hour, allow accurate warning information of a storm's location and probable progress. (NOAA/NESS photograph.)

again essentially complete global coverage. In the mid-1980s scientists from many countries are planning a major effort to analyze the cloud information obtained from the international satellite system to produce the most complete description available of the average cloudiness, height distribution of clouds, type of clouds, variation of these quantities throughout the year and variations from year to year. This work will form the basis of a continuing cloud watch, the results of which are needed to verify and improve numerical models of weather and climate, because the cloudiness itself could be an important index of any change in climate.

3.3.3 Winds

Knowledge of the motion of the atmosphere is one of the most important kinds of information needed to make useful, long-range forecasts of weather. Over land, and in some oceanic regions of the world where there are numerous islands, wind observations are obtained from tracking the motions of radiosonde or pilot balloons as they ascend. Over vast stretches of the oceans, however, it is impractical to obtain such observations using balloons. The satellite systems of the 1970s, however, make it possible to fill this information void, very usefully if only partially. The geostationary satellites are so accurate and stable in their timing and pointing that sequences of cloud images can be used to measure the winds by using certain types of stable clouds as tracers instead of balloons. Of course, the winds can be inferred only where there are clouds. Although suitable clouds are not present all the time and at all the desired altitudes, useful information is, nonetheless, available. Figure 3.13, for example, shows the kind of wind field that can be obtained by analysis of cloud motions. This example is over the Indian Ocean, at about the time of the onset of the monsoon in 1979, during an intensive international experiment to study the dynamics of the monsoon phenomenon in detail. Great advances are now being made in understanding how the monsoon evolves from the first northward

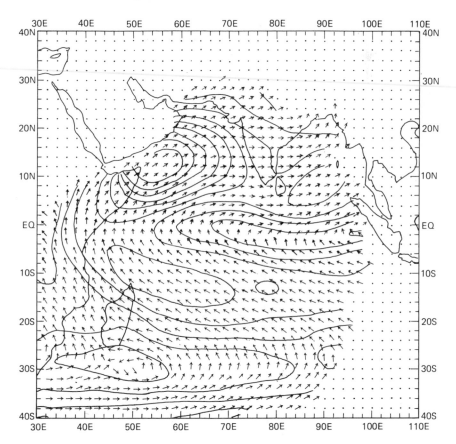

Figure 3.13 Computer derived wind field in the Indian Ocean region, during the onset phase of the 1979 monsoon. The cross equatorial air circulation can easily be seen. Such satellite observations are critical to mapping the change of the air flow that signals the beginning of the monsoon. (From Space Science and Engineering Center, University of Wisconsin.)

movement of moist air across the equator and why it behaves differently from year to year. The prospects of making useful monsoon forecasts within a decade or so are promising; such a prospect is important to hundreds of millions of people.

Even where there are no clouds, there is almost always some water vapor. Although it is invisible to the human eye, it is visible to the special detectors on satellites. Pictures of water vapor (Figure 3.14) can also be analyzed to obtain winds. In 1979, only METEOSAT contained such special detectors, but now it is planned that in the near future others will be so equipped for the purpose of increasing the amount of wind information that is vital for improved forecasting.

For future meteorological uses, in the 1990s, there is promise that a powerful laser ranging system on a satellite will provide wind information nearly everywhere and at a large number of different levels in the atmosphere. The laser signals reflected back to the satellite by scattering particles in the atmosphere, such as dust particles, water droplets, and possibly even molecules of the atmospheric gases, will make it possible to calculate the motions of these particles and, hence, of the winds. Such a laser system will require much technical development and perhaps new inventions to prolong the lifetime of the laser light

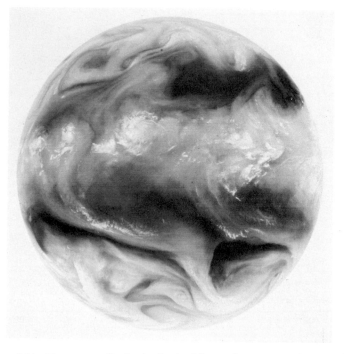

Figure 3.14 Water vapor distribution in cloud-free regions as seen using a special infrared channel on the European Space Agency geostationary satellite METEOSAT. Experiments have shown that it is possible to follow the movements of such patterns and thus derive wind fields in cloud-free regions. (ESA photograph.)

source, but plans are under way to attempt such developments. The impact will be great on daily weather forecasting should such developments be successful.

3.3.4 Temperature and Water Vapor

All substances emit electromagnetic radiation, the spectral characteristics of which depend on their temperature. In the atmosphere, carbon dioxide is a very stable gas; its annual rate of change of concentration is small and carefully measured. Because it is chemically stable, it is very well mixed with the other atmospheric gases. It emits long-wave infrared radiation in several bands, and these emissions can be measured quite accurately from satellites. Because much is known of the radiation characteristics of carbon dioxide and their variation with temperature, it is possible to infer from the measured radiation the temperature of the gas and thus of the atmosphere. Moreover, because the characteristics of the radiation change with the amount of gas between the emitting layer and the satellite, that is, with the pressure at the level of the emitting layer, temperature changes with height can be derived. Although the method of using satellite-measured infrared radiation to infer temperatures is not perfect, the results are especially useful over regions such as oceans where the information would otherwise be unobtainable.

Water vapor also emits radiation, the characteristics of which are well known. Since water vapor emits in the microwave part of the electromagnetic spectrum, microwave observations from satellites give useful information on temperature, water vapor amounts, and even liquid water in clouds. Combinations of instruments aboard satellites that observe infrared and microwave radiation are now used routinely to provide daily information to the world's meteorological services. This information is particularly valuable over regions where logistical or financial considerations make routine observations by any other means impractical or too expensive.

There are systems under study that will combine observations in several infrared and microwave bands to derive accurate temperature and water vapor profiles even in the presence of some cloudiness. The height of the separation (the tropopause) between the atmosphere's lower layer, the troposphere, and the middle layer, the stratosphere, should be obtainable. It is also thought that useful information will become available on the temperature of the very lowest part of the troposphere, where temperature inversions often occur. Such knowledge is important for predictions of the influence of the oceans on the atmosphere, and vice versa.

3.3.5 Earth Radiation Budget

The global balance between the absorption of solar radiation and the emission of longwave (infrared) radiation is pivotal to understanding the processes that govern the earth's climate. The climate at any particular time is described by a set of long-term averages of various variables. To be useful, the radiation budget must be averaged over approximately one month and observed over many years. Satellites are ideal platforms for observing the radiation budget because of their proximity to the top of the atmosphere and because a single instrument can make these measurements from space over a large number of locations for extended periods of time. The earth radiation budget comprises three parameters: the incident solar energy, the reflected solar energy, and the emitted long-wave energy.

The climate system can be visualized as a thermodynamic cycle in which energy is taken from a hot reservoir (the tropics and subtropics) and transferred to a cool reservoir (the polar regions), doing work in the process. From the observations to date, we have a reasonable picture of the spatial distribution of the absorbed solar energy and the emitted long-wave energy. The shortcomings of our existing data rest primarily with its marginal accuracy and the lack of correlative information to explain the interrelationships between the radiation budget and other climate processes. These two deficiencies are the subject of current and future research.

Much of the present understanding of the earth's radiation budget is based on data from the NIMBUS satellites, which carried five-channel Medium Resolution Infrared Radiometers (MRIR). However, each channel was limited in its spectral response to a small portion of the solar spectrum or the emitted spectrum. Since 1975 NIMBUS-6 and -7 have carried instruments specifically designed to measure the broadband parameters required for the radiation budget.

A new project, the Earth Radiation Budget Experiment (ERBE), has been designed to overcome the shortcomings of previous observations. The experiment utilizes sets of radiometers on three different satellites to adequately cover the earth. The radiometers will be placed on two sun-synchronous TIROS-N/NOAA meteorological satellites with differing equatorial crossing times and a special Earth Radiation Budget Satellite (ERBS) to be launched into a low 46° inclination orbit. The combination of ERBS and the near-polar orbiting satellites will provide a large number of samples uniformly distributed in space and time.

3.3.6 Atmospheric Pollution

The use of satellites for monitoring atmospheric smoke and dust, of natural or anthropogenic origin, is complicated by clouds, but satellite imagery nonetheless provides reasonably reliable data on particulate pollutants. Observations can be made of general levels of haze, and both meteorological and remote sensing satellite imagery show plumes or clouds of dust or smoke from extensive natural sources and, in some cases, from industrial point sources. The measurement of the concentration of gaseous pollutants is much more complicated and is as yet unresolved.

3.3.7 Ozone in the Stratosphere

Although this section concentrates on the lower region of the atmosphere, namely, the troposphere, which contains about 90 percent of the mass of the atmosphere and essentially all of the "weather" of direct importance to human activities, the next higher region, the stratosphere, cannot be neglected. Even though the stratosphere contains far less mass than does the troposphere, solar energy passing down to the lowest layers of the atmosphere must pass through it; the stratosphere can, therefore, influence the energy of the lower layer. The stratosphere is heated by the absorption of ultraviolet radiation from the sun, and many chemical species that are highly reactive are produced. It is stirred by strong winds causing considerable horizontal and vertical mixing, which strongly influences its chemical composition.

One gas produced in the stratosphere in this way, from oxygen, is ozone (O_3). Ozone is a very strong absorber of ultraviolet radiation that would otherwise be harmful to life on earth. In fact, life as it is now only developed after a small amount of oxygen was liberated from the early terrestrial life forms. The ozone formed by the solar ultraviolet then began to shield the lower layers from the strong and damaging ultraviolet radiation. Life forms, thus shielded, multiplied rapidly and developed to their present form.

Ozone in the atmosphere undergoes strong daily and seasonal variations, since its formation and destruction depend on the temperature and on air movements, that is, on the weather of the stratosphere. Also, very small amounts of certain comparatively inert gases released at the earth's surface can, after some time, reach the stratosphere. Some of these gases interact with the ozone, or with other gases which determine how much ozone is created or destroyed. Hydrocarbons, oxides of nitrogen, and chlorine containing compounds are some of the most important of these gases. It is strongly suspected that some human activities contribute to the introduction of these gases into the stratosphere which may change the ozone amounts. Thus, the chemistry and, indeed, the general behavior of the stratosphere must be better understood. On a global basis, the ozone itself and some of the bases which interact with it must be monitored, and satellites are the only practical way of doing this. In the presence of the daily and seasonal variations, many years of good observations are required to discern changes of only a few percent. Observations are especially needed over the large ocean areas. Over land, some types of ground-based or balloon-borne instruments can obtain some of the information needed; such ground networks also make possible cross calibrations with the satellite observations. Thus, man's concern over possible detrimental changes in the ozone layer requires a long-term satellite program and associated observations to provide the much-needed information on the behavior of the stratosphere.

There are other aspects of the stratosphere that are of current and strong meteorological interest, such as the sudden warmings that develop in northern polar regions signaling the early end of the wintertime circulation and the reversal of stratospheric wind systems in the tropics every 26 months or so. The stratosphere is also very sensitive to changes in the radiation balance of the anthropogenic influences on the atmosphere, such as chemical changes or temperature changes arising from the increased use of fossil fuels and the consequent release of carbon dioxide into the atmosphere. Satellites are the best tools for long-term monitoring of the state of the stratosphere.

3.4 DISASTER RELIEF

3.4.1 Disaster Prediction and Monitoring

3.4.1.1 Severe storms.   The use of satellites for the detection and monitoring of severe storms has advanced rapidly during the past 20 years. Severe storms include tropical and extratropical cyclones that cause heavy rain or snow and/or high winds and thunderstorms that produce excessive rains, high winds, hail, or tornadoes. Since violent weather patterns nearly always change rapidly and are usually localized, it is necessary to view these phenomena frequently and with relatively high spatial resolution. Geosynchronous satellites have become the dominant satellite platform for this purpose because of their ability to view the same scene almost continuously. However, low-altitude satellites are today providing valuable data at 6 to 12 hour intervals with a resolution not yet obtainable from geosynchronous spacecraft. It has been possible to monitor the three-dimensional extent and growth of the rate of convection in storms with visible and infrared data from the Geostationary Operational Environmental Satellites (GOES). This has been useful in developing means of assessing the strength of the convection associated with continental thunderstorms and therefore the nature of the severe weather that results. Geostationary satellite images have also shown the locations of fronts and thunderstorm outflow boundaries (as cloud lines), which are favored places for storm development. Monitoring of flooding which may result from these storms is discussed in Section 3.1.2.4.

Winds derived from observations of cloud motion have indicated regions of convergence in the lower troposphere where thunderstorms later form. Wind fields in the lower and upper troposphere can be determined for tropical cyclones by the same method. The lower tropospheric winds are used to determine the distance from the center for winds of various critical speeds. This information combined with vorticity estimates provides an indication of cyclone strength. The upper tropospheric winds can be used to assess the outflow of mass from the cyclone circulation, which is often a predictor of future intensity.

The next advance in geosynchronous satellite technology will be the Visible and Infrared Spin-Scan Radiometer Atmospheric Sounder (VAS) recently launched. This sensor will take the first temperature and moisture profiles from geosynchronous orbit and should make improvements in the measurement of surface temperature, cloud type and height, and winds. Better instrumentation in the future in both geosynchronous and lower orbits is expected to be coupled with advanced interactive computer systems that can effectively and rapidly combine data from numerous sources and will allow the selection of which data are to be analyzed.

3.4.1.2 Earthquakes.

As powerfully illustrated by the 1976 Tangshan, China, earthquake which caused over 650,000 deaths, earthquakes yearly kill tens to hundreds of thousands of people all over the world and cause extensive property damage. Most, but not all, of the really large quakes occur in conjunction with areas of active volcanism along well-defined crustal plate boundaries such as the "ring of fire" zone surrounding the Pacific Ocean. Scientists from many nations are currently engaging in extensive research attempting to understand and perhaps one day control earthquake forces. These efforts include studies of how the destructive forces are built up, stored, triggered, and dissipated and attempts to estimate where and with what repetition rate major quakes are likely to occur.

For studying or forecasting earthquakes, measurements of crustal stress and strain are essential. These measurements of very slow ground deformation or movements are obtained by repeated baseline measurements between fixed observing sites. For sites tens of kilometers apart, surface geodetic techniques suffice, but for sites separated by hundreds or thousands of kilometers, space techniques are invaluable.

The distance measurements used are of two types: laser ranging and very long baseline interferometry (VLBI). Laser ranging, as presently in use, measures the round-trip travel time of a laser pulse reflected off a satellite in a known orbit. This technique has been so perfected that the motion of the San Andreas Fault in California has been directly measured as 9 cm per year with an uncertainty of ± 3 cm. Current work is directed toward routine measurements over baselines of continental extent. A variant of this technique, now under consideration, involves placing the laser in an aircraft or satellite and ranging off a dense grid of ground-based reflectors. This technique will provide many more points of measurements over areas of thousands of square kilometers. The second technique, VLBI, makes use of the different times of arrival at two widely separated radio telescopes of a maximum or minimum in the waveforms from very distant celestial radio sources. This technique has been used to show that the North American plate is essentially rigid, i.e., averaged over a period of some four years, the distance between Massachusetts and California has changed by no more than 2.5 cm per year.

3.4.1.3 Volcanic eruptions.

On May 18, 1980, a dormant volcano of the Cascade Range in North America, Mount Saint Helens, erupted. Figure 3.15

Figure 3.15 Image taken by GOES-West satellite two hours after the eruption of Mount Saint Helens. (NOAA photograph.)

shows two images of this event from a U.S.A. meteorological geostationary satellite; the early phase, the dynamic effect of the blast wave on the cloud deck, and the dust plume are evident. In the second image (Figure 3.16), taken 24 hours later, the dust has spread more than 200 km and covered half of the state of Colorado. These images from space are not only of public interest, documenting a most dramatic natural event, but are also of scientific interest in studying the dynamics of the eruption and the dispersal of material which, if extensive enough, could have short-term weather and climatic effects. These would arise from the dust particles remaining in the upper atmosphere, which could affect the earth's radiation balance, as discussed earlier. Chemicals from the volcano, especially those containing chlorine (such as hydrochloric acid which is a common ingredient of volcanic gases), might have been released in a suffcent quantity to cause some effects in the ozone region of the stratosphere, as also discussed earlier. In fact, it is believed that major volcanic eruptions are related to some climatic changes. The Mount Agung eruption in 1963 and the Krakatoa eruption in the last century may have affected temperatures for several years after. It is uncertain that the Mount Saint Helens eruptions will have such long-lasting effects, unless, of course, the volcano continues to stay active for a long time and ejects a great quantity of fine particles and chemicals into the stratosphere. Satellite observations will continue to be one of the means by which its activity may best be studied.

Figure 3.16 Image taken by GOES-West satellite twenty-four hours after the eruption of Mount Saint Helens. (NOAA photograph.)

3.4.2 Emergency Communications

The disaster conditions that make relief programs necessary may also make them very difficult. For a major disaster, the magnitude of the relief effort requires national and often international mobilization. The availability of efficient communication links is essential if relief is to be provided in the right place at the right time. Small transportable earth terminals linked via geostationary communications satellites can provide rescue workers in the affected area with communications channels to each other, to national emergency headquarters, and to international relief organizations. This technology has been demonstrated in actual disaster situations. Light-weight communications devices can also aid emergency personnel such as forest fire fighters and mobile medical units.

Maritime communications and navigation satellites enable ships to determine their positions very accurately anywhere in the world and communicate their positions should an emergency arise. Navigation satellite systems can provide rapid position determinations accurate to about 300 m. One such system is the United States Transit Satellite System (TRANET) which is available for commercial use on a global basis. The system uses five orbiting satellites such that the time between satellite passes is about 80 min near the equator and about 40 min at high latitudes. A more advanced system, the NAVSTAR global air and sea positioning system, is now in a test and evaluation phase. When fully operational, it will provide continuous worldwide position determination capability with an accuracy of 50 to 100 m or better. Maritime communications are now provided on a global commercial basis by the United States MARISAT satellites. These satellites are now approaching the end of their useful lives, and the system is now being replaced by the international organization INMARSAT. These systems provide rapid, reliable, and high-quality voice and data communication links with the shore to ships equipped with satellite terminals.

Satellite systems are now being planned and developed for detecting and locating emergencies on land or at sea using dedicated search and rescue satellites. An international cooperative project called SARSAT (Search and Rescue Satellite Aided Tracking) is under way. The SARSAT concept makes use of a satellite system to receive distress signals from Emergency Locator Transmitters (ELT) on aircraft and Emergency Position Indicating Radio Beacons (EPIRB) on ships. These devices are currently carried on many aircraft and ships. In the case of an aircraft crash or a ship sinking, this battery-powered emergency transmitter is automatically activated and broadcasts an internationally recognized emergency radio signal via SARSAT to a ground station where the location of the distressed unit will be determined and rescue teams mobilized.

The SARSAT system will provide regional coverage for ELT signals at frequencies of 121.5, 243, and 406 MHz with immediate retransmission of the signals to ground receivers in the region. It will also provide worldwide coverage for the new experimental 406–MHz ELT/EPIRBs; the 406–MHz data will be stored on the spacecraft for subsequent transmission to a ground station when there is no ground station in the immediate region of the emergency.

The SARSAT system is an international cooperative effort involving Canada, France, and the United States, using the NOAA satellites and associated SAR ground stations. The USSR is equipping its COSPAS satellite with a SAR repeater which is compatible with the SARSAT system. Further, many other countries will be performing engineering tests of the system when it becomes available.

3.4.3 Conclusion

The use of satellites for developing and managing earth resources and for monitoring the environment has made great progress since the beginning of the space age 25 years ago. Many techniques, especially in the field of meteorology, are being used operationally, while others have been proved effective and await integration into operational systems. The development of new techniques, new satellite and ground systems, and new applications is continuing in a number of countries, and it seems safe to assume that progress will continue for the foreseeable future. By the end of this decade there will be satellite programs to manage valuable natural resources and monitor the changing environment.

While the benefits of earth observation satellite technology have been widely enjoyed, there are potential benefits that have not yet been realized, and there are countries which have not been able to avail themselves of these benefits. Problems of education and training, access to data, and acquisition and maintenance of equipment have limited the use of satellite data, particularly in developing countries.

The obstacles to the use of space technology are greatest for the smaller and less wealthy developing countries. If a country is to develop a capability for adapting technology to its needs and for developing new applications, it must have a group of scientists, engineers, and technicians working together. The creation of a viable research and development group of this sort may be beyond the means of a small country.

To overcome these difficulties, consideration is being given to the establishment of regional centers for training, applications development, and applications assistance, centers in which a number of countries pool their resources. Such centers can handle data from earth resources, and from meteorological and oceanographic satellites since much of the technology for receiving and processing the different kinds of data is the same or similar. Some economies might also be realized by combining satellite communications and broadcasting technology with earth observation technology to create a regional space applications program, but this might pose substantially greater political and organizational problems since communications and resource management tend to be organized differently.

The benefits of regionalization of receiving stations, data processing systems, applications development, and training seem clear. Whether a regional center might usefully undertake pilot projects or even operational applications is a more difficult question. Even if there are no political obstacles to transnational operations, the logistical difficulties in carrying out field surveys or acquiring data from other sources can be enormous. Whether operational programs should be regional or national might best be decided on a case-by-case basis; it might be feasible, for example, to regionalize meteorological, hydrological, and oceanographic surveys while keeping earth resource surveys on a national basis.

Regional projects are now underway in several parts of the world. In Africa, for example, regional training and applications development centers in remote sensing have been established in Ouagadougou, Upper Volta, and Nairobi, Kenya. Plans are in preparation for adding ground receiving stations and data processing systems to these centers, and additional regional training and assistance centers are planned for Ile Ife, Nigeria, Kinshasa, Zaire, and Cairo, Egypt. A regional center for hydrology located in Niamey, Niger, is one of a number of other regional centers which could use satellite data. These

African efforts to develop regional cooperation in the applications of space technology are encouraging, and similar programs are being planned elsewhere.

ACKNOWLEDGEMENTS

This chapter was prepared with the assistance of teams of experts organized by the International Astronautical Federation (IAF) and the Committee on Space Research (COSPAR) of the International Council of Scientific Unions (ICSU). The IAF contribution was compiled by Dr. F. O. von Bun (United States) and Mr. L. Jaffe (United States) from material provided by Dr. R. Allenby (United States), Dr. P. Castruccio (United States), Mr. C. Cote (United States), Dr. R. Curran (United States), Dr. L. Dottavio (United States), Dr. J. Gruau (France), Dr. G. Heath (United States), Dr. K. Ya. Kondratyev (USSR), Mr. L. Laidet (France), Dr. M. Oluic (Yugoslavia), Dr. A. Rango (United States), Mr. W. Shenk (United States), Dr. N. Short (United States), Dr. Y. Zonov (USSR), and Dr. J. Zwally (United States). The COSPAR contribution was edited by Dr. M. J. Rycroft (United Kingdom of Great Britain and Northern Ireland) and compiled by Mr. S. Ruttenberg (United States) from contributions by the following team members: Dr. T. Allen (United Kingdom), Dr. W. Alpers (Federal Republic of Germany), Prof. P. D. Bhavsar (India), Prof. W. Bohme (German Democratic Republic), Mr. W. D. Carter (United States), Dr. J. Garrett (Canada), Dr. P. Gudmansen (Denmark), Prof. K. Ya. Kondratyev (USSR), Dr. J. R. Lutjeharms (South Africa), Dr. P. McClain (United States), Prof. J. Otterman (Israel), Dr. S. I. Rasool (United States), and Dr. A. Shutko (USSR).

Chapter 4

IMPACT OF SPACE

ACTIVITIES ON THE EARTH

AND SPACE ENVIRONMENT

ABSTRACT

This chapter is the fourth of four general review chapters. It was prepared with the assistance of an international team of experts organized by the Committee on Space Research (COSPAR).

The purposes of this chapter are to examine some of the possible effects on the earth and space environment of space activities and to suggest means by which undesirable effects might be minimized. The report considers the possible impact of the introduction of new material, either deliberate or accidental, into the earth or space environment and the possible physical, chemical, or biological consequences.

INTRODUCTION

In the course of his day-to-day existence, man has always been obliged to submit himself to dangers of different kinds and to run the risk of being involved in accidents. However, since he is aware of the dangers and can often measure the probability of their occurrence, he normally takes precautions designed to eliminate accidents, or at least to limit the risk if complete elimination is not possible for economic or other reasons. In considering the dangers associated with space activities, it may be relevant to consider the way in which we deal with the risks associated with, for example, air and surface transport systems, or fire in buildings. Because of the very high exposure of populations to these dangers, many accidents actually occur and, in consequence, it is possible to identify their causes and to take appropriate steps to reduce the risk.

In space activities generally, the numbers of individuals exposed to the various risks are very small in comparison with those exposed to transport accidents. Hence the number of accidents has been small. However, it is possible to envisage accidents that have never occurred but that could conceivably occur even though they may be very unlikely. From the beginning of the space era, scientists and engineers responsible for the research programs have tried to assess the real and potential dangers and have taken, or have proposed, appropriate measures designed to eliminate or to minimize the risk of accidents.

Quite apart from the small risk of accidents to individuals engaged in space activities, there is also a potential risk of damage to the terrestrial environment. For example, certain gases emitted from space vehicles and their rocket engines could change the protective power of the atmosphere as a shield against harmful solar ultraviolet radiation, or they could modify the reflecting properties of the ionosphere which is an important element in radio communications.

Space activities could also have undesirable effects for other branches of scientific research. Very large numbers of artificial satellites, each emitting infrared radiation or radio waves or reflecting sunlight, would obviously be of concern to astronomers. Also, widespread contamination of the upper atmosphere and the magnetosphere would hinder studies of the natural state and composition of these regions and research on the origins of the earth itself.

It must be emphasized that in many cases it is not possible to assess with any accuracy the size of the risk. The precaution that needs to be taken to minimize a given task depends on the nature of the risk itself and its possible consequences, on the probability that the risk will actually lead to an accident, and on the economic consequences of the possible precautions. Of course, new developments may occur that would change the situation considerably.

4.1 POSSIBLE ENVIRONMENTAL EFFECTS OF LAUNCH AND RE-ENTRY

4.1.1 Launch Site

A rocket launching produces a so-called "ground cloud" consisting of exhaust gases, cooling water, sand, and dust (Figure 4.1). Generally, the resulting air pollution is limited both in extent and intensity, and the current level of satellite launchings poses no extensive danger.

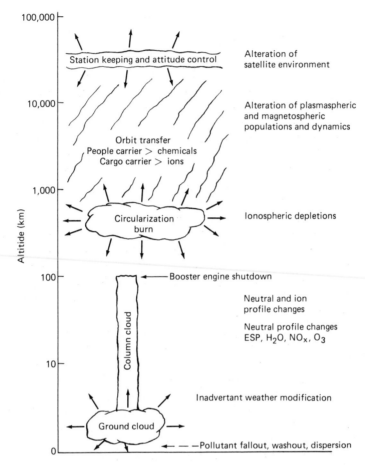

Figure 4.1 Summary of some potential atmospheric effects caused by rocket exhaust.

If launching activities increase greatly, however, the possible effects may have to be reexamined. In particular, if solar power satellite (SPS) systems were developed, consisting of perhaps 60 satellites, each with 50 sq km of solar panels, the cumulative launch effects on air and water quality around the launch site might be significant.

Given the unusual size of the SPS launching rockets, the risk of accidents and other undesirable effects is greater than for more conventional rockets. The fueling of the rockets will demand large volumes of liquid hydrogen and other explosive or inflammable liquids, and there are obvious risks during the transport of these materials to the launch site and during their transfer into the rocket fuel tanks. Spills of these liquids resulting from accidents could cause damage to life and local damage to ecosystems. The explosive potential of an SPS rocket will be about twice that of the Saturn V used for moon launches; this implies that structural damage could be caused to buildings at distances of the order of tens of kilometers in the case of a catastrophic explosion.

The noise levels generated during SPS operations would be high, and it seems possible that the 24 h average limits specified by the United States Environmental Protection Agency would be exceeded, not only because of the high noise produced by each rocket but also because there will be several launches per day. More detailed studies of this question are required. In addition to the noise generated by the rocket motors, there will also be sonic booms during the ascent and re-entry of the rockets. These are likely to startle humans and animals and might also cause structural damage.

4.1.2 Upper Atmosphere

More extensive atmospheric contamination is produced as the rocket engines continue to burn up to altitudes between 100 and 200 km. The magnitude of the contamination depends on the amount of propellant burned which depends on the mass of the vehicle and the altitude of the desired orbit. For larger rockets, the total amount of matter introduced into the atmosphere can be substantial; for example, during one launch of the Space Shuttle about 1 million kg of solid propellant are burned in the boosters and 750,000 kg of hydrogen/oxygen in the main engine.

At altitudes above 20 km in the stratosphere, and even above 60 km in the mesosphere, significant effects due to injection of water and nitrogen oxides could occur when very large vehicles are launched. Since atmospheric density decreases rapidly with height, the sensitivity of the atmosphere increases drastically with increasing altitude. Thus, not only exhaust during propulsion must be considered but also the outflow of gases from the combustion chamber after the engines have been shut off.

Because of deficiencies in our understanding of physical and chemical processes above 40 km, especially with regard to the effects of water vapor, all theoretical predictions suffer from a significant uncertainty. Potential effects of mesospheric water release on cloud formation, for example, have yet to be assessed.

Possible effects of stratospheric contamination include changes in the ozone (O_3) layer which shields the surface from harmful ultraviolet radiation. The amount of ozone in the layer is determined by chemical reactions that could be disturbed by changes in the concentration of gases already present such as carbon dioxide or by the introduction of other substances such as chlorine compounds.

The large solid-fuel rocket engines to be used in the Space Shuttle, for example, release chlorine and hydrogen chloride. Calculations, assuming a maximum of 60 launches per year, have indicated that the ozone content of the atmosphere would be reduced by less than 0.5 percent, an effect less than that due to other natural or man-made causes. Studies of other vapors released by the Space Shuttle suggest an insignificant effect.

A satellite in low earth orbit gradually loses energy and altitude due to the small resistance of the thin atmosphere. At an altitude of about 100 km, the resistance starts to rise rapidly, and the satellite descends quickly into the atmosphere. During this "re-entry" a very high temperature is reached at the leading edge of the vehicle, and generally the satellite disintegrates and vaporizes in the upper atmosphere. The resulting production of metal vapors and oxides could influence the balance of electrically charged constituents (ions) and, locally at least, change the ionospheric conditions that affect radio communications. Exotic elements such as beryllium and cadmium may be of particular concern. At present, however, it appears that meteorites, aircraft, and terrestrial sources introduce much greater quantities of material than satellite re-entry. The total meteoric mass entering the atmosphere, for example, is estimated at 10,000 kg/day. In the future, if large numbers of satellites were allowed to burn out on re-entry, the resulting contamination of the upper atmosphere might become important. Most future large vehicles, however, such as the Space Shuttle will be controlled during re-entry.

4.2 IMPACT ON THE ORBITAL ENVIRONMENT

4.2.1 Operational Releases

At orbital altitudes, above 100 km, the atmospheric density is so low that very small quantities of material may have significant environmental effects. The rocket engines have usually been shut down well before the satellite reaches altitude, but other releases occur as part of orbital operations. These releases include outgassing of spacecraft materials, propulsion for attitude control and station keeping, liquid from cooling systems, and, in the case of manned vehicles, leakage of vehicle atmosphere.

For the Shuttle Orbiter, detailed model calculations have been made concerning the different release sources and their importance. The results are summarized in Table 4.1. Water release by the evaporator, which might eject up to 140 kg/day, represents probably the most important impact on the atmosphere during a Shuttle flight. Cabin leakage can be expected to be harmless, and the thrust engines would only locally disturb the environment, with potential impact limited to a few on-board experiments.

Among the chemical reactions following the in-orbit releases, those concerning the ionosphere have been studied in detail. This is the region above 60 km where free electric charges (negative electrons and positive ions) exist in large quantities due mainly to the impact of ultraviolet and X-ray radiation from the sun. The peak charge density is reached at altitudes of about 250 to 400 km.

TABLE 4.1 Major Shuttle Orbiter Release Sources

Source	Duration	Flow Rate (Approx.)	Constituents
Outgassing	Continuous	10^{-4} kg/d	Hydrocarbon chain fragments, other volatiles
Offgassing	Continuous for first 100 h in orbit	10^{-3} kg/d	Water, light gases, volatiles
Evaporator	As required	10 kg/h	H_2O
Cabin leakage	Continuous	3 kg/d	O_2, N_2, CO_2, H_2O
Thrust engines	As required	0.04 kg/min	H_2O, N_2, H_2, CO, CO_2, H

While the numerical density of neutral species is much larger, the ionospheric electrons are of great importance for radio communications since waves in the high frequency (HF) range are normally reflected from the ionosphere back to the ground, so that by multiple reflection, in spite of the curvature of the earth, they may reach very large distances. Reflection also occurs at lower frequencies down to ELF (extremely low frequency). Waves of much higher frequencies (VHF, UHF) penetrate the ionosphere but suffer from changes in polarization and even direction. These effects decrease with increasing frequency, so that the importance of the ionospheric influence is very small in the UHF range. However, advanced systems for radio direction finding and navigation and, in particular, the so-called "very long baseline interferometry" (VLBI) which achieves enormous accuracy, still suffer most from the variable ionospheric influence, even when using UHF waves.

Therefore, the preservation of the ionosphere in its normal state has considerable practical importance. Under natural conditions, positive and negative charges recombine and so disappear in a rather complicated ion chemistry[1] which can be largely changed by the introduction of small amounts of interfering species not normally present in the upper atmosphere. Of particular concern are carbon and nitrogen oxides, most organic molecules, and unburned fuel, because these molecules become attached to electrons and hence form negative ions. Water vapor, if present, also removes a lot of electrons by catalyzing their recombination with positive ions. This increased removal of electrons results in a drastic change in the reflecting properties of the ionosphere for radio waves and for other radio-wave propagation phenomena. Thus, species produced in rocket exhausts and during re-entry may provoke large local changes, particularly in the lower ionosphere. Detailed observations were made with the 1973 SKYLAB launch of the National Aeronautics and Space Administration (NASA): The zone of depleted electron density covered an area 2,000 km in diameter lasting (at

[1] Chapter 1, "Current and Future State of Space Science."

peak altitude) for several hours. The phenomenon, which must justly be called a "big hole" in the ionosphere, had important effects on radio communications. The most important case studied in detail was the launch with an Atlas/Centaur thruster of the HEAO–C satellite in 1979. Usually the upper stage is shut down below 200 km, but in this case it was ignited only at 209 km and kept burning up to 466 km. Large deviations from normal radio propagation were observed, in particular with satellite beacons. Even larger thrusters are now being discussed, for example, in conjunction with the solar power satellite (SPS) project. These "heavy launch vehicles" would have 150 times the payload capabilities of the Atlas/Centaur combination. It must, therefore, be stated that each launch of a larger space vehicle produces a major regional disturbance in the ionosphere, which may last several hours. Ionospheric effects, though of smaller extent, are also expected during re-entry, as a consequence of the burnout and vaporization products that are produced.

The deliberate removal of satellites from orbit at the end of their usual lifetimes, as has been suggested to relieve congestion in orbit, might have substantial environmental consequences. The most effective means of removal, braking by retrofiring, would result in considerable fuel being burned at orbital altitude. At an altitude of 1,500 km, the required propellant mass would be about 10 percent of the satellite mass. "Clean" propellants would be most desirable, but no such solution is yet in sight.

In addition to the effect of releasing material, an inverse effect, namely, the collection of particles from the environment, is expected from satellites of larger size, in particular those with large metallic structures such as extended antennas, when they sweep through the outermost atmospheric regions where charged particles are predominant, i.e., in the plasmasphere and magnetosphere. They may also provoke artificial electric and magnetic fields as a result of motion relative to the earth's magnetic field and the accumulation of ions.

4.2.2 Release Experiments

4.2.2.1 Effects on chemical composition.

Artificial releases of chemically reactive substances are a valuable tool of space research. Light metal vapor, e.g., sodium, barium, or strontium, has often been released into the upper atmosphere as a "tracer." The vapor column, when illuminated by the sun, can be optically observed; and, by triangulation, winds in the upper atmosphere can be unambiguously determined. Barium and strontium are largely ionized by solar ultraviolet radiation, so that charged particles are formed which can also be identified by the specific color of the light they emit. Thus, the motion of charged clouds becomes observable. Electric fields in the upper atmosphere were so determined, and this information is of particular importance for understanding the phenomena occurring in the ionosphere at high latitudes. Outside the terrestrial environment, releases have been made to simulate natural comets.

Other releases are undertaken for studies of chemical reactions occurring naturally in the upper atmosphere, with the intention of using this particular environment as a laboratory where much higher cleanliness is achievable than in any terrestrial laboratory. Finally, releases have also been undertaken with the deliberate objective of changing the properties of the ionosphere, so as to improve or impair radio-wave propagation by the ionosphere.

Such artificial releases are certainly an important tool for scientific studies. But the question arises as to whether such activities might provoke unintentional changes that could not be reasonably foreseen, for example, long-lasting or extended deterioration of the natural environment, disadvantages for other scientific activities such as astronomy and geophysics, etc. Until now the ejected masses have been small in most scientific experiments undertaken; nevertheless, noticeable atmospheric changes have sometimes occurred, in particular in connection with the ion chemistry for the charged species. Since there is a tendency for the loads released to increase, some limitations on such experiments may be needed in the near future.

In order to estimate the potential detrimental effect of releases, the associated kinetics have been studied in great detail. A quantity of neutral gas, when released into a tenuous atmosphere, expands rapidly by the conversion of thermal into kinetic energy, and thus drops rapidly in temperature. The temperature decrease at the beginning of the release is particularly steep in a rarified atmosphere. The cooling can be so strong that condensation vapors and even droplets may occur for a short while. Condensation starts around any suitable "nucleus," e.g., dust particles, charged particles (ions), or clusters of gas ions. All this occurs during the first second after release. Subsequently, the expanded gas cloud is heated up by the ambient atmosphere with the evaporation of the condensed droplets. After about 10 s the ejected gas has reached about 1 km distance and diffuses steadily into the atmosphere, the speed of further expansion depending on the molecular weight of the gas and on the atmospheric density at the level of ejection. Particularly great diffusion speeds occur with hydrogen. As a consequence of gravity, the long-term expansion is larger in horizontal directions then downward. Gas clouds released at high initial velocity produce very asymmetric patterns and may cover a 100-km horizontal range in about 15 sec.

Kinetic effects dominate the release processes only during the first minute or so. The subsequent development depends on whether chemical reactions become important or not. If not, the released matter is further diluted by diffusion into the atmosphere with a propagation speed that is higher at greater altitudes. The removal of the injected material depends essentially on its downward propagation, which can be astonishingly slow even for constituents of large molecular weight. Of the light gases, only hydrogen is able to go upward and eventually leave the atmosphere in the direction of the magnetosphere. Most injected gases or vapors, though diluted, may remain for months in the upper atmosphere. In some cases, chemical reactions may determine the fate of the injected material. Above 100 km, highly reactive atomic oxygen is present and oxidation may occur extremely rapidly. The creation or transformation of unusual substances by unusual chemical reactions is a complex field in which much work is required before conclusions can be safely drawn.

Astronomical and geophysical observers are often interested in optical phenomena of extremely weak intensity which can easily be disturbed by unusual species deposited in the atmosphere; for example, the yellow emission of sodium light in the atmosphere has been used to estimate the natural overall meteoric activity. Releases, when frequently made, may introduce comparable amounts of sodium and so endanger such observations. Moreover, conclusions about the origin of our atmosphere may become invalid if the natural or artificial origin of some minor constituent is doubtful.

4.2.2.2 Effects on ion and electron concentration.

Detailed studies have been made of ion-chemistry reactions occurring after releases. Considerable changes in the natural conditions in the ionosphere were produced in systematic experiments carried out in the 1960s and 1970s. The release of various species has caused slight increases or, more commonly, substantial decreases in electron density. In contrast to the behavior of neutral constituents, the effects on electrons and ions normally disappear in hours or a day. Only at very high altitudes, above 2000 km, are longer lasting effects likely to occur.

If metal vapor is released, it is often easily ionized by sunlight so that free electrons and positive ions are produced. For certain species (like sodium, barium, or strontium) the energy needed for this transformation is much smaller than for normal air molecules. These ions therefore do not as easily recombine with electrons; thus they have a much longer lifetime and increase the total electron density.

Plasma depletion, i.e., large-scale reduction of the electron density in the ionosphere, can be provoked by the release of quite different gases. For example, water vapor, hydrogen, nitrogen dioxide (NO_2), and carbon dioxide (CO_2) all have a similar efficiency in decreasing the lifetime of O^+, the main ion in the upper atmosphere: a 1 percent concentration of these (compared with atmospheric nitrogen) decreases the lifetime by a factor of 10. Nitrous oxide (N_2O) is less efficient and the factor is only 3. A very strong effect on the vertical profile of electron density was obtained by a model computation for a release of 100 kg of hydrogen at 300 km altitude. The sequence of computed profiles reproduced in Figure 4.2 shows a reduction of the peak electron density by about a factor of 3, lasting for more than one hour. The ion temperature decreases, but the electron temperature is drastically increased, by 1500°K, 10 minutes after the release. At the same time, the hole created in the ionosphere (see Figure 4.3) is about 250 km wide. The disturbance propagates upward since hydrogen ions move up in a tube along the lines of the magnetic field of the earth.

Such changes strongly affect radio-wave propagation. In particular, the trajectories along which an HF wave travels can be seriously deformed. Waves otherwise reflected in the ionosphere may escape through the hole. For those

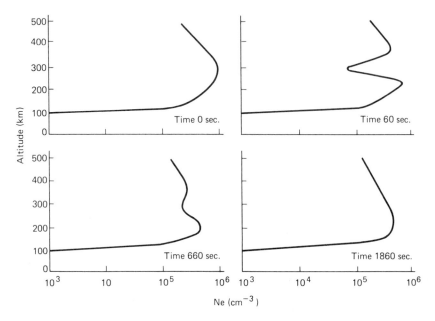

Figure 4.2 Daytime F. Region profiles after hydrogen injection.

reflected in the disturbed region, strong deviations from the usual "great circle propagation" may occur, with the result that direction finders give wrong indications. Waves may also be guided in the hole and so take a direction quite different from the usual one. This can, in particular, happen in the plasmasphere (above 200 km) for very low-frequency (VLF) waves. Such effects can be used for studying interactions between charged particles and very long waves—experiments that could never be carried out in a laboratory. A very particular "application" has been proposed in order to increase temporarily the intensities, at ground level, of frequencies used in radio astronomy, i.e., the use of the hole as a lens for radio waves, which would focus VHF radiation arriving from extraterrestrial sources.

The first successful depletion experiment in 1977 detonated 88 kg of high explosive at 261 km altitude, eventually producing water vapor and carbon and nitrogen oxides in large quantities. Rocket- and ground-based instruments recorded not only a decrease in electron concentration but also changes in the chemical composition of the ion population and an increased airglow. In order to produce an artificial "hole" in the ionosphere, 3,000 kg of rocket propellant will be burned during the second SPACELAB mission, probably in 1984. Scientifically, depletion experiments are an important tool for measuring chemical and diffusion coefficients directly in the "ionospheric laboratory," since they allow determinations to be made which, because of impurities, cannot be made in terrestrial laboratories. Further, such experiments allow particular effects in radio-wave propagation to be studied experimentally, e.g., ducting of waves in

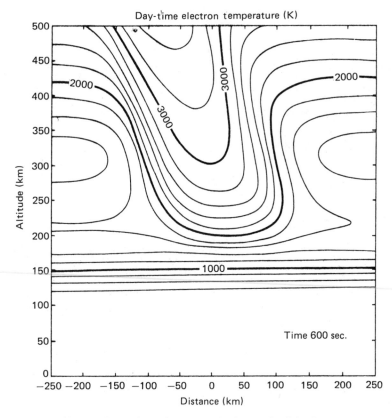

Figure 4.3 Isothermal contours 10 minutes after injection.

the HF or VLF ranges. The spectrum of the airglow artificially produced at the release can give information about the chemical and physical processes in the modified ionosphere.

Such depletion experiments have shown that, by this technique, drastic alterations of the natural conditions in the ionosphere can be artificially produced. Lowering of the electron density by 90 percent and raising the electron temperature by a factor of 10 can be achieved locally. Drastic effects on radio-wave propagation were observed during such experiments. Fortunately, there is a rapid "healing power" which, even for the plasmasphere, limits the duration of the artificial depletion to several hours.

A comparison of the benefits and the risks of release experiments is extremely difficult because the behavior of the upper atmosphere is not well understood and because comparisons of benefits to one discipline and risk to another is inevitably subjective. Nevertheless, competent scientific organizations should consider the establishment of limits on the releases of various materials in the space environment.

4.2.3 Nuclear Explosions

In addition to the nuclear explosions at or near ground level, explosions have been conducted in the upper atmosphere at altitudes around 400 km. Such experiments were conducted in the late 1950s and early 1960s with the intention of studying their geophysical consequences. The short-term effects were almost negligible on the ground, but they were large and extensive in the high atmosphere. In contrast to radiation from low-altitude explosions, which is absorbed within a short distance by the dense air, particles and X-radiation from high altitude explosions travel over large distances. As a result, the first large high-altitude explosion killed, by its radiation, the electronic devices of several satellites that were in orbit at the time. Charged particles produced in a northern hemisphere explosion, after traveling along the force lines of the magnetic field of the earth, produced an artificial aurora over New Zealand only seconds after ignition. Radio-wave absorption in the ionosphere was increased a few minutes later. Synchrotron radiation from 2 MeV electrons produced by the explosion was detected by radio astronomy observations at low latitude. Charged particles produced in the explosion largely increased the population of the natural radiation belt. Following another (1 kiloton) explosion at low latitude, an artificial radiation belt was created which lasted many years; similar explosions at high latitude produced belts that decayed rather rapidly.

4.2.4 Ion Beams

In recent years, for scientific purposes, electrons and ions have been ejected into the upper atmosphere in a controlled manner by activating suitable sources aboard space vehicles. Until now the importance of such "charged releases" or "beams" has been at or below the levels of naturally occurring processes. This situation may change as man's presence in space grows and as powerful ion engines are increasingly used for propulsion in the future, as is already intended in different projects now under discussion. Such engines will be required, for example, to propel large satellites from a lower-altitude parking and construction orbit into a geostationary orbit. In case of such activities, the effects of ion engines will no longer be minimal.

The most powerful electron beams that have so far been fired into the ionosphere produced an electric current of below 1 A supported by electrons of about 20 keV energy. The beam provoked an artificial aurora of very weak intensity, which could only be detected with sensitive instrumentation, and a variety of waves in the plasma, with weak intensities comparable to those of natural emissions. Radar reflections could be obtained from the beam, and bursts of radio noise were observed. The electron beams often provoked the creation of instabilities in the plasma and a hot plasma halo was formed around the rocket.

Injection of ion beams (cesium and xenon ions were used) also produced plasma waves and instabilities. These were studied in view of the natural occurrence of such phenomena and the search for a scientific explanation. Induced emission at the fundamental resonance frequency (of the ions in the local magnetic field) was observed with many harmonic frequencies up to twenty times the fundamental one. In other experiments it was confirmed that ions released in the upper ionosphere did not appear as contamination at much higher altitude in the magnetosphere.

Ion engines for propulsion, called ion thrusters, emit beams of particularly heavy ions. The currents now being used cause no harm, but more powerful engines are being planned. A proposed ion engine for transferring heavy structures from a lower into a geostationary orbit would deposit 1 million kg of argon into the magnetosphere, so creating an ion population about equal to the natural ion content of the magnetosphere, but of much larger energy. Such techniques could produce strong and long-lasting distortions of the outermost environment of the earth. Many questions are still open to discussion in this context, including the reaction processes involved and the time delays for the natural recovery after such operations. Preliminary experimentation with smaller ion thrusters in the magnetosphere could provide some answers to these questions.

4.2.5 Electromagnetic Waves

Electromagnetic waves are now also experimentally produced in space, for example, for feeding energy into natural or artificial particle populations. Such experiments are part of a new discipline in space research which uses the orbital environment as a laboratory providing conditions that cannot be achieved in a laboratory at ground level. Controlled experiments of this kind are normally sporadic and therefore of no serious consequence because with currently used power levels, recovery is quick in the ionosphere and even in the magnetosphere. However, emissions due to man's day-to-day activities may not be negligible in the outermost environment of earth. At present, these stem mainly from ground-based installations, viz., from powerful radio transmitters, including radars and, at extremely low frequencies, from the terrestrial power lines which emit harmonics of their 50- and 60-Hz fundamental frequencies. The effects of such emissions have clearly been demonstrated, but controversy still exists on their magnitude.

Suitable chosen waves in the HF range may provoke considerable heating of the electron and ion populations in the ionosphere. Installations at ground level must have extremely high power (of the order of megawatts (MW)) so as to produce strong fields at 100 km distance in the ionosphere. Similar experiments might be conducted in the future from larger spacecraft, but, of course, with much smaller transmitter power. The power density in the ionosphere, actually attained with the megawatt transmissions from ground level, is comparable with that reached at 1 km distance from a 20-kW transmitter. It has been shown that such rather modest energy fluxes are sufficient to modify conditions in the ionospheric medium and to create irregular structures ("cavities") with sizes of some meters. These have been shown to change considerably the usual conditions of radio-wave propagation in the ionosphere by scattering wave energy in unexpected directions. While in certain regions of the earth such phenomena occur naturally, they are rather rare at temperate latitudes.

Wave-particle interaction is a subject of some interest in space research and astrophysics and also for certain electrical technologies. Only in space can such phenomena be studied, since very long radio waves (in the VLF range) must be used. If a wave of suitable frequency is emitted in the magnetosphere, it can gain energy from the electrons "trapped" in the radiation belt and so be amplified. At the same time, the electrons are precipitated down into the ionosphere. It is a matter of some controversy how important human activity is, even at its present level, in provoking phenomena that were formerly considered due to natural causes. In short, it seems that radiations from ground level, in the VLF and other ranges, can to a certain extent affect the natural charged-particle population in the magnetosphere.

It seems probable that larger effects might be produced if, in the future, greater energy densities were generated in the magnetosphere through the use of transmitters aboard space vehicles.

Additional disturbances are introduced by powerful emitters of VHF radio waves. On-board emitters of HF and VHF waves generate electric and magnetic fields at large distances from the satellite. Usually, the aerials of these emitters are strictly directional and cause very large electromagnetic fields at their apertures. In the beam direction of the radiation pattern of such an antenna, the electric and magnetic effect of a satellite reaches the earth. In the most extreme case of an SPS so far designed, the enormous energy flux may significantly perturb the ionosphere and, when reaching the earth's surface, may even be harmful to the biosphere.

The energy fluxes in the sidelobes of a relatively poorly directional on-board antenna attain significant values and increase the electromagnetic effect of the satellite to a distance of about 1,000 km. Particularly dramatic is the case of an SPS, in that the desired directional properties of the VHF antennas cannot be guaranteed under such great powers; the energy within the sidelobes would be sufficient to disturb communications with other satellites and also to affect people and systems on the ground. Especially drastic would be the case were oscillations of a stabilized SPS to occur. This effect would lead to a sharp increase in the size of the disturbed area around the satellite and could alter the operation of other geosynchronous spacecraft, communications systems in particular. For an SPS with 1,000 MW, this disturbed region could extend over several thousand kilometers. For a detailed discussion of the area subject to electromagnetic disturbances by a satellite, see Appendix 4.2.

4.2.6 Material Sciences in Space

This new discipline uses the (near earth) space environment either as a laboratory in which gravity is extremely low (microgravity) or as an extremely powerful vacuum pumping system. Experiments made so far have used microgravity for investigating material processes (e.g., melting and recrystallization) in the absence of disturbances normally induced by gravity in terrestrial laboratories (e.g., gravity excited turbulence). The experimental structures themselves, if at all exposed to space, may modify the immediate environment of the vehicle by outgassing, but the masses involved will remain small (not more than 1 g per experiment). Moreover, the types of gases in question will be those occurring naturally in the high atmosphere, including, however, water vapor and carbon dioxide. Larger releases may take place from cooling systems, and for this reason, designers should use only non-aggressive species such as helium. For example, in the first SPACELAB mission, a release of 150 g of helium is envisaged.

The use of a windshield outside the vehicle for creating a vacuum of extremely high quality is now being considered. Here, too, care should be taken to avoid pumping vapors or gases that are aggressive to the natural upper atmosphere into the space environment.

4.3 POSSIBLE DIRECT EFFECTS ON HUMAN LIFE

4.3.1 Debris from Spacecraft

During the final re-entry of a spacecraft into the earth's atmosphere, the spacecraft gradually breaks up, leaving a trail of debris scattered over a fairly long, but narrow, track. Two recent events of this kind have received attention: the re-entry of SKYLAB over Western Australia and the fall of Cosmos-954 in Northwest Canada. In both cases, the breakup took place over uninhabited or very sparsely populated regions, and there were no casualties. Nevertheless, they demonstrate the risk and the need to take whatever steps are possible to reduce it.

The chance that a particular individual will be killed by a falling spacecraft is probably not greater than of being hit by a crashing aircraft. However, in view of the population of the world as a whole, the risk that someone may be hit is not completely negligible.

Several methods of minimizing the risk are available, and the most important is probably that of controlled re-entry. This implies that the spacecraft remains under control from the ground until its final orbit. Then, at an appropriate point in this orbit, a rocket motor is activated which slows down the spacecraft and allows it to fall into a predetermined area over the sea or an unpopulated land region. Other possibilities are the transfer of the spacecraft into a higher-level orbit, in which its lifetime will be greatly extended, or to arrange for its complete destruction before final re-entry. These methods are discussed in more detail in Section 4.4.

Particular problems arise in the case of satellites carrying nuclear power systems where there is the danger of radioactive material being introduced into the environment. The greatest risk arises when radioactive debris reaches the surface, but the introduction of radioactivity into the upper atmosphere through the disintegration and vaporization of the spacecraft may also have undesirable effects. Normally, however, the vaporization of a nuclear power system on re-entry would result in exposure to radioactivity at the surface well below the natural background and within the limits set by the International Commission on Radiological Protection (ICRP).

For low levels of electrical power (up to 475 W electrical power has been generated), radioisotope thermoelectric generators have been used on satellites such as LUNOKHOD (USSR) and VOYAGER (United States) where other power systems could not meet the mission requirements. Radioisotopes that have been used include plutonium 238 (half-life 87.7 years) and polonium 210 (half-life 138 days). Normally, these materials are enclosed in containers designed to survive re-entry and reach the surface intact. Radioisotope power sources are also used as heat sources, and small quantities of radioisotopes are used in some scientific instruments.

Higher levels of electric power can be generated by nuclear reactors using radioactive materials such as uranium 235 as fuel and generating a variety of radioactive products, most with rather short half-lives. Such systems have been used on the SNAPSHOT (United States) and COSMOS–954 (USSR) satellites. Since reactor power systems tend to be larger and more complex than radioisotope power systems, it is much more difficult to design the system to survive re-entry. Therefore, in cases of emergency re-entry, the reactors are designed to disintegrate and vaporize in the upper atmosphere.

Satellites carrying nuclear power systems are generally designed either to operate in high-altitude (long lifetime) orbits or, in the case of missions requiring lower orbits, to be moved to high orbits at the end of the working mission. Lifetimes of satellites cannot now be accurately predicted, but at an initial altitude above 800 km, the lifetime should be at least several hundred years, and above 1,000 km, the lifetime should be greater than 1,000 years. In 400 years, the fission product activity of a uranium 235 reactor should be about 1/1,000 of the activity one year after reactor shutdown.

The risks from nuclear power systems arise primarily from the possibility of the satellite's failure to reach high orbit combined with the failure of the mechanism designed to ensure the survival of the fuel containers, in the case of radioisotope systems, or to ensure the complete vaporization of the radioactive material, in the case of reactor systems. A few satellites carrying nuclear power systems have failed on launch or have re-entered the atmosphere prematurely:

1. TRANSIT-5-BN-3, launched on 21 April 1964, failed to reach orbit and reentered the atmosphere at an altitude of 121 km. The radioisotope system disintegrated and vaporized in the upper atmosphere as designed.

2. NIMBUS-B-1, launched 18 May 1968, failed on launch and fell into the ocean. The radioisotope system was recovered intact.

3. APOLLO 13, launched 11 April 1970, failed in circumlunar flight. The radioisotope system re-entered the atmosphere and was lost intact in the deep ocean.

4. COSMOS 954, launched 18 September 1974, failed in low orbit and re-entered the atmosphere on 24 January 1978. The reactor system disintegrated but did not completely vaporize, and some radioactive material reached the ground.

It is clear that great efforts have been made to protect the terrestrial environment from radioactive material from nuclear power systems, but it is equally clear that it is not possible to provide absolute guarantees against failure of the protection systems. It should be noted that the United Nations Committee on the Peaceful Uses of Outer Space has established a Working Group on the Use of Nuclear Power Sources in Outer Space to consider the technical aspects and safety measures relating to the use of nuclear power systems.

4.3.2 Balloons

Balloons are used for making measurements in the atmosphere at heights that are not accessible to satellites. Since these balloons must ultimately return to the ground, clearly some care is essential on the part of the users. In order to avoid the risk of collisions with aircraft, recommendations have been issued by the International Civil Aviation Organisation. At present, all short-term balloon flights are carried out in conformity with these recommendations. At the national level, local operational relationships are established between the balloon launching teams and the air traffic organizations.

The return of heavy payloads to the ground is controlled by a parachute, which reduces the rate of fall to a safe level. As for the large balloons themselves, they can be disintegrated into very low-density shreds by means of tearing devices operated after the release of the load.

4.3.3 Solar Power Satellites

Some proposals for solar power satellites envisage the construction in space of very large structures for collecting solar power and transmitting it to the ground by microwave beams. The satellites would use highly directional transmitting antennas in space for the transfer of energy to the ground at a radio frequency of 2.45 GHz. The very intense beam would be directed into special antennas on the ground, but there would also be some stray radiation, at a much lower level, to which the general population and the ecology would be exposed. Workers at the ground and space antenna sites would, of course, be exposed to somewhat higher levels of radiation.

Very little evidence is available on the danger to health and to the ecology of long-term exposure to weak microwave radiation. Some short-term experiments on animals and plants exposed to high-intensity radiation have been made, but the results are often contradictory and, in any case, the extrapolation of the results to long-term, low-level radiations is open to doubt. There is a need for an extensive investigation of these risks in order to determine whether the use of microwave beams for the transmission is feasible.

Mention should also be made of several other risks relating to the SPS. Because of the large quantities of gallium arsenide used in the solar cells, the workers handling the antenna components would be subject to risk of contamination. Those working on the antenna in orbit would also be exposed to space radiation, the effects of long periods of weightlessness, etc. Current experience is providing some information on the risks involved and on the protective measures that must be developed.

Since satellites require very large areas of solar panels (on the order of 50 sq km), a large number of powerful rockets would be required to transport the materials, first into low orbit for assembly and then into geostationary orbit for operation. The possible risks during launch are discussed in Section 4.2.

4.3.4 Laser Beams

The very narrow but intense beams emitted by lasers have many applications on the ground and in space. However, exposure to an intense laser beam can cause burns and skin damage. Precautions must be taken to avoid such accidents. There is a possibility that a laser beam emitted either from the ground or from a spacecraft may be accidentally directed toward an aircraft, but it should be pointed out that no accident of this kind occurred in the past 15 years when lasers were used in ranging satellites. Although a laser beam in a spacecraft could illuminate the earth's surface, the distance traveled by the beam would be so great that there would be no hazard.

4.3.5 Exobiology

It has been suggested that there are possible biological risks associated with space activities even though there is no evidence that any danger really exists. Three possible risks have been suggested:

1. The risk that terrestrial microorganisms carried by space vehicles may contaminate the surface of other planets.
2. The risk that extraterrestrial microorganisms carried by space vehicles returning with samples from other planets may contaminate the earth.

3. The risk that terrestrial microorganisms that have been exposed to the space environment may build up mutations which could be dangerous for man.

As far as the first of these risks is concerned, only the surface of Mars need be considered, since the surfaces of the other planets are so hostile to life that they would be totally unsuitable to the development of terrestrial organisms. Even on Mars, although the surface temperature and the atmosphere would not preclude the continued existence of the more resistant terrestrial microorganisms, their development would be highly unlikely because of the lack of free water, the scarcity of soil nutrients and organic matter, and the intense solar ultraviolet radiation. In any case, space probes which are intended to land on planets are sterilized and, taking everything into account, it is believed that the risk of contamination of the planets is negligibly small.

Those who envisage the second of the biological hazards suggest that the introduction of microorganisms from, say, Mars could be the starting point of an epidemic of some unknown Martian disease. However, this suggestion presupposes a high degree of functional overlap and chemical compatibility between terrestrial and Martian life forms. Even if the surface of Mars were more like that of the earth than it appears to be, the separate biochemistries of the two independently evolved systems imply that it is extremely unlikely that a hypothetical Martian parasite could thrive and develop in a terrestrial host. It should be added that no evidence has been found for the existence of plant or animal forms on Mars which might be considered a potential danger to terrestrial life.

The additional point has been made that terrestrial microorganisms that have been isolated for very long periods in high mountains, caves, or polar regions, and which share a common evolutionary heritage with other inhabitants of the biosphere, would be likely to present a far greater danger than samples imported from Mars.

It is probable that terrestrial organisms that have been exposed to space conditions will build up mutations. However, for many years genetic and radiation biological research has been in progress, and so far there has not been any reported case of the discovery of a mutant capable of infecting, or disrupting biogeochemical cycling, or of otherwise destroying parts of the biosphere.

4.4 PHYSICAL CONGESTION OF THE ORBITAL ENVIRONMENT

4.4.1 Number and Distribution of Objects

With the present population of satellites, problems of interference between satellites have been largely restricted to problems of radio-frequency interference, a subject that is handled within the International Telecommunication Union (ITU). Physical interference between satellites has not been a problem, but as satellites become larger and more numerous, the probability of collision and shadowing may become a problem.

Artificial objects in outer space can be divided into two groups, those that can be observed by radar, telescope, or other tracking device and those that cannot. Current systems are capable of tracking objects less than a square meter in cross section in low orbit. All launchings are announced through the United Nations and through COSPAR which also assigns international designations to trackable objects. Currently there are some 4,600 trackable artificial objects in outer space.

As Figure 4.4 shows, yearly numbers of launchings, after a sharp rise between 1957 and 1965, have been, with some fluctuations, almost constant with an average of 116 launchings per year. In a single launching, one or more satellites can be put into orbit while a certain number of nonfunctional objects, such as exhausted rocket stages, nuts and bolts, either fall back to the ground or follow the payload in approximately the same orbit. Only fairly large nonfunctional objects are tracked. Figure 4.5 shows the total number of trackable objects from 1974 to the spring of 1980. The increase between 1974 and 1978 indicates that the number of new objects exceeded the number of objects that decayed in the atmosphere. From 1978 to 1980, however, the total number of trackable objects decreased slightly. Since the launching activity remained about the same in that period, either the technology of launching improved by reducing numbers of nonfunctional objects or the maximum of solar activity in 1979–1980, which caused higher average densities of the atmosphere and thus shortened the lifetimes of space objects, speeded up the cleaning up of near outer space.

The debris of nuts and bolts and small fragments that escape tracking and detection move at a speed exceeding that of projectiles. If they collide with a satellite, they may damage it seriously. Two recent satellite malfunctions have, indeed, been ascribed to possible collisions with space debris. The numbers,

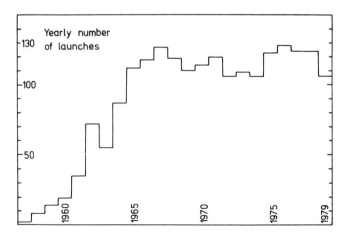

Figure 4.4   Yearly numbers of satellite launches 1959–1979.

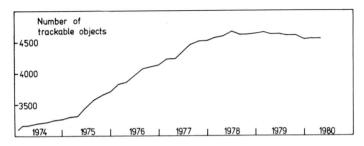

Figure 4.5 Number of trackable objects in outer space.

sizes, and orbits of untrackable debris can only be estimated. It seems that about twice as many untrackable objects as trackable objects are generated. Since the former are smaller, they decay faster and thus the actual number of untrackable objects can be estimated at about 5,000, i.e., not much different from the number of trackable objects.

Even more difficult than estimating the present situation is projection into the future. Some scientists maintain that new debris associated with the launch or breakup of a payload or a rocket are being generated faster than they can decay in the atmosphere. Once the collisional breakup begins—and it may have already begun—the number of pieces of debris would increase exponentially with time. It may quickly exceed the natural meteoroid flux and it may lead to the formation of a debris belt around the earth where only heavily protected spacecraft would survive. Other scientists point out the role of solar activity which, at its maximum, significantly contributes to the cleanup of near outer space. This mechanism is active for two to three years near the maximum of the 11-year cycle of solar activity.

The question of untrackable debris deserves further study extending over several cycles of solar activity, because the numbers and sizes of debris may be very important in considering the safety of future space missions.

It may be useful to compare numbers of artificial objects with the flux of meteors entering the earth's atmosphere. The estimated annual numbers of various sizes are:

$$
\begin{array}{ll}
1\ \text{mm} & 4 \times 10^{10}\ \text{per year} \\
1\ \text{cm} & 3 \times 10^{7}\ \text{per year} \\
10\ \text{cm} & 4 \times 10^{3}\ \text{per year} \\
1\ \text{m} & 40\ \text{per year}
\end{array}
$$

The shape and the orientation of a satellite's orbit in space are more or less variable. The orbital inclination is relatively constant, subject only to small and slow changes caused by the rotation of the atmosphere and by gravitational perturbations. Accordingly, the distribution of orbital inclinations is influenced only by new launches, since even an explosion or collision would leave most of

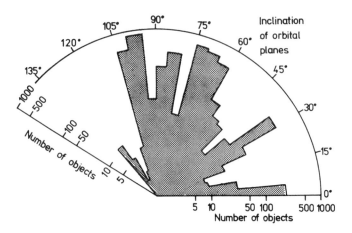

Figure 4.6   Distribution of inclinations of satellite orbital planes.

the fragments in the proximity of the orbital plane. Figure 4.6 shows the distribution of the inclinations of orbital planes and numbers of objects orbiting in those planes.

The maximum near 0° latitude belongs to equatorial orbits such as the geostationary orbit. The maxima around 30° and 60° reflect, at least partly, the location of launching sites. The highly eccentric orbits of some communications satellites favor inclinations above 60°. Satellites in orbits at about 80° overfly all inhabited regions of the world, while strictly polar orbits, at an inclination of 90°, do not seem to be that much in demand. Sun-synchronous orbits have inclinations between 95° and 105°. There are very few satellites in orbits at inclinations higher than that.

Spatial density is defined as the number of objects found on the average in a unit of volume. Here we use, as a reference volume, a cube with sides measuring 1,000 km. Due to the rapid motion of space objects, this spatial density exhibits large and rapid variations. The actual number of objects may at times differ by up to 50 percent from the average value. The spatial density depends on orbital properties since satellites spend more time near their apogees than near their perigees. Moreover, all orbits cross the equatorial plane, whereas at a given geographical latitude only those objects that have orbital inclinations at least equal to that latitude can appear.

The density (see Figures 4.7 and 4.8) varies with increasing altitude above the earth. From 100 km altitude, below which no space objects survive for any appreciable length of time, the spatial density increases with altitude. In the most densely populated region, between 500 and 1,000 km, almost 100 objects are found per reference cube, 60 out of that number being trackable objects, the rest untrackable debris. There are several peaks and valleys of density in that region. Another peak of relatively high density occurs at 1,450 km altitude. The density falls to three objects at 2,000 km and to one object per reference cube at 3,000

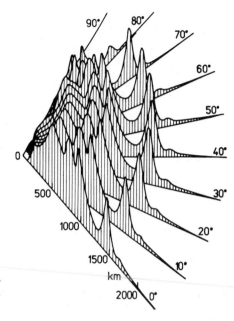

Figure 4.7   Height profiles of spatial density of earth-orbiting objects at various geographical latitudes.

km. Two small peaks appear at about 3,700 km and at the altitude of the geostationary orbit, 35,800 km. As can be seen in Figure 4.7, the density peaks are highest in the equatorial plane, becoming slightly lower at higher geographical latitudes.

The geostationary orbit is a special case in that the spatial density is rather high and growing and the relative positions of satellites are fixed. Large satellites can therefore cause prolonged shadowing of their neighbors. Most satellites and space stations use as a primary energy source solar radiation backed up by batteries. Should solar radiation be cut off for a longer period than that for which the batteries have been designed, some functions might be interrupted. The shadow of a space object can be more than 100 times as long as its dimension. A 20-km solar power station would throw a shadow extending over 2,000 km which, at the geostationary orbit, corresponds to almost 3° in longitude. A small communications satellite designed to work in the close neighborhood of a solar power station should have either the capability of steering out of the shadow or an alternate source of energy.

4.4.2 Collisions

Three parameters determine collision probability; the spatial density of objects, the cross sections of the objects, and their relative velocity. The spatial density is the most important factor, because the collision probability increases with the square of the density. Thus, the danger of collisions is greatest in most densely populated regions of outer space.

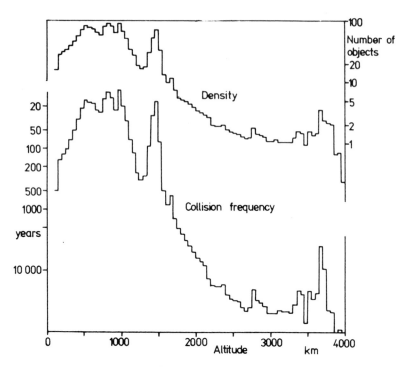

Figure 4.8 Spatial density of earth-orbiting objects near the equatorial plane (upper curve). The scale on the right-hand side gives the number of objects per reference cube, of side 1,000 km. The lower curve gives the collision frequencies at various altitudes. The scale on the left-hand side gives the number of years that will elapse before a collision can be expected.

The satellite cross section is not very important at present since most satellites are rather small bodies which present a small target area to the rapidly flying debris. The situation will be entirely different for large space stations with dimensions of an entirely different order of magnitude. At the time such stations are planned, the question of collision probabilities will have to be discussed in detail.

The velocity of a space object is given by laws of orbital mechanics, unless the body is being propelled by engines. The velocity is largest at the satellite's perigee; it is constant for satellites in circular orbits. An investigation of a representative sample of space objects yielded an average velocity of 7 km/s and a maximum of 14 km/s. An impact at that speed destroys the smaller of the two colliding bodies and ejects more than 100 times its mass from the larger body.

The results of a computation of collision probabilities, or collision frequencies, are shown in Figure 4.8. The left-hand scale gives the number of years within which a collision is to be expected at an altitude shown on the bottom scale. The highest probability of one collision in 20 years occurs in the most densely populated regions between 500 and 1,000 km altitude. Only one collision in a few hundred years is likely to occur at 1,200 to 1,300 km and the expectation is still lower above some 1,600 km.

The above data are valid for the present population of space objects. Should space activity markedly increase in the future, or should the number of pieces of debris build up during the coming minimum of solar activity, or should the next solar maximum be lower than the maximum of 1979–1980, such collision probabilities would have to be revised upward.

The probability of collisions was justifiably overlooked in the first decades of space activities. It may, however, become an important factor in the future if more and larger satellites and stations are launched into outer space. Since preventing all collisions is impossible, technical means might at least be used to reduce the probabilities of collisions. Such measures would be expensive; however, the sooner they are taken, the smaller might be the real cost. The most obvious way of reducing the probability of collision is to reduce the number of pieces of debris produced during the launching phase and during the lifetime of a satellite. Other options for removing satellites that are no longer active from frequently used regions of outer space are available and should also be considered.

4.4.3 Removal of Inactive Satellites

4.4.3.1 Natural forces. The density of the upper atmosphere decreases exponentially with height above the earth's surface and is only about 10 g/km^3 at a height near 350 km. However, a satellite moves at a speed of nearly 8 km/s, and its collisions with the air molecules are frequent enough to create a considerable drag force. If the orbit is noncircular, the air drag is much greater at perigee than at apogee. The satellite is thus retarded at perigee and does not fly out as far as expected at apogee. Hence, the orbit contracts with a tendency to become circular, as shown in Figure 4.9. If the orbit is initially circular, air drag acts continuously along the orbit, thus reducing the height of the orbit gradually.

For both circular and elliptic orbits, the drag rapidly increases as perigee height decreases. At perigee heights of about 100 km, the satellite can no longer remain in orbit and begins its final plunge into the lower atmosphere. It should be noted that, although air drag retards the satellite at perigee, the overall effect of air drag is to make the satellite move faster. Its orbital period decreases as its lifetime progresses and final decay usually occurs when the orbital period has decreased to about 87 min. A more complete discussion of the stability of satellite orbits is given in Appendix 4.1.

The braking force due to air drag depends on the density of the upper atmosphere, which varies widely and irregularly. Predictions of decay depend in particular on future variations of solar activity which are at present unpredictable in detail. Of 4,600 objects in orbit being tracked, about ten per week are slowed down to the point at which they re-enter the atmosphere. The majority of these are small fragments that burn up in the lower atmosphere, but an important proportion (about two per week) are large objects with a mass of more than a

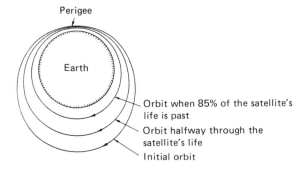

Perigee

Earth

Orbit when 85% of the satellite's
life is past

Orbit halfway through the
satellite's life

Initial orbit

Figure 4.9   Diagram showing the effect of air drag on a satellite orbit.

ton. Fragments of up to 10 kg in weight may reach the earth's surface when these large space objects decay.

The procedure for predicting satellite decay is first to assume that air density remains constant during the rest of the life of the satellite and then to calculate, either from theory or by numerical integration, the date at which the perigee will descend below 100 km. The observed current rate of decay of orbital period is used as a measure of the air drag. Then adjustments are made to this basic calculation to take into account variations in air density. In the absence of such variations in air density, the lifetime could be accurately predicted if the satellite retained a constant cross-sectional area. But, in practice, the air density varies, and lifetime estimates having an accuracy of ± 10 percent are about the best that can consistently be achieved.

If the perigee height is greater than about 500 km, the orbital lifetime is usually 20 years or more, but most satellites have perigee heights between 200 and 500 km. For these the orbital lifetime may be only a few days, or a few months, or a few years, depending on the exact orbit and the area/mass ratio of the satellite.

In predicting orbital lifetimes, the most uncertain factor is the great variation in air density at a height of a few hundred kilometers when the solar activity varies from sunspot maximum (1969 and 1979) to sunspot minimum (1976). At a height of 500 km, the density is about ten times greater at sunspot maximum than at sunspot minimum. Therefore, a satellite in a circular orbit at a height of 500 km might have a lifetime of about five years if launched shortly before solar minimum, but a lifetime of six months if launched at solar maximum. There is also a large variation in density between day and night, the density being higher by day than by night by a factor of up to about five at a height of 500 km. This variation must also be allowed for in predicting orbital lifetimes.

Unfortunately, at present, the progress of the solar cycle cannot be predicted accurately enough for good predictions of satellite decay. Solar activity also shows vigorous and unpredictable variations from day to day. There is normally a fairly steady outflow of plasma from the sun, termed the solar wind, but this is disrupted by shock waves when a solar disturbance occurs. When the shock waves reach the earth, the upper atmosphere responds strongly. At heights near 600 km, the density may increase by a factor of up to eight following a solar storm, and even at heights as low as 180 km the density may be doubled within a few hours. Thus predictions that a particular satellite will decay in a week or ten days can be seriously affected if an unpredicted solar disturbance occurs in the meantime.

The unpredictability of solar activity is the main difficulty in forecasting satellite lifetimes of several years or even of a few days. For lifetimes of intermediate length, i.e., between one month and one year, another partially unpredictable variation becomes more important; this is the semiannual variation in atmospheric density which exhibits maxima in April and early November and minima in January and July. Lifetime predictions can be in error by up to 30 percent if no allowance is made for this semiannual variation. Unfortunately, the semiannual variation itself changes appreciably from year to year. The future variation of the semiannual effect is not predictable with the present state of knowledge of upper atmosphere physics.

The prediction of satellite lifetimes is therefore likely to remain an inaccurate procedure. For lifetimes of between 1 and 20 years, the long-term forecasts of solar activity during a sunspot cycle are inadequate. For lifetimes between one month and one year, solar activity is still a major source of uncertainty (except near sunspot minimum), and so is the future course of the semiannual variation in atmosphere density. For lifetimes of less than one month, the day-to-day variations in density resulting from short-term solar disturbances are the major source of error. To achieve a lifetime prediction with an accuracy of ± 10 percent, in the light of these problems, calls for considerable skill and experience. Account has to be taken of the interaction of many factors, e.g., the future variations of perigee height due to both geometrical and dynamic factors and the synchronization of such variations with changing solar activity.

Although most satellites decay because their orbits steadily contract under the action of air drag, a small but important proportion of satellites have their orbital lifetimes reduced because the perigee is forced down into the lower atmosphere by luni-solar perturbations. Since luni-solar perturbations are accurately predictable and drag may not have an appreciable effect until the last few revolutions, the decay of such orbits is, in principle, quite accurately predictable. Unfortunately, however, such satellites are usually very difficult to track, because they move out to very great distances from the earth. Consequently, their orbits are not well determined, and the accuracy of the predictions is usually limited by the accuracy of the orbital parameters available.

4.4.3.2 Orbital transfers.

In some cases, it may become desirable to alter artificially the natural course of events. In low orbits, since the decay and, especially, the site of impact on the earth's surface are not predictable with sufficient accuracy, a controlled landing or a controlled decay in the atmosphere might be preferable. In some cases, the natural decay might have to be delayed by moving the satellite into a higher orbit with a longer lifetime.

A controlled landing or decay of a satellite in a high orbit might require prohibitive amounts of propellant. Such satellites could, however, be removed into disposal orbits, beyond the paths of active satellites.

In order to effect the removal of an active satellite from its orbit, some of its systems must be operational at that time, in particular the communications systems and the systems for satellite orientation and stabilization. The former are required to provide necessary commands and the latter are necessary for assuring the correct direction to the applied force. These conditions are satisfied in only a few satellites. Soviet orbital stations of the SALYUT-type and cargo-spacecraft of the Progress type are pushed into the earth's lower atmosphere before their active lifetimes expire. Similarly, INTELSAT communciations satellites are removed from their geostationary positions when a new satellite is launched.

The overwhelming majority of objects presently in orbit are inactive satellites (or debris) that cannot be removed from their orbits without some external assistance. Such satellites might be caught by the remote manipulator arm of the Space Shuttle, or some maneuvering rocket-powered unit could be attached to them in order to provide them with a maneuvering capability. However, better removal methods (in terms of energy conservation) can be found in making use of natural forces.

In principle, natural forces can also be used to effect rapid changes of the orbit. For example, a close encounter of a satellite with a natural celestial body might change its orbit profoundly. Or, the effective area/mass ratio of a satellite can be changed by the deployment of large "wings." This would increase the influence of nongravitational forces (air drag, radiation pressure) which, in turn; might cause faster decay. These methods, however, have not found practical application as yet.

In order to change the orbit of a satellite, it is necessary to change its kinetic energy, which is proportional to the square of its velocity. The theory of orbital transfers between circular orbits was elaborated as early as 1925 by W. Hohmann. He proved that the trajectory that connects two circular orbits in such a manner that it is tangential to both requires minimum velocity change and therefore minimum consumption of propellant. During such a transfer, the satellite travels exactly half the elliptical transfer orbit, called the Hohmann orbit. Transfer orbits, which are longer or shorter than Hohmann orbits, require larger velocity impulses. The orbital transfer involving two elliptical orbits can be treated similarly. Generally, it is preferable to perform a Hohmann transfer between the perigee of the initial orbit and the apogee of the final orbit. If the orientation of the planes of the two elliptical orbits is different, more than two impulses are necessary to accomplish the transfer.

The fundamental equation of rocket propulsion that is applicable to classical chemical rocket systems expresses the velocity of the vehicle in terms of specific impulse, which depends on the effective exhaust velocity of gases from the motor and the ratio of initial and final masses of the vehicle. Therefore, for a required change of velocity, using a specific propellant, one can compute the necessary amount of the propellant. As an example, for a typical geostationary satellite of the INTELSAT-IV type (mass 700 kg), 34 kg of hydrazine are needed to achieve a velocity change of 100 m/s.

Large velocity changes would demand amounts of propellant that would greatly increase the total mass of the satellite. Therefore, a careful analysis is necessary in order to minimize such maneuvers.

4.4.3.3 Planned decay.   The decay of a satellite in the atmosphere can be induced by pushing it into an elliptical Hohmann transfer orbit in which the satellite travels half of a revolution reaching the perigee at 100 km or preferably lower. The decelerating kick is applied in a direction opposite to the orbital velocity (see Figure 4.10), and its magnitude depends on the radius of the initial orbit. For altitudes up to the geostationary orbit, the higher the orbit, the greater the necessary retrokick (see Figure 4.11). If the satellite is in an elliptical orbit, the optimal retrokick is applied at apogee and its magnitude is always smaller than for a circular orbit at the same height. The lower the perigee, the smaller is the necessary retrokick. Obviously, if the perigee is already at 100 km, no impulse is required.

However, the use of a Hohmann orbit with a perigee at exactly 100 km is not a very safe method for removing a satellite from its orbit, because the satellite would travel horizontally at the perigee of the Hohmann orbit. It could happen that atmospheric drag would not be sufficient to "catch" the satellite. This situation is similar to the ballistic re-entry of spacecraft with high velocities (e.g., from a lunar mission), where the spacecraft "bounces" off the top of the atmosphere, i.e., "atmospheric skip" occurs if the angle to the local horizon is less than 3°. Therefore, when planning re-entry, it is advantageous to increase the retrokick for low orbits by a few percent and to increase the reentry angle to ensure atmospheric capture and a speedy decay.

Once in the atmosphere, the descending body moves according to the laws of aerodynamics. Its final fate depends on many factors, of which the most important are the entry velocity, shape and area/mass ratio of the body, and the thermal and mechanical characteristics of its structure. Usually, the object is destroyed at a certain altitude, but a few of the resulting debris are capable of reaching the earth's surface. Up to the end of 1979, from 6,733 objects which had already decayed, fragments surviving re-entry had been recovered on land in only 16 cases.

The conclusion could be drawn that, for low orbits (Figure 4.12), pushing the satellite into the earth's atmosphere is an effective means of removal. An application of the minimum necessary retrokick results in long trajectories and in small angles of atmospheric entry. Larger retrokicks increase the re-entry angles

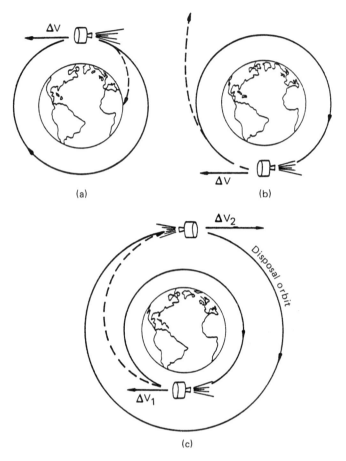

Figure 4.10 Three methods of a satellite's removal from orbit. (a) Into the atmosphere. (b) Out of the earth's influence. (c) Into a disposal orbit.

and for some values (which are used for re-entry of manned spacecraft), both thermal loading and mechanical overloading are relatively small, leading to a high probability for the survival of fragments which can consequently hit the earth's surface. For even greater velocities (and re-entry angles), mechanical overloading increases and destroys the object.

By timing the retrofiring carefully, a decay over uninhabited areas or over an ocean can be achieved. In such cases, resulting fragments have little chance of causing damage even if they reach the earth's surface.

4.4.3.4 Ejection from earth's orbit. According to Newton's law of gravitation, the gravitational acceleration produced by a celestial body decreases with the square of the distance between that body and the satellite. Consequently, the gravitational influence of a celestial body is important only up to a

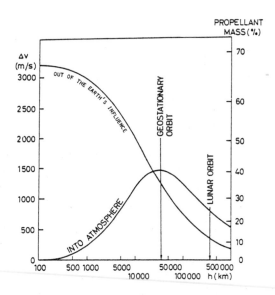

Figure 4.11 Mass of propellant and velocity necessary to push the satellite into the atmosphere or out of the earth's influence.

limited distance, where the gravity of some other body becomes stronger. The so-called "earth's sphere of influence" has a radius of approximately 1 million km. At greater distances, the sun's gravity prevails.

Should the apogee of an orbit lie beyond 1 million km, the satellite would travel toward the apogee but would not return back to earth. Such a satellite would leave the earth's sphere of influence on an escaping trajectory. The difference between the velocity of escape and the circular velocity for various

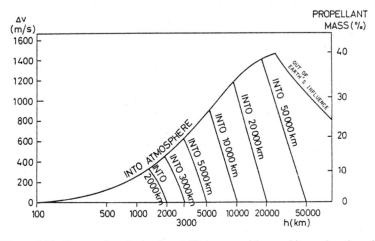

Figure 4.12 Proposed methods of a satellite's removal from orbit as a function of altitude.

heights above the earth's surface is shown in Figures 4.11 and 4.12 and labeled "out of earth's influence." In order to reach the escape velocity, a velocity kick is to be applied in the direction of the orbital motion of the satellite. For elliptical orbits, it is most economical to change the velocity at perigee.

What happens to the body after it leaves the earth's gravitational influence depends on the residual velocity after escape. If this velocity is directed contrary to the earth's revolution around the sun, the body will orbit around the sun inside the earth's orbit and touch the earth's orbit at the point of escape. If the residual velocity is directed along the earth's orbital velocity, the elliptical orbit of the body around the sun is outside the earth's orbit and touches it again at the point of escape. The probability of repeated entry into the earth's sphere of influence is very low, and therefore such orbits have also been considered for the disposal of nuclear waste in space.

In order to ensure that the removed body does not reappear within the earth's sphere of influence, it should either be placed in a disposal orbit around the sun which does not cross or touch the earth's orbit or be pushed out of the solar system completely. To change the 200-km high orbit around the earth into an orbit escaping the solar system requires a velocity impulse of almost 9 km/s. However, pushing the body directly into the sun demands an even greater impulse, a velocity change of 24 km/s. The velocity changes, ΔV, needed for a simple escape into a solar orbit are much smaller and are given in Figure 4.11.

4.4.3.5 Disposal orbits.   Still another method of removing a satellite from its initial orbit is to place it in a disposal orbit. By "disposal orbit" is meant an orbit that is generally higher than the original orbit and located at heights used by no (or only a few) active satellites.

For transfer from an initial circular orbit into a circular disposal orbit at a higher altitude, two velocity impulses are necessary. The first impulse changes the initial orbit into a Hohmann transfer, and the second impulse, half an orbit later, changes the Hohmann transfer into the final circular disposal orbit. Both impulses have to be parallel to the orbital velocity, increasing it by the necessary amounts. The first impulse is always larger than the second one since, in the former case, the satellite is nearer the earth's center of gravity. The change in velocity, ΔV, necessary for increasing the height of a given circular orbit by an increment, Δh, is shown in Figure 4.13. It can be seen that at greater heights even a small velocity kick could cause a significant orbital change (because the gravitational field of the earth is weaker there).

4.4.3.6 Summary.   The results of this section are summed up in Figure 4.12 which provides a comparison of the three methods for removing satellites from their orbits and shows which of the three methods would be most efficient and economical at a particular altitude.

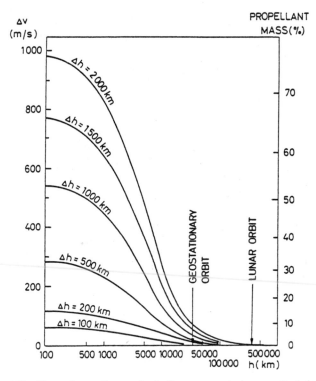

Figure 4.13 Mass of propellant and velocity necessary to increase the height of a given circular orbit.

1. *For low orbits* of a few hundred kilometers above the ground, dumping the satellite into the atmosphere is the most efficient method. The necessary velocity kick requires an amount of propellant that is only a few percent of the total initial mass of the satellite. Even at a relatively high orbit of 1,500 km, not more than 10 percent of the initial mass is required.

2. *For medium orbits,* between 1,500 and 10,000 km, either atmospheric re-entry or parking in a disposal orbit could be considered. Conveniently selected disposal orbits, not far from the operational orbits, would require only relatively small velocity impulses. Should the disposal orbits be far from the initial orbits, the required amount of propellant might become significant.

3. *For geostationary orbits,* at 35,800 km, the disposal orbit appears preferable to other methods. An orbit of about 500 km beyond the geostationary orbit would require an impulse of only 18 m/s, which is well within the limits of the station-keeping systems on board most communications satellites.

4. *For extremely high orbits,* above 40,000 km, the most economical method is to push the satellite out of the earth's gravitational sphere of influence altogether.

The above methods could be used if future satellites were designed so as to incorporate suitable propulsion systems or if the propulsion systems used for station-keeping and attitude control contained sufficient propellant.

Small satellites and debris already in orbit that are not equipped with propulsion systems cannot be removed from their orbits without a special collecting capability which might become possible in the future. For example, a manned space vehicle would be very convenient for this purpose, since its crew could reach the satellite with a remote manipulator arm and either collect it or attach it to a maneuvering unit.

In the era of frequent manned flights, possibly using reusable space vehicles, the number of inactive satellites might decrease because malfunctioning satellites could be repaired, returned to earth, or taken to the dense layers of the atmosphere for decay.

4.5 IMPACT ON ASTRONOMICAL OBSERVATIONS

4.5.1 General Considerations

For many centuries man has used optical telescopes for making observations of the planets, stars, nebulae, and other celestial objects. However, in the past 30 years, as a result of the application of remarkable new techniques, the astronomer now has at his disposal not only the optical window but also three important new windows through which he can look out into space; these windows are located in the parts of the spectrum which include radio waves, ultraviolet radiation, and X-rays. The radio window can be used either at observatories on the earth's surface or in high-altitude rockets or orbiting spacecraft. But since the terrestrial atmosphere almost completely absorbs ultraviolet and X-radiation, these two windows can be used only by installing the appropriate observing intstruments in spacecraft.

As a general rule, telescopes are instruments designed to collect radiation from weak sources in such a way that it can be recorded photographically, or by some other means, for subsequent examination. Since the radiation to be recorded by an optical telescope is weak, it is obviously necessary to site observatories far from large towns, where the street lighting would be a serious problem because of reflected light from clouds. Similarly, a radio telescope must be placed as far as possible from man-made sources of radio waves, and preferably in a site where the local topography helps to shield the telescope from terrestrial transmitters. However, even if all possible precautions have been taken in advance in siting the telescope so as to minimize terrestrial sources of interference, the astronomer can do nothing to prevent artificial satellites from passing across his telescope's field of view.

Satellites reflect sunlight in the same way as the planets in the solar system. Hence, a satellite that crosses the field of view of an optical telescope can destroy or distort the information that is being looked for. All satellites include radio transmitters, which are used both directly, in making scientific observations, or indirectly for transmitting down to the ground the observations made in the satellite. The radio emissions from satellites are very strong in comparison with the radio waves emitted by most radio stars or by the clouds of interstellar gas found in space. Hence, it is clear that the passage of a satellite can be a potential source of interference to a radio telescope.

4.5.2 Optical Astronomy

In order to assess the magnitude of the potential interference caused by satellites to optical astronomy, it is necessary to bear in mind such questions as the total number of satellites at present in orbit, the possible numbers in the future, their heights, the type of camera used, etc. For example, it has been calculated that, at the present time, the type of photograph taken by a Schmidt camera (which has a wide field of view) will almost certainly include a trace caused by the passage of an artificial satellite. For spectroscopic observations (using a small field of view), the chance of interference by a satellite is less than 1 percent. Satellite tracks on photographs of stars can usually be identified visually. However, automatic measuring machines are being increasingly used to reduce the manual effort necessary for the analysis of certain types of photograph, and there are difficulties in designing the machines to recognize and eliminate satellite tracks.

During astronomical observations that involve making recordings or other measurements over periods of tens of minutes or several hours, the passage of a satellite across the field of view of the observing instrument may not always be serious, provided that it is possible to identify the short-term disturbance caused by the satellite. Astronomers are also interested in making observations of certain types of transient phenomena which take place within a period of a few minutes. In these circumstances, it is possible that the passage of a satellite across the field of view of an instrument designed to record short-term phenomena may be misinterpreted as something else.

4.5.3 Infrared Astronomy

Artificial satellites have a temperature of roughly $300°K$, and hence they emit, spontaneously, infrared radiation which is very strong in comparison with the weak sources being investigated in infrared astronomical programs. Even if no strongly radiating satellite passes through the beam of an infrared telescope, large numbers of satellites will necessarily, in combination, lead to an increase in the level of background radiation with undesirable consequences. The particular case of the proposed solar power satellite must be mentioned because it is estimated that such a large object would be similar, so far as infrared radiation is concerned, to a full moon. It follows then that the only infrared observations that could be made would be those in which the effect of the moon is not important.

In principle, infrared astronomy could also be endangered by the radiation emitted by the exhaust gases from launching rockets, and especially from those gases that emit near the atmospheric windows used by infrared astronomers for their observations. However, given the present level of launches of satellites, this danger is not yet regarded as serious.

4.5.4 Radio Astronomy

Many of the astronomical observations that utilize radio telescopes are made at frequencies determined not by the choice of the astronomer but by the natural characterisitics of the emitting objects such as clouds of hydrogen in space, and other more complex gases, including those containing carbon which are of special interest since they may provide information about the origin of life. The very weak radio waves which reach the earth from these distant sources can be detected only with the aid of powerful specially constructed radio telescopes. These very sensitive instruments are susceptible to stray radiation from many man-made sources: broadcasting and television transmitters, radio communications stations, radio aids to air and sea navigation, etc. The frequencies allocated to each of these services are decided during World Administrative Radio Conferences which are held at infrequent intervals and are convened by ITU. Radio astronomy has been formally recognized by ITU as a radio service, and hence it is possible for the needs of radio astronomers for frequency allocations to be taken into account and for allocations to other services to be made in such a way as to minimize the interference likely to be caused to astronomical observations. Even when all possible precautions have been taken in siting a radio-astronomical observatory, so as to limit interference from terrestrial radiating sources, the instruments will not be immune to interference from satellites carrying radio transmitters of any kind. Therefore, special attention is necessary when allocating frequencies to satellites which radiate; their allocations must be as far as possible from those used by radio astronomers.

It is also important to bear in mind that a radio transmitter that operates nominally within a given band of frequency may, in fact, also emit radiation above and below this band. Unless special precautions are taken, it will also radiate at "harmonic" frequencies: that is, at frequencies that are two, three, four, etc., times the fundamental frequency. It follows that a satellite that emits harmonics can cause interference to radio astronomy at frequencies very far from the nominal operating frequency of the satellite. The limitation of these "unwanted" frequencies requires careful attention during the design stage of the transmitter if interference to radio astronomy, and indeed to other types of radio service, is to be avoided or at least reduced to a minimum.

The future survival of radio astronomy will depend on the degree of cooperation between the radio scientists on the one hand and the designers and users of satellites on the other. The reduction to a minimum of the "unwanted" out-of-band and harmonic radiation emitted by satellites will be of particular importance. These questions are considered in ITU and its International Radio Consultative Committee (CCIR).

The SPS also deserves mention here in view of the very large radio power it is proposed to generate (6,000 MW). Since spurious emissions are possible, they represent a potential danger to radio astronomy. The large physical size of this satellite also implies that it will be capable of acting as a passive reflector of radio waves emitted from the ground, or possibly from other satellites. Its reflecting power is estimated to be about a million times that of any other satellite now in orbit.

In this connection it should be noted that a small satellite at, for example, 500 km height would cause interference to radio astronomy if it were illuminated by a 3-kW transmitter at a distance of 2,500 km. Although this is not now a serious source of interference, with the rapidly increasing numbers of satellites and other passively reflecting objects in space, the problem could become serious in the future.

4.6 CONCLUSION

The current level of space activities does not appear to be having any extensive, long-term undesirable effects on the earth or space environment. This conclusion, however, must be tempered with reservations due to a limited understanding of many environmental processes.

Upper atmospheric processes in particular are still poorly understood and subtle changes in long-term balances cannot be excluded. It is not inconceivable that slow cumulative changes have been initiated that might continue for some time even if the activity causing the changes were stopped.

It does not appear that activities being planned for the next decade will significantly change the situation, but the development of large numbers of large space structures would require detailed study of possible environmental effects. In particular, the proposed solar power satellite system will require detailed study of the effects of the huge space transport system required for construction and of the high-power microwave beam used to transmit power to the surface.

Accidents, primarily arising from launch and re-entry, are and will continue to be a small but real risk. When nuclear power systems are involved, the risk becomes higher and more widespread. Clearly, every effort should be made to reduce accidents and limit the damage that can be caused.

APPENDIX 4.1: STABILITY OF SATELLITE ORBITS

A4.1.1 Forces Acting on a Satellite

A satellite in orbit around an isolated spherical planet with no atmosphere would follow an elliptical orbit, without variation, for thousands of revolutions. For the planet earth, however, this simple picture is greatly altered due to three different perturbing forces:

1. The variation of the earth's gravitational attraction that results from the flattening of the earth at the poles and other deviations from spherical symmetry, such as the "pear shape" of the earth.
2. The air drag caused by the rapid movement of the satellite through the tenuous upper atmosphere.
3. The forces due to the sun and moon, mainly their gravitational attraction, but also the effects of solar radiation pressure.

For most satellites, these are the three types of force that cause major changes in the orbits. Many other perturbations exist, but these do not cause major changes; they are caused by upper atmosphere winds, solar radiation reflected from the earth, earth tides and ocean tides, the precession of the earth's axis in space, resonances with the earth's gravitational field, and relativity effects. It should, however, be noted that these perturbations can become important for particular satellite orbits such as the orbits of ballon satellites.

A4.1.2 The Elliptic Orbit

The size and shape of the ellipse are defined by its semi-major axis, *a,* and its eccentricity, *e,* as shown in Figure A4.1. As the satellite, *S,* moves in its orbit,

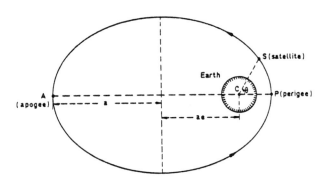

Figure A4.1 Diagram defining semi-major axis (a) and eccentricity (e) of an elliptic satellite orbit around the earth.

its nearest approach to the earth occurs at the perigee, P. The farthest distance of the satellite from the earth is reached at apogee, A.

The orientation of the orbital plane in space is defined by the right ascension of the ascending node, Ω, and the inclination to the equatorial plane, i, as shown in Figure A4.2, relative to a sphere with its center at the earth's center, C.

Finally, the orientation of the orbital ellipse within the orbital plane is specified by the argument of perigee, ω, which is the angle NCP in Figure A4.2.

These five parameters, together with a sixth parameter which specifies the position of the satellite in its orbit at a given time, provide a complete description of the motion of a satellite in an unperturbed elliptic orbit. The position can be specified by the angle PCS in Figure A4.1, denoted by $\ominus$ and known as the true anomaly.

The five parameters, a, e, i, Ω, ω, together with $\ominus$ at a given time, are known as orbital elements. More generally, orbital elements are any set of parameters that describe the orbit as fully as those given above.

A4.1.3 Perturbations of Satellite Orbits

A4.1.3.1 Earth's gravity.   The earth is not exactly spherical, for it is appreciably flattened at its poles: The equatorial radius is 6,378.14 km, and the polar radius is only 6,356.79 km. Because of this flattening of more than 21 km, the gravitational attraction of the real earth is slightly different from that of a sphere and, as a result, the elliptic orbit of a satellite suffers some changes.

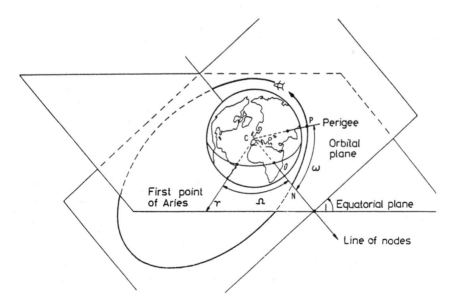

Figure A4.2   The position of satellite orbit in space showing definitions of the line of nodes, right ascension of node Ω, inclination i, and argument of perigee ω.

Although its size and shape remain almost the same (the changes in *a, e,* and *i* being small), the orbit is no longer fixed in space. The orbital plane rotates about the earth's axis, with the inclination remaining constant. If the satellite is heading eastward, the orbital plane swings to the west, as shown in Figure A4.3, and the right ascension of the node, Ω, steadily decreases. These effects are quite significant for satellites close to the earth.

More satellite orbits have inclinations less than 90° and thus travel eastward while their orbital planes swing from east to west. The rate of change of Ω depends on the inclination, *e;* for a near-equatorial orbit, the rate is about 8°/day. For an orbit at inclination 60°, the rate is 4°/day becoming zero for a polar orbit.

In addition to this movement of the orbital plane, the orientation of the orbit within the orbital plane also changes, so that the perigee latitude is continually changing. For a near-equatorial orbit, the perigee moves in the same direction in which the satellite is traveling, at a rate of about 16°/day; for a polar orbit, the perigee moves in the opposite direction from the satellite at about 4°/day. For an orbit at an inclination of 63.4°, often called the critical inclination, the perigree remains at a fixed latitude.

Because of these two effects, a satellite eventually passes through points within a toroidal volume bounded by its perigee and apogee heights, and a maximum latitude north and south, which is equal to the inclination, as shown in Figure A4.4.

Figure A4.3   The gravitational pull of the earth's equatorial bulge causes the orbital plane of an eastbound satellite to swing westward.

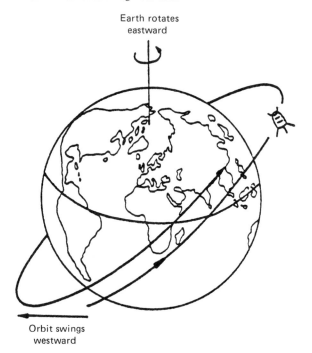

Earth rotates
eastward

Orbit swings
westward

Figure A4.4 The toroidal volume traversed by a satellite.

Although the earth's flattening is its greatest deviation from a sphere, there is also a slight asymmetry between the northern and southern hemispheres. Sea levels at the North Pole and the South Pole would differ by 44 m. This asymmetry is usually called the pear-shape effect. It causes slight changes in the shape of the orbit of a satellite, producing an oscillation in the orbital eccentricity, e, which leads to an oscillation in the perigee distance although the semi-major axis, a, remains constant. This oscillation can have an important effect on perigee height.

These effects are the most important perturbations for low-orbit satellites. But for satellites in higher orbits, such as geosynchronous orbits at a height of approximately 36,000 km, the effect of the earth's oblateness is very much smaller, and the variation of the gravitational field with longitude rather than latitude becomes more important.

A4.1.3.2 Air drag.   The density of the upper atmosphere decreases exponentially with height above the earth's surface, and is only about 10 g/km³ at a height near 350 km. However, a satellite moves at a speed of nearly 8 km/s, and its collisions with the air molecules are frequent enough to create a considerable drag force.

If the orbit is noncircular, the air drag is much greater at perigee than at apogee. The satellite is thus retarded at perigee and does not fly out as far as expected at apogee. Hence, the orbit contracts and has a tendency to become circular.

If the orbit is initially circular, air drag acts continuously along the orbit, thus reducing the height of the orbit gradually.

For both circular and elliptic orbits, the drag rapidly increases as perigee height decreases. At perigee heights of about 100 km the satellite can no longer remain in orbit and begins its final plunge into the lower atmosphere.

It should be noted that although air drag retards the satellite at perigee, the overall effect of air drag is to make the satellite move faster. Its orbital period decreases as its lifetime progresses, and final decay usually occurs when the orbital period has decreased to about 87 min.

A4.1.3.3 Luni-solar gravity.

Both sun and moon exert small gravitational attractions on satellites, and therefore perturb satellite orbits because these small attractions combine with the main attraction of the earth and change its amount and direction differently at different points of the orbit. In general, lunar gravity is about twice as effective as solar gravity in perturbing satellites in low earth orbits.

For a satellite close to the earth, luni-solar gravitational perturbations produce small oscillatory changes in all the orbital elements except the semi-major axis. For most near-earth satellites, the effects are small, displacing the satellite by less than about 2 km at periods ranging from ten days to more than a year.

The complete luni-solar perturbation is made up of many different terms, each with its own period. It is likely that one or more of these periods may be very long, perhaps several years, and perturbations building up for several years may become much larger than expected. For example, a perturbation may only amount to 10m/day, but if it continues for 500 days, it will amount to 5 km.

For satellites in higher orbits, luni-solar perturbations are of greater importance, being approximately proportional to the orbital period, for given eccentricity and inclination. For synchronous orbits, with orbital periods of 1,436 min, the effects are considerable.

An increase in eccentricity also increases the effect of luni-solar perturbations, and these are particularly severe for the many MOLNYA satellites having eccentricities near 0.7 and orbital periods near 720 min.

A4.1.3.4 Solar radiation pressure.

The pressure of solar radiation is very small ($4.6 \times 10^{-6} n/m^2$), but its effect on satellite orbits can be appreciable.

The acceleration of a satellite as a result of solar radiation pressure is directly proportional to the satellite's cross-sectional area divided by its mass, the area/mass ratio. A satellite constructed of metal usually has an area/mass ratio near $0.01 \, m^2/kg$, and the perturbations due to solar radiation pressure are very small, less than 1 km.

For a balloon satellite, however, the area/mass ratio and the perturbation can be a thousand times greater. For the ECHO 1 balloon, the area/mass ratio was about $10 \, m^2/kg$, and the main solar radiation pressure perturbation had a long period, about 20 months, with the result that the perigee height of ECHO 1 oscillated with an amplitude of 500 km.

Even for satellites with low area/mass ratios, the effects of solar radiation pressure can exceed the effects of air drag at heights above 500 km, though both effects are small at these altitudes.

A4.1.4 Special Orbits

A4.1.4.1 Geosynchronous orbits.   An orbit is called geosynchronous if the satellite moves at the same angular rate and in the same direction as the rotating earth. It completes one revolution in one sidereal day and its semi-major axis, a, has to be equal to 42,160 km. Viewed from the ground, such a satellite would be seen to complete a closed curve on the sky every day. The shape and dimensions of the curve depend on all orbital elements (see Section A4.1.2), with the exception of a, which is fixed. Generally, the smaller the inclination and eccentricity of the orbit, the smaller the apparent curve on the celestial sphere as seen from earth. In the extreme case, for a circular orbit in the equatorial plane (zero inclination and zero eccentricity) the curve shrinks to a point and the satellite is seen in a fixed direction. Natural perturbations, however, force the satellite out of the ideal geostationary orbit. Only by artificial "station-keeping" maneuvers can the satellite be kept reasonably close to the ideal geostationary orbit.

Because of the enormous importance of geostationary satellites for communications, meteorology, and other applications, Chapter 7, "Efficient Use of the Geostationary Orbit," has been devoted to this subject.

A4.1.4.2 Integer orbits.   By choosing suitable values for a, or, which is equivalent, for its mean altitude above the earth, a satellite can be made to pass daily over certain strips of the earth's surface, leaving other areas uncovered. These orbits are called integer orbits with N revolutions per sidereal day. Thus, for example, satellites at altitudes of 554 km, 881 km, and 1,248 km in polar orbits would complete 15, 14, and 13 revolutions per sidereal day, respectively.

A4.1.4.3 Sun-synchronous orbits.   By choosing suitable values of the altitude and inclination of the orbit, the movement of the orbital plane (see the first two paragraphs in Section A4.1.3.1) can be made to compensate for the annual movement of the sun in the ecliptic. Such a satellite would cross the equator always at the same local time. This was found very useful for earth observation satellites which then provide imagery of individual regions under nearly the same illumination by the sun. Thus, for example, satellites at altitudes of 565 km, 893 km, and 1,261 km and inclinations 97.6°, 99.0°, and 100.7° would complete 15, 14, and 13 revolutions per day, respectively.

Sun-synchronous orbits can also be achieved by using the attraction of the moon. A satellite proposed, e.g., for the study of the geomagnetic tail, which from the earth is opposite to the direction of the sun, is made to spend most of its time in the tail region by a double approach to the moon, as shown in Figure A4.5.

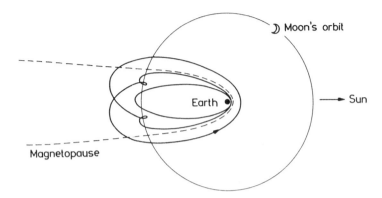

Figure A4.5   Proposed sun-synchronous orbit for geomagnetic tail mission.

A4.1.4.4 Highly eccentric orbits.   The laws of dynamics show that a satellite spends much more time near the apogee of a highly eccentric orbit than near its perigee. Also, its apparent movement in the sky is rather slow near the apogee. This has been used with advantage for communications satellites at high geographic latitudes. A typical orbit of this kind has an inclination of about 63°, a perigee at 400 to 500 km, and an apogee at 40,000 km, which is slightly beyond the altitude of geosynchronous satellites, and a period of exactly half a day. A system of four satellites has been designed in such a way that, from a ground station, one satellite is seen to rise slowly above the horizon and, after reaching maximum elevation, is seen to set slowly on almost the same path in the sky. Just before setting, another rising satellite almost meets the setting satellite; provision is thus made for easy tracking and ready switching of communications.

A4.1.4.5 Libration orbits.   In the earth–moon system, there are five important centers, called Lagrangian libration centers. A satellite placed at any of these centers would retain the same configuration with respect to the earth and moon. The first center, L_1, lies between the earth and the moon, at about 85 percent of the distance to the moon. The second center, L_2, lies beyond the moon, and the third, L_3, lies in a direction opposite to that of the moon. The equilibrium of a satellite in those centers would be unstable, i.e., it would require station-keeping. The most important centers, L_4 and L_5, lie at the apexes of equilateral triangles formed by the earth, the moon, and the satellite. A satellite at one of these centers would require no station-keeping. This is the reason why the last two centers have been proposed as suitable sites for the first space colonies or habitats.

A similar situation is presented by the sun–earth system. Here, the center, L_1, is 1.5 million km distant from the earth in the direction toward the sun. A scientific satellite has been placed near L_1 where it has a vantage point for observing solar activity and providing early warning of incoming charged particle streams from the sun.

APPENDIX 4.2: ELECTRICALLY AND MAGNETICALLY ACTIVE RADIUS OF A SATELLITE

A4.2.1 Concepts and Definitions

Since there is no exact definition, we suggest that the "electrically or magnetically active radius of a satellite" be understood as the maximum distance at which the electric or magnetic effects of the satellite can be observed. Good indicators of such effects are relative electron or ion density variations which are measurable down to about 0.1 percent. Therefore, this value was taken as the limit of observable electric effects. Another approach to that limit follows from the threshold of 0.2 gamma (nT) of recent measurements of natural variations of the local magnetic field at elevations of about 1,000 km. There are also other physical parameters that would serve the purpose of defining the limit of observable effects. Referring to electric and magnetic effects of one spacecraft on other satellites, the limit of observable effects would be defined by the sensitivity of the sensor and on-board receiving device and/or by the characteristics of the naturally occurring variations.

An important plasma characteristic, defining the electric radius of a satellite, is the Debye radius, D, i.e., the distance beyond which the ambient electric charges fully screen the electric field due to a given point charge. This is closely related to the penetration of the satellite electric field into the surrounding plasma, and it defines the thickness of the plasma sheath around the satellite. But the plasma sheath alters the electric field distribution around the satellite and the charged particle trajectories within the screening zone. It is important to consider the mutual dependence between the charge and potential distributions both over the satellite surface and in the plasma sheath.

In terms of the disturbing effects on terrestrial systems of satellite electric and magnetic fields due to on-board energy sources, the most important would be the effects of the proposed solar power satellites (SPS) on ground communications and on aircraft that might pass through the beams of electromagnetic energy between the spaceborne energy sources and the ground receiving stations. The documents of the CCIR of ITU have treated the present situation of satellite communications and radio systems excellently.

A4.2.2 The Plasma Environment

The electric (R_E) and magnetic (R_M) radius of a satellite, when there is no on-board electromagnetic source, is defined by the interaction between the satellite and the space plasma environment. For earth satellites, this environment is a partially or almost fully ionized multicomponent gas having large spatial variations of its characteristics. In space surrounding the earth, of interest to this study, the plasma density varies from about 5×10^6 cm^{-3} to about 1 cm^{-3}, and the particle thermal energy from 0.2 eV to 30–40 eV.

A4.2.2.1 The ionosphere. The ionosphere is considered to extend up to about 1,000 to 1,500 km, where transition to the plasmasphere occurs. The major ion of the ionosphere is atomic oxygen, with a density ranging from 10^3 cm^{-3} to $5 \times 10^6 cm^{-3}$ for quiet midlatitudinal conditions. Normally, the electron temperature exceeds $3,000°$K, while the ion temperature varies with altitude during the daytime from about $1,500°$K at 300 km to about $2,500°$K at 1,000 km.

A4.2.2.2 Plasma environment at geosynchronous orbits. The dynamic behavior of the plasma environment near geostationary orbit (altitude 36,000 km) is extremely complicated. The spatial and temporal variations cannot be defined either in a purely theoretical manner or based on experimental data from one satellite only. An overall model for the environment at geostationary orbit necessarily needs models for the electric and magnetic fields, plasma composition and density, charged particle fluxes and flux density, charged particle energy spectra, and other parameters. The typical parameters of the geosynchronous plasma environment depend strongly on geomagnetic activity. The number density varies between 0.5 and 150 cm^{-3}. A mean free ion path of 10^6 km, a Larmor ion radius of 1,000 km, and a Debye radius of several tens of meters to 1 km can be considered as representative values.

Solar ultraviolet and X-radiation also determine to a great extent the interaction between a satellite and the surrounding environment. Depending on the photoemission characteristics of the spacecraft surface materials, the sunlit part of a spacecraft can be charged to a high positive potential. Satellite charging to negative potentials occurs preferentially on non-illuminated surfaces or during an eclipse, as seen from the spacecraft, of the sun by the earth.

A4.2.3 Electric and Magnetic Satellite Radius

Satellite–environment interactions are either passive, when the environment affects the satellite and causes it to charge up, or active, caused by the operation of the spacecraft disturbing the environment.

Interaction between the environment and the satellite surface occurs via charged particle fluxes to and from the satellite surface and via solar illumination. The following particle fluxes contribute to the interaction: primary (magnetospheric) electrons and ions, photoelectrons, secondary electrons from electron and ion collisions with the surface, and backscattered electrons. As a result of this passive interaction, the spacecraft becomes charged. Problems of the degree of charging relate directly to the electrically active radius of the spacecraft.

A4.2.3.1 Spacecraft charging at geostationary orbits. The Debye radius in the outer magnetosphere is of the order of 1 km, and it is large compared to the satellite size. In the presence of high-energy electron fluxes, the dark shadow side of a satellite may be charged to a very high negative potential. If these electron fluxes are small compared to the photoemission current, insulated surfaces on the sunlit part may acquire a positive potential.

For a sufficiently large Debye radius and low photoemission current, it may be suggested that the total charge of the photoelectron cloud is small compared to the surface charge of the satellite. The behavior of the plasma sheath around the spacecraft is then complex. Experimental results on the degree of charging of the ATS-5 satellite show that the average surface potential is $-2\,kV$, and the maximum is $-10\,kV$. The ATS-6 satellite was charged, on average, up to $+4\,kV$ and even up to $+20\,kV$. The largest potentials were observed during solar eclipses; there is also direct correlation between the degree of charging and the geomagnetic activities.

A4.2.3.2 Spacecraft charging at low orbits.

For low-altitude satellites, the satellite velocity is much greater than the ion thermal velocity and much smaller than the electron thermal velocity. This results in a region of increased plasma density being formed in front of the satellite, while behind it there is a wake, a long region of low-density plasma. Due to their greater mobility over ions, electrons tend to fill the space behind the satellite. The quasi-neutrality of the plasma near the satellite surface is disturbed; a double layer with dimensions of the order of several Debye radii is formed as the potential over this distance drops from its value on the satellite surface to that of the nondisturbed plasma. Comparisons show that measured electron and ion densities agree well with theory. In the case of EXPLORER-31, it was found that plasma quasi-neutrality in the wake was attained within a distance as little as about twice the satellite radius ($2R_s$), at an altitude of 400 to 700 km. Measurements made during the docking of GEMINI and Agena give the number density in the wake half that of the undisturbed plasma, at a distance of 5 R_s at a height of 400 km. The difference between these two measurements is due to the fact that an increased density of H^+ ions much more effectively fill the region of negative potential behind the satellite.

A4.2.4   Electric and Magnetic Radius of Large Space Systems and Solar Power Satellites

When there are sources of electromagnetic disturbance on board a satellite, its electrically and magnetically active radius is defined by the superposition of the fields generated by the sources and the satellite body potentials. Solar panels, along with various types of ion engines, VHF antennas, electromagnetic radiators for research purposes, stabilization systems, etc., are systems which generate additional electromagnetic fields. Around the large panels of solar cells, used as a source of power for the satellite, there is a plasma sheath, the size of which depends on the Debye radius. At low orbits, the disturbing potential of the solar panel is screened at a distance of several Debye radii. The results of laboratory experiments show that, at a plasma density of about 10^6 cm^{-3} and panel potential of 30 V, the thickness of the disturbed layer is about 30 cm.

When large solar panels are designed for a solar power satellite at geostationary orbit, the potential distribution along their surfaces is the key problem. Computer modeling shows that the size of the disturbed zone around the solar panel, with a potential of 40 kV and dimensions of the order of 1 km, is commensurable with the panel size.

The use of ion engines will become more and more important for the propulsion and the stabilization of large space systems. These cause a plasma cloud around the satellite, which affects the satellite's potential. The efficiency of these engines depends on their orientation with respect to the earth's magnetic field lines and on the potential of the spacecraft. In many cases (at low orbits), a significant reverse current flows back to the satellite surface, due to the high plasma conductivity along the magnetic field lines; this results in a reduction of engine efficiency. For powerful energy systems at geostationary orbits, the use of ion engines at the ends of 500–m long booms is planned.

Additional disturbances are introduced by powerful emitters of VHF radio waves. At low orbits this results in plasma heating and a corresponding increase of the Debye radius. On-board emitters of HF and VHF waves generate electric and magnetic fields at large distances from the satellite. Usually, the aerials of these emitters are strictly directional and cause very large electromagnetic fields at their apertures. In the beam direction of the radiation pattern of such an antenna, the electrically and magnetically active radius of a satellite (which may, more appropriately, be named the "electromagnetic" radius) reaches the earth. In the most extreme case of an SPS so far designed, the enormous energy flux may significantly perturb the ionosphere and, when reaching the earth's surface, may even be harmful to the biosphere. The energy fluxes in the sidelobes of a relatively poorly directional on-board antenna attain significant values and increase the electromagnetic radius of the satellite to 1,000 km. Particularly dramatic is the case of an SPS, in that the desired directional properties of the VHF antennas cannot be guaranteed under such great powers; the energy within the sidelobes would be sufficient to disturb communications with other satellites and also to affect people and systems on the ground. Especially drastic would be the case in which oscillations of a stabilized SPS occurred. This effect would lead to a sharp increase of the size of a disturbed area around the satellite and could alter the operation of other geosynchronous spacecraft, communications systems in particular. For an SPS with 1,000 MW, this disturbed region could extend over several thousand kilometers.

Precise calculations of the electrically or magnetically active radius of the satellite would, in this case, require an accurate analysis of the VHF antenna design. In principle, the SPS and possible nuclear power stations at geostationary locations, because of their electric and magnetic effects, could affect world communications and navigation systems and may create disturbances to both air and surface transport systems.

The electromagnetic radius of the spacecraft is often determined by the telemetry and research emitters of HF and VHF radio signals rather than by the passive interactions between the spacecraft and the surrounding plasma. The boundary values of disturbing fields for these cases can be derived from the regulations of ITU (CCIR).

ACKNOWLEDGEMENTS

This chapter was prepared with the assistance of an international team of experts organized by the Committee on Space Research (COSPAR) established by the International Council of Scientific Unions (ICSU). For Sections 4.1, 4.2, 4.3, and 4.5, the team comprised Prof. K. Rawer (Federal Republic of Germany), Chairman; Dr. K. S. W. Champion (United States), Prof. A. P. Mitra (India), Prof. E. A. Müller (Switzerland), Prof. W. Riedler (Austria), Dr. M. J. Rycroft (United Kingdom), Prof. G. A. Skuridin (Union of Soviet Socialist Republics), Executive Members; Dr. R. E. Barrington (Canada), Dr. J. Baumgardner (United States), Dr. A. P. Bernhardt (United States), Mr. J. P. Chassaing (France), Dr. L. H. Doherty (Canada), Dr. R. D. Eberst (United Kingdom); Dr. J. C. Gille (United States), Dr. B. H. Grahl (Federal Republic of Germany), Prof. S. Hayakawa (Japan), Dr. F. Horner (United Kingdom), Dr. J. A. Klobuchar (United States), Dr. F. A. Koomanoff (United States), Dr. H. Martinides [European Space Agency (ESA)], Prof. M. J. Mendillo (United States), Dr. C. M. Minnis (United Kingdom), Dr. C. T. Russell (United States), Prof. S. M. Siegel (United States), Prof. F. G. Smith (United Kingdom); Prof. M. S. Vardya (India), Prof. R. Wielebinski (Federal Republic of Germany), Dr. P. Wilson (Federal Republic of Germany), Contributors. For Section 4.4, the team was composed of: Dr. L. Bankov (Bulgaria), Dr. Z. Dachev (Bulgaria), Dr. A. Gdalevitch (Union of Soviet Socialist Republics), Dr. M. Gogoshev (Bulgaria), Dr. D. G. King-Hele (United Kingdom), Dr. I. Kutiev (Bulgaria), Dr. P. Lála (Czechoslovakia), Dr. Yu. Matviichuk (Union of Soviet Socialist Republics), Prof. G. Moraitis (Greece), Prof. L. Perek (Czechoslovakia), Dr. I. Podgorny (Union of Soviet Socialist Republics), Dr. M. J. Rycroft (United Kingdom), Dr. L. Sehnal (Czechoslovakia) and Prof. K. Serafimov (Bulgaria). Apart from the writers, an even greater number of scientists have helped with advice or criticism.

Part II
CRITICAL ISSUES
IN SPACE TECHNOLOGY

Chapter 5

COMPATIBILITY

AND COMPLEMENTARITY

OF SATELLITE SYSTEMS

ABSTRACT

This chapter is the first of three concerning specific technical issues. It was prepared with the assistance of the World Meteorological Organization (WMO),
the International Telecommunication Union (ITU) and international teams of
experts organized by the Committee on Space Research (COSPAR) of the
International Council of Scientific Unions (ICSU), and the International Astronautical Federation (IAF).

The purpose of this chapter is to examine the possibility of increasing the
benefits that are realized through space technology by encouraging international
use of satellite services. In particular, the chapter will consider three applications
of space technology: meteorology, telecommunications and remote sensing. It
will also assess the potential benefits of making different national or international systems compatible and complementary. To the extent that compatibility
is feasible, users of one system could use other systems at relatively little additional cost.

It should be noted that WMO and ITU provide well-established
mechanisms for co-ordinating activities in meteorology and telecommunica-

tions. In the field of remote sensing, which is still in a pre-operational phase, co-ordination is being considered in the United Nations Committee on the Peaceful Uses of Outer Space as well as in bilateral and multilateral bodies, but no general international system of co-ordination has yet been established.

INTRODUCTION

The early years of the space age, following the launch of the first satellite in 1957, saw first the development of space research satellites, followed by a growing interest in experimental applications satellites for communications and meteorology. Since the satellites were experimental and the number of countries launching satellites was very small, co-ordination was minimal. As applications satellites became operational (for communications in 1965, for meteorology in 1966) and as more countries had satellites launched, the need for co-ordination and the potential benefits of co-operation grew. As of 1980, some 15 countries and three regional groups had satellites in orbit, and additional national and regional satellites are planned in the next few years.

Virtually every country in the world uses space technology to a greater or lesser extent. For applications such as meteorology and remote sensing, where the user countries may have access to more than one system, it would be beneficial to such countries if their interpretation equipment and training were applicable to data from all systems.

Satellite systems require corresponding ground systems, including facilities for data processing, data utilization, and training. Satellite systems can be called compatible if a ground system for one can be used for another with only minor adaptation. The question of using a ground system for more than one satellite system can arise either when two different satellites are operating simultaneously or when one satellite is being replaced by a more advanced model.

Satellite systems may be said to be complementary when one provides information or services that are particularly useful in conjunction with information or services provided by another. Complementarity therefore applies to satellite systems operating simultaneously and possibly requiring quite different ground systems.

Operational satellite systems have corresponding operational ground systems, utilization techniques, and training programs. Once such a system is created it changes rather slowly. Modifications to the satellite itself will generally be designed to minimize changes in ground systems since these generally represent a much greater investment than the satellite itself. Operational systems are created when the technology and applications have developed to the point where at least near-term future developments are reasonably predictable. It is in such operational systems that compatibility and complementarity become both feasible and beneficial.

Experimental systems generally have much smaller and more flexible ground systems. Neither the technology nor the applications are well developed, and their future development is unpredictable. The restraints which demands for compatibility place on systems are therefore likely to pose greater problems for experimental systems than for operational systems. Complementarity, however, can be highly valuable in comparing technologies and avoiding redundant experiments.

5.1 METEOROLOGY

5.1.1 Introduction

Meteorology is, by its nature, an international science. Weather patterns may extend over thousands of kilometers and move across a continent in a few days. Faced with the difficulties of observing such extensive and rapidly changing phenomena, meteorologists were among the first to appreciate the possibilities opened up by the first artificial satellites. In 1961 two leading meteorologists, V. A. Bugaev and H. Wexler, proposed the World Weather Watch (WWW) in which satellites would play a key role. This proposal was soon adopted by WMO.

The WWW followed this early planning for some time, but recently satellites have begun to play a far greater role than was anticipated in 1961. In addition to obtaining observational data, satellites now also collect and distribute other types of data to support a variety of meteorological, hydrological, and other related programs. It is generally recognized that satellites are indispensable for the success of the WWW, the Global Atmospheric Research Programme (GARP), the First GARP Global Experiment (FGGE), climate modeling, weather-modification studies, the Tropical Cyclone Programme, the Integrated Global Ocean Station System (IGOSS), the Operational Hydrology Programme, and other international programs.

The satellite subsystem of WWW, relying on the international cooperation that is traditional in meteorology, is supported by:

1. The European Meteorological Satellite Programme through its geostationary satellite METEOSAT, positioned at 0° over the equator.
2. The Meteorological Satellite Programme of Japan through its geostationary satellite GMS, positioned at 140°E over the equator.
3. The Meteorological Satellite Programme of the USSR through its polar-orbiting METEOR satellites.
4. The Environmental Satellite Programme of the United States through its polar-orbiting TIROS–N/NOAA satellites and geostationary GOES-type satellites, positioned at 75°W and 135°W over the equator.

Two more systems are to be added in the near future, the Soviet geostationary meteorological satellite (GOMS) at 70°E and the Indian national satellite (INSAT) at 74°E.

5.1.2 Meteorological and Environmental Satellite Systems

According to the WWW Plan and Implementation Programme for the period 1980-1983, meteorological satellites are divided into two groups: those in near-polar orbits and those in geostationary orbits. The different capabilities of the two groups complement each other and are necessary parts of the space subsystem of the Global Observing System. As well as taking observations directly, both are capable of accomplishing data-collection and data-dissemination missions.

Near-polar orbiting satellites are able to observe the whole globe twice a day, and hence two satellites in complementary orbits can provide the four sets of daily data required by WWW. The orbital altitude is determined by criteria of orbital stability, field of view, resolution, and coverage repeatability. The altitudes used to date range from 600 km to 1,500 km, lower orbits requiring excessive amounts of station-keeping propulsion to counteract air drag and higher orbits being less efficient with respect to field of view. The current generation of operational satellites uses orbits at about 850 km (TIROS–N) and 900 km (METEOR–2).

At an altitude of 900 km, the satellite has a period of 103 min and consecutive orbits are 2,860 km apart at the equator. To provide complete coverage, the satellite sensors must view a swath 2,860 wide corresponding, at the edges of the swath, to an oblique viewing angle 66° from the vertical. Some sensors have swath widths of 2,200 s to 2,600 km and hence provide complete coverage in mid and polar latitudes but incomplete coverage at the equator.

Near-polar orbiting satellites are generally placed in sun-synchronous orbits and provide data at the same local time every day. At an altitude of 900 km, the required inclination for a sun-synchronous orbit is about 99°. The TIROS–N system, for example, consists of two operational satellites, one crossing the equator southbound at 0730 and the other northbound at 1530 local time.

The payload of the near-polar orbiting satellites varies from system to system and from generation to generation, but in general the following missions are performed:

1. *Imagery mission.* All operational satellites carry radiometers operating in the visible (0.5 to 0.9 μm) and the infrared (10.5 to 12.5 μm) spectral range to provide image data with medium and high resolution (1 to 4 km in the visible and 1 to 7 km in the infrared along the subsatellite track).

2. *Sounding mission.* Some polar satellites are equipped with vertical profile radiometers that operate in the infrared (carbon dioxide and water vapor

channels) and in the microwave (oxygen and water vapor channels) spectral range to provide temperature and humidity measurements at different altitudes.

3. *Direct readout mission.* To support national and local meteorological activities around the globe, the polar-orbiting satellites provide a direct readout service which allows the reception of satellite data in real-time by ground stations within radio range of the satellite. Depending on the nature of the data-medium and high-resolution imagery or sounding data, they are broadcast continuously on VHF (136 to 137 MHz) or S-band (1,691 to 1,695 MHz) frequencies. The medium resolution image data are provided by automatic picture transmission (APT) and the high-resolution image data are provided by high-resolution picture transmission (HRPT). The direct sounding broadcast (DSB) is used for the transmission of the sounder data.

4. *Data-collection mission.* To close the gaps in surface observation systems, fixed and moving data-collection platforms (DCPs) are planned over data-sparse areas. Satellites are in general very well suited to collecting data from DCPs, and polar-orbiting satellites are specially tailored for the collection of data from moving DCPs which have no means of determining their geographic positions. For the up and down link, UHF frequencies are used.

Geostationary satellites orbit at an altitude of approximately 36,000 km. With this altitude the orbital period is about 24 hours and thus equal to the rotational period of the earth. If the orbital plane of the satellite coincides with the equatorial plane, the satellite remains fixed over the equator. From an altitude of 36,000 km a geostationary satellite provides nearly continuous information for an area within a range of about 50° (about 5,500 km) from the subsatellite point. As a consequence, five geostationary satellites are needed to provide full coverage around the globe within the range 50°N to 50°S latitudes.

The following missions are performed by geostationary satellites:

1. *Imagery mission.* The satellites are equipped with spin/scan radiometers operating in the visual and the infrared spectrum. With a spin rate of 100 rpm, the satellites scan the earth disk by pivoting the telescope of the radiometer stepwise from north to south, or vice versa, from rotation to rotation. By this technique sequential imagery data of the earth disk are obtained within about 20 to 30 min. The resolution of the image data at the subsatellite point is 0.9 to 2.5 km (depending on the type of spacecraft) in the visual spectrum and about 5 to 7 km in the infrared spectrum.

2. *Data-dissemination mission.* Since the raw data from the geostationary satellites can be received only by rather large central command and acquisition stations and, in addition, powerful computer centers are needed for

the necessary subsequent image processing (registration, rectification, gridding, annotation, sectorizing, etc.), one or two S-band transponders are installed aboard the satellites to relay pre-processed analog and digital image data to simple users' stations within the radio range of the satellite (76° geocentric arc from the subsatellite point at antenna elevation angles of 5° or greater). Of special importance is the dissemination of the information using the well-established APT-format (800 lines and 800 picture elements per line). This APT-type dissemination, called Weather Facsimile Service (WEFAX), allows the transmission of sectorized image data, as well as other meteorological information like analysis and forecast charts. The dissemination takes place at 1,691 MHz frequency on a fixed schedule.

3. *Data-collection mission.* Geostationary meteorological satellites are also capable of collecting information from a large number (up to 10,000) of fixed and moving DCPs of various types (meteorological, oceanographic, hydrological, etc.) and of relaying these data to central ground stations for further processing and dissemination.

Information obtained from near-polar orbiting and geostationary satellite systems is making a major contribution to routine operations and research in meteorology, hydrology, oceanography, and other related environmental activities, providing to users the following quantitative data and qualitative information; vertical profiles of temperature and humidity; temperatures of sea, land, and cloud top surfaces; wind field derived from cloud displacements; cloud amount, type and height of cloud tops; snow and ice cover; radiance balance data.

According to a recent study, the current status of experimental satellite meteorology and an assessment of future trends lead to the expectation that the following measurements will be made by means of satellites in the 1980s. By the mid-1980s, the parameters marked with an asterisk (*) are expected to be operational, the others experimental.

Atmosphere

Mean temperatures of isobaric layers.*

Total water-vapor content and its distribution by layers.*

Total ozone content and its distribution by layers.

Characteristics of aerosols.

Components of the radiation balance at the top of the atmosphere.*

Wind speed and direction in the troposphere (at two or three levels).*

Clouds

Spatial distribution and structure of cloud.*

Height and temperature of cloud tops.*

Phase composition (droplets, ice particles, or mixed) of the upper-cloud layer.

Total water content of clouds.

Location of precipitation areas and approximate intensity.*

Ocean surface

Temperature of the ocean surface.*

Location of the major ocean-surface currents.*

Degreee of roughness of the ocean (e.g., waves) and the related surface wind fields.

Ice conditions.*

Location of polluted areas (of certain types) on the ocean surface.*

Land surface

Temperature of the land surface.*

Degree of soil moisture.

Distribution of snow cover.*

Characteristics of the soil and plant cover.

Location of areas of melting snow and ice.*

5.1.3 Compatibility and Complementarity

5.1.3.1 Coordinating bodies.   All national and regional meteorological satellite programs participate in WWW co-ordinated through WMO. The last meeting of the World Meteorological Congress, held in April–May 1979, adopted the WWW Plan and Implementation Programme 1980–1983, the fourth such four-year plan. WWW comprises the Global Observing System, the Global Telecommunication System, and the Global Data-Processing System.

5.1.3.2 Orbits and sensors.   The wide variations in weather analysis and forecasting techniques and in the types of service performed in support of aviation, hydrology, agriculture, marine activities, etc., lead to large differences in observational requirements. It is, therefore, not to be expected that the various national systems will be identical in the foreseeable future. Nonetheless, the basic functions are the same, and a substantial degree of compatibility and complementarity of services exists.

Satellites in polar and geostationary orbits are complementary in the frequency and area of observations, collection and relay of surface observations, and dissemination of data by direct broadcast. With four or five geostationary satellites around the equator, images of clouds and storms in the tropics and mid-latitudes (up to about 60°) can be obtained at least once an hour. Atmospheric soundings with the same frequency over limited areas of the earth for use in mesoscale analysis should be possible from geostationary satellites by the end of the 1980s.

Three to four satellites in near-polar orbit will provide full global coverage every six hours, close to the synoptic times for global, large- and planetary-scale analysis and prediction. The convergence of these orbits near the poles provides more frequent coverage for regional and local use at high latitudes, complementary to the nearly continuous coverage by the geostationary satellites equatorward of about 50° latitude.

Polar and geostationary orbits are also complementary in providing frequent collection of surface observations from automatic or remote (e.g., ship, island) observing stations. While geostationary satellites can uniquely provide continuous contact with such stations, the polar-orbiting satellites have a unique capability for providing accurate position indications for mobile stations such as drifting oceanographic buoys.

From a system point of view, the polar-orbiting satellites will have a primary role in providing global sounding data and high-resolution surface observations, especially those required for specialized applications programs. The spacecraft in geostationary orbit will be used primarily for continuous monitoring of clouds for short-period and mesoscale forecasting and storm warnings and to obtain winds at two or three levels in the troposphere.

5.1.3.3 Data transmission.
Transmission frequencies for the Meteorological Satellite Service are set by ITU. The near-polar orbiting satellites use the 137-MHz band for medium-resolution imagery and the 1,690-MHz band for high-resolution imagery. The geostationary satellites use the 1,690-MHz band for image transmission and the 402-MHz band for transmissions from DCPs to the satellites.

5.1.3.4 Data dissemination.
WWW calls for reception, processing, and interpretation of both conventional and satellite data at three levels:

1. World meteorological centers (WMCs), whose primary job is to provide the products required for the analysis and forecasting of large- and planetary-scale processes.

2. Regional meteorological centers (RMCs), which primarily provide the products required for the analysis and forecasting of large-, meso-, and, to some extent, small-scale processes.

3. National meteorological centers (NMCs), which process data at the national level, using the products of the WMCs and RMCs, in conjunction with data which may be received directly from satellites.

In the past, the major role of satellites in WWW has been primarily to meet data requirements for the analysis and forecasting of large- and planetary-scale processes. This has led to the establishment of a system of data flow to WMCs and the subsequent transmission of low-resolution products on a hemispheric or global scale. At the same time, the use of the APT system has partially met regional and national needs for medium-resolution satellite images. The advent of new satellite techniques, which are increasing the types, frequency, and resolution of satellite observations, unquestionably calls for an improvement in the means of receiving, processing, interpreting, and disseminating satellite data at all three levels. Particular efforts in this respect should be directed toward meeting regional and national requirements for satellite data.

5.1.4 Conclusion

An effective mechanism for promoting compatibility and complementarity of meteorological satellites exists in the WMO World Weather Watch program. All existing national and regional meteorological satellite programs participate in this global program. For the foreseeable future the satellite component of the program will comprise national and regional satellite systems which are designed and built primarily to meet the needs of the operating countries or regions. Nonetheless, there appears to be little conflict between national and international priorities, and the level of co-operation is high.

The specific characteristics of satellites vary substantially between national systems reflecting in part the differences in national meteorological systems in general. The characteristics also evolve rather rapidly in response to new experiments and to experimental systems becoming operational. Even within a given operational series, the sensor package may change from satellite to satellite. It would appear that standardization of systems is neither possible nor desirable.

In the area of data dissemination, compatibility is both more feasible and more desirable. While the satellite systems differ in design, certain basic functions are performed by all operational systems. Since transmission frequencies are standardized by the ITU, ground receiving systems have a substantial degree of compatibility. A network of compatible and complementary data distribution centers makes the data internationally accessible.

Efforts to promote compatibility and complementarity will clearly continue to be necessary, in particular to increase existing levels of compatibility and to integrate new developments in a complementary way.

5.2 COMMUNICATIONS

5.2.1 Introduction

In contrast to meteorological and remote-sensing satellites which were first developed to meet national needs, communications satellites were from the outset designed for international use. The first operational communications satellite system to be established was the INTELSAT system which launched its first satellite in 1965, and which now links 111 countries. Since then communications satellites systems have been launched by the USSR, the United States, Canada, France, Federal Republic of Germany, Indonesia, and Japan. A number of other national and regional systems will be created during the next decade.

5.2.2 Communications Satellite Systems

5.2.2.1 Orbit. Most communications satellites use the geostationary orbit. These satellites, over the equator at an altitude of about 36,000 km, revolve about the earth with the same period as the earth's rotation and therefore appear from the surface to be stationary in the sky. An earth station can remain in continuous communication with a geostationary satellite and requires little tracking capability. The position of a satellite is designated by the longitude of the subsatellite point on the equator. A satellite can provide a communications channel between any two points within a range of about 8,000 km from the subsatellite point.

The geostationary satellite orbit, being unique, is limited, like any other natural resource, in its service to mankind for radiocommunications purposes. A geostationary satellite occupying a certain position will exercise an influence on neighboring satellites in the form of interference noise. This influence depends on many factors, the frequency being, of course, the major parameter. The efficient utilization of this orbit in its relation to frequency is obviously a subject for study. As a first approximation, which is also the simplest case, let us assume that all geostationary satellite systems are identical, and that all satellites, whether they belong to the same or different systems, are also the same. This constitutes what is called a homogeneous ensemble, and one can easily see that there is a minimum spacing between neighboring satellites beyond which the interference entering into the wanted system becomes unacceptable. Assuming this spacing to be 5°, one can see that the total number of satellites using a given frequency, if they are to be geostationary, cannot exceed 72.

In fact, however, satellite and earth stations differ in their characteristics and the minimum spacing between satellites varies from system to system. Since each satellite serves only a portion of the globe, and since the traffic demand is not distributed evenly, some segments of the geostationary orbit will be more congested than others.

Not all communications satellites are placed in the geostationary orbit. The MOLNYA system consists of a number of satellites at different positions in a highly elliptical orbit inclined at about 63°. Whereas geostationary satellites can communicate with earth stations up to 70° or 75° north and south latitude, the MOLNYA orbit provides communication links into the far north.

5.2.2.2 Communications technology.   Earth stations and satellites communicate through antennas that focus the transmissions in the desired direction and collect the signals coming from the desired direction. The angular width of the beam decreases as the diameter of the antenna and the frequency increase. Not all of the radiated energy, however, is directed into the main beam; some is radiated in other directions and can therefore cause interference with other systems.

Information is transmitted at a given radio frequency by the modulation of a carrier wave at that frequency. Modulation may be either analog, as in the case of frequency modulation (FM), or digital, as in the case of pulse code modulation (PCM).

In order to allow many earth stations to communicate via a single satellite, multiple-access systems are used. Multiple-access systems can be divided into frequency division multiple access, in which different earth stations use different frequencies, and time division multiple access, in which different stations transmit bursts of signals in separate time periods. Recently, single channel per carrier systems have been developed in which each carrier is modulated by one telephone channel. A third system, spread spectrum multiple access, makes use of a deterministic noise-like signal to spread the narrowband information over a relatively wide band of frequencies. The transmitter encodes the information using one of a series of complex codes, and the receiver uses the same code to extract the information from the wideband signal carrying many such messages using different codes. For direct communication between computer systems, pulse address multiple access uses bursts of data complete with address information, transmitted at a time determined only by the need for communication at the originating end and regardless of whether or not there may be another user using the transponder. The acceptable probability of interference between bursts determines the maximum data transmission capacity of the system.

5.2.2.3 Frequencies.   The quantity and quality of information that can be sent by a communications system depends on the environmental and propagation effects of the frequencies used and on the noise level in stations both on earth and in space at that frequency. The choice of frequency is therefore a primary decision in the design of a system.

The frequencies of major interest to space communications lie between the ionospheric MUF (maximum usable frequency for reflection by the ionosphere, about 30 MHz) and the first oxygen absorption line (about 60 GHz). Other "windows" between gaseous absorption lines above 60 GHz may also prove to be of interest. The attenuation characteristics of the earth's atmosphere are frequency dependent allowing energy at certain frequencies to pass through more readily than at others, and such frequency regions of atmosphere transparency are more useful for space communications. In the presence of rain, hail, snow, fog, and clouds, the attenuation increases with increasing frequency, especially at frequencies above 10 GHz, causing occasional deterioration of the signal, usually for brief periods. Furthermore, substantial attenuation will tend to increase the receiving system noise temperature of the earth station; in such situations the carrier/noise ratio for a received signal from the satellite will deteriorate more than from attenuation alone.

5.2.2.4 Communications satellite services. *The fixed-satellite service* relays signals between fixed and relatively large and complex earth stations. Since this service carries a very high volume of traffic and therefore requires a large bandwidth, frequencies in the gigahertz range are used. However, as the bandwidth used by a satellite increases, the power required to overcome noise and interference also increases. Current operational fixed-satellite service systems include INTELSAT (International), RADUGA (USSR), ANIK (Canada), PALAPA (Indonesia), WESTAR (United States), and SAKURA (Japan).

The mobile-satellite service relays signals between mobile ground stations or between mobile and fixed stations. Of the various mobile services, the technology is most highly developed for maritime communications. Until 1982, MARISAT (United States) satellites over the Atlantic, Pacific, and Indian Oceans are providing telephone, telex, facsimile, and data links with some 500 ships and off-shore facilities equipped with mobile terminals. In 1979 the International Maritime Satellite Organization (INMARSAT) was established to ensure the continuity of such services after the end of the working life of the MARISAT satellites. The INMARSAT system, initially using MARISAT satellites, will subsequently introduce three INTELSAT-V satellites with Maritime Communication subsystems, and two MARECS satellites launched by the European Space Agency.

The broadcasting-satellite service relays signals from fixed earth stations to large numbers of individual or community receivers. While much of the technology required is common to other services, particular requirements of the broadcasting service include the generation of high radio-frequency power, high-efficiency radio-frequency generators, effective methods of heat conduction and dissipation from these high-power radio-frequency sources, and the design and development of satellite antennas having low sidelobe levels and asymmetrically shaped beams.

The World Broadcasting-Satellite Administrative Radio Conference (WARC-BS) held in 1977, working on the basis of technical parameters and sharing criteria elaborated by CCIR, adopted a plan to assign frequency channels and orbital positions for the broadcasting-satellite service to countries in ITU Regions 1 and 3. A conference for Region 2 is scheduled for June–July 1983. The plan includes methods of reception (community or individual), reception quality, service area, coverage area, beam power, etc.

5.2.3 Compatibility and Complementarity

5.2.3.1 Co-ordinating bodies. The primary international coordinating body for communications satellite systems is ITU and its organs, the World Administrative Radio Conference (WARC), and the International Radio Consultative Committee (CCIR). CCIR studies technical and operational questions on all aspects of radio communication, issues recommendations based on those studies, and prepares the technical bases for conferences organized by ITU.

5.2.3.2 Orbits. The major problem of coordinating satellite communications systems relates to the geostationary orbit and the planning of the services using it. Since the orbit is unique, its capacity is limited. The usable capacity depends on available technology and the ultimate limit to the capacity is not known. Article 131 of the International Telecommunication Convention states:

> In using frequency bands for space radio services Members shall bear in mind that radio frequencies and the geostationary satellite orbit are limited natural resources, that they must be used efficiently and economically so that countries or groups of countries may have equitable access to both in conformity with the provisions of the Radio Regulations according to their needs and the technical facilities at their disposal.

Many countries do not now have the means to launch geostationary satellites for telecommunication purposes, and they are apprehensive that when they will have acquired this capability, there will be no more positions left in the geostationary satellite orbit for their use, or frequencies feasible for this purpose. Thus, they feel that orbital positions and frequency bands should be assigned to them by a conference to be convened in the near future, thereby guaranteeing equitable access regardless of the time until their capability to launch geostationary satellites can be acquired. Such a plan would have to be based on the technological possibilities foreseen at the time the plan is made. The present method of "tailor fitting" each new application into the orbit and frequency spectrum allows flexibility of design and facilitates maximum total usage of the orbit.

5.2.3.3 Interference. The radio-frequency spectrum used by communications satellites is also used by terrestrial communications services on the basis of frequency allocations by WARC. This sharing of the spectrum inevitably results in interference between systems. The relevant definitions of CCIR are:

Interference: The effect of unwanted energy due to one or a combination of emissions, radiations, or inductions upon reception in a radiocommunication system, manifested by any performance degradation, misrepresentation, or loss of information which could be extracted in the absence of such unwanted energy.

Interfering source: An emission, radiation, or induction that is determined to be a cause of interference in a radiocommunication system.

Permissible interference: Observed or predicted interference that complies with quantitative interference and sharing criteria contained in the Radio Regulations or in Recommendations of CCIR or in regional agreements as provided for in the Radio Regulations.

Harmful interference: Any interference that endangers the functioning of a radio-navigation service or of other safety services or seriously degrades, obstructs, or repeatedly interrupts a radiocommunication service operating in accordance with the Radio Regulations.

Note: It is recognized that, under certain circumstances, a higher level of interference than that defined as permissible may be accepted by agreement between the Administrations concerned without prejudice to other Administrations, but it is not considered possible to set down any precise values for this interference level—each case must be treated on its merits. This interference level may be called *accepted interference.*

Four parameters of particular importance for the efficient use of the geostationary satellite orbit are as follows:

1. The ratio between total internetwork interference and the maximum single-entry level, single entry being one network using neighboring satellites.
2. The increase of orbit utilization efficiency by reduction of inhomogeneity, i.e., dissimilarity of different satellite networks.
3. The limitation of earth station antenna sidelobe gain pattern and sidelobe radiation level.
4. The limitation of satellite antenna gain outside the service area.

Here one might cite two important examples of co-ordination procedures specified in the Radio Regulations. One deals with sharing between space and terrestrial services, the other between switching networks for the fixed-satellite service. Any Administration intending to build an earth station or to operate a space communication system using geostationary satellites must make preliminary examinations of the situation so that their systems do not receive or cause excessive interference.

CCIR studies of propagation data suggest appropriate frequency bands for various service requirements, thus laying the foundation of the ITU frequency allocation table. CCIR-elaborated sharing criteria and interference calculations for various radio services are essential to the International Frequency Registration Board (IFRB) in its daily work of technical examination of frequency assignments. The future work of CCIR will include studies related to resolutions of WARC-79:

1. One resolution sets up a world conference for the mobile-radio services, which will revise the Radio Regulations to harmonize some provisions for aeronautical, maritime, and land mobile services, to improve safety and distress signalization, and to initiate the maritime mobile-satellite service;

2. Another resolution asks CCIR to study the technical characteristics for a satellite sound-broadcasting system for individual reception by portable or automobile receivers and the feasibility of such a service sharing a band with a terrestrial service.

5.2.3.4 Conclusions. Communications satellites place much greater demands on the capacity of the radio-frequency spectrum and orbital space, particularly in the geostationary orbit, than do other types of satellites, and therefore, the need for co-ordination is correspondingly greater. ITU and its organs provide a mechanism for co-ordination, and the resolutions and recommendations of these bodies contain the agreed upon rules and procedures for co-ordination. For a more detailed discussion of the co-ordination procedure, see Appendix 6.1 in Chapter 6.

As the demand for satellite communications grows, the co-ordination function will become increasingly complex, and economic, technological, and political constraints will become more difficult to reconcile. Within the next few years a number of ITU conferences will be held to consider the need for new co-ordination procedures in the communications satellite services.

5.3 REMOTE SENSING

5.3.1 Introduction

Photographs taken during early manned space missions by both the USSR and the United States demonstrated the value of remote sensing of the earth from space. Automatic, unmanned weather satellites launched by several nations have been used to observe clouds and other meteorological conditions for several years. The LANDSAT (United States) series of remote sensing satellites has proven to be a phenomenal success. Remotely sensed photographic and other data from later manned flights such as the COSMOS and SOYUZ-SALYUT (USSR) series and the SKYLAB (United States) missions have also been of great utility. As a result of these developments, much of the world has come to recognize the benefits that can be derived through observation of the earth from space.

Several countries and organizations are now planning new satellites, and it is clear that several different systems will be in orbit from the mid-1980s onward. France and Japan have announced new satellite systems, SPOT and MOS 1, planned for 1984 and 1985, respectively, and the European Space Agency (ESA), India, and others are intending to do likewise. Co-operation among the remote sensing nations could produce the following benefits:

1. Complementarity between systems to improve diversity of observation, while enhancing continuity, availability, and timeliness of data.
2. Compatibility between systems to minimize capital and operating costs of ground systems to acquire, process, and use data from all available sources.

Equally important, close co-operation among the countries that will operate the space systems and the user countries could benefit the training and demonstration programs required to develop practical applications.

The possible benefits of complementarity and compatibility must be balanced against the costs due to the limitations that might be imposed on efforts to improve the performance and cost-effectiveness of systems, particularly in the current pre-operational phase. It should also be recognized that national programs are generally expected to give priority to domestic needs in the design and operation of space systems.

5.3.2 Existing and Planned Satellite and Ground Processing Systems

Several space systems have been used to gather earth resource and environmental data. Some of these are still operating. As the satellites designed for weather observations become more sophisticated, their use for observing resources and near-surface environmental conditions is increasing. Because of their large number, it is not practical to discuss them here. Moreover, because of the rapid developments in national and international planning for resource monitoring from space, the information included herein should be looked upon only as a representative sampling of the types of systems being planned.

Earth resource data acquired from space has been made widely available by the United States since the first photographs of the earth showing cloud cover from space were transmitted from VANGUARD 2 in 1959. For several years both the United States and the USSR have used many of their manned and unmanned space missions to gather earth resource data. The early systems served as forerunners to the large number of weather satellites now operated globally by these as well as several other countries. Aside from such atmospheric observations, most of the early earth resource data gathering was limited to that obtainable from photographic cameras carried on relatively short missions. The launch of LANDSAT–1 (United States) in 1972 demonstrated, for the first time, the feasibility of routine, repetitive earth observation using multichannel electro-optical scanners with much greater radiometric fidelity than can be achieved practicably using photographic film. SKYLAB later demonstrated the possibility of acquiring and distributing high-resolution data (10 m) from space. The reliability, long life, and high-image quality of the multispectral scanner system (MSS) carried by LANDSAT–1, 2, and 3 have encouraged resource managers to adopt techniques for including satellite remotely sensed data in their management systems. Furthermore, as a result of this success, added to the early work and the multitude of weather satellites, new earth resource satellites are being developed or planned by the United States, the USSR, Japan, India, France, Canada, and ESA.

LANDSAT–D, planned for launch in 1982, will carry the familiar MSS with four bands from the green region of the visible to the near infrared and with an instantaneous field of view (IFOV) of 80 m, corresponding to a spatial resolution of 100 to 200 m. In addition, it will carry a new scanner, the thematic mapper (TM) which will have seven channels. These will cover the same region of the spectrum as the MSS plus an additional band in the blue-green, and three new infrared bands (1.55 to 1.75 μm, 2.05 to 2.35 μm, and 10.4 to 12.5 μm— thermal infrared). The IFOV will be 30 m, corresponding to a spatial resolution of 40 to 75 m). The radiometric resolution will also be finer—256 levels compared to 64 for the MSS. The third LANDSAT carries a black and white television camera system, the Return Beam Vidicon, with 40 m resolution, the data from which can be used to obtain some estimation of the improvement that can be expected from the TM. A major cost for these improvements will result from the requirement to handle an order of magnitude higher data rate, which will be accommodated by satellite-to-ground transmission using X-band instead of S-band frequencies.

A serious limitation of the visible and near-infrared sensors is that they cannot observe the earth through cloud or in the absence of sunlight. For this reason, the United States launched the experimental SEASAT, which carried an L-band synthetic aperture radar (SAR).

As a result of experiments with the data from this satellite and those from airborne SAR, Canada is actively considering the development of an operational SAR satellite to provide all weather, year-round capability for earth observations. The system would be optimized for surveillance of Canada's arctic regions and in particular for ice observations to assist navigation and petroleum exploration in the northern ice-infested waters. The satellite would likely carry a C-band radar.

The NOSS satellite system is now being considered by the United States specifically for oceanographic observations. It would carry several instruments operating in the visible, infrared, and microwave regions to observe oceanographic variables such as surface wind velocity, temperature, and ocean color.

Another satellite being considered by the United States, MAPSAT, would provide variable spatial resolution (to 10 m) in the visible and near-infrared region and would have stereoscopic capability to permit topographic mapping with a 20-m contour. Data would be compatible with LANDSAT-1, 2, and 3. In addition, several experimental sensors, including radar, will be carried on various Space Shuttle missions.

For rapid, efficient dissemination of data, it is vital that a global network of ground data receiving stations be established. Such a network has been developed for LANDSAT, with more than a dozen receiving and processing stations in operation and more being planned. Many of these stations have plans to adapt their equipment to handle data from the new satellites being planned.

In the USSR, satellite sensor systems for studying earth resources are of two types: quick-look systems, using optical-mechanical scanners and direct transmission of data to study rapidly changing phenomena; and photographic film return systems for studying slowly changing phenomena. These sensors will be integrated into multi-sensor missions for studying earth, ocean, and atmospheric features. Photographic and digital products will be provided to both Soviet and foreign users.

The USSR has developed and operated quick-look systems consisting of four channel scanners with spectral bands similar to the LANDSAT MSS and with 1,000 to 1,700 m resolution and two channel scanners with visible and near-infrared bands and 240 m resolution. Data are transmitted at a frequency of 460–470 MHz to the ground receiving stations of the State Committee of Hydrometeorology and Environmental Protection. These systems are carried on METEOR satellites.

Other quick-look systems are being developed using multispectral optical-mechanical scanners transmitting data at X-band frequencies. A high resolution system with a field of view of 180 to 200 m will have eight spectral bands and a ground resolution of 50 m in the visible and 200 m in the infrared. The system will provide complete global coverage except for the polar regions, every 14 to 17 days. A medium resolution four-channel scanner with a field of view of 500 to 700 km will have a resolution of 150 to 200 m in the visible and 500 to 600 m in the infrared. The system will provide complete coverage every four to five days. In conjunction with these systems, VHF radar and radiometer instruments will be used for ocean and land surface studies, and a data collection system will relay information from ground-based and marine platforms. These systems are being developed on Cosmos and Intercosmos satellites and will transmit data at X-band frequencies.

For slowly changing phenomena, multispectral (MKF) and topographic cameras with three to six spectral bands in the visible and near infrared will be used. The field of view will be 120 to 240 km and the resolution better than 50 m. The film from these systems will be physically returned to the ground for processing. These images will be particularly useful for geologists and cartographers who require high resolution and accurate geometric control.

India's first satellite for earth observation (SEO) Bhaskara, was launched in 1979 with two television cameras operating in the 0.54 to 0.66 μm and 0.75 to 0.85 μm bands and two microwave radiometers. While the resolution was coarse, the experience has enabled India to begin planning for SEO-2 with higher resolution. Work has also begun on development of an SAR for future missions.

Satellites to date have carried photographic and television cameras and electro-mechanical scanners for observations in the visible and infrared regions of the spectrum. The SPOT satellite now being developed by France would be the first to use linear arrays of detectors for optical observations to maintain the radiometric advantages of discrete solid-state detectors without the problems associated with large moving mirrors. The SPOT sensors will have a selectable IFOV of 20 or 10 m and either three or one band in the visible to near-infrared. Two identical instruments will be carried. A unique feature is their pointability. This will permit more frequent coverage of selected ground sites than has been possible in the past. Stereoscopic coverage will also be possible. Many of the existing LANDSAT stations are developing plans to receive and process SPOT data.

The first Japanese marine observation satellite (MOS-1) will carry visible, near-infrared, thermal infrared, and microwave and radiometers. MOS-2 will also carry a microwave altimeter and a microwave scatterometer. MOS-1 is to be launched in early 1985. For geological and other land observations, an earth resources satellite (ERS-1) will use optical sensors and synthetic aperture radar (SAR).

ESA is developing plans to launch a series of earth observation satellites. Two satellite systems are now envisaged. The first, ERS-1, is planned primarily for ocean and coastal observation. The payload will likely include a radar altimeter, an active microwave instrumentation performing wind, field, and wave spectrum determination, and an all-weather imaging capability provided by a SAR. The second, LASS, will contain a high-resolution (30 m) visible and infrared radiometer using linear array technology and possibly a SAR. A launch date of 1987 is planned for the first satellite.

Other countries are actively considering the development of earth observation satellites. For example, Indonesia, in co-operation with the Netherlands, is considering a tropical earth resources system (TERS) in an equatorial orbit to provide optimum coverage of the tropical regions.

In addition to the earth observation sensors, many of the existing and planned satellites carry data relay equipment to facilitate the transmission of ground observations from remote, unattended locations to central processing facilities.

5.3.3 Compatibility and Complementarity

5.3.3.1 Co-ordinating bodies. Co-ordination activities have already taken place and are actively being pursued. For instance, the United States and France have, through a joint informal working group, ensured the compatibility of satellite-to-ground transmissions (X-band) of image data from the LAND-SAT-D and SPOT space systems. In November 1980, building upon this precedent, agencies in charge of developing and operating space systems for land observation, in an operational medium-term perspective, confirmed their agreement to work together toward the complementarity and compatibility of their systems. For this purpose, a special group, called co-ordination of land observation satellites (CLOS), was created. The group presently comprises representatives of the National Oceanic and Atmospheric Administration (NOAA) and the National Aeronautics and Space Administration (NASA) (United States), the National Space Development Agency (NASDA) (Japan) and the Centre nationale d'études spatiales (CNES) (France). A group of comparable scope for ocean observation systems planned to hold its first meeting in 1981.

Means of co-ordinating data acquisition, processing, and distribution have been initiated on a worldwide level by the United States through regular meetings of a LANDSAT Ground Station Operators Working Group. This approach is likely to be followed by some other agencies. France has announced that it will form a "Groupe des opérateurs de stations SPOT." Other agencies may choose other means for co-ordinating remote sensing experiments through their national systems. There is little doubt that the various activities and meetings of such groups will be co-ordinated in order to increase their effectiveness and reduce their cost.

Within the United Nations system, the Committee on the Peaceful Uses of Outer Space has been the forum for discussion of co-ordination of satellite remote sensing. This committee has promoted bilateral and multilateral efforts to make systems compatible and ensure that the benefits are as widely accessible as possible. The committee has particularly concerned itself with the question of international distribution of data.

5.3.3.2 Orbits. The characteristics of remote sensing missions depend on the interrelationship of launch capability, power supply and orbit control, sensor technology and reliability, ground system technology, user needs, and other factors. Frequently, the combination of these factors severely limits flexibility for achieving compatibility with other systems.

The choice of orbit is made during the design of the system. Because of different requirements, synchronization of orbit between different systems has not been achieved or even considered. The obvious exception has been the synchronization of the satellites with each other within multisatellite systems. However, co-ordination of orbits is a worthwhile goal for future systems, as it may enhance the diversity and timeliness of observation for areas of interest and reduce conflicts for ground facilities wishing to acquire data from two satellites. As a specific example of how mission characteristics are dependent on logistics, LANDSAT-D is to be placed into an orbit near 700 km in order to be compatible with retrieval by the Space Shuttle. This altitude is also compatible with the fact that the Delta launch vehicle is not capable of lifting the MSS and TM payload to the LANDSAT-1, 2, and 3 altitude of approximately 900 km. To launch to a 900-km altitude would require launch with an Atlas-class rocket which costs more and is not as reliable as the Delta-class vehicles. Once the 700-km altitude region is chosen, it limits the kinds of repeat, sun-synchronous orbits available. In particular, an 18-day repeat orbit, matching that of LANDSAT-1, 2, and 3, is not available. The current constraints on the orbital altitude of LANDSAT-D with the Thematic Mapper are 705 km + 12 km or −6 km. Eventually, the 705-km altitude, with a 16-day repeat cycle was chosen.

For the rest of this century, high-resolution earth observation will continue to be obtained from low earth orbits. The geostationary orbit, used at present for communication and meteorological programs, holds interesting possibilities, because it may allow the frequency of observation to be varied at will. Immediate and very frequent observation of specific areas is of great importance for mitigation of disasters. The geostationary orbit has, however, its limitations and drawbacks: It requires huge instruments for adequate resolution; several satellites are needed to cover the globe's longitudes and, thus, it may become overcrowded, and the polar regions are not covered. Low-altitude orbits have none of these limitations, but because the satellite cannot be everywhere at any one time, opportunities for observation of direct transmission of data are intrinsically less frequent. This limitation can be mitigated at the expense of putting up multisatellite systems and extensively using communications relay satellites. Other types of orbits, yet to be used for remote sensing, have a different mix of advantages and limitations. For instance, the use of non-equatorial orbits with a 24-hour period may be envisaged in order to extend high-altitude, daily observations to the polar regions.

Circular or low-eccentricity orbits are usually preferred in order to provide a relatively constant altitude above the surface and therefore a relatively constant image scale. For a given mean altitude, a circular orbit also results in less atmospheric drag and therefore less station-keeping propulsion. The altitude determines the area of the surface that can be seen at any time from the satellite and the area that can be covered by direct readout to a given ground station.

Orbital inclination to the equator is an important consideration that relates to launch capabilities, particularly for manned systems; it directly governs the range of the latitudes that can be observed. An inclination of about 63.4° has been used frequently for Soviet spacecraft in slightly elliptical orbits; at that particular inclination, the perigee–apogee line is stable, and closest observations always occur in the same latitude range. Higher inclinations, close to 90°, are required for truly global coverage. Of particular interest for missions of long lifetimes (whenever solar illumination controls the observation capability) are those orbits called sun-synchronous, where the rotation of the orbital plane in inertial space (caused primarily by the oblateness of the earth) precisely compensates the mean yearly rotation of the sun–earth line. These orbits have inclinations in the range 95° to 100°, depending on the altitude and thus require a slightly westward launch. They ensure that throughout a satellite's lifetime observation of an area will take place at approximately the same solar time, i.e., under similar conditions of illumination.

Within the continuum of (circular) orbital altitudes, there are discrete positions for which an integer number of revolutions take an integer number of days. If such orbits, called phased orbits, are maintained, the satellite will return to the same point above the earth (within a few kilometers) after each orbital cycle. Even within the family of sun-synchronous phased orbits that the LANDSAT, METEOR, and meteorological programs have been using, there are large numbers of possibilities (Figure 5.1). The choice bears directly on certain impor-

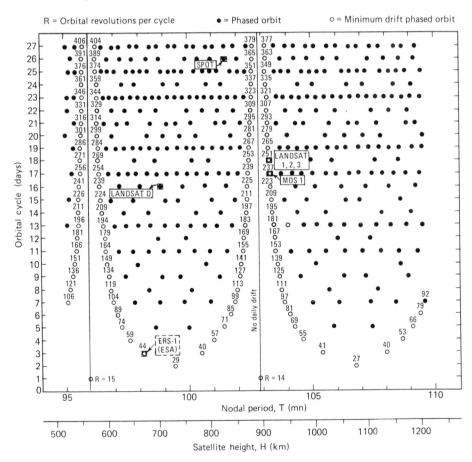

Figure 5.1   Low, short orbital cycle phased orbits.

tant mission aspects: minimum number of passes required to cover a broad area, space-time sequence of such coverage, minimum and maximum times between successive opportunities of observation, etc. Thus, there is good reason not even to try to standardize orbits whenever mission objectives are different; however, phasing two satellites in opposition on an identical orbit may be of particular interest for multisatellite missions in order to achieve a regular sequence of observations, while increasing their frequency, and has actually been used for LANDSAT.

5.3.3.3 Sensors. Within the constraints, and given generally broad mission objectives, the designer will study and select sensor systems, taking into account previous expertise and complementarity of measurements within the mission. For example, as in airborne photograpy, the geometric accuracy of space imagery may be enhanced by coupling an altimeter to the imager; the precision of ground reflectance determinations may be enhanced by a set of on-board atmospheric measurements.

For each of the sensors, there may be a choice between widely different concepts and techniques. The choice of a detector is usually crucial and has a strong impact on the instrument design. Arrays of electronically scanned detectors (charge coupled detectors) have recently been used for image acquisition in an experimental mode, both in the USSR (COSMOS-1208) and in the United States (DMSP satellites, 1979). They are likely to become standard practice within a few years.

Spectral band selection is also highly significant and mission dependent. Technical documents are available that discuss the advantages and disadvantages of various spectral bands. A major limiting factor for space missions is the availability of atmospheric "windows" within which to do remote sensing (for example, 0.4 to 0.7 μm, 0.75 to 0.91 μm, 1.5 to 1.8 μm, 2.05 to 2.40 μm). However, there is still a need for rigorously acquired, well-documented, and published material defining the advantages of various spectral bands, bandwidth variations within these bands, signal/noise ratios, effects of the atmosphere, etc. More data and sensor modeling results are needed to quantitatively assist in making sensor system trade-off studies and design selection. Experiments and simulations are being conducted in several countries in order to improve the assessment of atmospheric effects on sensor acquisition of radiation reflected or emitted by all types of ground covers.

It is already clear that optimal space missions for observing geology, vegetation, or water quality call for different spectral bands both in location and width, so different as to preclude, normally, the use of identical instruments. The spectral bands of sensors which have been flown repeatedly—MKF6 camera system designed in the German Democratic Republic in co-operation with the USSR, LANDSAT-1, 2, and 3 MSS—or which will soon be flown, are shown in Figure 5.2.

Other essential choices relate to radiometric performance (number of

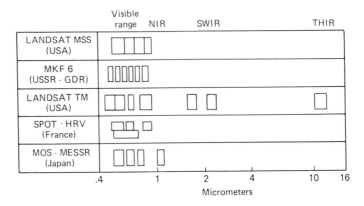

Figure 5.2 Spectral bands of various remote sensing satellite systems.

significant grey levels), the dimension on the ground of an elementary picture sample (pixel), and the ground breadth of observation (swath width); together these dictate the very high rates at which information has to be communicated to the ground and, later, processed.

5.3.3.4 Data distribution and coordination.
If maximum benefit is to be obtained from the use of remote sensing data, they must be made widely available to users the world over. Steps have been and are being taken in that direction: for example, the creation in 1966, and operation since that date, of the EROS Data Center in Sioux Falls (United States); construction of direct receiving stations in many countries for acquisition, pre-processing, and distribution of the data; creation of a distribution network in Europe by ESA (EARTH-NET); current plans to create efficient distribution entities prior to establishing fully operational systems (United States, France, and Japan); and the establishment of three regional data-receiving stations for Soviet remote sensing satellites (Moscow, Novosibirsk, Khabarovsk).

Such efforts must be encouraged, and working relationships between the distribution systems must be developed so that the end users can, upon request, get full information on, and easy access to, relevant data already required. ''Relevant data'' is not limited here to high-resolution imagery but may include data acquired by meteorological satellites or on the ground. A certain degree of standardization will have to be reached in order to facilitate direct data acquisition, access to existing data bases, data interpretation, and data application. Three broad areas are concerned: satellite-to-ground transmission; data product characteristics; and applications systems and procedures. Practical standardization has been achieved within meteorological space programs, in the overall framework of WMO. The experience gained will serve ongoing efforts in the field of remote sensing.

5.3.3.4.1 Data Transmission.   To maximize the use of data from several different satellites, it is desirable to be able to read out and process these data with the minimum amount of duplication of equipment. All unmanned earth resource satellite systems announced to date have a direct readout capability. Some also have on-board storage or satellite data relay capability. For the majority of the participants in satellite remote sensing programs, the most important means of access will be direct readout. In the future, because of the ITU Regulations, all remote sensing satellites are expected to transmit within the same general radio-frequency band (X-band) so that the same antenna, feed system, and RF receiver might be used. For several years to come, some satellites will also operate on the S-band, which has been used for several years for the LANDSAT series. Investigations are being conducted by some satellite operators to determine if certain signal parameters can be made the same from one satellite to another. Examples of these parameters include modulation, synchronization codes, and time codes. An example of successful co-operation in this area is the acquisition of data from geostationary meteorological satellites.

5.3.3.4.2 Data Products.   Significant progress has been made in developing standard data products by the operators of LANDSAT ground stations with respect to both end-user products and internal products intended primarily for intermediate storage of data, usually at a very high data density.

For LANDSAT, there are essentially five classes of user products: quick-look photographic images; standard photographic images; standard computer compatible tapes (CCTs); precision geometrically corrected CCTs; and high-density digital tapes (HDDTs). Within each class of product, there is scope for a great deal of variation; for example, a wide range of corrections may be applied. In some instances, the distinction between classes has become negligible. For example, in Canada, quick-look products, which originally had minimal annotation and radiometric correction and somewhat degraded resolution in comparison with the standard product, were available for shipment within hours of a satellite pass, while standard products took several weeks to process. Eventually, the quick-look processing system was upgraded in quality to the point where it became both the quick-look and standard product, still available within a few hours.

In examining both classes of photographic products, the LANDSAT Ground Station Operators Working Group (LGSOWG) decided that the only practical form of standardization would be "minimum annotation." Other possible standards with regard to format, scale, and quality were considered but rejected by the Working Group. It was felt that the user requirements on these matters varied so much from one area to another that it would be impossible to come to agreement on a set of standards in the near future. Undoubtedly, part of the reluctance to adopt any specific standards is the cost which would have to be borne by individual ground station operators if a significant product change were required. Nevertheless, the operators were willing to commit themselves to the expense of changing their annotation to conform to the minimum annotation requirement agreed upon. User concern seems to be focused on such questions as: What is the data source? What is the location on the ground? And when were the data collected? These concerns seem reasonably well treated by the annotation standards adopted.

Partly because the user cost associated with having to deal with a wide variety of formats for CCTs is high, a very versatile format was developed and adopted by LGSOWG for implementation for LANDSAT–D. In reality, a family of formats was developed to handle any conceivable image or grid form of data. Several of the LGSOWG members have indicated their intention to use the format family for all image-type remote sensing data, including airborne multispectral scanner and radar data. The format family has proven readily adaptable to non-image data as well. In Canada six major users of geographic data have adopted a format consistent with the LGSOWG format family for the interchange of data.

While the general data structure of precision geometrically corrected CCTs may be fully compatible with the standard CCT, there are several other areas of compatibility that have not yet been examined, such as pixel sizes, data orientation, and map projection. At present, the few facilities producing these products have adopted different philosophies regarding these parameters. The needs of the users in different countries are sufficiently diverse that no universal standard could be effective. Yet these differences may cause great difficulty to a user obtaining data from different ground facilities.

To date, there has not been any widespread external use of user high-density digital tape (HDDT) products. Examination of standards for such products has therefore been considered to be of low priority. Except for synchronization codes and frame identifiers, it is likely that any format standard would be essentially the same as that for the CCT format family.

Considerable effort was expended by LGSOWG in attempts to define standards for intermediate HDDT formats. These HDDTs of various types are used by the stations to record the satellite data between reception and processing or for archiving purposes. The desire for a standard was to permit the stations to interchange their data in a cost-effective manner. It was finally decided that the only standard should be at the station tape recorder level, such that the output of the station recorder should be the digital bit stream as transmitted by the satellite. This standard would not even guarantee that a tape recorded at one station could be read at another, unless both had identical recorders. Nevertheless, this was as far as the LGSOWG members felt that standardization should go for intermediate products.

Inherent in the satellite imagery data are radiometric and geometric errors that can be associated with the platform, the sensor, the atmospheric conditions, and the earth's surface itself. In most application areas, the full potential of the remote sensing imagery cannot be realized unless these aberrations are removed or greatly reduced. Even after it has been corrected, the task of integrating satellite imagery with other geographical information systems could present major difficulties because the two types of data do not necessarily use the same geographical references. From the user point of view, the conversion cost could be prohibitive. Also, in the case of multitemporal applications, each image from the same satellite would require its own conversion to account for orbit-dependent variations.

Given the situation of multiple satellites, the problem is complicated further by the orbit and sensor characteristics causing different image framing, field of view, and pixel sizes. The complexity of digitally registering data from complementary missions could constitute at the user's site a stumbling block which could deter and prevent new applications. With higher data rate satellites, many users would prefer subscenes to limit the volume of data that would have to be recorded digitally on computer compatible tapes. Finally, the concept of standard framing defined and fixed by the trajectory will become more difficult to apply with future orientable satellites.

To facilitate the utilization of data from multiple complementary missions and the integration with geographical data bases, some consideration is being given to the development of computer compatible tape and film products, which are platform and sensor independent in a cartographic projection. These products would have a subpixel geometric accuracy both in multitemporal registration and in absolute geodetic control. They might be offered in a subscene format to meet flexible user coverage requirements and in projections that are compatible with various geographical data bases. The pixels would be aligned with mesh conveniently with the projection grid. The pixel size of the corrected images might be such that it belongs to a natural hierarchy of sizes such as 5, 10, 20, 25, 50, 100, 1,000 m. In most cases, the chosen pixel size for the corrected products would constitute a slight oversampling by at least a factor of 1.4 with respect to the instantaneous field of view of the sensor.

5.3.3.5 Data interpretation and application.

Application research and development is taking place in many countries and is by no means limited to the satellite-operating countries. Procedures and media exist for the selection and broad dissemination of ideas and results; responsible organizations should be encouraged to publish or otherwise make available general purpose algorithms and computer programs. Dissemination of information is also served by numerous meetings and scientific symposia, for example, those organized or sponsored by COSPAR and IAF.

Introduction of remote sensing data into existing application systems usually calls for very specific methods that can be developed only by or with the proper authorities. Two suggestions could be formulated in this respect:

1. That the more experienced countries help the less experienced through specialist assistance and in-depth training programs, and that they coordinate further their efforts in the training field, as, for example, in the

regional remote sensing center in Ouagadougou, Upper Volta, under the aegis of the Economic Commission for Africa (ECA).

2. That satellite-operating countries, when initiating new programs, encourage, whenever possible, early involvement of scientists and potential users, the world over, to experiment at the earliest possible stage with data similar to or provided by the new systems, and disseminate the findings. The organization of regional technical information meetings could be seen as a significant step in this direction.

5.3.4 Conclusion

In the early 1980s, remote sensing is at a crossroads; the technology involved, in space and on the ground, is sophisticated and there is a justified drive to continually improve its performance. At the same time, a genuine potential for the application of technology to help understand and, hopefully, master important problems facing mankind on a worldwide scale is strongly perceived. Tapping such a potential requires, among other things, additional efforts and investments in establishing operational systems on the basis of stabilized technology. Several countries are proceeding with the establishment of such systems, but in view of the high costs, long lead times, and obstacles involved (such as cloud cover), good remote sensing data must itself be considered, for many years to come, as a limited resource.

A degree of co-ordination between operators of future systems is necessary to ensure that available systems can effectively combine their capabilities for the benefit of user communities. Co-operation is needed between the various entities concerned with acquisition, pre-processing, and distribution of data. Finally, interchange of ideas, know-how, and practices should be promoted between the scientific communities and the end users in each field of application. Work along these lines is already beginning; its success is as essential as the basic space technologies themselves if the promise of remote sensing is ever to be kept.

ACKNOWLEDGEMENTS

The teams from COSPAR and IAF, which assisted with the section on remote sensing, included the following: Dr. W. M. Strome (Canada, Co-ordinator IAF), Dr. G. M. Weill (France, Co-ordinator COSPAR), Mr. D. Duchossois (France), Dr. Yukio Hakura (Japan), Mr. C. Honvault (France), Mr. Mahson Irsyam (Indonesia), Prof. K. Ya. Kondratiev (Union of Soviet Socialist Republics), Dr. L. Marelli (Italy), Mr. E. Peytremann (France), Dr. S. I. Rasool (United States of America), Dr. V. Salomonson (United States), Dr. N. N. Shcheremetevsky (Union of Soviet Socialist Republics), Prof. K. Tsuchiya (Japan), Dr. A. B. Park (United States), Mr. Y. S. Rajan (India), Prof. K. B. Serafimov (Bulgaria), and Dr. H. Ricciardi (Argentina).

Chapter 6

FEASIBILITY AND PLANNING

OF INSTRUCTIONAL

SATELLITE SYSTEMS

ABSTRACT

This chapter, which is the second of three concerning specific technical issues, examines the problems involved in creating a satellite educational broadcasting system. The chapter considers the advantages and disadvantages of satellite systems and analyzes the decisions that must be made concerning orbit, transmitting frequency, earth station design, reception system design, equipment procurement, program production, maintenance, organization, management and evaluation. These analyses are based on the experience gained in the Satellite Instructional Television Experiment (SITE) in India, the Health Education Telecommunication (HET) experiment in the United States, and similar experiments in Canada, the USSR, and elsewhere. Because an educational system must be carefully adapted to its political, economic, social, cultural, and technological context, no attempt is made to define an ideal system. Rather, guidelines are presented that, it is hoped, will be of use to anyone attempting to plan such a system.

INTRODUCTION

The advent of satellite communications—and especially of satellites capable of broadcasting directly to augmented radio or television receivers—has made available a powerful new tool for education, instruction, and development. Not only is it possible to use this new technology for large-scale expansion of existing instructional systems, particularly to remote and inaccessible areas, but it has also become feasible to plan altogether new systems. Thus, satellite technology is capable of bringing about a qualitative change in the area of instructional systems.

The need for using a satellite-based system is not necessarily obvious; in fact, other means of delivery may be preferred in certain situations. A satellite does, however, enable a broadcasting and distribution system to be set up much more quickly than would be possible by conventional means. This is of particular relevance to developing countries which do not have extensive ground infrastructure, especially in rural and remote areas. Given the educational problems in most developing countries, time is a most crucial element. A satellite-based system can not only provide services more quickly and thereby attack the quantitative aspect of the problem, but it can also bring about a change in the quality of education by making the best teaching available to all. More innovative use of satellite capabilities can also bring about altogether new modes of education, especially for specialized instruction.

That this is not merely a vague and distant dream is borne out by the actual implementation of some experimental systems and the planned introduction of operational systems. Beginning with the simple point-to-point relay, first of voice and then of television signals, technology has advanced to the point where it is now possible to broadcast television via satellite so that it can be received directly in homes using slightly augmented sets.

Initially, television broadcasting via satellite could only be received by large, complex, and therefore expensive, earth stations from which the signals could be sent to conventional television transmitters for rebroadcast. Since reception of such signals does not require any augmentation of individual sets, this approach (the distribution mode) is still useful and economic in areas with high receiver densities (e.g., urban areas). Satellites have been used extensively for national distribution of programs in Canada, the United States, the USSR, and Indonesia. However, rapid advances in satellite technology soon made possible a "technology inversion," in which the spacecraft became larger, more sophisticated, and complex, enabling the reception equipment on the ground to become simpler, smaller, and cheaper. The first major example of this technology inversion was the ATS-6 spacecraft of the United States; using this, it became possible to broadcast directly to cheap and fairly simple ground receivers. In both the United States and India, where experiments in satellite broadcasting were carried out using this satellite, the television signals were received using 3-m diameter parabolic antennas and electronic converters. In India the cost of the augmentation was about twice the cost of the conventional television set.

More advanced technologies have since been proven, including the Canadian/United States CTS and the Canadian ANIK-B spacecraft, with broadcasts in the 12 GHz band. The ground receivers for these systems use antennas of the order of 1.2 m in diameter. Already, satellite systems are being planned that will use receiver antennas of 0.9 m and smaller. Another example of an advanced satellite system—operating at almost the opposite end of the frequency spectrum (702 to 726 MHz)—is the USSR system based on the EKRAN (Statsionar-T) satellites. Even more ambitious systems have been proposed, and conceptual studies of large space platforms, to be used for a variety of purposes, have been carried out.

While the focus has been almost totally on television broadcasting, there has recently been increasing interest in direct (as opposed to redistribution-type) radio broadcasting via satellite. A number of studies have been carried out by various organizations, including the European Broadcasting Union (EBU), the Indian Space Research Organization (ISRO), the European Space Agency (ESA), etc. The question of allocating a specific frequency band for such broadcasts is also being actively examined (see Appendix 6.1). Reception would, in this case, also require some augmentation to the conventional FM radio receiver (using AM does not seem feasible at this time). On a small scale, audio channels have been used for some very interesting and innovative experiments, especially in the South Pacific using the United States ATS-1.

Among the technological possibilities now offered by satellites are:

1. Radio and television distribution to broadcasting stations.
2. Direct television broadcasting to augmented sets.
3. Direct television broadcasting with two-way audio.
4. Two-way video and audio.
5. Two-way audio.
6. Slow-scan television with or without two-way audio.
7. Television with teletext.
8. Facsimile.
9. Access to computers, for computerized/computer-aided instruction, or for other purposes.

Despite all these advances in technology and the tremendous potential of satellite broadcasting, the early promise of great benefits in the instructional field does not seem to have materialized. Except for a few experiments, little has happened to fulfill the high hopes of the late 1960s, when satellite broadcasting was considered the ideal solution—particularly for the educational and developmental problems of developing countries. There have been few instances of large-scale use of instructional satellite systems, and even these have been only on an experimental basis. The biggest and most extensive so far has been the Indian Satellite Instructional Television Experiment (SITE) carried out for a year in 1975/76, using the ATS-6 satellite (1).[1] The same satellite was earlier used in the United States itself for experiments in the fields of health and education. Though the number of receivers in this case was smaller, several innovative experiments, including use of two-way video and audio, were conducted. More recently, the United States/Canadian communications technology satellite (CTS) has been used for a variety of experiments, including some in the instructional field (2). Some experiments in direct-to-home broadcasting are also planned in Canada, using its ANIK-B; however, the focus is basically technological and not instructional.

Operational satellite systems are already in place in the United States, the USSR, Canada, and Indonesia. None of these is known to be used extensively or exclusively for education. Basically, they are used for revenue-earning telecommunications/data traffic and for relay or exchange of normal television programs. The USSR, for example, has been using a satellite for television program distribution from the mid-1960s. However, satellites are not yet utilized for the extended rebroadcasting of instructional programs.

New communications satellite systems are planned or proposed in India, Europe, France, Federal Republic of Germany, the Nordic countries, Saudi Arabia, Latin America, the Arab countries, Japan, China, etc. While all these satellites will almost certainly have television broadcast capability, the main thrust is again likely to be in the telecommunications area. The extent to which the broadcasting capability will be used for educational and instructional purposes is still uncertain. Even so, the fact that so many systems are already being studied is an opportunity for policy makers, educators, and others to ensure—from the planning stage itself—that an instructional element is built in. For other countries, this may be an opportune moment to examine whether an instructional satellite system would be useful and economic for them. They would have to consider the type of system (national or regional), the services it should carry, the size of the ground network, the overall system design and a host of other factors. What is clear is that the feasibility of instructional satellite systems has been amply proven, not merely by studies but also by actual demonstrations. The moot point is the suitability and form for a particular country.

There are probably a wide variety of reasons for the slow progress in the field of instructional satellite systems. One obvious problem is the high investment required, not only in the spacecraft and its launch but also on the ground. Unfortunately, most governments look at education as an overhead rather than

---

[1] Numbers in parentheses refer to the References at the end of the chapter.

as a high-priority investment, probably because the returns are not immediately and concretely visible in hard quantitative terms. One solution to this is multipurpose satellites, in which the instructional element can, so to say, ride piggyback on other more "economically attractive" services like telecommunications; this is the case with the Indian INSAT, due to become operational in early 1982, and the proposed ARABSAT. Most countries now planning their satellite systems seem to favour this approach. A second solution is for the investment in the spacecraft and launch to be shared by a number of countries. This is a particularly attractive approach for smaller countries, even affluent ones. Such a "regional" approach is being considered by a number of countries, including the Arab countries, the Nordic countries, the ESA countries, the Latin American countries, some African countries, etc.

Another problem may be the complete lack of familiarity with satellite technology (and often with any technology at all) on the part of the instructional designers and educators almost all over the world. In contrast, telecommunications personnel have always been technology oriented, possibly accounting for the quick and easy entry of satellites in this field.

The third possible problem relates to questions about the effectiveness and the development of instructional television. Oleg Belotserkovskii points out that although rapid development of instructional television began in the USSR in the 1960s, it has slowed down now. He attributes this to a complex of unsolved sociological and pedagogical problems.

A more basic problem with regard to the setting up of instructional satellite systems may be structural. It is well known that the integration of a major new element in any system requires a change in the mode of functioning, and often in the structure, of the system itself. It therefore seems that satellite broadcasting can be added to the existing instructional systems in a meaningful manner only when certain operational and structural changes are made. The type and extent of change required will obviously vary from situation to situation and from country to country. However, the "stability" or inertia of the educational system is a fact of life in almost all countries, and therefore the difficulties in bringing about changes should not be underestimated.

At the same time, the magnitude of the educational problems, especially in the developing nations, is so great that conventional technologies and approaches are unlikely to even create a dent in the problem. This has been brought out clearly and very well in the study by the United Nations Educational, Scientific and Cultural Organization (UNESCO) on tele-education in South America (3), carried out at the request of the Spanish-speaking South American countries. The origin of this study was the recognition that the traditional system was inefficient and needed transformation. This study has recommended a new system (SERLA) that, in a phased manner, will move toward the use of a satellite.

Special situations even in developed countries may require a satellite-based system. For example, the Educational Communications Authority of Ontario (TVO) in Canada has said that it will use the ANIK-C satellite to distribute its existing programming. Also in Canada, the British Columbia government has expressed interest in the educational use of satellites. Similarly, the USSR has, for many years now, been using satellites for distribution of instructional television programs.

As mentioned earlier, the desirability (in certain circumstances) and the feasibility of instructional satellites systems have not merely been a conclusion of many studies; this has been actually demonstrated and proven in practice. This chapter seeks to provide information and to discuss briefly some of the major issues involved in planning and establishing an instructional satellite system. It necessarily draws heavily upon the practical experiences so far, especially those of India, which has conducted the largest such planned experiment (4). This also means that this chapter is "television-oriented," since there is little experience of large-scale satellite instructional radio broadcasting to date. While much of the discussion is universally applicable, the focus is on developing countries, where the need (and impact) of instructional satellite systems could probably be greater.

Technical details have been kept to a minimum, though they are not always avoidable. Similarly, in the interest of brevity and conciseness, we will not dwell in detail on program production or other aspects of programming.

6.1 SYSTEM PLANNING AND OPERATIONS

6.1.1 Introduction

An instructional satellite system consists of a large number of component systems working in unison to provide a broadcasting service to a target audience. The objective of this section is to discuss the role of various components and the manner in which their planning and operation need to be dovetailed so as to provide a trouble-free system. Once the technical system is established, the actual use of the system in terms of education, training, etc. can proceed. An important point to be noted is that the system must be planned around the projected usage. It is, therefore, necessary to have a clear conception of the usage at the start. It would be disastrous to establish a system first and then try to plan its utilization. At the same time, it would not be desirable to stick very rigidly to blueprint systems and usage plans. A certain amount of flexibility in the system and its usage should be built in, to ensure that the system can respond to changes and fresh ideas.

The focus here is on the ground segment, since more countries are likely to be interested and involved in this than in the space segment.

6.1.2 Instructional Satellite Systems and Their Operational Goals

Expressed in simple terms, the operational goal of an instructional satellite system is to provide a broadcasting service—television or radio or both—with a particular level of reliability to a target audience. The level of reliability is defined as that percentage of the total broadcast time for which the service is available to the target audience. It is clear that the reliability, as defined here, only applies to the existence of a broadcast channel from the source to the receiver. What goes on the channel is not being considered; that is dealt with in Section 6.4.

An instructional system must also have an operational goal related to feedback from the audience to the source. The critical parameters of this will be the speed with which feedback is available and the accuracy of feedback. This aspect is also dealt with in Section 6.4.

To illustrate the above requirements, SITE in India had as its goal the provision of a television broadcasting link to 2,400 villages in six clusters spread over India. The target villages were those which were electrified and accessible for 80 percent of the time by road from a nearby town. The former requirement was waived for about 180 villages in Orissa, where battery-operated sets were used on an experimental basis. The target audiences were children in the 5- to 12-year age group in the morning and adults in the evening. A total of four hours of program time was available per day. The goal for channel reliability was fixed at 80 percent.

Other examples are that of TELESAT Canada which provides television/radio services via satellite for communities of 200 or more people in Canada's rugged northern areas. Eighty stations were served by ANIK-A of TELESAT Canada. Canada has also experimented with direct broadcasting in the 12 GHz band using the CTS satellite to provide tele-education and tele-medicine as well as broadcast services on an experimental basis. Following this, Canada is now using the 14/12 GHz transponder on-board ANIK-B to experiment with television direct broadcast services. One hundred and ten terminals are involved.

Another example of an instructional satellite system is the Health Education Telecommunications Experiment (HET) (1974/75) using ATS-6 and covering Appalachia, the Rocky Mountain states and the north-west, and Alaska in the United States, where terrestrial television networks had not yet reached. In Appalachia, the target audience was 1,200 teachers for college courses and ten Veterans Administration hospitals for tele-consultation. In the Rocky Mountain states, 15,000 students in junior high school were served. In Alaska, links were provided with Washington University for the benefit of medical students, and hospitals were linked to remote clinics. Apart from this, primary and secondary school programs and community services were also provided (5).

In the USSR, instructional television programs have been broadcast since 1964. Satellites have been used for this purpose since the launching of MOLNYA-1 in April 1965. The programs are aimed at secondary schools, students intending to enter institutes, correspondence students, managerial personnel of ministries, offices, and enterprises, experts, and so forth. The target audiences are located across the length and breadth of the country.

The Indonesian system using the PALAPA satellites is also capable of television distribution. Thirty-four VHF stations are linked via 40 PALAPA earth stations in a national television network (6).

6.1.3 Components of an Operating System

As mentioned earlier, an operating system for an instructional satellite system consists of several parts working in unison to provide a broadcast channel. Five main components can be identified:

1. Program production.
2. Program transmission.
3. Program reception.
4. Feedback and evaluation.
5. Organization and management.

The program production component consists of studios and the associated production staff. This would include fixed studios, mobile studios (outside broadcast vans), etc. The program transmission component includes satellite earth stations and their feeder links, the satellite itself, and terrestrial transmitters or cable systems. The program reception component consists of the broadcast receivers and the program utilization elements. The receivers could be of two types: conventional television sets or specially augmented satellite receivers. The feedback and evaluation component consists basically of social research staff. The organization and management segment is the nerve system that organizes, directs, and co-ordinates the operation of the other four components. An important part of this is the satellite management system which is responsible for satellite operations, housekeeping, and station keeping.

The SITE operating system consisted of five program production centers equipped for field production as well as studio-based programming. One of the innovations tried at one center was field production of programs using inexpensive "consumer" grade video equipment.

The program transmission component consisted of the ATS-6 satellite loaned by the United States and located in the geostationary orbit above 35°E; the main earth stations and the backup earth station; two receive-only stations; a mobile receive-only station; conventional television transmitters and their associated feeder links.

The program reception component consisted of 2,400 direct reception television sets located in six clusters in northern, eastern, and southern India and conventional sets located in and around the terrestrial television transmitters.

Feedback and evaluation were provided by a team of about 100 social scientists working in ISRO. In addition, evaluation studies were also carried out by other institutions (see Section 6.5).

The satellite management was by the United States National Aeronautics and Space Administration (NASA) from its station at Buitrago, Spain. A 24-hour dedicated voice and teleprinter link was maintained between the SITE Operations Control Center of ISRO and the ATS Operation Control Center of NASA. SITE was responsible for control of the ground segment and liaison with the space segment control; NASA managed the space segment. SITE provided ranging support to NASA from the Ahmedabad Earth Station.

The TELESAT Canada system established with ANIK-A and B consists of three A series and one B series spacecraft in geostationary orbit. Satellite control is from Ottawa through earth stations at Allan Park, Ontario, and Lake Couichan, British Columbia. Television and radio services are provided by the two main earth stations and eight network earth stations which have transmission and reception facilities. In addition, there are about 80 reception-only stations. The television distribution to the individual receivers is effected by retransmission over conventional channels. In addition, seven stations are equipped for radio transmission and reception and 14 for radio reception only. There are also two mobile television transmission earth stations (7).

The CTS experiment used several receive-only terminals using antennas from 0.6 to 1.6 m in diameter located for various periods for specific experiments or demonstrations in Lima, Peru; Ontario, Labrador, and Saskatoon, Canada; and Australia.

The ANIK-B 12-GHz direct broadcasting experiment involves the use of 110 terminals in northern Ontario and British Columbia. There are three types of terminals: 1.2-m antennas for use in high-signal regions, 1.8-m antennas for low-signal regions, and 3-m antennas for feeding into cable television or rebroad-casting stations.

The HET experiment using the ATS-6 2.5-GHz transmitter and the ATS-1 and 3 VHF transponders involved eight transmission terminals and 119 receivers, 51 with two-way audio links. The HET Network Control was located at Denver, Colorado, and linked to ATS Operation Control and two other earth stations. The Veterans Administration experiment deployed ten television receivers in hospitals in Appalachia to receive programs originating from Denver. Two-way audio was provided for teleconferencing over conventional telephones. The Appalachian Education Satellite Project involved programs originating from the University of Kentucky at Lexington and received by 15 television receivers. One-third of the receivers had VHF two-way audio capability via ATS-3 to the originating point. The other two receivers were linked to this receiver by landlines. The Rocky Mountain experiment catered to 56 receivers and 12 Public Broadcast Service rebroadcast stations. Twenty-four receivers also had VHF two-way audio via ATS-3. In the USSR, 85 Orbita earth stations receive programs from the MOLNYA series of satellites as well as the RADUGA (STATSIONAR) series for rebroadcasting by terrestrial transmitters located within 10 km of the earth station. While the MOLNYA system provides only one television channel, the RADUGA system provides three channels which can be used for different time zones. A new series of satellites called EKRAN (STATSIONAR-T) has a high-power UHF FM transponder that allows the use of simple low-cost earth stations. Each station, coupled to a limited range rebroadcast transmitter, serves a small community. At present 60 such stations are being used experimentally.

6.1.4 Planning for Operations

The planning activity for system operations begins from the stage where the system has been defined and consists of the following:

1. Identification of hardware items and suppliers, including indigenous development options.
2. Identification of manpower requirements, including recruitment and training.
3. Site selection for various components of the system.
4. Acceptance testing of hardware items.
5. Deployment of equipment, scheduling, and logistics.

6.1.4.1 Identification of hardware items and their suppliers. Once the system configuration has been decided, the next job is to look for system components and suppliers.

6.1.4.1.1 The Satellite. This can be either a leased service or purchase of a complete system. While INTELSAT transponders can be leased for telecommunications or television distribution, as of now there are no operational international satellites offering broadcasting service to users. However, several domestic and regional systems are planned. The choice of leased satellites is necessarily limited to those located in the portion of the geostationary orbit visible to the country. This is further narrowed depending on the characteristics of the available satellites and the possibility of entering into a bilateral agreement with the country or countries operating the satellite. If leasing proves impossible, the alternative is to develop a new satellite. This would involve a detailed process of drawing up specifications, releasing a request for tenders, evaluating proposals, negotiating the contract, etc., before an order can be placed. Those countries that have the capability may also design, develop, and build their own satellites. One has also to plan and arrange for the launch of the satellite and its management in orbit.

The satellite used for SITE was the NASA applications technology satellite, ATS-6, loaned to India for a period of one year from August 1975 to July 1976 under a bilateral agreement. ISRO provided the complete ground segment. Under the agreement, the results of all experiments were to be shared. There was no transfer of funds. The same satellite was used earlier for the HET experiment in the United States.

TELESAT Canada has its own satellites which are used for telecommunications services as well as radio and television distribution. Similar arrangements exist in the USSR and Indonesia.

6.1.4.1.2 Earth Stations.   Today, earth stations for satellites can be purchased from a wide range of suppliers. The choices are:

1. To buy the earth stations on a turnkey basis.
2. To buy subsystems and put together the earth stations.
3. To buy some subsystems and build others.
4. To design and build the earth stations from the subsystems up.

Of these, the first is the easiest approach, but it may not always be cost effective. Even if it is apparently "cheaper," it limits the development of a tailor-made system designed to meet the specific needs of the country and may therefore in reality turn out to be more expensive. The next two alternatives would ensure that an optimized system is procured but would require personnel having extensive background in subsystem engineering and overall system integration. The last is probably the most desirable but also the most difficult. Essentially, the choice is dictated by the resources available to the country, its technological capabilities, and its future plans.

For SITE, the main earth station at Ahmedabad was an existing station set up for experimentation with ATS-2 in 1967 with assistance from the United Nations Development Programme (UNDP). That station was extensively modified and augmented for SITE. This included complete redesign and fabrication of the communication chains, including the high-power amplifiers. These were fully indigenous efforts. Only the parametric amplifiers were purchased. For ranging support, NASA provided a complete set of ATS ranging equipment. The original antenna was retained and used after alignment. The secondary earth station at Delhi was a totally indigenous effort. It was designed as a low-cost, tailor-made limited steerability earth station meant for use with geostationary satellites; consequently, it could do without elaborate tracking gear for the antenna. All other earth stations were of the receive-only type and were indigenous.

In the field of earth stations, SITE followed the third alternative by building almost all subsystems, buying a few and putting them together. For its operational INSAT system, India is following a similar approach though, due to various considerations, more subsystems are being imported. The INSAT system will have over 30 earth stations with antenna diameters of 11 m, 7.8 m, and 4.5 m. In addition, one transportable and two air-liftable stations are likely to be part of the network.

The TELESAT Canada system uses four types of earth stations for television and radio distribution. The heavy-route main terminals use 30-m fully steerable antennas and redundant electronics with power backup. The network terminals have 30-m limited steerability antennas, redundant electronics, and power backup. There are two types of television receive-only terminals: The Remote terminals have 8- or 4.6-m limited steerability antennas; the Frontier terminals have a 4.5- or 3.6-m antenna.

The USSR Orbita system has been using 12-m tracking antennas. However, the EKRAN system uses phased array antennas for the reception terminals.

6.1.4.1.3 Terrestrial Transmitters, Microwave Feeder Links, and Studios.
These are conventional items and chances are that they are already in use locally as part of the terrestrial broadcasting system. If so, then a suitable adaptation could be planned.

For the studios, it would be necessary to look at the economics of operation, technical quality desired, planned usage, and scope for future expansion in order to arrive at the system configuration. Here again, we have choices for turnkey contracts or subsystems procurement and integration.

During SITE, existing terrestrial transmitters were used, except in the Ahmedabad area where a special rural television transmitter was located at Pij, in a dairy-farming district adjoining Ahmedabad. Cable links from studios to earth stations had to be laid at Ahmedabad and New Delhi. A unique feature of the transmitter at Pij was that its antenna array shared the steel tower on which the microwave link dish was mounted.

The SITE system used five regional studios. Three were located in the regions where the receivers were deployed, a fourth was located in New Delhi to provide national level programs as well as regional programs, and the fifth was located at Bombay for the production of children's science-education programs. Almost all programs were recorded on tape and transported by air in special containers to the main studio at Ahmedabad. However, the national programs from New Delhi were directly fed to the New Delhi earth station. All the studios were newly set up for SITE and almost all the equipment was purchased from foreign sources.

The TELESAT Canada, HET, and Orbita systems use existing program production facilities and are hence not described here.

6.1.4.1.4. Receivers.
The reception system consists of direct reception television sets (receiving signals directly from satellites) and conventional sets. While conventional sets are in ample supply, the direct reception systems are not yet commercially available. However, the direct receivers can be made up of a conventional set with an add-on system for direct reception. Thus, the problem is

reduced to one of procurement of the add-on system. A word of caution needs to be sounded here: Conventional television sets are designed for a fairly controlled environment; community sets are likely to be subjected to severe environmental conditions and poor handling. Therefore, in the long run, specially designed receivers for both direct reception and conventional reception would be more economical in terms of reducing maintenance costs for community reception.

The SITE direct receivers consisted of a specially designed antenna and an electronic system called the front-end converter to receive signals from ATS-6 and generate audio and video signals for feeding into the television receiver. The receiver was a commercial unit modified to interface with the converter. The antenna and converter were new designs specially made to reduce costs and ease maintenance in the field. The antenna was in kit form that could be easily erected in the field by two persons using simple off-the-shelf tools. The television set was basically a commercial design but modified to tolerate rough "field" usage. The major modifications were an enlarged cabinet to protect the picture tube, particularly during rural transportation, a power supply that could withstand the $+20$ percent to -15 percent voltage fluctuation expected in rural areas (as against the ± 5 percent of urban areas), and a patching system whereby the set could be converted from direct mode to conventional VHF mode. About 100 sets were "ruggedized" further through the use of superior quality electronic components. Also, 200 sets were designed to work off 24-V batteries and deployed in areas having no electricity.

The conventional television receivers for terrestrial transmissions were commercial sets available in the market.

Direct broadcast receivers were used in the CTS and the HET experiments. Here the antenna and front end were designed and fabricated by commercial firms. These were linked to commercial television sets or television monitors.

6.1.4.1.5 Maintenance Equipment and Spares.   The maintenance of the system will require equipment, spare parts, and trained manpower. Here it is necessary to control costs without degrading the reliability of the system. This could be achieved as follows:

1. The number of different types of subsystems should be minimized.
2. As far as possible, "off-the-shelf" items should be used.

For all purchased items, it is useful to order 10 percent spares. The manufacturer should identify a critical spares list and these could be ordered at a level of 15 percent. All spares should be included in the purchase contract. For items built by the project, it is better to keep the total spares inventory at a 15 to 20 percent level. The higher end is preferable when components are imported or in short supply.

The test equipment needed for maintenance also needs to be carefully selected. Here again, the aim should be to standardize on a specific set of equipment rather than a large variety of test equipment. Some equipment will need to be custom-built for the system, for example, "satellite signal simulators," battery-operated test equipment for field use, etc.

For SITE, the earth stations and studios had adequate spare subsystems as well as spare components. Since almost all earth station subsystems were built by the project, provision was made for 1:1 replacement. Studio equipment was procured from sources abroad and spares were procured simultaneously. The manufacturers also provided a preferred spares list. Spares for the direct receivers were procured from a local supplier at a 10 percent level, which was later raised to 15 percent to reduce system downtime due to lack of spares.

Most of the test equipment for maintenance consisted of standard items such as signal generators, spectrum analyzers, etc. For the direct receivers, two special pieces of equipment had to be made: one was a satellite signal simulator which was built, for such items are not sold by any manufacturer; the other was a 230-V inverter required to convert a storage battery output to the AC mains voltage and frequency in order to run the receivers and test equipment in the event of a power failure during a service call. (Power failures are common in rural areas.) The storage battery used was the battery fitted in the jeep.

6.1.4.2 Manpower.
The operation of an instructional satellite system would require trained manpower. Requirements would be in the fields of:

1. Operation and maintenance of studios.
2. Operation and maintenance of earth stations, transmitters, and microwave links.
3. Operation and maintenance of community receivers.
4. Operation and maintenance of support facilities, e.g., power systems, air conditioning, transport, etc.

The manpower requirements for research and evaluation and for management are not elaborated here.

The background required of personnel depends on the nature of the work. Usually, supervisory staff should have a degree in engineering and the technicians should have an engineering diploma. Experience in maintenance is an additional requirement. It is necessary in all cases to have these people trained on the equipment they are to operate and maintain. One way of doing this is to include training as a part of the procurement contract. Further, these persons can be present during acceptance testing at the manufacturer's facility; this will give them a better understanding of the equipment. Some manufacturers also hold courses in operations and maintenance.

As SITE was an experiment, the manpower for the operation and maintenance of earth stations and studios were drawn from those who had been responsible for their design, fabrication, and installation. The personnel consisted of engineers at the supervisory level and technicians at the working level.

Supervisory staff also had had training in the manufacturer's facility during fabrication and acceptance testing of the studio equipment. Earth station staff had had experience in the installation of an INTELSAT earth station as well as the design and development phase of the equipment for the various SITE earth stations. Working-level staff were trained on the job. In some cases, special training courses were organized by the suppliers of equipment, e.g., video equipment operation and maintenance.

The direct receiver maintenance staff were initially given a brief exposure to the receivers by the project, followed by a month at the manufacturer's facility to get familiar with circuit details, alignment procedures, etc. Later they were given a specific installation and maintenance course before being deployed in the field.

The receivers had to be operated by villager custodians who had to be given basic training on how to set the controls, make minor adjustments to the antennas, and report faults accurately if and when they occurred.

The HET experiment used personnel with fairly widely varying backgrounds. They had been trained in equipment operations and care, operations protocol, and routine testing.

6.1.4.3 Site selection.

Once the equipment has been selected and the areas where the various facilities are to be located have been chosen, the next job is to select specific sites within the various areas. The requirements of facilities such as studios and earth stations are discussed separately from the requirements of the direct reception system.

Earth stations have to be located in a (radio) noise-free environment. Site selection therefore entails a detailed noise survey of candidate sites. Other factors to be considered are possible interference with other installations such as line-of-sight links, radars, etc. The earth station is linked to its studio via a coaxial line or a microwave link. The provision of this link also has a bearing on the site. Other requirements are roads for accessibility, electric power for running the station, and communication links (telephone/telex).

Terrestrial transmitters could be located in urban or rural areas. Their siting would also depend on the factors mentioned above, though, in general, the problem of radio noise is minimal in rural areas.

The studios could be either located in towns near the earth stations they serve or co-located with the earth stations.

The site selection for the community receivers is perhaps the most time-consuming process. It starts by first selecting the target villages where community sets are to be located. This is followed by a visit to each of the villages to determine the following:

1. Accessibility of the village.
2. A suitable building for the community receiver.
3. Nearest electricity point (except in the case of battery-operated sets).

Accessibility (or lack of it) has a direct impact on maintenance. Though in principle all villages should be served, inaccessible villages will severely strain the maintenance system and hence are best left unserved till the situation improves.

In community viewing situations in developing countries, the suitability of a building to house the community set is decided by three criteria: It should be a public place capable of housing 200 to 300 persons; it should provide security to the installation; and it should have near it a suitable site for locating the antenna.

Although the building selected may already have electricity, it may be necessary to do some additional wiring to bring electricity to the set location. In the case of a building having no electricity, it will be necessary to run wires from the nearest available point.

The problem of power, however, may not be critical if the installations can be operated on batteries or from alternate power sources such as windmills, solar cell panels, etc. During SITE, about 180 sets were operated on batteries.

The major site selection activities for SITE were related to the standby earth station, the rural television transmitter, studios, receive-only earth stations, and, of course, the 2,400 direct receivers. The standby earth station was located at New Delhi to permit national-level program dissemination. However, finding a (radio) noise-free location near New Delhi turned out to be a major problem. Fortunately, an excellent location was eventually found close enough to the studio. Its location in the midst of a "mini-forest" turned out to be ideal, for the trees served as good absorbers of radio noise.

The studios should be located in cities in or near the areas they are intended to serve so that regional nuances can be catered to and local people (producers, actors, subject specialists) could be used for program production. This was done in SITE to the extent that practical constraints and finances would allow. In the operational Indian system, it is intended that studios will be set up within each of the areas served.

The sites for the direct receivers for SITE had to be selected very painstakingly. The target areas were selected on the basis of backwardness and available developmental infrastructure; within these areas, candidate villages were listed. Only electrified villages within a radius of 40 km from a district town (which would serve as a maintenance center) were chosen. These were visited to ascertain accessibility, existence of a suitable building for the receiver, and extent of electrification. More than 10,000 villages had to be visited before the final list of 2,338 villages was selected. In many cases, criteria like year-round accessibility and 40 km distance from the nearest town had to be dropped. Criteria of maintainability had to be compromised if a sufficiently large population was to be served. Many villages had electricity, not in the chosen building, but kilometers away in the fields to power irrigation pumps. A special crash program had to be mounted to have the authorities connect these points to the chosen buildings. At times, this involved installation of transformers and major changes in the state's rural electrification schedule.

As mentioned earlier, SITE had 180 battery-operated sets installed. These were full-size (24 in.) sets fitted with a AC/DC converter to enable the set to operate off a 24-V storage battery pack. The pack was sized so as to give ten days of operation on one charge, and batteries were recharged every seven days to ensure that delays would not result in a dead battery. The battery-operated sets showed certain interesting features as compared to the mains operated sets. The rural mains voltage fluctuations proved to be a major problem and several failure modes were attributable to this factor. These failure modes were naturally not present in battery-operated sets. Further, since the battery sets were visited once a week for battery replacement, the technician resorted to preventive maintenance which led to better reliability figures as compared to the mains-operated sets. This indicates that battery-operated sets, in spite of their higher initial cost, may be preferable in rural areas in developing countries.

Another interesting experiment was a solar-cell powered television set. Two sets of solar panel power sources were obtained on loan from NASA. These were used to recharge the batteries on site. One such installation was in a heavy rainfall area and the other in a dry area. Both installations worked well. This also could be a future candidate for power sources for community receivers.

6.1.4.4 Acceptance testing.

It has already been mentioned that acceptance testing at the manufacturer's facility should be part of a procurement contract. The test specifications, parameters to be tested, test conditions, criteria for rework/rejection, etc., need to be clearly spelt out and accepted by both parties.

The equipment should again be checked on receipt at the installation site to detect any damage resulting from transportation and handling. After the installation work is over, the full system has to be checked out to ensure that all system specifications are met. For the community set, such detailed checking may not be possible. Here the final checkout would be the reception of signals from the satellite after the set has been installed in the village.

During SITE, equipment testing and evaluation formed a major activity. Purchased items were tested at manufacturers' facilities. Items built by the project were evaluated by independent experts. The direct reception system was tested by NASA, in the United States, using ATS–6 since, during this time, the satellite was not within range of India. All items procured from Indian manufacturers were subject to previously agreed specifications and test procedures. The testing involved 100 percent checks on critical items and selective (10 to 20 percent) checks on others. Testing was repeated on receipt of items and again after installation.

6.1.4.5 Deployment.

The deployment activities cover the movement of the various hardware items from the manufacturers' facilities to their locations for installation. The major deployment activity is that of the direct receivers. This involves, for village community sets, movement in two stages: the first from the factories to staging points; the second from staging points to the village locations. The villages where receivers are to be deployed could first be grouped into clusters around a town or city. These towns form the staging points. Each of these needs to have warehousing facilities and a deployment center office staffed by technicians and provided with transport.

After the receivers are tested at the factory, they are packed and sent to the staging points. Large items like antennas are sent in kit form. At the staging points the items are inspected and repacked. Special teams then take these to the villages and assemble them in the selected locations. The assembly consists of preparing the antenna mount, assembling and mounting the antenna, connecting the antenna to the receiver, and testing the system, using either a satellite simulator or the satellite signal itself, if available.

During SITE, 2,400 direct receivers had to be installed in villages. Installation consisted in preparing the antenna site, erecting the antenna, installing the television set, and handing over the entire system to the custodian. There were four deployment teams, consisting of four persons each, operating in each of the six clusters. It was estimated that each team could complete one installation a day, thus requiring 100 working days or nearly four months to complete the job. In order to avoid a situation in which some installations would remain idle for several months with the attendant problems of damage by pests or improper handling, the deployment was done in two phases. The first phase concentrated on siting, erecting, and pointing the antennas (this was the most difficult part). The antenna pointing initially was a major task since each installation had a unique look angle. It was soon realized, however, that within the tolerance of pointing and the beam width of the antenna, a large cluster of antennas could be pointed in one direction without any danger of any of the antennas being misaligned. The pointing method was specially developed. The ephemeris of the sun was used to find the true north after which the satellite azimuth and elevation were marked and the reflector pointed in the direction of the satellite. Marks were made on the structure for reference in case the reflector got disturbed. The pointing equipment used was adapted from the equipment used by surveyors for plane table surveys. During this phase a parallel effort was mounted to electrify the building designated to house the television set. The second phase was the installation of the television set and a quick check of the whole system. Though the installation phase was conducted by government staff, there was no dearth of helpers. Villagers volunteered help and actively took part even in antenna assembly. The installation thus quickly became "their" installation.

The four teams were quartered in four towns of each cluster and each team was responsible for deployment in villages around the town in which it was located. The receivers were dispatched by road from the manufacturer's facility to the towns. This delivery schedule was co-ordinated by a procurement team located at the manufacturer's facility, the headquarters staff at Ahmedabad and the cluster offices. This was perhaps the most finely tuned activity. Since the field offices had limited storage space, the equipment could not all be stored and then deployed. The *modus operandi* was to get a consignment and begin deployment, and, as this consignment was nearing exhaustion, a fresh consignment would arrive from the manufacturer.

Obviously, problems with regard to deployment are likely to be minimal in developed countries. However, most developing countries will probably face situations similar to those in India, and the SITE experience would be especially meaningful to them.

Deployment will need to be supported by a logistics group. Apart from the movement of personnel, this group will also have to plan the movement of equipment from manufacturers' facilities to intermediate staging points and to the final locations. The schedule for movement should also take account of the schedule of completion of site selection, site preparation, etc. In SITE, these activities were supported by a very efficient logistics group. This group formed the vanguard—proceeding to each point, arranging for accommodation, setting up offices, organizing transport facilities, establishing links with local authorities, recruiting and training administrative personnel, etc.

6.1.5 System Operations

Operation of an instructional satellite system will consist of the following: (a) scheduling; (b) maintenance; (c) performance monitoring.

6.1.5.1 Scheduling.

Broadcasts are scheduled according to the time zone and the target audience. In general, programs for audiences located in the eastern part of a large country have to start earlier than those farther west. School programs generally have to be broadcast during the morning while general programs could be broadcast in the evening. Thus, a fairly large country (the USSR is an excellent example) could keep a satellite busy most of the day. However, there are periods when the satellite will be required for ranging and station-keeping. Eclipse periods will also require reduced operation to conserve on-board batteries.

When earth stations are not transmitting programs to the satellite or receiving programs for rebroadcast, time has to be set apart for earth station alignment, periodic performance checks, and overhauls. The prime earth station is required to work round the clock.

The studios provide two services: production of programs and origination of programs being transmitted. Program production involves both studios and mobile facilities made available to different users. The origination of programs requires mainly playback facilities, announcers, and standby facilities for transmission. Scheduling has to allow for all this and for routine maintenance and performance checks.

The schedule for SITE involved the use of the ATS-6 satellite for four hours a day—$1\frac{1}{2}$ hours in the morning for schools and $2\frac{1}{2}$ hours in the evening for general audiences. In addition to this, about half an hour was available for system checks before the program transmission. The program production centers were geared to provide about 20 hours of material a week and a week's program material was always on standby to meet any disruption in the production schedule. Since the studios and earth stations were dedicated to SITE alone, their scheduling posed no major problem. The maintenance of the village reception sets, however, had to be carefully planned so that at any time 80 percent of the sets were working and so that individual sets would not be allowed to remain "dead" for longer than 15 days.

6.1.5.2 Maintenance.
The maintenance of major installations such as earth stations and studios is routine and need not be dealt with at length. However, it is important to note that periodic checks and preventive actions are important for reducing downtime.

The maintenance of the receivers is a major activity as it involves effort spread out geographically. It is also an extremely important and difficult one, especially in developing countries. In general, the maintenance has to be done in the field so that the downtime of the receivers is reduced. The activity is hampered by problems such as the limited test equipment and spares which can be carried and also the uncertainty of rural power supply. A suitable procedure is to do module or card level repairs in the field. Here the technicians carry with them a satellite simulator and spare modules for the set. They repair the set by module replacement. The modules are then taken to a laboratory located in a town where they are repaired for re-use. In order to meet the desired system reliability figure for SITE, a large number of field service centers were necessary. If each service center had required a comprehensive set of test equipment and inventory of spare components, the costs would have been prohibitive. To meet this clash of requirements, a two-tier maintenance system was selected.

The first tier consisted of a fully equipped central service laboratory capable of dealing with detailed fault-finding and repair down to individual component level. This laboratory served a large number of field service stations and was equipped with test equipment and a spare components inventory. The second tier consisted of the field service stations which dealt with fault location and repair at a module level. These stations were equipped with simple test equipment and they carried an inventory of spare modules.

The hub of each group of 100 villages was the field service station, so there were four field service stations per cluster of 400 villages. One of these stations was designated as the cluster headquarters (CLHQ) and was the control center for the maintenance activity in the cluster. The central laboratory was staffed and equipped by Electronics Corporation of India Limited (ECIL), the contractors for the supply of direct reception equipment. This laboratory serviced the television sets.

Maintenance of the front-end converters required special test equipment, but from cost considerations it was not possible to equip each cluster headquarters for this. Therefore, servicing of converters was provided at only two locations: the New Delhi earth station where the necessary equipment was available and the factory at Hyderabad where the converters had been fabricated. Figure 6.1 shows the SITE field maintenance system.

The rate of failure of direct receivers was expected to be about two per 100 sets per day. It was felt that this workload could be handled by one technician. Thus, each field station had four field technicians, one for each 100 village installations. In addition, one extra field technician was posted in the cluster headquarters as a leave-relief man. The cluster headquarters was controlled by an engineer-in-charge assisted by a senior technical assistant. The converter maintenance center was staffed by one engineer and two technical assistants.

Each field station was equipped with a jeep for field maintenance and for transporting unserviceable modules and sets to the cluster headquarters. Transportation of faulty converters from the headquarters to their maintenance centers was by air and railway.

The fault reporting system consisted of simple prepaid postcards showing

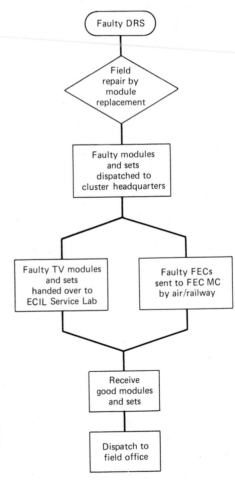

Figure 6.1 Flow diagram for DRS maintenance in a cluster. DRS stands for direct reception system and FEC stands for front-end converter.

illustrations of likely failure situations, e.g., antenna damaged, no electricity, or antenna O.K., electricity O.K. but no sound or no picture, or a totally dead television set. Appropriate legends were included in the local language. The set custodian was required to keep a log of set operation.

The maintenance system, in spite of detailed planning, faced severe problems. Although the sets were designed to withstand voltage fluctuations of ± 15 percent, the rural electric supply fluctuations occasionally exceeded these figures. At times, problems such as a "floating neutral" incapacitated the sets. There were even cases of fire due to excessive high voltage. High winds and heavy rainfall during the monsoon season resulted in damage to the antennas. Roads became impassable to the extent that jeeps got stuck in the mire. The spares stock turned out to be inadequate and had to be increased substantially. Incorrect operation of the sets also led to set failure. Often, the brightness control was set too high, which, in spite of limiting circuitry in the sets, led to set failures. A television program on how to correctly set the controls had to be repeatedly transmitted before this problem was brought under control.

In spite of all these problems, the receiver reliability was raised from a low 60 percent at the start of SITE to about 85 percent in about three months. This indicated that once the true problem was understood, remedial measures solved the problems rather quickly.

6.1.5.3 Performance monitoring.

In any operating system, periodic monitoring of performance is the only way to assure the health of the system. Gradual degradations can be corrected so that only sudden failures have to be handled on an emergency basis. The performance checks of the transmitting system were divided into two classes: daily warm-up checks and periodic (fortnightly or monthly) specification checks. The daily warm-up checks are functional checks designed to ensure that the system components are operating nominally. These include a few voltage and input–output signal transfer checks and are done prior to the operation of the equipment. The periodic checks, which require a few days, are more in the nature of subsystem and system specification checks, e.g., signal-to-noise ratio, video bandwidth, etc.

The receiver situation is a little different since such checks are not possible at every location. Here a few sample sets located at the maintenance centers could be monitored. The sets located in villages would be checked during the periodic visits to service the sets.

As an indirect indicator of the health of the sets in the villages, two techniques can be adopted: records of sets that have failed and have not been repaired and monthly records of set operation from the village set operator.

Since SITE was an experiment, a very elaborate system of performance monitoring was used. Each earth station and studio subsystem had its own operations log. The overall performance of each earth station and studio was also recorded for each transmission. Some individual system and subsystem performances were checked daily, while others were done once a week or fortnight.

Apart from this, all system links, including the links to the direct receivers, were checked out daily. The performance of the maintenance system was monitored through a system of service record sheets and logs. Thus, each service call was completely documented. This data was later used to evaluate the system. The overall reliability of the village sets was also monitored by two independent sources every day. Reports came in once a week in one case and once a month in the other. This provided a very accurate picture of system reliability.

The performance was found to be as per the specifications. Picture quality was good for the community sets and excellent for the earth stations. Audio was also very good and the two-channel system worked well. The earth stations had reliability figures in excess of 99 percent and the receiver reliability was 85 percent after an initial low of 60 percent in the first month of operations.

The HET system receivers also provided good to excellent picture quality and reliability of about 99 percent.

The ANIK–B 12 GHz experiments have shown that relatively simple terminals could yield picture quality better than any received from retransmission by terrestrial relay stations.

6.2 OVERALL SYSTEM DESIGN

6.2.1 Introduction

Although there are a number of constraints and regulations under which a satellite system has to operate, even within these constraints there is a very wide range of choices to be made with regard to a number of parameters. It is these choices that result in tailoring the system to the specific needs and resources of a country.

The system design of an instructional satellite system is a complex process involving a large number of variables and trade-offs. The parameters involved include those relating to the satellite, to the ground hardware, to the program design and production system, to the maintenance system, and to the overall organizational structure. While some of the trade-offs involve straightforward technical economic choices, others are qualitative and can be resolved only on the basis of "subjective" preferences. In order that such "subjective" choices be made, it is necessary to define a value frame that would enable relative preferences to be established.

System design has to take account of political, social, and cultural factors in addition to the more obvious technical and economic factors. This section outlines the major parameters to be considered in optimizing system design and provides some real-life examples of the choices involved. Since the process is necessarily iterative, a serial description has its limitations; therefore, the attempt is merely to highlight the critical variables and steps.

6.2.2 Choice of Service Areas

A basic parameter in designing a satellite system is the choice of the areas to be served. This could be parts of a country, a whole country, or a region including two or more countries. A satellite system can, in theory, serve an area as large as one-third of the earth's surface. However, this implies that power is transmitted from the satellite in a very wide beam, leading to a low power flux density on the ground and therefore requiring large, sophisticated and expensive receiving systems. Such wide-area coverage does have its advantages and uses— for international telecommunications, for example. The INTELSAT system is an example of such wide-area coverage.

Focusing the satellite beam to a smaller area helps to reduce the complexity and cost of individual receiving systems, but it increases the cost of the satellite, requiring a trade-off to arrive at an optimal solution. But there are other considerations with regard to the service area. For example, is the satellite to be shared by a number of countries? If so, this would determine the total service area. There are then further choices to be made about whether a common program is desired or whether country or area specific programs are to be broadcast. If the latter, the question arises as to whether or not these can be staggered over a period of time. If they cannot, one has to either plan for a multichannel satellite system or for separate "spot beams" to service each area. Similar choices and decisions have to be made with regard to serving different areas within a country. Given the magnitude of investments in a satellite system, it is clear that a very small service area will not be economical. For this reason, many of the smaller countries are thinking in terms of a regional satellite, i.e., a satellite that will serve many countries. Studies for various regional systems have been carried out, including those for South America, the Arab countries, the Nordic countries, Africa, and Europe. Until recently, regional systems could be considered only for those countries using the same language and where reception of a common regional program was politically acceptable. However, technological advances have made possible the concept of "spot beams" allowing a single satellite to provide separate coverage to different areas or countries through independent beams. Thus, not only can each area have separate dedicated channels, but the program of one area will not cover another area (except as unintended spillover).

A country must therefore first examine whether it would like to be part of a regional satellite system or have a dedicated satellite for its own use. The decision variables here would include:

1. Size of the country.
2. Potential number of receivers and audience.
3. Single-mission or multimission satellite (this has a major effect on national cost allocation and is addressed separately in Section 6.3).
4. Social, cultural, and political factors.

While a detailed analysis will have to be performed for each situation, it is clear that dedicated satellites are more suitable for large countries having large populations and that a multimission spacecraft would reduce the cost allocated to the instructional system.

Ideally, in national systems one would like to utilize the satellite system to serve the whole country. But economic, logistic, and other constraints often necessitate selection of only some areas within a country for satellite service. Even if the whole country is to be served, there is the matter of timing, and therefore priority areas will have to be chosen. Such choices are not easy, especially since it means favoring one area over another. It is, however, possible to lay down certain criteria for selection. These criteria could be based on the following variables:

1. Present instructional and communications infrastructures.
2. Extent of "backwardness" (or underdevelopment) in relative terms. Use of various socioeconomic and educational indicators may be in order.
3. Physical infrastructure (roads etc.). This is important from the point of view of maintenance.
4. Potential and feasible number of receivers, taking into account the number of villages/towns, extent of electrification, etc.
5. Likely receiver density, so as to determine the most economic mode of serving the area—satellite direct broadcast, satellite rebroadcast, or terrestrial.
6. Cultural and linguistic homogeneity/heterogeneity within the area and with adjoining areas.

The above are more or less self-explanatory, but the last one needs some elaboration. An area having considerable cultural heterogeneity needs a number of different channels. If such an area is itself small, then satellite direct broadcasting is not likely to be the best choice. One example of such an area is the northeastern part of India, which consists of seven distinct areas in an area of 255,000 sq km. Each of these areas uses a different language (and has many dialects) and is culturally distinctive. Therefore, in planning for INSAT, it has been recommended that this region should be served by terrestrial television transmitters broadcasting their own programs. These could, of course, be linked to the satellite through reception terminals for other programs whenever desired.

An example of choosing areas for satellite direct broadcast is provided by India's SITE. The criteria used for SITE were: (a) backwardness, (b) continuity of television service in the area after SITE, (c) maximum common agro-economic conditions, and (d) availability of matching facilities and infrastructure on the ground so that the aspirations or expectations created through television programs could be satisfied.

SITE was an experiment; it may therefore be interesting to note what criteria were used for selecting areas for the operational Indian (INSAT) system. Though this is as yet tentative, the criteria presently being considered are: (a) remoteness, with poor communications infrastructure; (b) support to ongoing extension programs, especially in backward areas; (c) former SITE areas not already covered; (d) contiguity of selected clusters; and (e) logistics.

In both the above instances, it is important to note that the number of areas had to be limited. The constraints were:

1. Satellite availability.
2. Audience availability
3. Extent of audience/area diversity.
4. Number of channels available.
5. Desired minimum duration and frequency of broadcasts.

In the case of SITE, the ATS-6 satellite was available to India for four hours a day ($1\frac{1}{2}$ hours in the morning and $2\frac{1}{2}$ hours in the evening) and had only one video channel with provision for two simultaneous audio channels. Six areas were chosen, involving the use of four different languages. Since there was also a "common program" in the national language (Hindi) for 30 minutes each evening, some of the areas received only 30 minutes of programs in their language in the evening and $22\frac{1}{2}$ minutes in the morning.

INSAT will be available on almost a full-time basis. However, when one considers audience availability, the evening transmission will have to be limited to about three or four hours. Even with two video channels, this constrains the number of areas to be served, especially if one assumes a daily transmission of a minimum of 45 minutes to one hour duration per day to each area. Thus, even in the INSAT system, it is unlikely that more than six areas will be served by direct broadcasts.

However, as a counterpoint to this, it should be mentioned that concepts like frequency re-use, spot beams, bandwidth reduction, large space platforms, etc., may make it possible to broadcast a large number of channels from a satellite, thus removing many of the constraints mentioned above. In fact, many of these technologies are already in use; examples include the ANIK–B system of Canada and the proposed Nordic system. The latter proposes, as one alternative, the use of two spot-beam satellites (plus one in-orbit spare), each of which can provide five television channels to the eastern sector (Denmark, Finland, Norway, and Sweden) and three channels to the western sector (Iceland, Greenland, and the Faroe Islands). Each channel will be accompanied by stereo sound and a number of commentary channels in mono. A teletext system will be used for subtitles.

6.2.3 Choice of Medium

Once a target area has been chosen, it is necessary to decide on the medium of communication and on the program content. As already mentioned, these are basically iterative and serial tasks. However, one can mention the various factors that go into the decision regarding the medium. Of course, in this context we are considering only the various electronic means and not interpersonal communication, mail, etc. Thus, the choice is basically restricted to one (or a combination) of the following: (a) radio, (b) television, and (c) interactive means.

Some studies have been done comparing radio and television as a means of education. There is some controversy about the relative cost effectiveness, and the assumption that television—being both visual and aural—not only has greater impact but is also more successful in attracting larger audiences may not necessarily be true. However, while experiments have been carried out with direct television broadcasting via satellite, there is as yet no satellite for direct radio broadcasting, though studies on radio broadcasting via satellites have been carried out (8), and, in fact, the recent World Administrative Radio Conference (WARC) has also recommended certain frequency bands for further studies in this field. It is possible to conceive of satellite radio broadcasts playing an important role in future instructional satellite systems.

In discussing radio, it should be mentioned that there is worldwide a tendency to greater localization in radio, i.e., radio stations serving small areas and even special audience groups. Such a trend would seem to run counter to the concept of satellite radio broadcasting. At the same time, there is increasing interest in using satellites for network distribution of radio programs. For example, India proposes to use INSAT to distribute programs to all its 82 radio stations by installing receiving terminals at each of the stations. Capability for transmission to the satellite will be provided at four stations. Another use of audio channels which has evoked a great deal of user interest is interactive learning, in which a one-way broadcast (television or radio) is followed by audio interaction between the learner (receiver) and the teacher (or source). This is achieved by installing equipment with "talk-back" capability at some (or all) receivers. Such experiments have been tried out in the United States, in the South Pacific, and in Canada. One example is the experiment on curriculum sharing by digital television carried out between Stanford University (United States) and Carleton University (Canada) using CTS. The satellite provided two-way television between the two institutions and used digital compression techniques on narrowband channels. In one mode of operation, two-way audio and one-way video made it possible for students in remote locations to participate in classes. In a second mode, two-way audio and two-way video made teleconferencing possible between the two universities.

Choice of a particular medium or means of communication depends on various parameters, including the following:

1. Size, level, and type of audience.
2. Messages to be conveyed and objectives.
3. Mode of reception: community or individual.
4. Existing infrastructure for television, radio, and telephony in the area.
5. Investment constraints.
6. Physical constraints in terms of power, viewing and listening space, etc.
7. Maintenance.

It must be stressed that while it may not always be possible to put all the desired services on a satellite, on the ground one need not be restricted to choosing only one medium. This is an important consideration, since even limited experience has indicated the great advantages of a multimedia approach for education and instruction. Thus, one can conceive of systems in which television programs are received via satellite and radio support is provided through conventional terrestrial radio stations. In fact, exactly such an approach was used with considerable success in India during SITE for training of schoolteachers. The Ontario Educational Communications Authority (Canada) used the commercial telephone network for audio interaction while using one-way television via CTS as part of its Tele-Academies project. Therefore, one needs first to look at the most appropriate multimedia combination in a given situation and then to examine the best means (terrestrial or satellite) for providing each of the media services.

Nevertheless, it must be mentioned that such an ideal approach is rarely feasible. First, there is no known formula for arriving at an optimum media combination. Second, the desirable combination (if one could be precisely identified at all) may vary from area to area. Therefore, one has to look at the total service area of the satellite and decide on a best-approximation basis what capabilities the satellite should have. Clearly, a great deal of cost analysis is also required before a decision can be made; but, with current costs and technologies, a satellite dedicated to direct radio broadcasting does not seem to be an economic proposition. Almost all instructional satellite systems would therefore include television; in addition, they may or may not have radio and/or talk-back capability.

6.2.4 Definition of Message Content

In any instructional or learning/teaching system—whether satellite-based or not—a clear definition of the program content is an essential primary requirement. The message may vary depending on the choice of the area and will greatly influence the choice of the media to be used because certain media are best adapted to certain types of material.

In this context, it is necessary to note that almost all instructional systems can carry a variety of program material. Thus, even if a system is initially designed for a single purpose, the very existence of the system will invariably lead to its usage for dissemination of other material. Despite this, it is well to define the basic purpose of an instructional satellite system in terms of program content and target audience. Such clear definition is especially important for the ground segment decisions.

Thus, a system meant as an in-school teaching aid will obviously require that the receivers be placed in the schoolroom, while a system aimed mainly at adults can be more flexible with regard to receiver placement (assuming broadcasting basically to community sets). For example, the television sets for the Kheda rural television project in India are almost all in the community center or milk co-operative buildings since broadcasts are made only in the evenings and are not directly school-linked. During SITE, on the other hand, most of the sets were installed in schools, since $1\frac{1}{2}$ hours each morning were devoted to broadcasts meant specifically for schoolchildren. In the evenings the adults (and children) generally brought the set out onto the verandah of the school for viewing by the large audience sitting outside in the open. In Canada the primarily technical experiment in direct-to-home satellite broadcasting (using the ANIK-B) has 110 terminals located in small communities (for rebroadcast) and in private homes.

The program content therefore must be determined and spelt out. Important considerations in doing this are:

1. National priorities and policies.
2. Needs of the target area.
3. Potential and limitations of the selected media.
4. Existing infrastructure.

Using these considerations as a basis, one can determine the basic content of the instructional messages. This would include a clear definition of the fields and topics to be covered and the target audience for each message. Obviously, a great deal of research is necessary to arrive at proper answers in this field, and this research, in order to contribute usefully, must begin way ahead of all other plans.

Incidentally, program definition may affect not only the choice of hardware for program production but even the design of the studios. It will also affect the type of training to be imparted to producers and to "utilization assistance" people in the field.

6.2.5 Number of Channels

In simplistic terms, the number of channels depends on the information to be carried, the capacity of the channel, and the time available. In real life, however, a number of other variables have to be considered. Among the more important ones are:

1. Effectiveness of the particular channel (medium) for the given message and audience.
2. Extent of linguistic, agro-climatic, cultural, and other heterogeneity in the target area.
3. Audience and reception contraints, e.g., availability of audience only at a particular time of the day.
4. Extent of repetition required because of message complexity or other reasons.
5. Technical and economic considerations.
6. Support available through other media or means.

Some of these are not easy to assess and only practical field experience in the given situation can help to define the particular variable. For example, it is not possible to assess the first factor if the particular medium is being introduced to the audience for the first time. But it may be feasible to plan explicitly for the last, i.e., to ensure a specific type and amount of support through other media.

Technical and economic considerations will also play a prime role in determining the number of channels. Television channels are much more expensive than radio channels, but the beneft may also be greater. The only satellite used extensively for instructional broadcasting, ATS-6, had only one television channel for the SITE project in India, though two audio channels could be simultaneously broadcast with the picture. Since such broadcasts require transmission of high power from the satellite, one major constraint until now has been the limited power available on the satellite. However, with larger structures being launched into space and with increasingly efficient solar cells, generation of sufficient power is no longer a serious contraint.

A more important constraint on the number of television channels is the availability of radio-frequencies. Radio-frequencies have been and are being allocated (or planned) by the International Telecommunication Union (ITU) through its administrative conferences. While new technologies of various kinds have made it possible to put more channels within a given bandwidth, there is necessarily a limit to this. Thus, frequency allocation itself could be a constraint on the number of channels. For example, in the proposed NORDSAT system, the number of channels in the western sector is limited by international frequency allocations. Also, one has to keep in mind the increasing cost and complexity of a multichannel ground segment.

While SITE used only one television channel, the Indian operational satellite INSAT will have two channels. The experimental CTS satellite had two television channels, and the Japanese had one channel for their experimental broadcast satellite. ARABSAT is presently configured for two channels. Most currently operational satellite systems (PALAPA in Indonesia, the United States domestic satellite system, and INTELSAT) do not have any direct broadcast capability for radio or television. All of them do have capability for network distribution and program exchange for television. The Canadian ANIK-B will be used for experiments in broadcasting directly to small terminals.

As mentioned earlier, use of satellites for instructional radio programs has so far been minimal, and none of the systems planned at present will use it in a major way. However, two-way radio capability has been used in some experiments. In India, use of a single video channel with two-way audio capability for instructional purposes will be tried out, using the experimental APPLE spacecraft launched in June 1981. Examples of the use of audio only are those by the University of the South Pacific project (using ATS-1) and the Wa-Wa-Ta satellite radio project in Canada (using CTS). A large number of other experiments using two-way audio in conjunction with one-way or two-way video were also carried out using the HERMES/CTS.

Thus, although large numbers of channels from spacecraft may be desirable for instructional purposes, practical considerations (including technical and economic considerations) constrain the number. As a result, all existing systems have only a small number of channels. While advances in technology will make it possible to increase the number of channels considerably, each country will have to take account of its own needs and constraints to decide on the number that are required. It is clear that with new (foreseeable and feasible) developments, the constraint on the number of channels will no longer be technological. In fact, there are already proposals for a satellite with 1,500 television channels, along with interactive audio and even limited two-way video, for education.

In this context, one major consideration not mentioned earlier (and one that may well be the major constraint) is the program-production capacity. Experience in all countries indicates that the production of high-quality instructional programs is a very difficult time and resources-consuming task. This is true of radio as well as of television programs. Therefore, apart from the considerations mentioned in the preceding paragraphs, the decision on the number of channels will be influenced in a major way by the program production capacity and capability.

6.2.6 Intensity of Coverage

One final variable that is an essential part of the overall system design is the planned intensity of coverage in terms of the proposed number of receivers within a given service area. As mentioned, this number plays a crucial role in overall system planning and is directly related to the trade-off between the relative complexities of the space and ground segments. All other things being equal, a system intended to serve a large number of receivers can afford a more expensive space system than a system intended to serve only a small number of receivers.

The intensity of coverage will also have an impact on the organizational structure of the program production and receiver maintenance elements, besides, of course, on the support and utilization activities. In determining the intensity of coverage, two basic strategies can be followed. The first involves a "saturation approach" in which the receivers are deployed in as concentrated an area as possible. The second strategy can be characterized as an "extensive-equitable" one in which the receivers are dispersed widely so as to cover all states or areas. In reality, the ability to implement either of these extreme strategies is limited in most countries. In developing countries, physical constraints such as availability of roads (for maintenance of receivers), electricity, and schoolrooms or public building for the community set, etc., are major limiting factors. Thus, a full-saturation approach does not seem feasible. At the same time, some of these very constraints (e.g., transportation problems) make mobility and thus widespread maintenance difficult, thereby negating very wide dispersal of sets. As a result, a more or less middle path has to be followed.

In India, SITE adopted a mixed mode by selecting (for experimental reasons) widely dispersed areas, but then installing the television sets in "clusters." Thus, there were areas of concentration, but these areas were widely dispersed. Of course, even each "cluster" area was far from saturated—only one village in ten in the selected area had a television set. This was less by intent than by force of circumstances, due to the low level of electrification, limited availability of public buildings, etc.

In the operational INSAT system, the philosophy will be similar, with the aim being to locate television sets in "clusters," which themselves are dispersed. In each area, saturation will be attempted to the extent feasible, i.e., within the limitations imposed by availability of electricity, public buildings, roads, etc.

In considering wide dispersal, it has to be noted that in many situations this means catering to different linguistic and cultural groups, often with different agro-climatic, educational, and health conditions. Therefore, instructional programs to each of these areas may require different channels or time slots. For example, SITE catered to six different areas of India, requiring broadcasts in four languages and resulting in each area's getting only a fraction of available satellite time for its programs. Fewer areas and more saturation (if at all feasible) could have resulted in greater broadcast time per area. Of course, since SITE was an experiment, there were obvious advantages to selecting a number of areas so as to study the impact under different cultural, linguistic, and other conditions.

The actual choice about intensity obviously has to be made on the basis of the situation in each country, and so the above discussion can only serve as an indication of the factors that may be relevant.

6.2.7 Management and Organization

As in any venture, the most careful planning can come to nought without proper management and organization. In an instructional satellite system, the very nature of the system and the sophisticated technology involved require exceptionally good management and organization. Therefore, these should be treated as critical elements of the overall system design.

Social and cultural factors are prime determinants in the design of management systems and organizational structures. Therefore, just as the hardware should be custom-configured for the needs of each country, the management and organization structure too has to be designed uniquely for each situation.

In an instructional satellite system, the major agencies generally involved are:

1. The agency dealing with space (a separate agency in many countries; in Canada the Department of Communication is responsible for communication satellites).
2. The "education" sector (including agriculture, health, etc.).
3. The broadcasting agency/agencies.

All of these agencies should be involved in a co-ordinated manner in the various phases of planning and implementation.

An important aspect of the organization of an instructional satellite system is the extent and mode of interaction between the education/instruction system and the satellite system. In no country so far has the overall management of the satellite system been vested in the education sector; therefore, means of interfacing have to be found and built into the system. It is imperative that education/instruction experts be involved right from the planning stage of an instructional satellite system. Since a composite organization including these experts and hardware experts does not seem feasible (no country has attempted this), such interaction can only be brought about through joint planning teams.

Some of the organizational options are:

1. To create, within the agency responsible for space activities, a group that understands developmental and educational needs.
2. To develop an understanding of satellite technology in the education sector in each of the key agencies.
3. To create a group that represents the users.

The preferred option necessarily depends on the infrastructure, expertise, etc., available. Thus, in the United States there does not exist a state-owned operational instructional satellite system; the proposed Indian INSAT will be completely state-owned. INSAT will basically follow the second option in terms of organization. Interestingly, in Canada there exists a group of the type mentioned in (3) above: the Federation of Canadian Users of Satellites (FOCUS).

Most of the existing and planned satellite systems that include broadcasting have telecommunication as their primary, revenue-earning payload.

Therefore, the ministry of communication or the common carrier plays the pivotal role in the overall organization. Of course, in countries where it is already involved (e.g., India), the space agency itself plays a major role. Here, planning the system can be done only through the active involvement of the users. In the United States some users got together to set up complete satellite systems (e.g., the Satellite Business System), but these are not primarily intended for instruction or education.

Maintaining some kind of joint team for monitoring hardware development and operation seems desirable, but it may not be practical. Experience seems to indicate that education experts do not have much interest in the development of the hardware component. Therefore, this phase may best be carried out by the agency responsible for hardware (satellite, earth stations, program production centers, receivers).

Apart from the space segment, there is a great deal of hardware (and, of course, software) required on the ground for any instructional satellite system. An important parameter in organizing this is the extent of decentralization desired.

Co-ordination between the various elements involved is very important, the main elements being:

1. Program content planning and production, studio operation, and maintenance.
2. Program dissemination, including the satellite and terrestrial transmitters.
3. Receiver operation and maintenance.
4. Evaluation and feedback.

In most cases, not all these elements will be handled by a single organization. Thus, proper interfacing and co-ordinated functioning have to be achieved through appropriate organizational and managerial mechanisms. Committees, working groups, and other groups have been the most common means of achieving this. However, as stated at the outset, since both organizational structure and managerial methods are necessarily influenced a great deal by sociological and cultural factors, each country will have to evolve structures and methods best suited for its environment. There can be no prescriptive ideals.

6.3 TECHNICAL AND ECONOMIC CONSIDERATIONS

6.3.1 Introduction

This section considers the technical and economic aspects of the feasibility and planning of instructional satellite systems. First, the many variables that have to be considered in planning a system are enumerated and explained. Then the feasibility of implementing a system with available and projected technology is examined. Finally, the planning of a complete system is illustrated with a few examples of planned and operational systems.

6.3.2 Classification of Systems

Instructional satellite systems could be classified into three broad categories. These are:

1. Direct broadcast systems.
2. Rebroadcast systems.
3. Hybrid systems which are a combination of the above.

A direct broadcast system enables simple "inexpensive" receivers to receive signals directly from a satellite. The receivers themselves could be at individual homes or at public places for community reception.

In a rebroadcast system the signal received from the satellite is rebroadcast by conventional terrestrial means for reception by individual or community sets. Rebroadcast can be done by transmitters or television and cable distribution systems. Both direct and rebroadcast systems can be used for radio.

6.3.3 System Considerations

The major parameters to be considered are the signal characteristics, modulation scheme, radio frequency, satellite orbit, coverage area, ground terminal (receiver) size, satellite transmitting power, etc. All these factors are interdependent and one may have to trade off each factor with respect to others to arrive at an optimum technical solution. Also, there is an interplay of cost, available resources, requirements, and technology. An optimum system definition has to take all these factors into consideration.

Signal characteristics and quality objectives. The instructional programs are transmitted as video or audio signals. The video signal could be in the form of normal television images or in the form of alphanumeric or sketch displays on a video monitor or a facsimile recorder. The selection of the particular signal format depends on the program design. The technical characteristics of the system will be determined by the signal formats chosen.

Once the signal characteristics are determined, one has to examine the quality objectives for the system. At the international level, several standards have been in existence, particularly for program exchange. Similarly, in many countries, there are national standards as well. Several studies linking the subjective quality as graded by viewers and listeners to the technical parameters have been conducted.

The major technical parameters that need to be defined are the signal-to-noise ratio and the fidelity (bandwidth, distortions, etc.). These, along with other parameters, could then be used to completely characterize the system. The quality objectives should be derived from the end use. Most of the existing standards are meant for high-quality program exchange or for entertainment. Though a system meant mainly for instructional purposes should have fairly good quality, it could do away with many of the frills associated with entertainment programs. These aspects should be objectively considered as part of system planning.

For example, the signal-to-noise ratio in a television program for exchange purposes is normally taken to be 51 db. But in a direct broadcast system, the quality objective could be based on a subjective assessment as perceived by a viewer, since the system caters directly to the ultimate user. That this value could be considerably lower, say around 42 db or so, is borne out by experience in both India and Canada.

Also, the normal signal bandwidth employed for television transmissions is 5 MHz. But the energy content at high frequencies in a television signal is fairly low and the subjective quality of a picture, as observed by a viewer, may not be adversely affected even if the transmission bandwidth is somewhat less than 5 MHz. Similarly, if the main objective is the transmission of graphic information along with instructional voice messages, slow-scan systems that require fairly narrow bandwidths of the order of only 100 KHz could be employed. The impact of restricting the signal bandwidth on the total system is considerable. Since many countries adopt certain standards for their terrestrial systems, compatibility with these standards must be kept in mind.

Modulation scheme. The most common modulation scheme used for terrestrial radio and television broadcasts is amplitude modulation (AM). AM transmissions, by nature, require very high powers (about 20 times that required for FM) to be transmitted from the satellite if the receivers are to be inexpensive. Nevertheless, AM transmissions are definitely feasible, at least in certain frequency bands.

In a frequency modulation (FM) scheme, one trades off bandwidth for power. Because of the large bandwidth used, an improvement in signal-to-noise ratio becomes feasible. Because of the power constraints on board a spacecraft, the FM system is favored in almost all cases. Because of the complexity and cost, digital techniques are yet to be tried out for large-scale applications. But in specific cases, particularly in the context of rebroadcasting, digital transmission techniques might be feasible. Digital techniques allow for considerable signal processing in terms of bandwidth compression, redundancy reduction, etc., and this may come in handy in reducing the spacecraft power requirements or in reducing ground terminal size.

The transmission of the audio signal associated with the television or slow-scan television signal also needs to be considered. The commonly used methods are to transmit the audio in the form of a subcarrier above the video band or to include it as a separate narrowband carrier. A method of inserting audio channels in the horizontal or vertical intervals of the television signal is also available. The best method will depend on the application.

Frequency selection. The operating frequency of the system is one of the most important parameters in planning an instructional satellite system. The total system cost, the technological requirements, many software related factors, etc., are strongly influenced by frequency of operation. The ITU Table of Frequency Allocations, as disussed in Appendix 6.1, contains several bands for radio and television broadcasting satellites, including bands around 700 MHz, 2.5 GHz, and 12 GHz. Specific allocations for sound broadcasting around 1.5 GHz are under consideration. Choice of an appropriate frequency band will depend on the particular application, and a detailed study of the many technical factors and constraints needs to be carried out in planning specific systems. Some of the main factors that must be considered with regard to frequency selection are: international regulations on power flux density and spillover, propagation phenomena at various frequencies with particular reference to the climatic conditions, technology (particularly for ground receivers), and co-ordination with terrestrial systems.

In the 700 MHz and 2.5 GHz bands, which are shared with terrestrial systems, the maximum power flux density that can reach the surface of the earth is specified in the ITU Radio Regulations. In the 12 GHz band, allocations specifying satellite power (EIRP), coverage, bandwidth, etc., for different countries have been worked out by WARC-77. These allocations and constraints on power flux density determine the ground reception terminal size and complexity and to an extent, albeit indirectly, define the receiver.

One important factor to be considered in choosing the higher frequencies such as the 12 GHz band and the 23 GHz band is the propagation phenomena and the rain attenuation at these frequencies. The higher loss of the signal necessitates greater margins in the signal-to-noise ratio depending on requirements for continuity of service. The margins to be allowed in the case of tropical countries are somewhat higher due to higher rainfall rates and longer rainy seasons.

The receiver technology in the 700 MHz and 2.5 GHz bands are well established. The 12-GHz band technology is rapidly developing, and already simple receivers for satellite broadcast applications are available. The status of receiver technology, as well as the ability of a particular country to adopt the required technology (based on its infrastructure), should be important considerations in selecting the operating frequency.

All three frequency bands mentioned above are shared with other terrestrial systems. Hence, there is a need to carefully analyze the potential interference problems to and from the terrestrial systems and if necessary to adopt co-ordination procedures. In the 12 GHz band there is less terrestrial communication traffic than at 700 MHz or 2.5 GHz, so co-ordination may be easier, though in the lower bands there are also no insurmountable problems.

Orbits. The geostationary orbit is best suited for an instructional satellite system since it enables the ground systems to be simplified and permits continuous service with a single satellite. It is suitable for most countries except those in the extreme north and south latitudes (the USSR, for example, has this problem in its northern areas). However, a rebroadcast system could be implemented making use of a highly elliptical 12-hour orbit with at least three satellites phased in orbit in such a manner that at least one is visible to the ground station at all times. Also, the ground stations must have the capability to follow the satellites in their orbit and to change over from one satellite (as it sets) to the other (as it comes into view from the station). Such systems have been successfully used for many years in the USSR, where they are essential for communications between Moscow and Kamchatka and Chukotka.

For geostationary satellites, position in the orbit is very important. The position is determined by the coverage requirement, interference potential to and from other satellite systems, etc. If the area to be covered is small, then the range of possible orbit positions becomes wider. For regional coverage, if the orbit position is properly chosen, a single satellite could be profitably shared between areas in different time zones. If the satellite position is chosen to the west of the area to be covered, the satellite will experience "eclipse" (i.e., the satellite enters the earth's shadow and the power generated from the solar cells is reduced) after local midnight. This will lead to certain advantages in spacecraft power system design because the instructional system will not usually operate after midnight.

Coverage. The main advantage of a satellite system is its geographic range. A country or an entire continent can be covered with a single system. Smaller areas could also be covered by using narrow-beam antennas on board. Use of several narrow beams in a multibeam configuration permits coverage of different regions by independent beams. In all cases, the spillover to other regions not in the service area has to be carefully considered from technical and other points of view.

Satellite power and ground receiver system complexity. It is almost axiomatic that receivers should be as simple as possible so that the instructional system could have a large number of receivers and thus cater to more people. One basic requirement for this is that the power transmitted from the satellite be high enough so that small and inexpensive antennas could be used to receive this signal. There are two limitations on increasing the received power on the ground: (a) the technological limitations on generating power on board and converting it into useful transmitted energy and (b) power flux density limitations imposed due to sharing of frequency bands with other services. Interference between satellite systems is also a consideration.

Spot beams can concentrate all available power on to a small area; but this involves use of large antennas on the spacecraft. The ground receiver complexity again is a function of frequency, the number of channels to be received, satellite power modulation scheme, etc. Integrated system planning taking into account the cost and maintenance aspects of the ground system will be required for arriving at an optimum configuration.

Reliability and availability. System reliability depends on the construction of the spacecraft and ground receiver hardware and on the maintenance organization for the ground hardware. It also depends on the margins that have been built into the system to handle unforeseen problems. For instance, it is not always possible to maintain the orientation of the antenna to the accuracy desired. Also, rusting of connectors may cause increased noise temperature in the receiver. Since maintenance of unattended or almost unattended reception sets will at best be periodic, these small but inevitable failures should not result in complete breakdown in the reception system. An instructional satellite system meant for use in remote and less developed areas should be capable of degraded operation, and the degradation should be "graceful" (i.e., gradual). But providing margins in a satellite system is expensive and, if not properly balanced, self-defeating.

Interactive system. Whether audio or video, instructional systems can realize their full potential only if they are interactive. The provision of an interactive channel calls for additional hardware. Whereas an interactive two-way video system could be very expensive, a fully interactive system may be realized more economically by providing just an audio channel from the viewer to the program originating center.

The interactive system must be designed with respect to the number of return channels required, type and number of accesses required, whether a conferencing facility is to be included, etc. The larger the total system in terms of its reach, the more complex will be the interactive system.

6.3.4 Spacecraft Configuration and Hardware

This section briefly reviews the status of the satellite and launch vehicle technology which has made instructional satellite systems possible. In the earlier days of satellite communication there were tremendous constraints on the spacecraft weight, power, and volume. Two basic developments have removed these constraints to a large extent. One is the development of large launch vehicles that can put very heavy satellites weighing a few tons in orbit. The other is miniaturization and the use of special light-weight materials permitting the packing of enormous performance capabilities into a relatively low weight and volume.

As stated earlier, the higher the transmitted power from the spacecraft, the less expensive the receivers. Transmission of high radio-frequency power requires a large antenna, a high-power transmitter, and a capacity to generate high power on board the satellite. One option for generating high power could be use of nuclear energy. However, use of solar cells in geostationary orbit is a safer and time-tested method, and present-day solar-cell technology allows generation of adequate power.

Depending on the number of channels provided, extent of coverage, etc., the power generation requirement could be from 500 W to several kilowatts. This requires the use of solar-cell arrays that are stowed during launch and are deployed after the satellite is launched. If the solar arrays are oriented toward the sun all the time, the power generation efficiency is improved considerably. The satellite is body-stabilized on all three axes for this purpose. High-efficiency batteries are used to cater to the power needs of the satellite during the short periods during which the satellite is eclipsed from the sun by the earth.

Large antennas on the spacecraft help to concentrate the energy into a narrow beam and direct it toward the target area. While the effective transmitted power is thereby increased, the area covered is also reduced. Thus there is a trade-off between antenna size and the coverage area. Many average-sized countries could be covered with narrow beams of between 1 to 3°. To give an illustration, for radio broadcasting from a satellite in the 1.5 GHz band, a 1.5° beam that could roughly cover an area of 1,500 km across will require a 10 m antenna on the spacecraft. Such antennas have to be stowed during launch and deployed or unfurled after the spacecraft reaches its orbit. This again requires three-axis stabilization. For covering larger areas, one would require broader beams and hence smaller antennas. It is also possible to generate several beams pointing toward different areas using the same reflector. This concept of multibeam formation will be useful in the context of regional satellite systems or for very large countries.

Techniques have also been developed to shape the antenna beam in such a way that any required coverage contour on the ground can be obtained. This procedure maximizes the power transmitted to the desired area and reduces the spillover to adjacent areas. However, it should be emphasized that some spillover to adjacent areas is unavoidable in any system.

The transmission of relatively high radio-frequency power is another particular requirement of instructional satellite systems. Typically, the transmitted power ranges from 50 W to a kilowatt or more. This compares with the 5 to 20 W of power normally employed in telecommunications satellites. The transmission of high-power levels requires, apart from large solar panels, a proper thermal control system, electromagnetic interference control, etc. Generation of transmitted power up to 200 W or so has been demonstrated. While solid-state transmitters are feasible in the 700 MHz and 2.5 GHz bands, power tubes must be used at higher frequencies.

Use of narrow-beam antennas along with high-power transmitters calls for precise station-keeping and control of the spacecraft attitude. This is because any shift in the spacecraft position shifts the area covered by the beam. The narrower the beam, the more accurate should be the station-keeping and attitude control. Station-keeping and attitude control to a fraction of a degree is now feasible. With high-power transmissions, precise station-keeping is also required from the point of view of interference with other systems.

Another important technological achievement is the increase in the in-orbit useful "life" of the satellites. Since a satellite is an expensive piece of hardware, frequent replacement is out of question. While a life of seven years has become routine, even 10 to 15 years can be easily achieved. The useful life of a satellite is essentially governed by the amount of fuel it carries for corrections in its orbit. The reduction in the power generation capability of the solar cells as time passes is another important factor. Use of highly efficient (high specific impulse) propulsion systems, solar cells with low rates of degradation, solid-state electronics with improved component reliability, etc., have all contributed to increase satellite life. However, apart from the increased expense of achieving long life, it should be kept in mind that because technology changes rapidly, too long a life (and hence continuity of one technology) may not prove to be very attractive in the long run.

The launching of satellites is another important point to be considered. Only a few countries in the world have a capability for launching heavy satellites into geostationary orbit. Satellites weighing around 1.5 tons are commonly launched. The advent of new space transportation systems will allow building and launching of much heavier satellites and probably assembly and repair of satellites in space. Many of the present-day constraints in terms of weight, volume, power, etc., will also be removed and one may think in terms of very large space systems that serve many purposes and many countries simultaneously. The advanced concepts of satellite clusters and large space platforms, while potentially beneficial, have to await further developments, both in technology as well as international co-operation. However, it needs to be stressed that with the available, proven technology, it is fully possible to implement viable instructional satellite systems.

6.3.5 Ground Segment

The ground segment of the instructional satellite system consists of the program production facilities, namely, the radio or television studios, the transmitting earth stations, and a large number of receiving stations. An interactive system will, in addition to the above, have associated transmitting equipment with the receivers.

The basic approach to the planning of instructional satellite systems ought to be the simplification of the ground hardware, particularly the receivers. Since the number of receivers is likely to be very large, their cost will largely determine the total cost of the system. The lower the cost of each receiver, the more likely is the prospect of a large audience for the instructional program. Apart from this, complex equipment is difficult to maintain and service and hence not of much use for the areas with the greatest need for instructional systems. In an ideal situation, the receivers should be comparable in cost and complexity to the radio and television sets that are commonly available. But invariably the satellite system will require additional equipment, and every effort must be made to minimize this and keep the cost low.

The allocated operating frequencies are different from the frequencies that are presently used for ground-based broadcasting. Moreover, these allocations are generally in higher frequency bands, which make a straightforward adoption of conventional receiver technology difficult and, in most cases, impossible. Moreover, even with the use of the highest permissible power from the spacecraft, the satellite receivers have to be somewhat more "sensitive" than conventional receivers.

The ground receiver design should be amenable to mass production. One way of achieving this is to incorporate a front end that converts the signal received from the satellite into a form that could be fed to the standard broadcast receivers. The design should also take into account the environments in which the equipment has to work. In most cases, the receivers have to work in extreme temperatures and humidity, with little or no maintenance. This is particularly true of the antenna and antenna mounted electronics, which will be required of all satellite systems. Further, the antenna may have to withstand high wind speeds. Apart from the physical environment, consideration has to be given to the social and economic environments also.

Reliability and maintainability have to be built into the system by proper design, selection of components, quality control, inspection, and testing during production. This is particularly necessary since no redundancy can be provided in the system. Adopting a modular concept for easy replacement of parts will help reduce downtime. In some cases, the receivers may have to operate on batteries. Depending on the climatic zones, use of solar cells and wind-powered generators could be other alternatives.

Use of large-scale production techniques will certainly bring down cost, probably without a decrease in technical standards. But this may not be applicable in all situations; availability of local talent for maintenance and operation of the system is essential for success, especially in less developed areas. Installation and maintenance could become extremely simple if locally available materials are used.

In addition to the above, two other technical parameters that must be addressed in receiver design are:

1. The trade-off between receiver antenna size and the front-end noise figure (sensitivity). Good, inexpensive solid-state devices (transistors and diodes) with fairly low noise figures are available for operation in the frequency bands of interest, and imaginative use of these would permit minimization of the antenna size. Also, the larger the antenna, the more complex will be the mount and the pointing system, and the greater the logistics problems. The antenna could be an indigenously made item in many countries, but the latest low-noise devices would almost certainly have to be imported.

2. Selection of the receiver type, i.e., the choice between frequency conversion-type receivers and direct demodulation-type receivers. The choice will depend on the number of channels that the receiver should be capable of

receiving, the potential for adjacent channel interference, etc. Present-day technology and interference environment favor the first option.

The program production studios and the transmitting earth stations could be relatively more sophisticated. But even here, in a situation where a broad-based instructional system requires a large number of production units, one has to take recourse to low-cost technologies. Use of transportable terminals in conjunction with mobile studio facilities will enhance the program production capabilities. For instance, a television studio facility could consist essentially of simple video cameras and video tape recorders, basic editing facilities, audio and video processing facilities, and appropriate mixers. All the paraphernalia connected with filming could be dispensed with. The technical quality achieved will be adequate for the purpose of instructional systems.

Similarly, the transmitting earth stations (fixed or transportable) could also do away with some of the frills normally associated with earth stations. A no-break power supply system, which is an essential part of a telecommunications earth station, could be dispensed with and only a backup diesel generator might be provided. In the event of a power failure, the generator could take over within a minute or so. Because of the nature of service, a short break may be acceptable.

If the transmitting earth stations could be located close to the program production facilities, the feeder links to connect them could be avoided. However, the program production facilities may have to be located in urban centers, and the need to avoid interference with the various terrestrial systems may force the earth stations to be located at a considerable distance from the program production facility. This may not be a serious handicap if the production studies are limited in number.

6.3.6 Orbit and Spectrum Sharing

Right now the fixed-satellite service (FSS) is threatened with overcrowding in geostationary orbit. But the broadcasting satellite service (BSS) may also face the same situation soon since many developed and developing countries are now planning broadcasting satellites. For the sake of orderly growth of instructional satellite systems and for providing reasonable opportunities for countries and regions that may decide to launch such systems in the future (as technology and need mature), some sort of planning in orbit and spectrum sharing will be essential. The assignments in the 12 GHz band made for ITU regions 1 and 3 at WARC-BS (1977) is an example. However, there are several technical aspects that need to be considered in such planning. One such factor is the receiving antenna size and its technical characteristics, particularly the sidelobe specifications. The larger the ground station antenna, the better will be its resistance to interference leading to a better utilization of the orbit since the spacing between spacecraft operating in the same band could be reduced. But the impact of such a

TABLE 6.1 Technical Parameters of Some Satellite Systems

	JAPAN (JBS)	INDIA–USA (SITE)	USA (HET)
Satellite designation	YURI	ATS-6	ATS-6
EIRP per channel (dBW)	58	51	51
Satellite antenna	37 dB gain	30 ft	30 ft
Frequency of operation	12 GHz	860 MHz	2.5 GHz
Number of beams	1	1	2
Bandwidth per channel	50 and 80 MHz	40 MHz	40 MHz
Satellite DC power	1 KW	550 W	550 W
Attitude control	Three axis stabilized	Three axis stabilized	Three axis stabilized
Station keeping accuracy	± 0.1° N–S and E–W	± 0.1° N–S and E–W	± 0.1° N–S and E–W
Lifetime	3 years	2 years	2 years
Satellite weight (in orbit)	355 kg	1,360 kg	1,360 kg
Launch vehicle	Thor-Delta 2914	Titan-3C	Titan-3C
Date of launch	April 1978	May 1974	May 1974
Used for	TV broadcasting	TV broadcasting	TV broadcasting
Satellite position	110° E	35° E	95° W
Exclusive/multipurpose	Exclusive	Multipurpose	Multipurpose
Nature	Experimental	Experimental	Experimental
Program status	Ongoing	Completed August 1976	Completed June 1975
Ground receivers			
Antenna used	1 m	3 m	3 m
G/T	—	− 6 dB/°K	8 dB/°K
Noise figure/temperature	—	6 dB	4.5 dB

step on the individual or community receivers may be to make them uneconomical. Similarly, certain bandwidth-efficient digital systems which conserve bandwidth may require more complex receiver systems and hence should be approached with caution.

6.3.7 Typical Planned and Operational Systems

Many of the points discussed earlier may become clearer if they are illustrated with examples of actual operating systems. In examining these systems from a technical standpoint, one could adopt two broad classifications:

CANADA (CTS)	USSR (EKRAN)	INDIA (INSAT)	Consortium of Arab countries (ARABSAT)	INDONESIA (PALAPA)
CTS	EKRAN	INSAT-1, 2	ARABSAT 1, 2	PALAPA 1, 2
59	57	45	43	35
2.5° beam	33.6 dB gain	1.5 × 1.6 M	28 dB peak gain	1.5 M
12 GHz	702–726 MHz	(2.5 GHz)	2.5 GHz	4 GHz
2	1	1	1	1
85 MHz	24 MHz	40 MHz	30 MHz	26 MHz
1,365 W	—	1,186 W	—	319 W
Three-axis stabilized	Three-axis stabilized	Three-axis stabilized	—	Spin stabilized
Within 0.2° E-W	0.5 to 1.0°	±0.1° E-W and N-S	±0.5° E-W	±0.1° N-S and E-W
2 years	—	7 years	—	7 years
346 kg	—	620 kg	—	562 lbs EOL
Thor-Delta 2914	—	Thor-Delta 3910-PAM or STS	—	Thor-Delta 2914
January 1976	October 1976	Early 1982	1982	July 1976
TV and radio broadcasting	TV broadcasting	TV broadcasting and radio distribution	TV broadcasting and distribution	TV rebroadcasting; Telecommunications
116° W	99° E	74° and 94° E	19° and 26° E	83° and 77° E
Multipurpose	Exclusive	Multipurpose	Multipurpose	Multipurpose
Experimental	Operational	Operational	Operational	Operational
Completed	Ongoing	1982 start	Planning	Ongoing
0.6 m	23 dB gain yagi	3.6 m	6 m	10 m (TV)
—	—	8.2 dB/°K	13 dB/°K	—
4.3 dB	450°	4.5 dB	4.5 dB	—

1. Systems broadcasting primarily to direct reception community receivers or individual home receivers. Under this category, one may include India's SITE project, the United States HET experiment, the Japanese broadcasting satellite systems, Canada's CTS experimental system, India's INSAT-1 system, and the USSR's EKRAN network.

2. Systems operating with low power using relatively more complex community receivers and rebroadcasting. Examples of this include Indonesia's PALAPA system, Canada's ANIK-A system, the USSR's Orbita network, the system adopted by several African countries in using the SYMPHONIE satellite, etc.

Table 6.1 summarizes the major parameters of the satellite and ground receivers used or planned. While the technical feasibility of instructional satellite systems has been amply demonstrated, the operational systems tend to combine the broadcast services with other revenue-earning services, particularly telecommunications. This may also represent more efficient use of spacecraft and launch vehicle capacities.

Many of the United States domestic satellite programs, the proposed heavy satellites of Europe, and the Franco-German SYMPHONIE satellite have not been listed here though all of them were, are, or will be used for public broadcasting and instructional purposes to a lesser or greater extent.

6.4 PROGRAM PRODUCTION AND UTILIZATION

6.4.1 Introduction

While all elements of an instructional satellite system are—like links in a chain—equally important, the creation of programs can probably be considered "more equal" because it is the programs that will create the final impact. At the same time, this is probably the most difficult element in the total system, especially because it cannot (or rather should not) normally be "purchased" from other, more experienced countries. Programs must necessarily be area and culture-specific to be effective and beneficial.

Programs can also be used to promote better understanding of cultural integration among a group of countries. Some satellite systems are being planned with this specific purpose in mind, e.g., ARABSAT, NORDSAT, SERLA, etc. At the same time, concern has been expressed about programs from one country being beamed to another without its consent. In Latin America six countries signed the Andres Bello Convention in 1970 reaffirming each country's right to determine "on a basis of full sovereignty, liberty and equality, the content, pattern, production and supervision of the educational programs to be offered to their respective peoples via satellite." The countries also requested, through UNDP, that UNESCO and ITU carry out a study of the feasibility of a satellite communications system for the Andean region with the management, administration, and supervision of the satellite and the system shared on an equal basis.

This section discusses some of the aspects connected with program production and program utilization. The discussion is not limited to in-school instructional programs, but includes "instructional" programs for development.

6.4.2 Program Production

The basic feature of a satellite broadcasting system that must be borne in mind while designing the instructional software system is its ability to cover large and possibly widely separated areas quickly and simultaneously. This feature tends to centralize both program production and transmission. It also exerts tremendous strain on the program production system. The astonishingly hungry monster must be fed regularly with enormous quantities of programs. If the software planning and execution do not keep pace with the hardware implementation, a mad rush for program production results.

In this situation, quality becomes the first casualty and quantity becomes the end-all. The program production authority tends to feel satisfied and even proud if the production targets are miraculously met and time slots filled. Thus, the messages, hastily prepared and transmitted, lack specificity of purpose. Producers, being too busy churning out programs day in and day out, have neither the time nor inclination to respond to feedback from the field.

An advanced satellite broadcasting system demands an advanced software system. It is assumed that the primary target of the instructional programs in developing countries will be the socioeconomically backward population living in remote and isolated rural areas which are usually not well served by agencies responsible for education and development. Program design is discussed here in this context.

The wide coverage of the satellite leads to physical separation of the designers and the receivers of programs, sometimes by more than 1,000 kms. But perhaps a more serious problem is the perception gap between the designer's plans and the real needs, aspirations, background, and beliefs of the rural audience. More often than not, the design team is likely to have very special ideas and orientations concerning the role of the medium itself. While this may not be a serious problem in program production for urban audiences, it becomes significant in the case of rural audiences. The perception gap will considerably influence the producer's attitude and approach in terms of topics and their treatment.

Development itself has different connotations for urban program designers and their rural target audience. For many producers trained to make feature films for mass entertainment, the medium is likely to be more important than the message. Though handicapped by their background and training, the producers, researchers, writers, and experts (who constitute the information source) must be willing to learn and develop an empathy with the rural audience. Unless this happens, the source will lack credibility.

It is here that "formative research" has to play a major role. The aim of this research—carried out by appropriate social scientists (sociologists, anthropologists, psychologists, communication researchers, etc.) in the area to be served—is to sensitize the producers and to help them develop empathy with their audience. Production, at least of instructional and developmental programs, has therefore to be carried out in a "team mode," with each team consisting of a producer, a formative researcher, a "content expert," and a script writer. Practical experience in working in this mode for educational programs exists in many places, including the United States and India.

The program designers will have to acquire background information about the target groups, which should include their psychological makeup, socio-economic conditions, beliefs, superstitions, customs and practices, etc. For this purpose, detailed "audience profiles" will have to be prepared after extensive field work. The audience profile should contain information on climate, population, social structure, customs, beliefs and superstitions, current informational and educational level, economic activity, agricultural practices, animal husbandry, health and family welfare, cottage and small industries, planning and development, social life, etc. The next step would be to establish the perceived and the real needs of the different socioeconomic categories of people in the instructional areas. There should be special emphasis on the relevance of instruction, its utility, practicability, importance, timeliness, and area specificity. Prioritization of topics is another equally important input required for program design.

If the full potential of an instructional satellite system is to be realized, one must view the software system (its design, development, and utilization) in the context of the process of development itself. In developing countries, one has to view it in the context of large-scale illiteracy among adults. Even though primary and secondary education has received very high priority and millions of children are going to school, they are yet untouched by modern methods of education. A satellite could bring scientific methods of education to school teachers to bring their teaching skills up to date. It is important to recognize that while financial and physical inputs like fertilizers, seeds, water, and power are all essential for progress, they nevertheless are not sufficient. One must provide information inputs based on perceived needs of the villagers as well as the motivation of the individual farmer. Ultimately, it is the motivation of the individual that is going to count. It is therefore necessary to learn how to make persuasive programs that will trigger and catalyze development activity and then assess their real impact. The objective is to acquire the ability to use programs for the real needs of development, for uplifting the mind and heart of man, taking him away from poverty—that is the biggest challenge.

If the satellite is going to bring radio or television for the first time to the viewers, then considerable additional effort would be required in preparing guidelines for the producers. In the case of television, information will also have to be obtained to establish the level of visual literacy of the audience. For example, one will have to examine whether or not flashback and flashforward techniques would work. One will have to check and test out whether the instructional messages could be effectively communicated through song, dance, and drama or whether villagers prefer instructional messages without "sugar coating." Thus, pretesting of the programs, before they are broadcast on a large scale, is desirable and even necessary.

It is logical to assume that in economically backward rural areas there will be hardly any individually owned television sets and it will be necessary to have community receivers. Experience in India has indicated the importance of appropriate locations for these community sets; they must be placed in a public building such as a school or community hall accessible to the entire village community. The program designers must also be conscious of the viewing conditions. In the daytime, for school programs, one may have an essentially captive audience and a teacher who ensures that there is minimum disturbance. During the evenings, one may have a composite audience. With such a mixed audience of children and adults, there are bound to be some instructional programs which would be of little interest to children and they are likely to make noise. The designer of the program must depend largely on the visual rather than the audio for the impact of his message. It must also be recognized that even though programs are directed at special audience groups (e.g., students), others may also be viewing or listening to the programs—this is the intrinsic character of a mass medium. Similarly, one must be sensitive to the viewing situation while preparing programs on family planning. In this context, the importance of female researchers in the feedback team cannot be overemphasized if one is to get true and intimate reactions of women in the audience. The viewing conditions in different seasons must also be kept in mind in deciding the duration as well as the frequency of transmissions. One should be aware that the audience may drop to almost zero in the peak agricultural seasons and when conditions are not comfortable for viewing. Data on the real-life situation can be acquired only by proper research on these issues; here also the researcher's role is of great significance.

Feedback from the audience to the program designers is of crucial importance in any communications system. It is of even greater importance in a satellite-based system in which the program makers and the audience are likely to be separated by substantially greater distances than in conventional systems. Ideally, the feedback should be instantaneous (through two-way video or audio); but since this is not always possible, it may be necessary to have teams of professional researchers in the field whose job is to provide feedback on all aspects of the programs, their utility, the viewing situation, utilization, etc.

Since the upper-class people in villages may not like to mix with the disadvantaged class every day to view television programs, it is expected that the community viewing situation is likely to draw, apart from the ever-present children, a majority of viewers from the low socioeconomic strata of society. This has been the experience in the community viewing situation in India. It is, therefore, imperative that a significant proportion of the programs be specially designed for this strata. This is likely to lead, in the long run, to the closing of the communication gap. The socioeconomically disadvantaged viewers are also likely to have more than an average proportion of illiteracy and ill health. The program designers must therefore place emphasis on the use of literacy-independent visuals. They should set aside a sizable fraction of the transmissions for programs on topics such as health, nutrition, family welfare, etc.

The lower strata of society usually consists of those who are engaged in subsistence agriculture—small farmers, the landless laborers, and the unemployed. These sections of the audience may be without work for long spells during the year and could profit from instructional programs that would create awareness about self-employment and income-raising activities. Such programs should be backed up by ensuring availability of resources, raw materials, and a market for such activities. Even then the chances of these deprived people readily coming forward to put income-raising recommendations into practice are rather meager. This is often due to lack of self-confidence and a fatalistic attitude. To combat these drawbacks, it would be necessary to launch a series of programs aimed at building their self-confidence and changing their attitudes. Such a campaign would be a most challenging proposition, the magnitude and complexity of which should not be underestimated.

But even before a program can be designed, one must have a "credo" that will guide the production groups. However, while the credo could be a touchstone—a validating point—the program design will be considerably strengthened if the people in the audience identify the medium as their own, a tool to which they have access for ventilating their grievances, for closing the information gap, and for establishing two-way communication between them and the decision-makers. Their involvement and participation in the very process of program design leads to improvement in their self-image and self-confidence. Such a program design system will not only earn credibility but it will also acquire the status of a friend, philosopher, and guide to turn to in times of emergency, crisis, or need.

The creation of an experimental communications laboratory for ascertaining the effectiveness of various programs under field conditions would be essential. A few programs should be designed and developed on a pilot basis. Then a complete series could be produced after the evaluation process. The laboratory would also serve as a facility for conducting in-depth studies on the specific aspects of communication and development.

This slow experimental process would lead to the development of writers, artists, producers, researchers, and experts from the target area. The target area could then take over the role of the information source. Programs are designed by the people to suit their needs—they will rarely be irrelevant, their language will be completely comprehensible, and their impact will be greater.

If one takes the view that a developing country in a state of economic and social backwardness should deploy the most powerful techniques at its disposal, then different agencies responsible for education, agriculture, animal husbandry, and health should use instructional broadcasting (radio and television) as an extension tool in their promotional effort. If they really succeed in using these media effectively as instruments for development, then one can regard the money spent on them as an investment rather than overhead. To promote their large-scale use by the agencies, a conscious effort will have to be made to bring down the cost of equipment and production of programs, especially in the case of television. It is also necessary to demystify the television medium for potential users. In this context, appropriate technological choices must be made. This may require some indigenous research and developmental efforts.

During SITE (1975–76) for example, 1-in. video tape recorders (VTRs) were used for the first time, resulting in very considerable cost savings in comparison with the then conventional 2-in. VTRs. A start was also made in this period in the use of "amateur" half-inch VTRs. After some modifications and with the use of a digital time base corrector, portable half-inch VTRs are now regularly used for recording broadcast-quality programs. Three-quarter-inch video cassettes are also in use. As a result of these developments, it has been possible to reduce studio cost by an order of magnitude.

An important consequence of these developments is that it is now possible to make programs much more field- and locale-based, and thus much more participatory. This increases relevance, interest, credibility, and impact.

Another consequence of this is that lower costs enable more studios to be set up. This decentralization of program production makes it possible to carry out production closer to the audience and also provides for greater variety and thereby higher quality. It is especially important in the satellite era, since the new technology makes it feasible to broadcast programs to a whole large country with just one production facility and one earth station.

Choice of equipment is an important element in the total software design. An expensive, large studio with sophisticated but immobile equipment will perforce lead to a high degree of studio-based programs. A decentralized production system with a number of small studios equipped with portable equipment will almost necessarily result in a substantial amount of field-based and participatory programs. It will also encourage extension agencies to take up some programming responsibility themselves, rather than to merely serve as "experts" or advisors.

In most countries, considerable effort would be needed to create interest and awareness about the potential of television as an extension tool among the user agencies. Initially, it may be difficult to get them involved because television is often looked upon as an expensive and difficult medium to handle. But recent technological developments have made it possible to use inexpensive and easy-to-operate portable equipment and move out of the suffocating atmosphere of costly equipment-studded studios. The new developments enable the extension agency to do extensive television coverage of its work in the field and bring it to the viewers. This would facilitate the viewers' understanding of the subject and may lead to improvement of their skills and practices.

When the extension agent becomes the program designer for programs in his special field of expertise, there are certain built-in advantages. First, he is in constant and close touch with his clients and knows the actual status of the know-how of the users and what is required to advance their knowledge. Second, the specialist-producer will have the confidence and faith of his clients, since his recommendations will have already produced fruitful results for them. When such well-known experts generate programs, the credibility of the source is established beyond doubt. Another aspect is the timeliness of the information. An agriculture or health expert, for example, knows precisely what topic he should handle at a given point of time in a year. A producer who is not a specialist in health or animal husbandry will have to consult the experts to establish the priority of topics. It is at the same time important for a specialist to realize that he cannot promote extension work via television programs in isolation; he must understand that there is a totality about development and that what is of primary importance is concern for the audience.

The program expert in a particular extension agency also has a better chance of getting proper assessment of the impact of his program since he has his own network of contacts in the field and the feedback on messages could be easily arranged. There will be times when programs have to be produced quickly in response to urgent needs of the audience or in response to emergency situations. It is much easier for the extension agents to undertake this work quickly if they are producing the programs themselves. Extension agents can also easily decide upon the frequency of repetition of the message, because they know the field situation and the viewers' needs and level of comprehension.

6.4.3 Utilization

The use of a medium by an extension agency greatly facilitates the proper utilization of the information. Thus, the extension agency can ensure a proper tie-up between information about farming techniques and the availability of physical inputs recommended. Similarly, in the case of "persuasion" or change programs, the agency concerned can try to ensure the availability of experts in at least a few locations to answer questions and clear doubts. Supplementary printed material like posters can be made available in time. If the extension agency is not directly involved, co-ordination of all this becomes a major and virtually impossible task. As exemplified by the Indian experience in SITE, such co-ordination is especially difficult in a satellite-based system serving far-flung areas. SITE also experimented with providing preparatory and follow-up material to the schoolteacher in connection with the school broadcasts. While this proved very useful, the problems of logistics, in terms of translation and distribution, were tremendous. From the USSR too there are examples of difficulties in ensuring full utilization of instructional programs especially by part-time students. Without proper organization at the viewing end, the programs cannot be of much help to a considerable majority of students taking part-time or correspondence courses.

Another form of utilization assistance that has been tested widely is the post-program discussion group. This was tried in many countries with radio and subsequently with television. During SITE, an experiment was carried out not only to organize such discussions but also to evaluate their impact. The results are very interesting: It was found that when the pre-exposure (i.e., before the program) knowledge was low, most of the later gain was due to the program itself and the contribution of the discussion was minimal. But when the pre-exposure knowledge was high, further gain in knowledge was due only to discussion. The implications of this finding for crucial fields like agriculture, health, etc., are immense.

Utilization assistance should cover not only provision of support material and the organization of discussion groups but also set operation (who operates? should he be paid? by whom?), training of the set operator in operation and possibly simple maintenance, stability of the power supply, determining the optimum set location and viewing situation, etc. As such, utilization assistance is of prime importance, and it is desirable that each field team have not only researchers for feedback and formative work but also a utilization expert.

While the experts in education and in extension work can assume the role of program-makers and thus combine the role of producer, researcher, expert, and even scriptwriter, it is important to broaden the base of program designers. In this context, the role of the various voluntary agencies and people interested and concerned about education and social work must be recognized. It is important to create facilities to allow such people as well as university and school teachers to try their hand at the medium so that those who have special talent for program design can be recruited. Through such conscious effort, the instructional television network would be enriched and would be able to cater to the wide-ranging educational and extension requirements of the people.

6.5 EVALUATION

6.5.1 Introduction

Evaluation has to be an integral part of any system. In the case of instructional systems, it has obvious importance and relevance since periodic assessment of the learner's progress is essential to any educational system. In addition, two other kinds of evaluation are important, especially for satellite-based systems: an economic evaluation because of the very large financial investment and a technical evaluation because the technology is still under development. This section discusses these three forms of evaluation and gives some real-life examples.

6.5.2 Economic Evaluation

In the planning of a total instructional system, two economic questions arise in relation to satellites:

1. Is a satellite-based system more economical than a conventional one?
2. If so, how does one design an optimal instructional satellite system?

Both questions require an economic evaluation. One technique for such an evaluation is cost effectiveness, which involves the determination of cost and the definition of measures to determine effectiveness.

Basically, there are two distinct ways of distributing television in a given country or region. They are:

1. Through conventional television transmitters and connecting links (satellite or terrestrial).
2. Through a direct broadcast satellite.

In addition, combinations of the above are possible. The factors that would generally go into the choice of a particular system are described below.

6.5.2.1 Size of the coverage area.

When the area to be covered is large, the number of terrestrial transmitters needed is high. Building such a system is both expensive and time-consuming. Hence, for coverage of large areas, a satellite is likely to be more cost effective. Studies carried out in India and Canada have concluded that it would be almost impossible to cover these countries in their entirety without the help of a satellite. In particular, if conventional television transmitters are to be linked in a network so as to transmit simultaneously a single program, then satellite methods become even more cost effective, especially in large countries. In the USSR, for example, the use of satellites has permitted the All-Union Central Television Program to reach 93 percent of the population.

6.5.2.2 Number of receivers, audience size, and beneficiaries.

The number of receivers plays an important role in determining the configuration. The cost of a normal receiver (for terrestrial broadcasts) was about one-third of that of a direct-from-satellite reception set for the SITE project in India in 1975. The projections for INSAT (1982) indicate that this ratio would be around one-fourth or less. At the same time, use of conventional receivers requires a large fixed investment for each transmitter, while satellite broadcasting requires an investment of a larger magnitude but in only one satellite. For a given area, the costs of direct reception and terrestrial broadcasting can be compared, based on the expected density of receivers. Wherever receiver density is low, direct reception would probably be more economical. A graph of the costs of the two approaches will have a cross-over point indicating the minimum density of receivers for which a terrestrial broadcasting is more economical.

In Canada, though more than 99 percent of the population has access to at least one television channel through the terrestrial network, vast areas of the country have no television service at all. It is extremely expensive to extend this coverage solely by terrestrial means, and studies have shown that "direct-to-home satellite broadcasting is the only cost-effective means of providing a substantial improvement in television service to rural and remote areas of Canada."

The USSR too had the problem of providing television signals to a number of places with low populations in the north and in Siberia. Setting up the Orbita-type earth stations would be very expensive and time-consuming. Therefore, simple receiving systems operating with the new high-power EKRAN satellites have been set up and the received signal is rebroadcast (after processing) by a 1-W television transmitter.

6.5.2.3 Geographical features.

The coverage of a terrestrial transmitter is dependent on the geographical features of the particular region. For example, hilly terrain can reduce the range of a transmitter considerably. In dispersed islands, such as in Indonesia, the number of transmitters required is quite high. In such areas, extension of conventional communications facilities, especially television, is difficult. Satellite service, however, is not affected by geographical features, and therein lies its advantage.

6.5.2.4 Existing infrastructure.

The satellite-based system becomes advantageous primarily in areas where existing infrastructure is inadequate. In less developed countries where the investment in existing communication links is low, the use of a satellite makes it possible to greatly extend the coverage very rapidly. However, it has been established that even in an advanced country a satellite service has definite advantages, as observed in the case of the Public Broadcasting Service (PBS) in the United States. After several years of study of the available alternatives to terrestrial links between public broadcasting stations, the satellite emerged as having significant advantages over all competing alternatives. In addition to meeting the goals set for local automony, flexibility, quality, reliability, growth, and coverage, the satellite system also proved cost effective because there was (a) the prospect of reducing operating costs for the entire system, (b) the possibility of preventing costs from rising sharply over time, and (c) the opportunity for long-range financial planning.

6.5.2.5 Other services available on the satellite.

In a multimission satellite, the cost allocation for the instructional service could be small, making the system more attractive. In addition to the instructional television service, the satellite can also provide telecommunications, computer interconnection, teleconferencing, remote computer batch-processing, automated remote information retrieval, etc. So far, instructional satellite systems have been implemented or are being planned by large countries (Australia, Brazil, Canada, India, Indonesia, United States, USSR) and by regional groupings (Europe, South America, Arab countries, etc.). Almost all the existing and planned systems include a considerable amount of telecommunications capability as a revenue-earning component. Thus, most are multimission satellites in which the major share of the cost is allocated (actually or notionally) to services other than the instructional one.

The cost effectiveness of a satellite system seems to be particularly great for a large country that has a minimal infrastructure but an urgent need to establish an instructional television system. In other cases, the factors mentioned above have to be fully considered before a decision can be made. The initial investment in a satellite-based system is high. However, if the economics of the total system, which may include other services, is considered, the "return" very often justifies the investment.

Once the decision on a satellite-based system is made, there are other factors that have to be considered in arriving at a system configuration. There are trade-offs to be evaluated between the satellite complexity and the receiver simplicity (relating to factors such as power, pointing accuracy, frequency, etc., of the satellite and ruggedness, low power consumption, maintainability, cost, etc., of the receivers). Whether a given area should have direct reception or rebroadcast has to be determined on the (expected) receiver density, future growth, and other factors. It is also necessary to identify the extent to which local autonomy is essential since the satellite-based system generally tends to centralize the production and transmission of programs. This is an important factor that can sometimes offset cost considerations. However, a satellite can be effectively used for providing program exchange and outside programs to autonomous, local stations. This involves software and policy, rather than economic, decisions.

The limitations on the use of the cost-effectiveness technique for decision-making lie in the measurement of "costs" and "effectiveness." A number of questions and ambiguities arise and it is entirely possible to interpret the same set of data in two (or more) ways and reach contradictory conclusions. This applies not merely to the subjective elements of cost or effectiveness but even to very quantitative ones. For example, a change in the discount rate in a discounted cash flow analysis can alter the ranking of alternatives. Since most of the system components serve other purposes as well, the allocation of costs to "instructional television" tends to be arbitrary. The measurement of effectiveness is heavily based on assumptions and time scales. Despite these limitations, the concept of cost effectiveness is a powerful tool for aiding decision making. It helps to determine optimum system configurations and system parameters and has been extensively used to compare satellite-based systems with conventional terrestrial systems.

6.5.3 Technical Evaluation

Though communication satellites have now been in use for almost two decades, the technologies involved have been changing very rapidly. There are now large satellites with very high power capable of broadcasting television programs directly to augmented receivers. At the same time, new modes of usage—slow-scan television, two-way audio or video—have become available. Clearly, for each new mode and new technology, an evaluation of the technical performance is required, and measures have to be established to define what is a satisfactory grade of service. It is only through such evaluations that improvements and refinements in the system (and especially in future systems) can be made, leading toward an optimum configuration.

In this context, it needs to be noted that the "optimum" technical parameters will vary from country to country, depending on a host of factors. For example, the attenuation of the signal in the atmosphere is a function not only of frequency but also of local environmental conditions; thus, signals in the 12 GHz broadcasting band suffer great attenuation in regions subject to high rainfall rates. Similarly, and more obviously, since ground hardware performance depends on the environment (temperature, humidity, dust, rainfall), the results in one country are not directly applicable to another. Hardware performance may also depend on the organization and management structures, which vary from culture to culture and country to country, since these determine the efficiency of equipment maintenance. Thus, while simple technical results do have global validity, in the complex real-life situation, each country must carry out a technical evaluation of its system.

A technical evaluation of an instructional satellite system should cover the following most important factors:

1. *Reliability* of the total system and of each major component: satellite, transmitting stations, receiving stations, redistribution transmitters, direct reception systems, studios, etc. In the direct reception system, and especially in developing countries where power shortages are not uncommon, a measure of system reliability can be the fraction of the total broadcast time for which the signal is actually available to the audience.

2. *Signal quality* received at various receiving points (this can be defined as the signal-to-noise ratio) and transmitted from various points (the studio, transmitting earth station, rebroadcast transmitters).

3. *Link margins.* Any system design has certain assumptions and certain margins. The evaluation should help to test the assumptions and to determine the adequacy of the margins.

4. *System sensitivity.* In order to arrive at the optimum system parameters, it is necessary to establish the sensitivity of quality, reliability, etc., to changes in input or environmental variables. Thus, one may want to determine the change in signal-to-noise ratio due to changes in transmitting power, bandwidth, increase in receiver noise figure, etc. The effect of changes in maintenance procedures on system reliability is also an important factor.

A proper technical evaluation involves not only the measurement of various parameters but also controlled experimentation for evaluating factors like the system sensitivity mentioned above. Therefore, the technical evaluation of the system has to be planned in advance, and measurements must be made over appropriate periods of time.

While technical evaluations have been very much a part of most of the satellite systems so far, they have been particularly intensive for the experimental projects. For example, the United States HET experiments using ATS–6, the Indian SITE (again using ATS–6), and the Canadian experiments using HERMES/CTS all had very substantial technical evaluation components. From these has emerged a wealth of data on many crucial variables. For example, both the Indian and Canadian experiments have indicated the feasibility of getting subjectively good pictures at much lower signal-to-noise level than the generally accepted minimum. The Indian experience has proven the feasibility of achieving reliability of the order of 85 percent for direct reception television sets even in remote and inaccessible areas. Both ATS–6 and CTS have demonstrated the practical feasibility of building and operating high-power, high-technology spacecraft for direct television broadcasts to augmented sets. A number of other important technical findings in fields such as slow-scan, video-conferencing, audio feedback, etc., have emerged from the ATS–6 experiments in the United States and the CTS experiments in Canada and the United States. The USSR too has developed a vast reservoir of technical experience through its many years of working with communications satellites.

Despite this wealth of data, on which, of course, any proposed system must draw, there is need for continued technical experiments and evaluation. In addition, evaluation is essential to determine whether specifications and performance standards are being met, an important consideration when equipment is bought. This kind of evaluation may have to be done even before a project begins—at the stage of acceptance testing when purchased equipment is tested to determine whether or not it meets the agreed specifications.

Similar pre-project evaluation of project-developed equipment is also necessary and should be undertaken. It may sometimes be necessary to check the compatibility of different pieces of equipment in advance. Thus, for example, the Indian-developed direct reception system was taken to the United States for tests with ATS–6 before SITE began. Similarly, a "spacecraft simulator" sent from the United States was used to verify the compatibility between the Indian earth stations and the spacecraft.

6.5.4 Social Impact Evaluation

In any system, and especially in an instructional system, evaluation of the social impact is proof of the final efficacy of the total system. Obviously, such an evaluation is meaningful only for projects taken up on appropriately large scales and over reasonable periods of time. It would be meaningless, in terms of statistical significance, to study the impact of instructional programs on an audience of a dozen people over a week or two.

SITE included broadcasts meant for schoolchildren and those for a general audience. The latter included programs of "instruction" in agriculture, health, nutrition, etc. Since it was an experiment, evaluation was included from the beginning as an important component of the total system design. The social research for SITE included contextual research (e.g., audience profiles), formative research for program production and process research or feedback, as well as the impact evaluation. Among the techniques used for impact evaluation were:

1. Anthropological studies carried out by anthropologists living in the selected villages for the duration (12 months) of SITE plus three months before and after.
2. Large-scale sample surveys of adults carried out before, during, and after the experiment.
3. Observation studies on student–teacher interaction.
4. Various smaller studies (on sample or in-depth basis) on children and adults.

Apart from the impact of the programs, two important studies covered the impact of a multimedia package for teacher training and the effect of intensive utilization assistance activity in fields like agriculture.

Much of the impact evaluation of SITE was carried out by ISRO, the major participant in the experiment. In general, however, it is advisable that impact evaluation be done by an independent agency, not so much because of the danger of bias, but so that it is seen to be unbiased. The other types of research, especially formative research and feedback, are most effective when done by the program production agency itself.

For impact evaluation, the need and importance of each country doing its own research is obvious. Clearly, social effects and impact are a function of culture, of the socioeconomic condition, and even of the political environment. As such, extrapolation of research findings in one culture or country to another are not feasible. In fact, even research methodologies may require change, or at least adaptation. This is a field in which the extent of scientific findings is still limited. Oley Belotserkovskii reports that in the USSR the examination of the problem "development of scientific foundations of instructional television" is under way. He also notes that for a long time there prevailed a rather narrow-minded approach to instructional television. As a result, a whole complex of important sociological and psychological questions were excluded from consideration.

There are so many environmental and intervening variables that direct cause-and-effect relationships cannot easily be established in this field. In this sense, it is a challenging and virgin field, in which each country, whether developed or developing, has great scope to make its own contribution.

6.6 CONCLUSION

This chapter relates to an admittedly challenging and difficult area. It is also an area of tremendous implication to the future of mankind. Education and instruction involve human interactions and human interfaces. Technology can be only an intermediary. This chapter brings out the great capabilities of the existing and newly emerging space technology for education and instruction. Specifically, the following points are clear:

1. Telecommunications can be used to provide channels for instructional use. Space technology undoubtedly increases the range, density, and flexibility of telecommunications and decreases the costs for reaching some of the thin route, nonrevenue-generating, distant areas where educational inputs are meager.
2. Space broadcasting systems are feasible and have been demonstrated in fairly realistic environments, both in developed and developing countries.
3. It has also been demonstrated that given the right amount of effort in program creation, space broadcasting can be used very effectively for instruction, developmental education, and national integration.

Progress in setting up specially dedicated systems for education has been somewhat slow. This is natural. First, because even the use of conventional telecommunications and broadcasting for education is not as widespread as had been hoped a few decades ago. Second, the education and telecommunications organizations in most countries work under their historical constraints and limitations, while the use of space technology for education is somewhat difficult without new organizational structures. Third, the increased range and reach of satellite signals present genuine problems and challenges for devising educational curricula to suit the needs of different cultural and language groups. Fourth, new mechanisms have to be devised for debiting relatively large initial costs of space systems which, instead of being used as an additional investment in education, can also be used for revenue-earning telecommuncations services.

These are genuine problems, but all of them can be solved. The revolutionary development of space systems could ultimately find an important place in the instructional activity on this planet. Perhaps the beginning should be made by using systems whose cost is shared between the revenue-earning telecommunications services and the education sector. This would also enable these two well-organized sectors around the world to evolve structural relations necessary for exploiting the full potential of space systems for education.

Finally, an explanation. This chapter contains a great deal of information arising out of the SITE project. This is mainly because SITE was the largest satellite instructional experiment conducted so far—and that too in a developing country. It is felt that some of the successes and travails of that experiment would provide useful insights to those, particularly in developing countries, who might be considering using space technology for education and development.

ACKNOWLEDGEMENTS

This chapter was prepared with the assistance of an international team of experts headed by Prof. Yash Pal of India. Other members of the team were Academician Oleg M. Belotserkovskii (Union of Soviet Socialist Republics), Dr. Anna Casey-Stahmer (Canada), Dr. Delbert D. Smith (United States), and Dr. Teofilo M. Tabanera (Argentina). Additional assistance was provided to Prof. Yash Pal by Mr. E. V. Chitnis, Mr. A. R. Dasgupta, Mr. K. Narayanan, Mr. N. Sampath, and Messrs. Kiran Karnik and Narender K. Sehgal, who traveled around the world in order to co-ordinate the submissions of the various contributors.

APPENDIX 6.1: INTERNATIONAL REGULATIONS AND CONSTRAINTS

The radio-frequency spectrum is a limited natural resource, and hence maximum benefit can only be derived by equitable and efficient utilization. The instructional systems will draw heavily upon this resource, particularly because very often these systems have to employ high-powered satellites and low-cost small reception terminals to be effective. Issues such as interference between satellite networks, coexistence of satellite and terrestrial networks, etc., are international problems and have to be dealt with as such. Because of this, over the years, several international conventions and regulations have been established to govern the use of the radio-frequency spectrum and a consideration of these is important in planning a satellite system.

The International Telecommunication Union (ITU) is one of the specialized agencies of the United Nations and one of the oldest international bodies. ITU and its various organs have played an important role in evolving the various international regulatory mechanisms in relation to the use of the frequency spectrum. The matters that are of permanent nature such as legal relations between the contracting nations and the basis for fixing of tariffs, etc., are placed in the conventions adopted by the Plenipotentiary Conference which is the supreme organ of the Union. Technical matters are governed by the regulations which are adopted by the Administrative Conferences. The International Frequency Registration Board (IFRB), the International Radio Consultative Committee (CCIR), and the International Telegraph and Telephone Consultative Committee (CCITT), which work within the framework of ITU, provide the technical support to ITU and the Administrative Conferences.

The various Administrative Conferences have established a table of frequency allocations. Conformity to these allocations is compulsory for all members of ITU. The table of allocations is arrived at after careful consideration of various service requirements, the technical merits of using certain frequency bands for specific services, probabilities of interference between systems, equip-

451

ment, technology, etc. At present, allocations cover the frequency bands up to 275 GHz. Tables A6.1 and A6.2 give the frequency bands allocated for the broadcasting satellite and fixed-satellite services. ITU has divided the globe into three geographical regions and the frequency allocations are made on the basis of regions.

The allocations are also split into two distinct categories, including the fixed-satellite service (FSS) and broadcasting satellite service (BSS). The fixed-satellite service generally refers to point-to-point or point-to-multiple-destination telephone, telex, or data communication, as well as television distribution. The broadcasting satellite service provides television and radio programs for direct reception by individual home receivers or by small community receivers. Both services could be used for instructional purposes, though the broadcasting satellite service may be more useful from the point of view of its reach.

Apart from the frequencies listed in Tables A6.1 and A6.2, WARC–79 has recommended that experiments and studies in the 0.5 to 2.0 GHz range be carried out for satellite sound broadcasting. The resolution adopted also makes a specific reference to the band around 1.5 GHz for this application.

Most of the frequency bands are shared with terrestrial services, and to

TABLE A6.1   Broadcasting Satellite Service Frequency Allocations (WARC–79) Below 35 GHz

Satellite to earth	ITU regions	Remarks
620–790 MHz	1, 2, and 3	Notes 1 and 4
2.5–2.69 GHz	1, 2, and 3	Notes 2 and 4
11.7–12.5 GHz	1	Notes 3 and 5
12.1–12.7 GHz	2	Notes 3 and 5
11.7–12.2 GHz	3	Notes 3 and 5
12.5–12.75 GHz	3	
22.5–23.0 GHz	2 and 3	

General remark: All of these frequency bands are shared with other satellite services and/or with terrestrial services.

Note 1:   Only frequency modulation may be used. The satellite shall not produce a power flux density in excess of the value -129 dBW/m² for angle of arrival less than 20° within the territories of other countries without the consent of the administrations of those countries.

Note 2:   Defines in technical terms the power flux density limitations.

Note 3:   For regions 1 and 3, WARC–77 has defined the frequency and orbit assignments. For region 2, a regional conference in 1983 will plan the assignments.

Note 4:   The up-link requirements are to be met out of FSS allocations in S and C bands.

Note 5:   The up-link requirements (feeder) links will be decided in a forthcoming radio conference. 14 GHz and 18 GHz bands are the possible options.

TABLE A6.2 Fixed-Satellite Service Frequency Allocations (WARC–79) Below 35 GHz

Satellite to earth	ITU region	Remarks
2,500–2,535 MHz	3	Note 1
2,535–2,655 MHz	2	Note 1
3,400–4,200 MHz	1, 2, and 3	Note 2
4,500–4,800 MHz	1, 2, and 3	Note 2
7,250–7,750 MHz	1, 2, and 3	Note 2
		Note 3
10.70–11.70 GHz	1, 2, and 3	Note 4
11.70–12.30 GHz	2	Note 5
12.50–12.75 GHz	1, 2, and 3	Note 5
17.70–21.20 GHz	1, 2, and 3	—
Earth to satellite		
2,655–2,690 MHz	2 and 3	—
5,725–5,850 MHz	1	—
5,850–7,075 MHz	1, 2, and 3	—
7,900–8,400 MHz	1, 2, and 3	Note 3
10.70–11.70 GHz	1	Note 6
12.50–12.75 GHz	1	—
12.70–12.75 GHz	2	—
12.75–13.25 GHz	1, 2, and 3	—
14.00–14.50 GHz	1, 2, and 3	Note 6
14.50–14.80 GHz	1, 2, and 3	Note 7
17.30–17.70 GHz	1, 2, and 3	Note 8
27.00–27.50 GHz	2 and 3	—
27.50–31.00 GHz	1, 2, and 3	—

General remark: All of these frequency bands are shared with other satellite services and/or terrestrial services.

Note 1: Limited to national and regional systems (Further, the note defines the power flux density and limitations.)

Note 2: Defines power flux density limitations.

Note 3: May also be used by the mobile satellite service.

Note 4: Defines power flux density limitations.

Note 5: Defines power flux density limitations.

Note 6: This band is available for BSS feeder links subject to coordination.

Note 7: The band 14.5 to 14.8 GHz is limited to feeder links for BSS for countries outside Europe.

Note 8: Limited to feeder links for BSS on global basis.

protect the terrestrial services, restrictions have been imposed on the power flux density at the surface of the earth. These constraints determine the maximum power that can be radiated from a satellite and hence the size of the ground terminals.

The power flux densities are not the only constraint in designing a satellite system. Though, in principle, a satellite in any orbit could be used for communications, the geostationary orbit (approximately 36,000 km above the surface of the earth and over the equator) is ideally suited for satellite communications. This is because a satellite positioned in the geostationary orbit moves around the earth in the same direction and with the same period as the earth around the polar axis and hence appears to be located at a fixed point in the sky when viewed from the earth. Thus, over the region covered by the satellite, round-the-clock communications can be established with a single satellite. Moreover, after initial acquisition of the satellite, there is little or no need to follow the satellite by moving the ground antenna.

Because of the growing number of international, regional, and domestic satellite systems, the geostationary orbit is rapidly getting crowded with respect to orbital positions and use of the radio-frequency spectrum. The limitation comes not so much because of physical interaction between satellites, but primarily because of the radio-frequency interference between satellite systems. The interference could be due to a satellite or ground station receiving unwanted signals originating from nearby satellites or ground stations. Several modes of interference are possible. Consideration of the interference modes is a significant constraint on the system planning.

As mentioned earlier, the establishment of a satellite system is subject to the regulations adopted by ITU. These regulations pertain to technical factors such as the maximum permissible level of interference from one satellite to another, station-keeping tolerances, satellite and ground station antenna radiation patterns, etc.

Another aspect of the regulations pertains to the co-ordination procedures to be followed in the establishment of a satellite system. Since, for communication systems operating in most frequency bands, orbital positions and operating frequencies are currently assigned on a first-come first-served basis, an elaborate co-ordination procedure that guarantees interference-free operations of various satellite systems has been evolved by ITU. According to this procedure, a country intending to set up a satellite system should notify its intention to all other countries through what is known as the advance publication of information (API) giving all relevant technical parameters of the system. It cannot be filed earlier than five years from the date of bringing the system into use. API enables other countries to examine whether or not their own operating and planned systems will be affected by the new system. If potential for interference is detected, the affected countries can call for co-ordination and, after mutually accepted parameters have been evolved, IFRB will register the frequencies and the orbit location in favor of the country proposing to establish the system. The steps involved in notification and the required procedure are briefly summarized in Table A6.3. The co-ordination procedure can be a time-consuming and involved process, especially if a large number of countries are affected. However, it should be noted that establishment of a satellite system itself is a long process, taking anywhere from five to ten years from decision to operation.

Of late, there is a perceptible desire and trend amongst many nations to

TABLE A6.3   Steps and Procedure Involved in Notification

No.	Step	Remarks
1.	Advance publication of information	The technical characteristics of the planned satellite network as listed in appendix 1B to Radio Regulations should be published not earlier than five years and preferably not later than two years before the date of bringing into use.
2.	Comments on published information	Other administrations will provide their comments/objections with respect to interference from and into their existing and planned satellite systems.
3.	Resolution of difficulties	The concerned administrations should resolve the difficulties by exploring all the possibilities.
4.	Requirement of co-ordination	In the case of a planned satellite network using the geostationary orbit, the administration requesting co-ordination should submit the technical parameters of the planned satellite system as listed in appendix 1A to Radio Regulations.
5.	Notification of frequency assignments	After resolving all co-ordination difficulties, the administration should submit notification to IFRB for registration of frequencies not earlier than three years before the date of bringing into use.

move away from the first-come first-served basis for the assignment of orbital positions and frequencies and to adopt a structured plan that will lead to an ordered and regulated use of the limited natural resource, namely, the geostationary orbit. While rapidly advancing technology can lead to more efficient use of the orbit and spectrum, it is necessary to consider the economic impact. Technical efficiency alone cannot be the objective, and one has to think in terms of maximum benefit to the maximum number of people at minimum cost. If this is accepted, then any meaningful regulation has to take into account the potential requirements of all nations and the economic viability of systems, particularly in reference to the developing countries.

This basic concern led WARC–77 to adopt a detailed (if somewhat rigid) plan for orbit and frequency assignments for broadcasting satellite systems in the 12 GHz band for regions 1 and 3. Under this plan, various countries were assigned channel frequencies, orbit locations, coverage area for each beam, transmitted power [equivalent isotropically radiated power (EIRP], and polarization. Table A6.4 reproduces a portion of these assignments as an example and Table A6.5 shows channel numbers and assigned frequencies. A plan for region 2 (the Americas) will be decided by a regional conference in 1983.

The Final Acts of the Conference also developed procedures for the modification of the plan, notification of frequency assignments being brought into use, co-ordination procedures to be applied for terrestrial stations, and criteria for interregional sharing.

TABLE A6.4   Extract from the Broadcasting Satellite Plan
in the Frequency Bands 11.7 to 12.2 GHz in Region
3 and 11.7 to 12.5 GHz in Region 1

Country symbol and IFRB serial number	Nominal orbital positon[1]	Channel number[2]	Bore sight[3]		Antenna beam width[4]		Orientation of ellipse[5]	Polarization[6]	EIRP[7]
URS 060 A	23.0	4	41.5	57.4	3.08	1.56	153	1	66.7
ZAI 322 A	− 19.0	4	22.4	0.0	2.16	1.88	48	1	64.7
AFG 246 B	50.0	5	64.5	33.1	1.44	1.40	21	1	63.4
AUS 005 B	98.0	5	133.5	− 18.8	2.70	1.40	76	2	64.3

[1] Longitude in degrees east (+) or west (−).
[2] See Table A6.5.
[3] Longitude/latitude in degrees for center of beam.
[4] Width in degrees of major/minor axis of beam.
[5] Angle in degrees of major axis of beam with respect to equator.
[6] 1 = direct; 2 = indirect.
[7] Equivalent isotropically radiated power at center of beam.

TABLE A6.5   Channel Numbers and Assigned Frequencies
in the 12 GHz BSS Band

Channel No.	Assigned frequency (MHz)	Channel No.	Assigned frequency (MHz)
1	11 727.48	21	12 111.08
2	11 746.66	22	12 130.26
3	11 765.84	23	12 149.44
4	11 785.02	24	12 168.62
5	11 804.20	25	12 187.80
6	11 823.38	26	12 206.98
7	11 842.56	27	12 226.16
8	11 861.74	28	12 245.34
9	11 880.92	29	12 264.52
10	11 900.10	30	12 283.70
11	11 919.28	31	12 302.88
12	11 938.46	32	12 322.06
13	11 957.64	33	12 341.24
14	11 976.82	34	12 360.42
15	11 996.00	35	12 379.60
16	12 015.18	36	12 398.78
17	12 034.36	37	12 417.96
18	12 053.54	38	12 437.14
19	12 072.72	39	12 456.32
20	12 091.90	40	12 475.50

Some of the forthcoming conferences dealing with other issues of interest to those desirous of planning instructional satellite systems are:

1. World Administrative Radio Conference on the use of the Geostationary Satellite Orbit and the Planning of Space Services Utilizing It (1985/87).
2. Region 2 Broadcasting Satellite Planning Conference (1983).
3. Conference to draw up agreements and associated plans for feeder links to broadcasting satellites operating in the 12 GHz band in regions 1 and 3.

These conferences will have far-reaching implications on the future of instructional satellite systems.

Direct broadcasting from one country to another has been another area of considerable discussion and debate in international forums. Principles regulating such broadcasts are currently under discussion, particularly in the United Nations Committee on the Peaceful Uses of Outer Space.

REFERENCES

1. A large number of reports exist on SITE and on its impact. A good general overview is provided in K. S. Karnik and N. Sampath (eds.), *UN–UNESCO Panel Meeting on SITE Experiences,* Space Applications Center, Ahmedabad, 1977.
2. *Hermes/CTS Consolidated List of Papers and Reports* lists the very large literature available on this. A brief, but interesting, document on the experiment is D. H. Jelly (ed.), *Hermes Experimenters Debriefing,* Ottawa, 20–21 September 1977.
3. *Feasibility Study of a Regional System of Tele-Education for the Countries of South America* (in two volumes), UNESCO, France, 1975.
4. Romesh Chander and Kiran Karnik, *Planning for Satellite Broadcasting—The Indian Instructional Television Experiment,* UNESCO, 1976.
5. For further information on ATS-6 and the HET experiments, see W. N. Redisch, ATS-6 Description, EASCON 75, pp. 153-A-E, and A. A. Whalen, *Health Education Telecommunications Experiment,* EASCON 75, pp. 154-A-C.
6. J. Sutanggar Tengker, *Indonesian Domestic Satellite System,* EASCON 76, pp. 11-A-U.
7. *Telesat Canada-A Technical Description,* TELESAT Canada, September 1979.
8. Direct radio broadcast studies have been carried out by various groups. The studies include: *Feasibility of a Satellite Sound-Broadcasting System for a National Service with Portable Receivers,* European Broadcasting Union—CCIR Study Group SPM (WARC-79), Geneva, 1978; *Feasibility Study of a System for Sound Broadcasting by Satellite,* ESA-CCIR Study Group SPM (WARC-79), Geneva, 1978; "Preliminary Study Report on Indian National Sound Broadcast Satellite," Space Applications Center (ISRO), Ahmedabad, 1980 (unpublished).
9. Department of Communications, *A Satellite Delivered Direct-to-Home Television Pilot Project,* Ottawa, May 1980.

Chapter 7

EFFICIENT USE

OF THE GEOSTATIONARY ORBIT

ABSTRACT

This chapter is the third of three on specific technical issues. It was prepared with the assistance of an international team of experts organized by the International Astronautical Federation (IAF). It should be noted that two international organizations share the primary responsibility for co-ordinating international use of the geostationary orbit: the United Nations, through its Committee on the Peaceful Uses of Outer Space, has the responsibility for general co-ordination of space activities; ITU, through the World Administrative Radio Conference (WARC), the International Radio Consultative Committee (CCIR), and the International Frequency Registration Board (IFRB), has responsibility for co-ordinating use of the radio-frequency spectrum. Both of these organizations have dealt with the question in a number of their reports and publications ((1) to (5)].[1]

The purpose of this chapter is to examine the geostationary orbit as a limited natural resource, to define the nature of the limitations, and to examine

---

[1] Numbers in parentheses refer to the references at the end of the chapter.

the extent to which the limitations can be overcome by technological means. The geostationary orbit is defined and described, and its uses and potential uses are discussed. The limitations on the capacity of the orbit are described and the technological means for overcoming these limitations are discussed. Finally, a general assessment is made of the current situation and of the foreseeable evolution of the use of the orbit.

7.1 CHARACTERISTICS OF THE GEOSTATIONARY ORBIT AND THE RADIO-FREQUENCY COMMUNICATIONS SPECTRUM

7.1.1 The Geostationary Orbit

Earth satellites orbit about the center of the earth with a period determined by the radius of the orbit. At a radius of some 42,165 km, corresponding to an altitude of 35,787 km above the earth's surface, a satellite has a period of 23 hours 56 minutes and is therefore synchronous with the earth's rotation. If the satellite is moving in the same sense as the earth, from west to east, and if the satellite orbit is over the equator, the satellite will appear to an observer on the earth's surface to be stationed at a fixed point in the sky, i.e., geostationary. The geostationary orbit can therefore be defined as a circular earth orbit in the plane of the equator at an altitude of some 35,800 km. In terms of geographical coordinates, the position of a geostationary satellite is defined by its longitude. The advantage of the orbit is that a geostationary satellite has a constant view of a large area of the earth and is constantly visible from any point within that area. A fixed ground antenna need not be continually reoriented to track the satellite.

Orbital variations. The above is an ideal definition since natural forces change the satellite orbit. Forces due to the ellipticity of the earth's equator, the gravitational attraction of the sun and moon, and solar radiation pressure cause the satellite to drift in longitude and to move north and south of the equator in a figure 8 pattern. "Station-keeping" systems on board the satellite are used to counteract these forces and maintain the satellite in the desired position within the orbit. Recent geostationary satellites can maintain position to an accuracy of $\pm 0.1°$ both in longitude and latitude, corresponding to a square 150 km from north to south and 150 km from east to west. The satellite will also vary in altitude by about 30 km. In practice, the geostationary orbit, rather than being a line in space, is actually a ring 150 km from north to south and 30 km thick. A more detailed description of the orbit and the forces acting upon it can be found in reference (1).

Geosynchronous orbits. The geostationary orbit belongs to a family of orbits called geosynchronous. A geosynchronous satellite has the same period as the earth's rotation, but moves in an orbit that is elliptical and/or inclined with respect to the equator. From the earth, such a satellite will appear to describe a single or double loop about a point on the equator once every 24 hours. Although such orbits are not the subject of this chapter, they do have a potential for relieving some of the congestion in the geostationary orbit, as will be discussed later.

Geographical coverage. The area of the earth's surface that is visible from a geostationary satellite is a circle of radius 9,050 km about the subsatellite point on the equator, in other words, a circle extending in latitude from 81.3° N to 81.3° S and in longitude 81.3° east and west from the subsatellite point. Conversely, the satellite is visible from every point in this circle, appearing at the zenith from the center and on the horizon from the circumference. In practice, because of atmospheric attenuation, the satellite must be above the horizon from reliable communication, and a minimum elevation of 10° will be used here, corresponding to a circle of radius 71.43° or about 7,952 km. This circle is called the "area of visibility" for the satellite and has a width that depends on latitude as follows:

0° (equator)	15,900 km
30° N or S	13,200
45°	10,000
60°	5,620
70°	1,640
71,43°	0

It would be useful here to introduce two other definitions: the "coverage area," which is the area effectively covered by the satellite's communications antennas or sensors, and the "service area," which is the area in which the earth stations are located. If the width of the service area is equal to the width of the area of visibility, e.g., 10,000 km for an area at 45° N, then the satellite can only be located at the central longitude of the service area. If, the service area is much smaller than the area of visibility, then the satellite can be located anywhere within a range of longitudes about the central point. Thus, satellites serving large areas, e.g., satellites for intercontinental communications, require quite specific positions within the geostationary orbit, while those serving small areas, e.g., domestic satellites for smaller countries, can be positioned more flexibly. This flexibility can be important in avoiding interference between satellites. The positioning of space research satellites may be constrained not only by the location of the earth station or stations but also by the phenomenon to be studied.

Due to the constraints on satellite position, the geostationary orbit is not, and most likely will not be in the future, uniformly populated with satellites. In particular, portions of the orbit that serve large areas of heavy communication traffic will become congested while other portions are sparsely populated.

Eclipses. Satellites in the geostationary orbit are subject to eclipses of the sun by the earth twice a year around the spring and fall equinoxes. The satellite is eclipsed for 44 consecutive nights around each equinox for a period of up to 72 min. This will occur before midnight if the satellite is east of the service area and after midnight if the satellite is west of the service area. Since satellites generally rely on solar cells for their power supply, this can result in interruptions of service. There are several possible ways to resolve this problem: The satellite can carry batteries to provide power during eclipses; the satellite can be positioned such that the interruptions occur at acceptable times such as after midnight; or two satellites with different eclipse periods can be used.

Solar interference. Interruption of service to an earth station will occur whenever the satellite, as seen from the earth station, passes in front of the sun so that radio noise from the sun overwhelms the satellite signal. This happens twice a year around the equinoxes, and communication is interrupted on four consecutive days, for up to 6 min. If communication must be maintained, a second satellite must be used. Since most operational communications satellite systems have spare satellites in orbit to protect the system against satellite failure, this is often a feasible solution.

7.1.2 The Radio-Frequency Spectrum

Satellites use radio communications for command and control of the satellite and for transmission or relay of information. Since limitations on the use of the radio-frequency spectrum present the greatest constraint on use of the geostationary orbit, some relevant properties of that spectrum will be discussed in this section.

The radio-frequency spectrum is defined by ITU to be that part of the electromagnetic spectrum whose frequencies are lower than 3,000 GHz (corresponding to wavelengths greater than 100 m). Frequency bands within this are allocated to various services, including satellite services, by the WARCs of ITU the last of which met in 1979. A large number of frequency bands with frequencies up to 275 GHz have been allocated to various satellite services. To date, most satellites have used frequencies below 10 GHz, a few operational and experimental satellites have used frequencies up to 14.5 GHz, and a few experimental satellites have used frequencies up to 31 GHz. As the lower frequencies become congested and as high-frequency technology develops, more satellites will use the higher frequencies. However, at frequencies above 10 GHz, attenuation of the signal due to rainfall and, at higher frequencies, to the constituent gases of the atmosphere, becomes a problem.

Bandwidth. Information to be transmitted by radio communication occupies a certain range of frequencies or bandwidth of the radio-frequency spectrum. The bandwidth depends on the amount of information to be transmitted and on the way in which the radio waves are modulated to carry the information. On a typical communications satellite, a telephone or voice channel will occupy about 40 kHz of bandwidth and a television channel will occupy about 40 MHz. The voice channels are grouped together and relayed by transponders, and a typical transponder with a bandwidth of 40 MHz carries either about 1,000 voice channels or one television channel. The most commonly used frequency band for transmission from a satellite to an earth station is the 3.7 GHz to 4.2 GHz band, a band 0.5 GHz or 500 MHz wide. A satellite might have 12 transponders of 40 MHz bandwidth each within this band for a total capacity of 12,000 voice or 12 television channels.

Polarization. Electromagnetic waves can be polarized in two ways: linearly, such that the electric and magnetic fields are oriented in fixed planes, and circularly, such that the fields rotate. In either case, two signals can be transmitted and received independently at the same frequency; in the linear mode, for example, one can be polarized vertically and the other horizontally. Use of the two polarizations can therefore double the amount of information transmitted in a given bandwidth.

Antenna beamwidth. The antennas used for satellite communications, for reception and transmission, on the satellite and on the ground, have a beamwidth or field of view that depends on the frequency of the radiation and the geometry of the antennas. The beamwidth for a typical antenna is given by the formula:

$$\alpha = 70\frac{\lambda}{D}$$

where α = the angle of divergence of the beam in degrees;

       λ = the wavelength of the radiation;

       D = the diameter of the antenna.

As the formula indicates, the beamwidth decreases as the antenna diameter increases and as the wavelength decreases (i.e., as the frequency increases). Some examples of beamwidths calculated by the formula are as follows:

Freq.	λ	D	α
6 GHz	5 cm	2 m	1.72°
		10 m	0.34°
		30 m	0.11°
14 GHz	2.1 cm	2 m	0.74°
		10 m	0.15°

An antenna on a satellite may have a very wide beam in order to cover the entire area of visibility, or it may have a very narrow or "spot beam" in order to cover a small geographical area. While a circular antenna will normally transmit a circular beam, shaped beams can be generated either by shaping the antenna or by using several antenna feeds to generate several overlapping spot beams that combine to give the desired coverage. If separate beams are directed at well-separated geographic areas, information can be transmitted independently on each beam using the same frequency, thereby increasing the information that can be transmitted in a given bandwidth. The INTELSAT V Atlantic satellite, for example, uses separate beams for the two sides of the Atlantic, thus doubling the utilization of a large bandwidth in the 6/4 GHz band. Since ground antennas are normally intended to communicate with a single satellite, the beamwidth is kept small to avoid transmitting to or receiving from adjacent satellites.

Antenna sidelobes. A perfect antenna would radiate energy only into the beam as defined above. In practice, antennas transmit significant amounts of energy in other directions, and in particular into the sidelobes adjacent to the main beam. The angular distribution of power radiated by an antenna is therefore divided into the main beam, in which the power is sufficient for reliable communication, the sidelobe area, in which the power is insufficient for communication but may interfere with communication, and the rest of the circle, in which the power level is sufficiently low to avoid interference. Thus, at a given frequency the minimum distance between satellites, or between earth stations communicating with different, but closely spaced, satellites, is defined not by the beamwidth but by the power levels in the sidelobes and the sensitivity of the system to interference. The satellite spacing can be reduced either by reducing the sidelobe levels of earth station antennas or by reducing the system sensitivity to interference. Currently, in the 6/4 GHz band, the distance between satellites is typically 3 to 5°.

7.2 USES OF THE GEOSTATIONARY ORBIT

7.2.1 Communications

The major use of the geostationary orbit, both in terms of number of satellites and radio-frequency bandwidth used, is communications. The geostationary orbit is uniquely capable of providing continuous communication between earth stations via a single satellite. Communications satellite systems are divided into three categories depending on the type of earth station served.

Fixed satellite service refers to communication between fixed earth stations. Since the antennas can be large and the stations complex, the satellites can be relatively simple; hence, this was the first type of satellite communications system to be developed. The earliest operational systems were for international communications (e.g., INTELSAT), but national systems (e.g., ANIK, STATSIONAR, WESTAR, PALAPA) have also been established, and regional systems (e.g., ARABSAT, ECS) are now being planned. The fixed-satellite service carries television, telephone, telegraph, and telex traffic now, and new capabilities are being introduced, including:

1. High-speed transmission of documents (text, graphs, pictures, newspaper print).
2. High-speed data transmission between computers.
3. Audio conference with visual aids.

4. Video conferences (carried by live television transmissions between different sites).

   Mobile satellite service refers to communication with earth stations located on aircraft, ships, or land vehicles. Since there are large numbers of mobile stations widely scattered, and since there are constraints on the size and complexity of these stations, the satellite must be correspondingly more powerful and complex. For this reason, a mobile satellite service has been slower to develop than the fixed-satellite service. Maritime mobile communications systems, in which the constraints on the earth stations are least, have been using satellites (e.g., MARISAT) for several years now and the INMARSAT system became operational early in 1982. Aeronautical mobile communications satellites are still in the development stage. Since mobile communication is primarily voice and low-speed data, the bandwidth requirements are less than for the fixed service, and the volume of communications is considerably lower.

   Broadcasting satellite service refers to communication of television or radio from a fixed earth station to a very large number of small inexpensive receiving stations, serving large and small communities, and even individual homes. As with the mobile service, since the receiving stations must be simple and inexpensive, the satellite must be relatively powerful. Experimental broadcast satellite programs have been carried out in Canada, Brazil, India, Japan, the United States, and the USSR, using the ATS-6, CTS, BS (YURI) and EKRAN satellites, and a number of national or regional broadcast systems are now being planned. Direct sound broadcasting is also being considered and WARC-1979 encouraged experiments at frequencies around 1.5 GHz. The ground receivers would be very small and might include receivers for motor vehicles.

7.2.2 Meteorology

   Meteorological satellites, unlike communications satellites, generate information by means of on-board sensors, and transmit this data to fixed ground stations. There is a very small volume of command and control data sent to the satellite and a much larger volume of meteorological data transmitted from the satellite. The geostationary orbit allows the satellite to make frequent observations of the earth's atmosphere, unlike lower-altitude satellites which provide coverage only once every 12 hours. The WMO World Weather Watch (WWW) plan and implementation program for the period 1980–1983 includes six operational geostationary satellites in the following positions: 140° E (Japan), 74° E (India), 70° E (USSR), 0° (European Space Agency), 75° W, and 135° W (United States) [(6) and (7)]. This system will provide global coverage from about 60° S to 60° N and can generate images as frequently as once every 20 to 30 mins with a ground resolution between 1 and 7 km. The polar regions cannot be observed from the geostationary orbit.

   Since raw data from the geostationary satellites can be received only by rather large central command and acquisition stations and, in addition, powerful computer centers are needed for the subsequent image processing, one or two S-band transponders are installed aboard the satellites for the transmission of pre-processed analog and digital image data to simple users' stations within the radio range of the satellite (76° geocentric arc from the subsatellite point at antenna elevation angles of 5° or greater). Of special importance is the dissemination of information using the well-established automatic picture transmission (APT) format (800 lines and 800 picture elements per line). This APT-type dissemination, called Weather Facsimile Service (WEFAX), allows the transmission of sectorized image data as well as other meteorological information like analysis and forecast charts. The dissemination takes place at 1,691 MHz frequency on a fixed schedule.

Geostationary meteorological satellites are also capable of collecting information from a large number (up to 10,000) of fixed and moving data collection platforms (DCPs) of various types (meteorological, oceanographic, hydrological, etc.) and of relaying these data to central ground stations for further processing and dissemination.

Meteorological satellites generate a volume of data that is very low compared with that carried by communication satellites. On METEOSAT, for example, the data collection system uses about 200 kHz of bandwidth at a frequency of 400 MHz while the image transmission and relay and other communications with the central ground station are carried within a 20 MHz bandwidth (between 1,675 and 1,695 MHz). Since the six satellites of the current system provide complete coverage, and since the currently allocated bandwidth seems adequate, there is little prospect for congestion of the geostationary orbit due to meteorological satellites.

7.2.3 Space Research

In addition to the communications and meteorological satellites currently in the geostationary orbit, there are two operational space research satellites: the international ultra violet explorer (IUE), an astronomical observatory, and the geodetic satellite GEOS-2, a magnetospheric monitor. For such satellites, the advantage of the geostationary orbit is that it allows continuous contact with the ground station. Since these satellites produce relatively small volumes of data, and since the number of satellites being planned is rather small, space research satellites are not likely to contribute significantly to the congestion of the geostationary orbit.

7.2.4 Proposed Geostationary Satellite Systems

Tracking and data relay satellites in geostationary orbit have been proposed as a solution to the problem of maintaining continuous contact with satellites whose function requires them to be in low earth orbit, for example, remote sensing satellites. Two geostationary satellites could track low-orbit satellites almost continuously and could relay data to a single central ground station. Such a system has been proposed by the United States for launch in 1982.

Remote sensing satellites in geostationary orbit have been considered as a means of studying rapidly evolving ground phenomena, such as soil moisture, and of monitoring disaster areas. Because of the difficulties in obtaining sufficiently high ground resolution from the geostationary orbit, no such systems are yet being implemented.

7.2.5 Proposed Large Space Structures

A large space structure can be defined as one that requires two or more launches to put the material in orbit, followed by assembly in space. No such structures have yet been assembled in space, and low-orbiting structures will certainly precede any in geostationary orbit. Any such structures would require advances in the technology of lightweight materials, transport into orbit, and assembly in orbit. It is possible that large structures could be in low orbit by about 1990 and in the geostationary orbit in another decade.

Solar power satellites have been proposed as an efficient means of harnessing solar energy. Such satellites, which might be located in geostationary orbit, would be designed to receive full sunlight almost continuously and would transmit the power to terrestrial receivers by various forms of electromagnetic radiation—microwaves, infrared radiation, or visible light. On the ground, the energy would be converted to electrical power and distributed, providing an essentially continuous and constant power supply. Among the possible techniques that have been considered are the following:

1. Photovoltaic panels in space generating electricity which in turn generates microwave power which is beamed to earth and reconverted to electricity
2. Photovoltaic panels in space powering lasers which transmit infrared radiation to earth for reconversion to electrical energy
3. Mirrors in space reflecting and focusing a high-intensity beam of sunlight to earth for photoelectric or solar-thermal conversion to electricity.

Such systems have been analyzed extensively and appear to pose no insurmountable scientific or technological obstacles, but it is not yet clear which, if any, of such systems might eventually become economically and operationally feasible. A great deal of additional research and technological development would be required to assess these factors and eventually to implement such a system. Furthermore, depending on the technology used, consideration would have to be given to possible environmental problems due to atmospheric and surface effects of high intensity beams, high frequency rocket launches, and electromagnetic interference with communication systems.

Multimission platforms have been proposed as being more efficient than separate satellites. A platform would have a common system for power generation and distribution, precision station-keeping, and stabilization systems and could carry a variety of data gathering sensors and telecommunications relay systems connected by a central switching unit to a variety of antennas including large antennas with several precisely shaped beams. Because of the large number of antennas such as platforms might carry, they are sometimes referred to as antenna farms. The possibilities for switching signals between systems and the fixed geometrical relations between systems might allow for more efficient use of the radio-frequency spectrum from geostationary orbit.

Manned space stations in the geostationary orbit could provide a means for assembling other large structures and for maintaining systems in orbit. The repair, elimination, or replacement of nonfunctioning satellites could reduce the risk of physical or radio-frequency interference between satellites.

7.3 LIMITATIONS ON THE USE OF THE GEOSTATIONARY ORBIT

There are two factors that limit the number of satellites that can use the geostationary orbit: physical interference between satellites and radio-frequency interference between systems. It is not meaningful to express the limit to the capacity of the orbit in terms of a number of satellites since satellites vary greatly in their characteristics. The limitations are much more usefully expressed in terms of cross-sectional area subject to collision and radio-frequency bandwidth subject to interference.

7.3.1 Collision Probabilities

As noted earlier, geostationary satellites require active station-keeping to maintain their designated positions. Since most current satellites are able to maintain their position within $\pm 0.1°$ of longitude, there are 1,800 "slots," each $0.2°$ wide, in the geostationary orbit such that there would be no risk of collision between functioning satellites. If two or more satellites are positioned at the same nominal position, there is a risk of collision which depends on the size of the satellites. A recent study (8) concludes that two satellites, each of 100 m² cross section, would have a probability of collision of 9×10^{-7} per year. If ten such satellites were assigned the same location, the probability would increase to 4×10^{-5} per year; that is, there would be an average of one collision every 400,000 years between active satellites.

A more serious risk of collision arises between active position-keeping satellites and the collection of former geostationary satellites that have ended their useful lifetime and are drifting. It has been noted above that natural forces cause the orbits of satellites to vary in longitude, inclination, and eccentricity. Variations in the earth's gravitational field will cause a satellite to oscillate in longitude about the nearer extension of the minor axis of the earth's equator, i.e., about either 75° E or 105° W. The forces due to the gravitational forces of the sun and moon and the radiation pressure due to the sun will cause the orbit to vary in inclination from 0 to 14.6° with a period of about 52 years and to vary in eccentricity causing a daily variation in altitude up to perhaps 60 km. An inactive satellite will therefore generally be in a geosynchronous orbit that crosses the geostationary orbit twice a day. It is during these passages of inactive satellites through the ring occupied by the active satellites that the risk of collision arises. Collisions between inactive satellites are also possible, but are relevant only in that they may create additional objects passing through the geostationary ring.

The study cited above has calculated the probability of collision for projected populations of active and inactive satellites for the next two decades. The results are summarized in Table 7.1. It is assumed that the growth in demand for the use of the geostationary orbit will be largely met by increasing the size of satellites rather than increasing the numbers, and the cross-sectional area is a determining factor in the collision probability. The risk is expressed as the probability of a single collision during the five-year period. Two calculations are made for the period 1996–2000, one with no solar power satellites (SPS) and one with four SPS of 50 km² each (8).

The figures in Table 7.1 indicate that the probability of collision will remain low but not negligible for the foreseeable future unless SPS or other structures of comparable size are launched. Given some 300 smaller inactive satellites crossing the geostationary orbit, a population of SPS with a total of 200 km² of solar panels would suffer an average of one hit every five years. If these large satellites were allowed to drift after their active lives, then they would collide with active satellites with a similarly high frequency.

Since the main risk of collision to active satellites arises from inactive satellites drifting in geosynchronous orbit, a possible solution to the problem lies in the removal of satellites from the geostationary orbit at the end of their useful lifetimes. Since it appears that inactive satellites drift in altitude by less than 100 km, the raising of satellites into stable circular orbits 100 to 200 km above the geostationary orbit would eliminate the risk of collision. Such a change of orbit would require some 0.2 to 0.4 kg of hydrazine propellant per 100 kg of satellite mass. The International Telecommunications Satellite Organization (INTELSAT) has already removed a number of its communications satellites from the geostationary orbit at the end of their lifetimes.

TABLE 7.1 **Probability of Collision for Projected Populations of Active and Inactive Satellites, 1980–2000**

Period	Satellites		Average Area	Probability	
	Active	Inactive			
1981–1985	95	65	13 m²	0.3×10^{-4}	
1986–1990	135	140	21 m²	1.2×10^{-4}	
1991–1995	115	250	124 m²	5.5×10^{-4}	
1996–2000	110	310	168 m²	9.7×10^{-4}	no SPS
1996–2000	120	310	1.7 km²	0.98	4 SPS

On the basis of a proposal made by India at CCIR in 1978 [see (3), document No. 318], CCIR has acknowledged [see (3), p. 5, 165] that it would be desirable to study the question of removal of satellites from the geostationary orbit and to agree on measures for such removal if necessary. It should be noted that the failure of the station-keeping propulsion system before the planned end-of-lifetime would prevent removal, but since this should be relatively rare, the risk may be minor. The removal of inactive satellites by other manned or automatic devices would appear to be a rather costly operation.

7.3.2 Radio-Frequency Limitations

The limitations on the use of the radio-frequency spectrum are both natural and man-made. The natural limitations include attenuation due to the constituent gases of the atmosphere and the more localized attenuation due to precipitation. The man-made limitations are due to interference between satellite communications systems and other uses of the radio-frequency spectrum and to interference between satellite systems.

7.3.2.1 Atmospheric attenuation.   The attenuation of electromagnetic radiation by the constituent gases, most importantly by water vapor, generally increases with frequency. Superimposed upon this general increase are narrow bands of very high attenuation, e.g., around 22 and 183 GHz due to water vapor and around 60 and 118 GHz due to oxygen. Above 500 GHz attenuation begins to become prohibitively high, but it drops again to potentially usable values above 10,000 GHz.

Attenuation due to precipitation and clouds is highly variable both in space and time. It is generally negligible at frequencies below about 10 GHz and increases with increasing frequency above 10 GHz. The attenuation due to both the constituent gases and precipitation increases as the elevation angle to the satellite, or site angle, decreases. An example of the attenuation due to rainfall as a function of frequency and site angle is given in Table 7.2. In humid zones, earth stations that are far from the subsatellite point (i.e., have a low site angle) will be much more susceptible to interruption of service by rainfall than stations with high site angles. Possible ways of avoiding interruption include using larger antennas with more power and higher gain and using two or more earth stations

TABLE 7.2   Attenuation Factor of 50 mm/Hour Rainfall

| | Elevation Angle | |
	30°	10°
11 GHz	3 (4.7 dB)	6 (7.8 dB)
30 GHz	1,200 (31 dB)	160,000 (52 dB)

some distance apart ("site diversity"), since the probability of rain at the two stations simultaneously is less than the probability for a single station.

7.3.2.2 Interference.

Interference refers to a degradation of performance of a communications system due to unwanted signals. Interference may arise from signals intended for a different service area or from signals intended for a different frequency. Since all communications systems radiate some energy outside the intended service area and outside the intended frequency band, interference cannot be eliminated entirely, but it can be minimized and can be allowed for in the design of the systems. The geographic area covered and the frequency range occupied by a communications satellite therefore include a minimum service area and a minimum bandwidth necessary for the satellite to transmit its signals to the earth stations and a bordering geographical area and frequency bandwidth over which the transmitted energy is reduced to a level that will not interfere with other systems. The capacity of the geostationary orbit can be increased both by reducing the bandwidth necessary for transmitting a given amount of information and by reducing the extended area and bandwidth required to avoid interference.

One mechanism for avoiding interference between communications systems is ITU and its subsidiary bodies, CCIR, IFRB, and WARC. Among other functions, these bodies developed standards for communications systems and allocate frequency bands to the various communications services. The most commonly used frequencies allocated to the fixed satellite service are given in Table 7.3. The up and down transmissions are at separate frequencies so that the large amount of energy being transmitted does not interfere with the small amount of energy being received.

In recognition of the increasing demand for fixed satellite communications, the most recent WARC held in 1979 increased the bandwidth of the 6/4 GHz band by about 600 MHz and that of the 14/11 GHz band by 500 MHz. In addition to these bands, a bandwidth of 3,500 MHz remains allocated in the 30/20 GHz band (17.7 to 21.2 GHz down and 27.5 to 31 GHz up), but this band has only been used experimentally to date because of the technological problems

TABLE 7.3 Most Commonly Used Frequencies Allocated to the Fixed Satellite Service

Band	Bandwidth	Direction	Frequency
6/4 GHz	500 MHz	down[1]	3.7–4.2 GHz
		up[2]	5.925–6.425 GHz
14/11 GHz	500 MHz	down[1]	10.95–11.2 GHz
			11.45–11.7 GHz
		up[2]	14–14.5 GHz

[1] Down refers to transmissions from satellite to earth station.

[2] Up refers to transmissions from earth station to satellite.

involved in working at these high frequencies. Additional allocations have been made at various frequencies up to 241 GHz, but there are no current plans to use these frequencies because of the technological problems and atmospheric attenuation (9).

The frequencies in Table 7.3 are not allocated exclusively to the fixed satellite service but are shared with other services, in particular with fixed and mobile terrestrial communications. This poses particular problems in the 6/4 GHz band which is widely used for terrestrial microwave transmissions. Since an earth station transmits very high-power signals to the satellite, and receives very weak signals from the satellite, there is a potential for stray earth station radiation to interfere with nearby microwave receivers or for terrestrial microwave beams to interfere with earth station reception. The location for the earth station must therefore be selected with care and may have to be at some distance from major cities. This interference is less of a problem at the higher frequency bands such as the 14/11 GHz band since there is much less terrestrial traffic at these frequencies. It will be easier therefore to locate user antennas in cities.

Communications satellites must avoid interfering with other satellites as well as with terrestrial communications. While most communication satellites are now in the geostationary orbit, it should be noted that nongeostationary communications satellites such as the Molnya system share the same frequency allocation. Between geostationary satellites, interference is avoided by separating the satellites in space or in frequency. As noted above, the current spacing for satellites operating in the 6/4 GHz band is typically 3 to 5°. Prior to WARC-79, the bandwidth within this band allocated to the fixed-satellite service was 500 MHz. Since most operational communications satellites use essentially all of this bandwidth, a maximum of about 90 such satellites with global beams could be placed in the geostationary orbit. Since certain areas of the world have more communications traffic than others, a small number of satellites could create congestion in popular parts of the orbit. Fixed service satellites are placed in the geostationary orbit according to need, with co-ordination under the auspices of IFRB, essentially on the basis of "first come, first served." As demand grows, the job of co-ordinating the use of communications satellites will become more complex. A WARC to deal with this question will be convened by ITU in two sessions, the first in June 1985 and the second in September 1987.

For broadcasting satellites, a plan was adopted by the WARC held in 1977, assigning frequencies, orbital positions, and service areas on a country by country basis. This plan does not cover the Americas for which a regional conference will be held in June 1983. For the other regions, 40 television channels are designated in the 11.7 to 12.5 GHz band, and 34 satellite positions separated by 6° are designated in the geostationary orbit between 37° W and 170° E. Generally, five channels are allocated to each country, and the direction, size, shape, polarization, and power of each antenna beam and the technical characteristics of the transmission are specified. Although the shape of the irradiated areas has been designed to cover each country, there are inevitably adjacent areas that will be covered as well, a phenomenon called "spillover." In some cases, the same position and frequency have been assigned to two countries that are sufficiently separated geographically to avoid interference between the beams (10).

7.4 DEMAND FOR GEOSTATIONARY SATELLITE SERVICES

Since the first geostationary satellite was launched in 1963, 126 satellites have been placed in geostationary orbit, including 96 (76 percent) communications satellites, 17 (13 percent) reconnaissance satellites, 10 (8 percent) meteorological satellites, and 3 (2 percent) scientific research satellites. In the decade 1970 to 1980, the number of geostationary satellites increased at an average rate of about 18 percent per year. At the beginning of 1980, approximately 80 percent of these were operational, with the rest inactive and drifting across the geostationary orbit. To date, there have been few problems in accommodating these satellites in the orbit. If, however, use of the orbit continues to grow as it has in the past, parts of the orbit will become congested in the near future. Projections of the increase in satellite numbers and size and the resulting increase in the probability of collision were discussed in Section 7.3.1 and will therefore not be reconsidered in this section. Since satellites for purposes other than communications form a small percentage of the total and use an even smaller percentage of the frequency spectrum, it does not appear that they will significantly limit the use of the geostationary orbit in the foreseeable future. This section will therefore consider only the problems posed by an increase in communications satellites.

In order to discuss the growth of satellite communications, it is useful to consider a typical satellite using essentially the entire 500 MHz bandwidth in the 6/4 GHz band by means of 24 transponders (12 in each of two orthogonal polarizations), each using 40 MHz bandwidth and capable of carrying 1,000 telephone channels or one television channel. The transmissions use global beams covering the entire area of visibility. Given a 4° spacing between satellites to avoid interference, a total of 90 such satellites could operate simultaneously. It should be noted that national satellites using smaller beams would permit frequency re-use, thereby increasing the possible number of satellites.

There are at present about 40 operational communications satellites in the geostationary orbit that use the 6/4 GHz band and hence many positions are still available. However, the portions of the orbit serving regions of the highest volume of communications already have satellites as closely spaced as possible. Thus, the orbital arcs from 49° E to 90° E (over the Indian Ocean), from 135° W to 87° W (serving North America), and from 1°W to 35° W (over the Atlantic Ocean) are virtually full with respect to the satellite described above. Any growth in traffic in these areas must therefore be accommodated by either more efficient use of bandwidth in the 6/4 GHz band, a reduction in satellite spacing, or by the use of other frequency bands.

The local congestion described above poses two problems, one involving growth of traffic on the routes that are already heavily used, and the other involving access to the orbit by regions with relatively low traffic. Some of the most recent satellites in orbit have been designed to reduce congestion, in the first case by frequency re-use through polarization, and in the second by directional antennas that can permit two or more satellites in the same orbital position to serve different regions without interference.

As noted above, the number of satellites in orbit has been growing by about 18 percent over the last decade. Launches in the last few years and announced plans for the next few years indicate that this growth rate will decline somewhat from 18 percent in the near future. The annual meeting of INTELSAT in June 1980 concluded that the demand for INTELSAT international telephone circuits would double by 1984, corresponding to an average increase of about 18 percent per year. The INTELSAT-V and the planned INTELSAT-VI satellites are designed to meet this demand through frequency re-use and more efficient use of bandwidth.

Table 7.4 gives a list of satellites currently in geostationary or geosynchronous orbit. The main frequency band has been indicated, but many of these satellites use more than one band. A nominal longitude is given, but the position of satellites is sometimes changed, often as new satellites of the same system are launched. Lifetimes of the satellites are variable but are generally shorter for meteorological satellites (three to seven years) than for communications satellites (typically seven years now but may increase to ten years in the future).

Table 7.5 gives a list of planned geostationary satellite systems. The list does not include additions or replacements to the systems included in Table 7.4. The list has been compiled from various sources and may omit some planned systems. The dates are also to be taken as indicative since experience indicates that it is not uncommon for such systems to be delayed for three or four years.

Figure 7.1 shows the distribution of the satellites listed in Table 7.4 as a function of longitude along the geostationary orbit. The four peaks are largely due to North American domestic service, transatlantic service, trans-Indian Ocean and USSR domestic service, and transpacific service. It should be noted that congestion in these services does not necessarily mean a lack of satellite positions for other regions using the same arcs. For example, if the North American and USSR domestic services use directional antennas, they can avoid interfering with South American or Asian services using satellites in the same arc.

The question of assessing and predicting the degree of congestion of the geostationary orbit is complex and cannot be solved reliably and quantitatively. The evolution of the demand and supply of services depends on general economic growth, growth of telecommunications, technological advances, costs of existing and new technologies, and other factors. Furthermore, these factors

TABLE 7.4 Geostationary Satellites

Designation	Name	Country	Launch Date	Nominal Longitude	Service	Main Frequency Band (GHz)
1963–004A	SYNCOM-1	USA	14 Feb.	Drift	COM(E)	Inact
031A	SYNCOM-2	USA	26 July	Drift	COM(E)	Inact
1964–047A	SYNCOM-3	USA	19 Aug.	Drift	COM(E)	Inact
1965–028A	EARLY BIRD	USA	6 Apr.	Drift	COM	Inact
1966–110A	ATS-1	USA	7 Dec.	149W	MULTI(E)	MULTI
1967–001A	INTELSAT-II-F2	INT	11 Jan.	Drift	COM	Inact
026A	INTELSAT-II-F3	INT	22 Mar.	Drift	COM	Inact
094A	INTELSAT-II-F4	INT	28 Sept.	Drift	COM	Inact
111A	ATS-3	USA	5 Nov.	105W	MULTI(E)	MULTI
1968–063A	BMEWS-1	USA	6 Aug.	Drift	RECON	Inact
081C	ERS-21	USA	26 Sept.	Drift	RES	Inact
081D	LES-6	USA	26 Sept.	85W	MOB(E)	Inact
116A	INTELSAT-III-F2	INT	19 Dec.	Out	COM	Inact
1969–011A	INTELSAT-III-F3	INT	6 Feb.	Drift	COM	Inact
013A	TACSAT-1	USA	9 Feb.	Drift	MOB	Inact
036A	BMEWS-2	USA	13 Apr.	Drift	RECOM	Inact
045A	INTELSAT-III-F4	INT	22 May	Out	COM	Inact
069A	ATS-5	USA	12 Aug.	70W	MULTI(E)	MULTI
101A	SKYNET-A	UK	22 Nov.	Drift	COM	Inact
1970–003A	INTELSAT-III-F6	INT	15 Jan.	Out	COM	Inact
021A	NATO-1	NATO	20 Mar.	Drift	COM	Inact
032A	INTELSAT-III-F7	INT	23 Apr.	Drift	COM	Inact
046A	BMEWS-3	USA	19 June	Drift	RECON	Inact
069A	BMEWS-4	USA	1 Sept.	Drift	RECON	Inact
1971–006A	INTELSAT-IV-F2	INT	26 Jan.	3W	COM	6/4
009A	NATO-2	NATO	3 Feb.	Drift	COM	Inact
039A	IMEWS-2	USA	5 May	Drift	RECON	Inact

095A	DSCS-1	USA	3 Nov.	Drift	COM	Inact
095B	DSCS-2	USA	3 Nov.	Drift	COM	Inact
116A	INTELSAT-IV-F3	INT	20 Dec.	21W	COM	6/4
1972–003A	INTELSAT-IV-F4	INT	23 Jan.	179E	COM	6/4
010A	IMEWS-3	USA	1 Mar.	Drift	RECON	Inact
041A	INTELSAT-IV-F5	INT	13 June	57E	COM	6/4
090A	ANIK-1	CAN	10 Nov.	104W	COM	6/4
101A	BMEWS-6	USA	20 Dec.	Drift	RECON	Inact
1973–013A	BMEWS	USA	6 Mar.	Drift	RECON	Inact
023A	ANIK-2	CAN	20 Apr.	106.5W	COM	6/4
040A	BMEWS	USA	12 June	Drift	RECON	Inact
058A	INTELSAT-IV-F7	INT	23 Aug.	56E	COM	6/4
100A	DSCS-3	USA	13 Dec.	Drift	COM	Inact
100B	DSCS4	USA	13 Dec.	54E	COM	8/7
1974–017A	COSMOS-637	USSR	26 Mar.	Drift	COM	Inact
022A	WESTAR-1	USA	13 Apr.	99W	COM	6/4
033A	SMS-1	USA	17 May	113W	MET	468M
039A	ATS-6	USA	30 May	140W	MULTI(E)	MULTI
060A	MOLNYA—1S	USSR	29 July	Drift	COM(E)	Inactive
075A	WESTAR-2	USA	10 Oct.	123.5W	COM	6/4
093A	INTELSAT-N-F8	INT	21 Nov.	174E	COM	6/4
094A	SKYNET-2B	UK	23 Nov.	Drift	COM	Inactive
101A	SYMPHONIE-1	FR/FRG	19 Dec.	49E	COM(E)	6/4
1975–011A	SMS-2	USA	6 Feb.	75W	MET	468M
038A	ANIK-3	CAN	7 May	114W	COM	6/4
042A	INTELSAT-IV-F1	INT	22 May	19W	COM	6/4
055A	BMEWS	USA	18 June	Drift	RECON	Inact
077A	SYMPHONIE-2	FR/FRG	27 Aug.	11.5W	COM(E)	6/4
091A	INTELSAT-NA-F1	INT	26 Sept.	25W	COM	6/4
097A	COSMOS-77S	USSR	8 Oct.	44E	COM	Inactive
100A	GOES-1	USA	16 Oct.	90W	MET	468M
117A	RCA SATCOM-1	USA	13 Dec.	135W	COM	6/4
118A	—	USA	14 Dec.	Drift	COM	Inact
123A	RADUGA-1	USSR	22 Dec.	Drift	COM	Inact

(cont.)

TABLE 7.4 *(continued)*

Designation	Name	Country	Launch Date	Nominal Longitude	Service	Main Frequency Band (GHz)
1976-004A	CTS-1	CAN	17 Jan.	142W	COM(E)	Inact
010A	INTELSAT IVA-F2	INT	29 Jan.	27.5W	COM	6/4
017A	MARISAT-1	USA	19 Feb.	15W	MAR	1.6/1.5
023A	LES-8	USA	15 Mar.	110W	COM(E)	8/7
023B	LES-9	USA	15 Mar.	100W	COM(E)	8/7
029A	RCA SATCOM-2	USA	26 Mar.	119W	COM	6/4
035A	NATO-3A	NATO	22 Apr.	18W	COM	8/7
042A	COMSTAR-1	USA	13 May	128W	COM	6/4
050A	—	USA	2 June	Drift	—	Inact
053A	MARISAT-2	USA	10 June	176.5E	MAR	1.6/1.5
066A	PALAPA-1	INDO	8 July	83E	COM	6/4
073A	COMSTAR-2	USA	22 July	95W	COM	6/4
092A	RADUGA-2	USSR	11 Sept.	85E	COM	6/4
101A	MARISAT-3	USA	14 Oct.	73E	MAR	1.6/1.5
107A	EKRAN-1	USSR	26 Oct.	99E	BRC	6/0.7
1977-005A	NATO-3B	NATO	28 Jan.	135W	COM	8/7
014A	ETS 2-(KIKU-2)	JAP	23 Feb.	130E	COM(E)	2.1/1.7
018A	PALAPA-2	INDO	10 Mar.	77E	COM	6/4
034A	DSCS II-7	USA	12 May	13W	COM	8/7
034B	DSCS II-8	USA	12 May	175E	COM	8/7
038A	—	USA	23 May	Drift	RECON	Inact
041A	INTELSAT-IVA-F4	INT	26 May	19.5W	COM	6/4
048A	GOES-2	USA	16 June	105W	MET	1.7
065A	GMS-1	JAP	14 July	140E	MET	1.7
071A	RADUGA-3	USSR	24 July	35E	COM	6/4
080A	SIRIO	ITAL	25 Aug.	15W	COM(E)	18/12
092A	EKRAN-2	USSR	20 Sept.	99E	BRC	6/0.7
108A	METEOSAT-1	ESA	23 Nov.	OE	MET	1.7
118A	CS SAKURA-2	JAP	15 Dec.	135E	COM	6/4

016A	FLTSATCOM-1	USA	9 Feb.	100W	MAR	240
035A	INTELSAT-IVA-F6	INT	31 Mar.	63E	COM	6/4
039A	BSE YURI	JAP	7 Apr.	110E	BRC	14/11
044A	OTS-2	ESA	11 May	10E	COM(E)	14/11
058A	—	USA	10 June	Drift	RECON	Inact
062A	GOES-3	USA	16 June	135W	MET	1.7
068A	COMSTAR-3	USA	29 June	87W	COM	6/4
071A	GEOS-2	ESA	14 July	0-29E	RES	2.2
073A	RADUGA-4	USSR	19 Nov.	35E	COM	6/4
106A	NATO-3C	NATO	19 Nov.	50W	COM	8/7
113A	DSCS-II-9	USA	13 Dec.	130W	COM	8/7
113B	DSCS-II-10	USA	13 Dec.	175E	COM	8/7
116A	ANIK-B1	CAN	16 Dec.	109W	COM	6/4
118A	GORIZONT-1	USSR	19 Dec.	13.5W	COM	6/4
1979-015A	EKRAN-3	USSR	21 Feb.	53E	BRC	6/0.7
035A	RADUGA-4	USSR	25 Apr.	35E	COM	6/4
038A	—	USA	4 May	75E	MAR	240M
053A	—	USA	10 June	Drift	RECON	Inact
062A	GORIZONT-2	USSR	5 July	13.5W	COM	6/4
072A	WESTAR-3	USA	10 Aug.	91W	COM	6/4
086A	—	USA	1 Oct.	—	RECON	—
087A	EKRAN-4	USSR	3 Oct.	53E	BRC	6/0.7
098A	DSCS-II-13	USA	21 Nov.	175E	COM	8/7
098B	DSCS-II-14	USA	21 Nov.	135W	COM	8/7
101A	RCA SATCOM-3	USA	7 Dec.	Drift	COM	Inact
105A	GORIZONT-3	USSR	28 Dec.	53E	COM	6/4
1980-004A	FLTSATCOM-3	USA	18 Jan.	23W	MAR	240M
016A	RADUGA-5	USSR	20 Feb.	80E	COM	6/4
049A	GORIZONT-4	USSR	14 June	—	COM	6/4
060A	EKRAN-5	USSR	15 July	—	BRC	6/0.7
074A	GOES-4	USA	9 Sept.	95W	MET	1.7
081A	RADUGA-6	USSR	6 Oct.	—	COM	6/4
087A	FLTSATCOM-4	USA	30 Oct.	172E	MAR	240M
091A	SBS 1	USA	15 Nov.	122W	COM	14/11
078A	INTELSAT-V	INT	6 Dec.	24.5W	COM	6/4, 14/11
104A	EXRAN-6	USSR	26 Dec.	—	BRC	6/4
1981-018A	COMSTAR-4	USA	21 Feb.	127W	COM	6/4
027A	RADUGA-7	USSR	18 Mar.	—	COM	6/4

TABLE 7.5 Planned Geostationary Satellite Systems

Name	Country	Service	Approximate Date	Frequency (GHz)
MARECS	ESA	MAR	1982	1.6/1.5
GALS	USSR	COM	1982	8/7
VOLNA	USSR	MOB	1982	1.6/1.5
LOUTCH	USSR	COM	1982	14/11
GOMS	USSR	MET	1982	1.7
STW	CHINA	COM	1982	6/4
INSAT (IND. NAT. SAT.)	INDIA	MULTI	1982	MULTI
ISCOM (APPLE)	INDIA	COM(E)	1982	6/4
ARABSAT	REGION	COM	1984	6/4
TELECOM	FRA	COM	1983	14/11
NORDSAT	REGION	COM		14/11
EBS (EUR. BRC. SAT.)	REGION	BRC	1982	14/11
ECS (EUR. COM. SAT.)	REGION	COM	1982	14/11
LEASAT	USA	COM	1984	8/7
TDRSS (TRACK & DATA RELAY)	USA	ISS	1983	15/13
TV-SAT	FRG	BRC	1985	19/12
TDF	FRA	BRC	1985	19/12
L-SAT	ESA	BRC	1986	14/11
PRC	CHINA	COM	1986	6/4
ZOHREH	IRAN	COM		14/11
CBSS	CANADA	BRC		14/11
SATCOL	COLUMBIA	COM		6/14
BRASILSAT	BRAZIL	COM		6/4
ITALSAT	ITALY	COM		30/20
CONDOR	PERU	COM		6/4

COM   = Communications
RECON = Reconnaissance
MULTI = Multi-purpose
RES   = Scientific Research
MOB   = Mobile
MET   = Meteorological
MAR   = Maritime
BRC   = Broadcasting
(E)   = Experimental

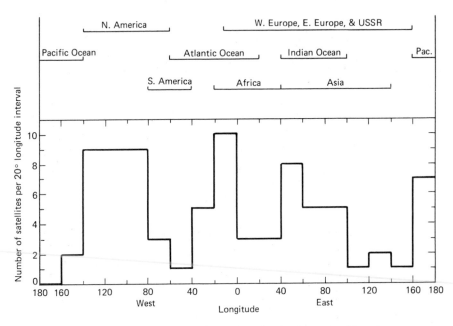

Figure 7.1 Geographic distribution of geostationary satellites.

interact with each other to make predictions even more hazardous. Some examples of these interactions are the following:

1. As demand grows and as existing systems become saturated, new technology will be developed.
2. If the new technology requires a rapid increase in price, the growth in demand will be slowed.
3. New technology for terrestrial communications, for example, fiber optics, may make terrestrial channels more competitive, thereby reducing the demand for satellite services.
4. Efforts to conserve energy may include replacing travel by communications, creating, for example, a demand for video conferences by satellite.
5. Advances in space transportation systems may make possible large space platforms that may provide economies of scale.

7.5 REVIEW AND PROJECTION OF TECHNOLOGY

As noted above, the demand for communication channels via satellite is expected to increase rapidly, with a doubling time in the near future of perhaps four years. Satellite capacity can be increased either by increasing the number of satellites that can use the orbit or by increasing the efficiency by which satellites use the radio-frequency spectrum. Some of the techniques that can be used for increasing capacity are discussed below and are summarized in Table 7.6.

Frequency allocations. At the WARC-79, the bandwidth allocated to the fixed satellite service in the 6/4 GHz region was approximately doubled, as was the bandwidth allocated in the 14/11 GHz region, although there are restraints

TABLE 7.6   Potential Utilization Improvement Factors

Purpose	Technology	Change	Approximate Improvement Factor	Notes
Reducing intersatellite spacing	Antenna beamwidth reduction	Larger antennas 30 m vs. 11 m	2	Increased earth station cost
	Antenna sidelobe reduction	7.5 dB reduction from CCIR standard	2	New antenna technology
	Station-keeping	$\pm 1°$ to $\pm 0.1°$	1.8	Recent satellites have $\pm 0.1°$ capability
	Optimum positioning	Multilateral co-ordination vs. bilateral co-ordination	1.5	Complex co-ordination and increased satellite cost
	Modified orbits	Inclined and eccentric orbits	2–4	New station-keeping technology
Increasing effective bandwidth	Available frequencies	Bandwidth increase from 1,000 MHz to 5,500 MHz	5.5	New communications technology
	Beam separation	Global coverage to 6 spot beams	6	Limited by concentration of traffic
	Polarization	Unpolarized to dual polarization	2	Implemented on some recent satellites
	Up-link/down-link assignments	Alteration of up-link/down-link frequencies	1.8	Interference and co-ordination problems
	Intersatellite links	Elimination of double-hops	1.1	New satellite technology
Increasing communication efficiency	Modulation and multiple access	FM/FDMA to DSI/TDMA	3	Increased system cost
	Data compression	Voice and TV signal compression	3–6	New communications technology

on the use of these frequencies in some countries. These allocations will take effect as of 1 January 1982. The 3,500 MHz bandwidth in the 30/20 GHz band, previously allocated and currently being used experimentally, was not changed, giving a total of over 5000 MHz bandwidth in these bands. This total allocation amounts to 11 times the bandwidth of the currently allocated 6/4 GHz band that is used by most general communications satellites today. Additional frequencies are allocated at 50/40 GHz, 80/70 GHz, 105/95 GHz, 210/150 GHz, and 270/240 GHz, a total bandwidth of 34 GHz, but these have not yet been used experimentally and there are problems resulting from atmospheric attenuation.

Up link/down link reversal. In each of these bands, equal bandwidths at different frequencies are designated for satellite to earth station transmissions and for earth station to satellite transmissions. A CCIR report (4) concludes that it may be feasible to place a satellite using the frequencies in the reverse sense in between two satellites using the frequencies in the currently designated sense, thus possibly increasing the number of satellites that could use a given portion of the orbit by 70 to 90 percent. This, however, raises new problems of interference between satellites on opposite sides of the orbit, of interference between earth stations, and of interference between earth stations and terrestrial communications systems.

Polarization. As noted in Section 7.1.2, electromagnetic radiation can be polarized in two independent modes, thus doubling the communication capacity compared with unpolarized transmissions. A number of satellites now in orbit use this technique to get an effective 1,000 MHz bandwidth in a 500 MHz bandwidth band. A problem arises at frequencies above 10 GHz, however, because of the depolarization of the transmitted signals by precipitation.

Beam separation. If separate beams as described in Section 7.1.2 are used for satellite transmissions, the same frequency can be used independently in each beam, thereby increasing the capacity of the system by a factor equal to the number of beams. The INTELSAT-V satellite uses beam separation to double the capacity of the 6/4 GHz band, in addition to the doubling achieved by dual polarization, to achieve fourfold frequency re-use for the first time. Beam separation is advantageous when the demand is widely distributed geographically, but it cannot increase capacity within a small region of high-traffic density. Since traffic is currently concentrated in certain regions of the world, the practical gains that could be obtained by multiple spot beams is rather less than the theoretically possible increase. In theory, for example, if each beam were 1.6° in diameter, corresponding to 1,000 km at the subsatellite point, and if the beams were 3.5° apart to avoid interference, about 20 beams could be used independently to serve different parts of the visible surface.

Satellite spacing. The distance between satellites in the geostationary orbit must be sufficient to prevent a ground station using one satellite from transmitting to or receiving from a neighboring satellite. The distance therefore depends on the station-keeping ability of the satellites and the shape of the beam of the ground station antennas. The WARC-79 reduced the station-keeping tolerance for the fixed satellite service to $\pm 0.1°$, thereby reducing the spacing necessary for avoiding interference. The beamwidth of the ground station transmissions and therefore the minimum distance between satellites decreases as the antenna diameter and the frequency are increased. Thus, the INTELSAT standard B antennas (11 m) require a larger satellite spacing than the standard A antennas (30 m), and given the same size antennas, satellites operating in the 14/11 GHz band could be more closely spaced than those operating in the 6/4 GHz band. However, satellite spacing depends not only on the width of the main beam but also on the level of radiation in the sidelobes. Advances in antenna design to reduce sidelobe radiation could permit reduced satellite spacing with the same size antennas. If the level could be lowered by 7.5 dB from the CCIR standard directivity for the sidelobe level, the capacity of the orbit could be about doubled.

In the Broadcasting Satellite Service the situation is somewhat different since relatively small and inexpensive receiving systems are essential to the system. Since these small stations do not transmit any signals, there is no problem of interference with other satellite systems; but since the satellites transmit relatively high-power signals, there is a problem of the satellites interfering with each other's broadcasts or with other services. This problem can be reduced by improving the performance of the receiving system, thereby permitting a reduction in the satellite's transmitted power level.

Modulation and multiple access technology. Satellite communications have generally used the frequency modulation (FM) system also used by terrestrial systems. In order to allow a number of earth stations to use the same satellite, a frequency division multiple access (FDMA) system has been used in which different earth stations use different frequencies within the assigned frequency band. The FDMA system, however, becomes less efficient as the number of earth stations increases. Time division multiple access (TDMA) systems using digital signals are now being developed to improve the efficiency of multiple access communications systems. The conversion of signals to digital form would also allow satellites to act as switchboards, switching transmissions between frequencies and between antennas with a flexibility that would provide an increase in efficiency. A third modulation system, the spread spectrum system, has been proposed for mobile communications systems serving very large numbers of stations for short periods of time.

Data compression. For many types of communications, in particular for voice and television which constitute most of the traffic, digital data compression techniques allow the essential information to be transmitted more efficiently. Thus, compared with the FM/FDMA system, a TDMA system with digital speech interpolation (DSI) can transmit voice three times as efficiently, and other digital data compression techniques can transmit high-fidelity television three times as efficiently.

Optimization of satellite positioning. The ITU Radio Regulations specify the allowable interference noise power induced in a satellite system in order to regulate mutual radio interference when a number of satellites are positioned in the geostationary orbit. From the standpoint of the utilization of the orbit, it would in principle be most effective to rearrange satellite positions so that each satellite would have an overall interference noise power below the allowable level. This could be achieved by readjusting the positions of all satellites whenever a new satellite system joins them. In reality, however, since it is difficult to readjust the positions of all satellites, satellite networks are co-ordinated such that the interference level of a new satellite network with other individual satellite networks is kept below the allowable noise power. In this bilateral co-ordination procedure, which is based upon the exchange of technical data as prescribed by the ITU Radio Regulations, the value recommended by CCIR as single-entry interference noise power is used as a limit of interference between two satellite networks. Co-ordination of this kind is likely to result in less effective utilization of the geostationary satellite orbit as compared with that implying the rearrangement of positions of all the satellites concerned. For instance, for a given amount of traffic, the required total orbital arc can be larger by some 50 percent if the former co-ordination method is used instead of the more efficient latter one. It will, of course, still be possible for new satellite networks to find room in the orbit by means of this bilateral co-ordination as long as congestion of the orbit is not serious. But it is obvious that this situation will change sooner or later.

As a result of a growing concern about possible shortage of capacity in the orbit, the WARC held in 1979 resolved that a WARC should be convened (in June 1985 and September 1987) with the objective of guaranteeing countries equitable access to the geostationary satellite orbit. CCIR has initiated preparatory studies on technical aspects of the question to determine the most suitable method to achieve this objective. One possible method to guarantee equitable access to all countries is the long-term planning of the orbit in a way similar to the broadcast satellite plan adopted by the WARC in 1977. In the case of long-term planning, the satellite spacing is determined assuming certain technical standards of the satellite and of the earth stations. Advances in technology, however, may render them obsolete and could consequently result in a somewhat less efficient utilization in the future. It is expected that ITU will establish an appropriate method to overcome the disadvantages of a rigid plan, and at the same time to guarantee equitable access to all countries without sacrificing the considerable potential for more effective utilization of the orbit, principally with regard to the radio-frequency spectrum.

It is shown in the CCIR study that in locating satellites in the geostationary orbit, the effective utilization factor of the orbit is highest for satellite systems having similar characteristics (homogeneous systems). Therefore, it may be desirable that satellites falling under different categories in respect of antenna diameters, receiving sensitivities, etc., be planned to be positioned in different sectors of the orbit or be operated on different frequency bands.

Nonstationary geosynchronous orbits. If satellites are placed in eccentric or inclined geosynchronous orbits, the satellites will move through single or double loops about a geostationary position with a period of 24 hours. A number of satellites could be placed at intervals on such a loop with the required minimum spacing between satellites. The number of satellites in such a system could be two to four times the number of regular geostationary satellites that could be placed in the orbital arc occupied by such a system. The system would require steerable ground antennas and very precise satellite position control mechanisms.

Intersatellite links. A geostationary satellite can provide a communication channel between two ground stations only if both are within the area of visibility of the satellite. If the ground stations are more than about 8,000 km apart, they can only communicate by way of an intermediate ground station and a second satellite, thereby using twice as much of the geostationary orbit/spectrum capacity as a channel via a single satellite. The efficiency of the system can therefore be improved by using direct intersatellite links to avoid "double-hops." Such links could use frequencies of high atmospheric attenuation that are not useful for satellite to ground transmissions and that are resistant to interference with terrestrial systems.

Multipurpose satellites and platforms. Combining two or more missions onto a single satellite having common power, orientation, and population systems could provide some reduction in the probability of collision through reduction of the total cross section, and in the case of multimission communication platforms, more efficient use of the radio-frequency spectrum through more flexible switching between ground networks, between geographic areas, and possibly through direct intersatellite or interplatform links. A single satellite serving several countries might be less expensive and could contribute less to congestion of the orbit than satellites for each country, particularly in the case of countries having a low level of demand. The Indonesian satellite PALAPA, for example, has spare capacity which has been offered to countries of the Association of South-East Asian Nations (ASEAN). Multimission systems, however, present substantial problems of co-ordination and of technical design. It is therefore not clear whether or not they would offer any net benefit.

7.6 CONCLUSION

Despite the great difficulty in predicting demand, it would appear that certain conclusions can be drawn regarding the use of the geostationary orbit in the next two decades. These conclusions are based on the following assumptions:

1. That the rate of growth in demand will not greatly exceed current rates.
2. That most of the technologies that now appear feasible will prove to be so and will be developed and used.
3. That the users of the geostationary orbit will co-ordinate their programs.

Given these assumptions, it would appear that foreseeable technological advances will permit the geostationary orbit to accommodate both the growth of existing systems and the introduction of new systems by new users for at least the next two decades.

This conclusion should be tempered by the following considerations:

1. Competing requests for specific positions using the most desirable frequency bands have started to appear and will appear more frequently in the future.

2. Future systems desiring access to the orbit, especially in the congested arcs, may have to use more advanced, and probably more expensive, technology.

3. Advanced technology being developed and used on a voluntary basis will gradually become mandatory.

4. The technology and the planning required to improve the efficiency of utilization of the orbit will exact a price, and the improvements realized will be closely related to the resources invested.

The burdens imposed by these considerations will fall most heavily on the developing countries, because of both their limited resources and their later entry into the orbit. Special measures may have to be taken in their favor.

It is possible that the assumptions made above may prove invalid. If the demand grows very rapidly and technological advances are not realized, conflicts could arise rather soon, especially in certain portions of the orbit and in the most widely used frequencies (6/4 GHz).

If radically new systems make large demands on the orbit, in particular very large systems such as solar power satellites, the situation might be altered profoundly, and solutions that would satisfy all demands for geostationary services might be very difficult to find.

In light of the problems of bilateral co-ordination and the advantages of multilateral co-ordination, and the increasing complexity of co-ordination as demand increases, existing regulatory mechanisms should be reviewed. It appears that these mechanisms could be revised to the mutual benefit of all parties concerned. In addition to the purely technical questions of efficient utilization, special attention might be given to the question of equitable access.

While, at present, the probability of collision between active and inactive satellites is low, it will grow continuously with time and will become very great if large structures such as solar power satellites are placed in geostationary orbit. To avoid this problem, efforts should be made to provide all geostationary satellites with the means to remove themselves from the geostationary orbit at the end of their active lifetimes.

Because of uncertainties in forecasts and of the rapid evolution of technology, the whole situation should be reviewed periodically (e.g., every five years). The reviews should take into account all aspects of utilization of the orbit, including technical, economic, legal, and political considerations. It is noteworthy in this respect that ITU will organize a conference on this subject in 1985 and 1987.

ACKNOWLEDGEMENTS

This chapter was prepared with the assistance of an international team of experts organized by the International Astronautical Federation (IAF) and including the following team members: Dr. E. Peytremann (Switzerland), Co-ordinator, and Dr. W. Arismunandar (Indonesia), Dr. K. Miya (Japan), Prof. G. S. Narimanov (Union of Soviet Socialist Republics), Dr. J. P. Singh (India), Ing. T. M. Tabanera (Argentina), and Dr. H. Weiss (United States of America). The team had discussions with and received comments from a large number of other people. The draft report was sent to the International Telecommunication Union (ITU) and the World Meteorological Organization (WMO) for comments, and the comments received have been incorporated into this chapter.

REFERENCES

1. Physical Nature and Technical Attributes of the Geostationary Orbit (A/AC.105/203 and Add.1-3).
2. Views of Member States on the Most Efficient and Economical Means of Using the Geostationary Orbit (A/AC.105/252 and Add.1).
3. ITU (CCIR): *Technical Bases for WARC-1979,* Special Preparatory Meeting, Geneva, 1978 (in particular Section 5.3.5, pp. 5.133-5.155).
4. ITU: *Recommendations and Reports of the CCIR,* Kyoto, 1978. Cf. in particular, Vol. IV, Report No. 453-2, p. 181: Fixed Service Using Communications Satellites.
5. ITU: *Final Acts* (WARC, Geneva, 1979). See resolution BP, R.13-18, relating to the use of the geostationary orbit and to the planning of space services utilizing it.
6. WMO (World Weather Watch): *Global Observing System-Satellite Subsystem* (publication No. 411 and supplements).
7. WMO (World Weather Watch): *Guide on the Global Observing System* (publication No. 488).
8. M. Hechler and J. C. Van der Ha, "The Probability of Collisions on the Geostationary Ring," *European Space Agency Journal,* Vol. 4, No. 3, 1980.
9. ITU: *Final Acts,* WARC, Geneva, 1979.
10. ITU: *Final Acts,* WARC-BS, Geneva, 1977.

Part III
SOCIAL AND ECONOMIC
ASPECTS OF
SPACE TECHNOLOGY

Chapter 8

RELEVANCE OF SPACE

ACTIVITIES TO ECONOMIC

AND SOCIAL DEVELOPMENT

ABSTRACT

This chapter is one of two concerning the socioeconomic aspects of space activities. Its purpose is to examine the contribution that space activities are making and could potentially make to economic and social development, particularly in the developing countries. Examples are given of results that some developing countries have achieved through the application of space technology. Also considered are the obstacles to the development of space activities in developing countries and the social and economic problems that can arise due to the introduction of new technology. Certain subjects that are relevant to economic and social development are discussed in detail in other chapters and are therefore only summarily included here. Current and future space technology and the technical aspects of its application are considered in Chapter 2 (technology), Chapter 3 (remote sensing and meteorology), and Chapter 6 (instructional satellite systems), and the needs of developing countries for education and training are considered in Chapter 9.

8.1 TECHNOLOGY AND DEVELOPMENT

8.1.1 The Introduction of Technology

Through the exploration of outer space, man acquires knowledge of the earth, the space environment, the sun, and other planets and stars. This knowledge gives us a new perspective on the nature of man and his home planet and on the interactions between the atmosphere, the oceans, the land surfaces, the subsurface, and man's activities. This perspective may prove to be vital to our growing population whose economic and social well-being will depend on our understanding of our complex planet and our relationship to it.

Without losing sight of the future, we must deal with immediate problems. In varying proportions and degrees of urgency for different countries and regions, we need food, education, health care, jobs, energy, and raw materials. To the extent that these needs can be met, the earth will be a richer and more stable place to live. Clearly, space technology cannot, by itself, solve these problems, because of both the limitations of space technology and the limited resources that can be made available for space activities. Nonetheless, it now seems clear that space technology can and will have an effect on all of these needs.

New technology can improve man's standard of living, but it can also exert a disruptive effect on his social and cultural systems. The development of technology in what are now the developed countries caused problems as it was introduced, grew, and was adapted over a period of centuries. This same technology can now be introduced full-blown into societies to which it has not been adapted. The potential disruption is much greater than anything faced by the developed countries, but the knowledge and understanding that can mitigate the disruptive effects are also more highly developed than they were in the past.

A society can maximize the benefits and minimize the disruption of space technology only if the decisions regarding its introduction and use are made through the decision-making processes of that society with full understanding of the implications of those decisions. This condition can be met only if technologically advanced countries cooperate with developing countries to ensure full access to information, satellite data, and the equipment and training necessary to participate in space activities.

8.1.2 Contribution to Human Needs

In order to discuss the contribution that space activities can make to economic and social development, it is essential to define what we mean by economic and social development. There have been many efforts to define the factors that constitute development and a variety of factors and priorities have been proposed. Traditionally, the per capita gross national product (GNP) has been used as an index, both to measure progress within a country and to compare the levels of different countries. While the per capita GNP may be a useful rough index, it does not accurately reflect the quality of life or the degree to which the basic needs of all of the people are met.

If economic development is difficult to measure, social development is even more difficult. While social services such as education and health care are universally considered desirable, there is no agreement on measures of their quality. Problems arise particularly concerning the distribution of social services as a factor in economic development. It is much easier to ensure access to such services by educated urban populations than by uneducated or rural populations. Even the wealthiest countries have difficulties in meeting the basic needs of all of the people.

This chapter does not attempt a quantitative assessment of the actual or potential economic benefits of space activities. Rather, the chapter focuses on two questions: What are the contributions that space activities can make to meeting people's basic needs and what are the potential effects of space technology on the distribution of economic and social services? People's basic needs are both physiological—food, clean air and water, health care, shelter—and social—education, communication, transportation. Table 8.1 gives a very rough indication of the extent to which basic needs are not being met in the world today.

This chapter will not attempt to survey all human needs or review all possible contributions of space activities but will discuss what currently appear to be the most important and direct contributions. The general approach outlined in this section is intended to be applicable to all countries while the specific examples given in the following sections will emphasize the needs of developing countries.

Communication is not only a basic human need itself but it also contributes to meeting all other needs. Economic and social development generally implies increased mobility of populations, and communications systems can help people maintain important relationships. Trade and commerce, whether national or international, depend on communications. Emergency communications in disaster situations can prevent or reduce suffering. Communications is the most

TABLE 8.1   The World's Poor 1974

Total world population	4,000 million
Undernourished	570 million
Illiterate adults	800 million
Children not in school	250 million
No access to effective medical care	1,500 million
Income less than $90 per year	1,300 million

highly developed of space technologies and most countries now use satellites for international communications. A few countries now use satellites for domestic communications, and the number will grow rapidly in the next decade.

Transportation, like communication, serves personal and economic needs. Access to transport facilities enables people to maintain and develop human relationships, exchange goods, expand their opportunities and help each other in time of need. Satellite remote sensing has contributed to route planning and construction nationally, and communications satellites provide operational navigation assistance to international shipping.

Education is both a prerequisite and a benefit of development. An educated person is better able both to provide for himself and to contribute to his society. In a modern industrial and technological society, only the educated person can fully enjoy the benefits or fulfill his responsibilities. Education should be understood here to include not just formal schooling but also the general learning process that continues throughout life. Broadcasting satellites have been used experimentally by a few countries for educational purposes, and they will undoubtedly be used more extensively as they become operational in the 1980s.

Food is essential to human well-being, and food production and distribution are important economic activities in all societies. Failures in food production or distribution are an important cause of human suffering and social instability in the world. Information from meteorological satellites can contribute to planning of planting and harvesting, and satellite remote sensing has been used experimentally for gathering information on crops and livestock. As remote sensing satellites become operational, agro-meteorological crop and livestock monitoring systems will help in agricultural planning.

Water is essential for drinking and other domestic uses, for agriculture, and for industry. The vagaries of the hydrological cycle, however, are such that water is frequently overabundant or very scarce. Water management is therefore essential if large variations in water supplies, food production, and industrial production are to be avoided. Meteorological satellites are being used operationally in many countries for monitoring the weather, and remote sensing satellite data are now being used experimentally for monitoring surface water and aiding in the search for subsurface water. The development of new sensors in the 1980s for monitoring soil moisture could make an important contribution to the efficient use of water resources.

Health care and safety contribute to general well-being and to economic productivity. Providing health services to isolated areas, however, is difficult and expensive. When disasters strike, the need for health services becomes much greater, as does the difficulty in providing it. Communications satellites can assist in providing expert medical assistance to local health workers, both in normal and emergency situations. In emergency situations, such as floods, data from remote sensing and meteorological satellites can be used to assess the extent of damage and to plan relief operations.

The factors of development discussed above exhaust neither the list of basic human needs nor the potential contribution of space technology. Rather, they constitute a range of needs and a range of technologies that provide a basis for discussing the role of space technology in development. There are two other aspects of space activities that should be mentioned here although they are outside the focus of this chapter. An important application of remote sensing is the search for commercially exploitable natural resources. The development of these resources can provide the material and financial means for meeting the human needs discussed above. Finally, the use of satellites for scientific research in such fields as geophysics, atmospheric physics, and astronomy can stimulate intellectual and educational activities and can, in some cases, lead to more direct economic and social benefits in the future.

Many of the services that can be provided by space technology can also be provided by other means; telecommunications can be by terrestrial microwave links, resource surveys can be done from aircraft or on the ground. A decision to use space technology will therefore generally imply two decisions: a decision to provide the services in question and a decision that space technology is the best means to provide those services.

A country can adopt space technology either on an *ad hoc* basis—with each agency deciding whether existing space technology is cost-effective for meeting its requirements—or by establishing a space program to develop technology to meet national needs. With the exception of a few countries which have been involved in the development of space technology since the beginning of the space age, most countries have initially adopted space technology on an *ad hoc* basis, using existing satellite systems for international telecommunications, for meteorology, or for remote sensing. Some of these countries have subsequently established space programs to develop or adapt new technology to their needs.

The establishment of a space program should be based on an assessment of the country's development priorities, its technical capabilities, and its financial resources. This assessment could be conducted by a multidisciplinary group including scientific and technical personnel familiar with space technology, economists, social scientists, and others interested in ensuring the proper utilization of space technology. To ensure a balanced assessment, it is important to have people who are not already committed to space technology by virtue of their positions.

For the assessment, the following methodology may be appropriate:

1. The group should study the possible applications of space technology within the country based on the existing technological capabilities in the world.

2. A detailed survey should be made of the areas of priority in the country taking into account economic plans or other documents which establish priorities.

3. The group should set down the selection criteria and rationale for the evaluation.

4. The group should carry out a methodical study of the impact of similar space activities on similar situations in other countries.

5. The group should then make a detailed analysis of how the space activities can contribute directly to meeting basic human needs.

6. An assessment of the indirect benefits should then be made separately.

7. A detailed study should be made of the infrastructure and other support facilities that will be required for the implementation of the given program, and also of what is involved in building these facilities.

8. As assessment should be made as to how the same objectives could be met by the use of more conventional technology already available either in the country or outside.

9. A detailed examination of the advantages, if any, of using conventional technology should be made (for example, the utilization of more manpower in a country where unemployment is a serious problem).

10. Finally, a comparison should be made between the use of space technology and alternative technology.

Such technological assessments often have a tendency to highlight abstract indirect benefits rather than the direct contribution to meeting human needs which is more difficult to realize and to assess. The temptation to justify the program on the abstract benefits and minimize the difficult process of assessing concrete results should be avoided. Possible harmful consequences, such as the effect of imported capital intensive technology on employment, must also be carefully considered.

The social implications of the technology should be carefully studied. The introduction of advanced technology into a low-technology society may increase the separation of the wealthier, educated population from the poorer, uneducated population. It is not only the total economic benefit that should be assessed, but also the distribution of benefits among the various economic, social, and cultural groups in the population.

Modern science and technology have made possible accelerated development on a large scale. However, there has been a tendency, in some cases, for the use of modern technology and expert assistance from advanced countries to result in the creation, in the developing country, of an internationally oriented technical elite that is insufficiently responsive to national needs. If space technology is to achieve maximum benefit for all the people of a country, the technology and organization of the application system must be integrated with local technology and local organization practices. This will generally require an adaptation of the technology to the local industrial, economic, social, and political environment.

In the developed countries, the cost of labor is relatively high, so the development of technology has emphasized automated systems, generally requiring a large investment per worker. Most developing countries are poor in economic resources and have a large force of unskilled and semiskilled labor. Developing countries therefore require technology based on machines that can be built locally with a minimum of specialized labor or else imported machines that provide work for large numbers of workers.

It may be useful to compare the transfer of technology to developing countries with the changes in technology that took place in the developed countries during the nineteenth century. In the first half of the century, the introduction of mechanization and the spread of commerce tended to increase poverty among the poorest segments of society, in some cases in spite of rising average wages. The extent of increased poverty depended on the balance between the elimination of existing economic activities and the creation of new activities, but the process of displacement and reabsorption was hardest on the poorer strata of society. Eventually, the poor did benefit from the economic expansion associated with the Industrial Revolution, but the benefits came some time after the initiation of the process of technical change.

The problem of maximizing the benefits of technological change while minimizing the harmful effects is complex and will probably be different for every country. If the benefits of the new technology are not apparent to all economic and social strata, the technology may come to be perceived as being introduced by the elite for their own benefit and as irrelevant to the real needs of the people. If this experience is repeated, a general opposition to all modern technology may arise and persist.

The integration of modern science and technology in a developing country and the participation of all strata in that process require great efforts toward information and education at all levels, from basic literacy for all to specialized technical training for a small number. The magnitude of the problem is indicated by Table 8.2.

TABLE 8.2   Educational Expenditures by Economic Groups of Countries (1974)

	Low-Income Countries	Lower-middle-Income Countries	Upper-middle-Income Countries	High-Income Countries
Population	1,341,300	1,145,400	470,600	1,057,000
Per capita GNP	$152	$338	$1,091	$4,361
Literacy rate	33%	34%	65%	97%
Per capita expenditure on education	$3	$10	$28	$217
Education as percent of GNP	2.0%	3.0%	2.6%	5.0%

8.2 APPLICATIONS IN DEVELOPING COUNTRIES

In most developing countries there is very little infrastructure on which applications of space technology can be based. The contribution that space technology can make therefore depends greatly on the simplicity or complexity of the application. At one end of the spectrum are applications requiring little or no specialized equipment or techniques, for example, the use of satellite images for surface water mapping using photo interpretation techniques. At the other end of the spectrum are applications requiring advanced electronic equipment and extensive ground systems, for example, a satellite-based crop statistics system or a broadcasting satellite system for rural areas.

By and large, both developing and developed countries have so far tended to apply space technology to functions that are not being performed by other technologies. When existing systems are replaced by satellite systems, problems can arise due to both an existing financial investment being made obsolete and people's experience and interest being made obsolete. Since developing countries have fewer established technologies, there may be less resistance to the introduction of space technology.

The infrastructure in the developing countries, in technical skills, managerial abilities, and data handling is, at present, inadequate to the needs of space technology. Most countries have had to and will continue to have to depend largely on commercial firms in the advanced countries to build turnkey systems. Some countries, such as India, Brazil, Indonesia, have been using space technology for many years now and have established substantial infrastructures. Even in these countries, however, cooperation with developed countries is still essential.

The educational requirements of a country wishing to adapt and integrate space technology include a range of programs from brief and concentrated introductions to policy questions for senior administrators to intensive technical training for engineers and technicians. While administrators may acquire the necessary knowledge through participation in meetings, symposia, and workshops, the training of technical personnel will require a greater effort. The general shortage of technically trained manpower in developing countries is particularly severe in technology related to space activities.

Training programs for technicians and engineers might take from six months to four years, although on-the-job training might replace training courses for the shorter programs. Technicians with high school diplomas might be recruited and trained for six months to two years in specialized courses. Engineers and scientists would normally be educated in university degree programs. Personnel with technical training but without exposure to space technology could benefit from short courses in the uses of space technology relating to their discipline.

Training the technical personnel required to ensure effective integration of space technology into development programs poses an enormous problem.

Space technology can also contribute to the training programs by enabling countries to make more effective use of their limited educational resources. However, even with the most efficient systems available, the overall problem of education for development remains formidable.

Some examples of the application of space technology to development will be discussed in detail in the following sections. There are examples from other developing countries as is indicated by the large number of countries that have made use of satellites for communications, meteorology, and remote sensing.

8.2.1 Education: The SITE Experiment

India conducted the Satellite Instructional Television Experiment (SITE) for one year (1975/1976). This experiment carried instructional and entertainment television programs to some 2,400 villages scattered over the country and was a collaborative effort between the National Aeronautics and Space Administration (NASA) and the Indian Space Research Organization (ISRO). The satellite used was ATS-6. Programs were transmitted to the satellite from stations in Delhi and Ahmedabad, were broadcast by the satellite, and were received by small receiving antennas in the villages. Technical information on the experiment is given in Chapter 6. The experiment sought to demonstrate the use of direct television broadcasting by satellites for rural development programs. It was a very complex mass-communication experiment of a magnitude never attempted before.

As a system test of satellite broadcasting technology in a country like India, the experiment was a success. All the ground equipment designed and built by ISRO performed well. The reliability of the earth stations was higher than 99.8 percent, and the community sets received a picture quality better than the normal VHF receivers in cities. The reliability of the receiving equipment coupled with the efficiency of the maintenance system (which required travel of as much as 70 km to repair one set) was sufficient to keep 90 percent of the sets working at any one time. The research and development capability generated during this experiment was a valuable spinoff. The project produced television programs in many rural areas and solved problems of equipment compatibility and logistics of tape movement for programs recorded in different parts of the country.

The average audience for community receiving sets settled to about 100 after the initial novelty had worn off. The number of viewers was largely limited by available space and not by the quality of programs. Many of the viewers of SITE programs were first-generation mass-media participants in the sense that they had never been exposed to radio, newspapers, or cinema. Most of the first-generation mass-media participants were illiterates and came from the poorer sections of rural society. On the basis of continuous feedback from 22 villages where 210 interviews were conducted every day, it was established that the audience favored instructional programs to sociocultural programs.

A large-scale longitudinal survey has shown evidence of statistically significant and unexpectedly large gains in information, awareness, and knowledge in areas such as health and hygiene, political consciousness, overall modernity, and family planning. The gains were greater for underprivileged sections of the rural society such as females and illiterates, and the gains increased with the degree of television viewing. In the area of agriculture, there is definite evidence in the holistic studies of a large number of innovations triggered by the television programs. Some case histories of these innovations indicated that farmers adopted only those new practices that did not demand additional expense or infrastructure. They were also secretive about their intentions until they achieved success. Either for these reasons, or because of lack of specificity in the questionnaire, the large longitudinal survey did not show statistically significant gains in agricultural production. On the other hand, a smaller survey using a different method suggested high gains in agriculture.

A survey of children exposed to the morning programs showed substantial gains in the area of language development and in seeking knowledge and information from sources other than conventional classroom teaching. Since the television programs for children were not syllabus-oriented, it is not surprising that there was no significant change in traditional scholastic achievement. The observed fact that the school attendance or drop-out rates were not affected by introduction of television in schools indicates that these factors depend primarily on social and economic parameters and not on the attractiveness or otherwise of the school curriculum; the children probaby do not have a choice in this matter. It was noticed, however, that during the television lessons the number of children present was greater than the number of registered students. Children learnt new stories and songs, and activities such as making of models and toys became popular in most of the schools. The experiment was particularly successful in focusing programs such as those involving teacher training and field worker training. It might be mentioned that 50,000 rural teachers were exposed to a multimedia package for teaching science and mathematics.

A number of lessons have been learnt about the efficacy of various types of programs, about the use of inexpensive portable equipment for decentralized participatory program production, and about the problems of programming for mixed rural and urban audiences. SITE was an exhilarating experience for everyone involved. It brought a large number of scientists, engineers, sociologists, and programmers close to the rural reality. It provided a practical example of how a large number of agencies with different basic disciplines can work in close partnership. SITE has given a permanent rural orientation to Indian broadcasting.

The above assessment of SITE emphasizes the technical achievements and some indicators of economic and social impact. Demonstrating actual economic and social change due to the educational programs is, of course, more difficult and is perhaps not to be expected from a program of limited duration and scope. The very success of the experiment highlights the difficulties involved in measuring the effect of any one activity on economic and social development.

The ATS-6 satellite was available to India for one year. Subsequently, a large fraction of the villages covered by SITE have been served by terrestrial transmitters specially set up for this purpose. It is largely because of the SITE experience that India has decided to have two television broadcast transponders on its operational satellite due to be launched early in 1982. SITE has also highlighted the importance of paying much more attention to the program production aspects and to the involvement of the education segments and other user agencies.

Extending educational satellite broadcasting from a controlled experiment in 2,400 selected villages to a system covering all of India would be an enormous task. The problems of recruiting and training personnel, maintaining receivers in remote areas, and producing the required volume of educational programs will be difficult and expensive. Another difficulty is posed by the diversity of languages in India, necessitating either dividing the satellite time between linguistic regions or providing several sound channels for each video channel.

The educational broadcasting satellite project in Brazil, using the ATS-6 satellite, also successfully demonstrated the feasibility of using satellites for education in developing countries. The extension of these two experiments into large-scale operational systems will undoubtedly pose problems as great as those overcome so far. The use of the ATS-6 satellite for the experiments was provided by the United States at no cost to the user countries. There are no plans currently for existing or planned broadcasting satellites to be made available to developing countries on a comparable basis. Future experiments may therefore require buying or leasing satellite capacity and may therefore be substantially more expensive. India and Brazil and a few other very large countries can perhaps afford national systems of this sort. For most developing countries, regional systems may be required for economic feasibility, making the organization of the system more complicated. In any case, if the potential of educational satellite broadcasting is to be realized, the next step will be the establishment of a pilot project using an operational satellite system.

8.2.2 Communications: The PALAPA System

In July of 1976, the PALAPA-1 satellite was launched making Indonesia the first developing country to have its own domestic communications satellite. Service was inaugurated in August 1976, providing telephone and television links between the various parts of the country, in most cases for the first time. A second satellite, PALAPA-2, was launched in March 1977. Services have been gradually expanded since 1976, and satellite channels have been leased to the Phillipines, Thailand, and Malaysia.

The Indonesian population of about 130 million is spread over hundreds of islands extending a distance of about 5,000 km from east to west. Prior to the establishment of satellite service, five islands were connected by microwave links and a sixth by a troposcatter system providing telephone, telegraph, and telex services. Some other areas are linked by high-frequency radio, but generally not on a 24-hour basis. The difficulty of communicating with remote areas was a great obstacle to national development, particularly since the development of the more remote areas has been a national priority. Many major development projects have had to operate with inadequate communications services. Extension of conventional links to new areas would have been a slow and expensive process. Since 1969 Indonesia has had reliable international communications via the INTELSAT Indian Ocean Satellite, but the domestic network has not been able to meet the demand. To compensate for the limitations of the national telecommunications network, some governmental and private agencies have established their own high-frequency radio links resulting in duplication, inefficient utilization, and frequency management problems.

The domestic satellite system is not designed to replace the good-quality terrestrial circuits in the areas of high demand. Rather, the objectives are the following:

1. To provide growth capacity for telephone, telegraph, and telex services throughout the country, particularly in the outer islands and as a back-up to the terrestrial microwave network in Java and Sumatra.
2. To extend television service to all of the provinces.
3. To introduce educational television on a national basis.

The satellite system is far more flexible than a terrestrial system since even the most remote area can be easily integrated into the system and the capacity of the system can be allocated to any distribution of demand that may evolve. Satellite technology also permitted a much more rapid introduction of the new links, the period between signing the contract and the inauguration of the system being 18 months. For these reasons, a satellite system was considered more cost effective than any alternative for meeting the needs of national development.

The system includes two satellites built and launched by the United States under contract to the Indonesian government. The satellites operate in the 6/4 GHz frequency band and are located over the equator, in geostationary orbit, at longitudes 83°E and 77°E. Each satellite carries 12 transponders, and each transponder has a capacity of 500 telephone circuits or one television channel. The first satellite is currently carrying the Indonesian domestic traffic and the second, intended primarily as a spare, provides the transponders leased to the Philippines, Thailand, and Malaysia.

The satellite system has permitted the radio links to be phased out. In 1978, of the estimated demand for some 3,200 long-distance telephone circuits, terrestrial links could provide 2,100 good-quality circuits, leaving a requirement for about 1,100 satellite circuits. One television channel is also required. Most of the satellite capacity is therefore available for growth. Currently, 40 percent of the circuits of five transponders are permanently assigned to 19 cities with the greatest demand, one operational and one backup transponder are shared by 21 light-traffic towns and a number of small terminals throughout the archipelago, one operational and one backup transponder are assigned to television transmissions, and three transponders are reserved for spares or for future growth.

The earth stations are of four types:

1. Master control station (1) for satellite control and all types of communications including television transmission.
2. Main traffic stations (18) for telephone, telegraph, and telex service and television reception; one of these has television transmission.
3. Light-traffic stations (21) for telephone, telegraph, and telex service, 15 with television reception.
4. Small terminals, 4.5 m in diameter, for telephone and telegraph service.

The small terminals are being installed as needed throughout the country. A staff of three assures 24-hour availability of one telephone and one telegraph circuit.

The satellite system permits a truly national television service. Television programs transmitted from Jakarta or Surabaya, the two earth stations with television transmission capability, can be received by the 32 earth stations with television reception capability and rebroadcast by a local VHF television transmitter for reception by ordinary television sets. A major objective of the television service is to support the development of the educational system. With the current system, educational programs can be used by any educational institution within range of one of the rebroadcasting stations. With future generations of PALAPA satellites, direct broadcasting from the satellite to special television receivers in schools may be possible.

8.2.3 Transportation: Road Building in Africa

Development of rural areas requires the development of a network of roads. Since the total length of a large network will be very great, and since initially the traffic will be rather light, low-cost unsurfaced gravel roads made of local materials may be the most cost-effective construction technique. Once established, these roads can be upgraded as required by the growth of traffic. In Upper Volta, a program to upgrade 500 km of secondary roads was carried out with the assistance of satellite remote sensing data from the LANDSAT satellites.

Topographic maps of the area at a scale of 1:500,000, compiled from aerial photographs taken in 1956, provided basic topographic information, but some important features have changed since that time. The remote sensing data and the topographic maps provided complementary data. The objective of the study of the remote sensing data was to identify the bridges and culverts that would be required, to identify lakes or reservoirs along the route that could provide water for construction, and to identify sources of road building materials.

In this semi-arid climate, with highly seasonal rainfall patterns, cloud-free satellite data, like aerial photography, are generally acquired in the dry season when the stream beds are dry. The width of the streams had to be deduced from the vegetation patterns due to the residual soil moisture early in the dry season. In general, the topographic maps provided the most complete information on the number and locations of stream beds to be bridged, and the satellite imagery provided the best information on the maximum width of the streams. In performing ground surveys of the route, it was found that orientation was significantly easier with the satellite images than with the maps. In some cases, streams as narrow as 1 m were visible on the satellite imagery because of the much greater width of the vegetation affected by ground water. A comparison of satellite images with the maps indicated that some streams had changed courses along roads that had been built previously without bridges or culverts.

An important factor in the cost of road building was the need for water for compaction and other construction purposes. Almost all of the water bodies in the area are small man-made reservoirs which have been built since 1956 and therefore do not appear on maps. A series of satellite images acquired at different times of year provides a rough estimate of the volume of the reservoir and its variation through the dry season. A number of reservoirs were identified along the route to provide water at minimum cost.

The basic road construction material in the region is laterite which is found in scattered outcrops. The satellite imagery indicated outcrops located along the route which could provide material at minimum cost. LANDSAT data have also been used in Libya to locate convenient sandstone deposits covered by surface sand. In the Kalahari region of Botswana, outcrops of bedrock are rare, and satellite imagery has been used to assist in the location of calcareous material known as calcrete for low-cost road construction.

Development of rural roads in developing countries requires fast, efficient surveys and low-cost construction using local materials. Satellite remote sensing can make substantial contributions to both of these requirements. In many cases, as in the Upper Volta Study, the images used were standard LANDSAT products, photographically reproduced and interpreted in the country. The cost of the remote sensing information was therefore very low.

8.2.4 Food and Water: Potential Benefit Studies

The food supply of a country depends on food production and food management. While space technology, in particular remote sensing and meteorology, has important potential contributions to make to food production, for example, through selection of planting and harvesting dates, soil moisture monitoring, and irrigation scheduling, it is in the field of food management that applications of space technology are more advanced. Much theoretical and practical work has been done on the use of remote sensing and satellite meteorology as part of agricultural information systems. Such information systems are essential to efficient management of transportation and storage of food and planning of imports and exports. In times of crop failure, the early availability of accurate estimates of food shortage can help prevent suffering and death.

Various methods have been suggested for measuring the economic benefits of accurate food production estimates. The benefits are generally assumed to be of three types:

1. Accurate estimates allow food to be allocated so that essential needs are met first followed by other uses according to priority, the priorities being set either by the market or by government planning.
2. Accurate estimates allow national reserve stocks to be maintained at a level that ensures continuity of food supply while minimizing storage and transportation costs.
3. If planting and harvesting occur throughout the year, accurate estimates allow planting to be planned taking into account existing crops.

In a case study of rice crop forecasts in Thailand it was envisaged that satellite remote sensing data could be interpreted to provide more accurate estimates of total acreage, cropping patterns, and intensity of cultivation. From this information, estimates of the acreage devoted to each of the three principal types of cultivation could be made. An economic model was used to calculate the economic gains that could be realized through more efficient allocation of food supply due to more accurate crop forecasts. With different assumed values for economic parameters such as price elasticity, the calculated benefits ranged from about $1 million annually to $49 million.

In a case study of range land management in Kenya, benefits are attributed to more efficient allocation of grazing land for domestic and wild animals. In Kenya, cattle and wild life are the major sources of local income. Cattle are raised both on a subsistence basis by certain ethnic groups and on a commercial basis. Cattle compete for range land with the abundant wild life which generates tourism which accounts for a major share of Kenya's foreign exchange earnings. The study considered the use of satellite remote sensing data as part of a range management program that would maximize economic benefits from commercial ranching and tourism while assuring adequate range resources for the pastoral ethnic groups. If use of satellite data could increase the range carrying capacity by 5 percent, the economic benefits that could be realized would be $8 million over a 20-year period, with future benefits discounted at 20 percent per year. The benefits were due 30 percent to ranching and 70 percent to tourism.

Satellite remote sensing data have been used in water resources management studies in Botswana and Thailand. In Botswana, repetitive satellite images of the Okavango Delta provided information for the development of a mathematical forecasting model which will have applications to agricultural, ranching, and mining activities. In Thailand, repetitive imagery providing a synoptic view of drainage and flooding patterns in the Mekong river basin has proved to be essential to understanding the complex hydrological processes in the region.

8.3 NEW TECHNOLOGY FOR DEVELOPMENT

8.3.1 Communications

One of the major areas of impact of space activities has been the advance in worldwide communications via satellite. The first communications satellite, SCORE, was launched in December 1958 and lasted 13 days. This was followed by ECHO-I, a passive communications satellite in 1960 and later on by TELSTAR, an active repeater satellite in 1962. The launch of INTELSAT-I in April 1965 started a new era which has dramatically improved communications throughout the world. The MOLNYA and GORIZONT satellites have provided international communication links through the INTERSPUTNIK network, and a number of systems have been launched to serve national networks. While this revolution is still in progress, the concept of communications has undergone a tremendous and permanent change. Transoceanic communications were earlier limited by the availability and capacity of submarine cables and high-frequency radio circuits. Communications satellites now provide the capability for information to be reliably transmitted to any place on earth where a satellite earth station has been located. This leads to the possibility that even small countries can be linked with the world communications network for relatively low costs. Even in the case of communications over land involving distances of more than 800 miles, communication satellites offer several advantages over cables and microwave radio relay systems.

In the coming decade, satellites with a capacity of 12,000 simultaneous telephone conversations will become operational as in the case of INTELSAT-V, the first of which was launched in December 1980. In the foreseeable future, satellites with 100,000 or more telephone circuits providing television and data channels to both large and small users will be a matter of reality. The concept of direct broadcast to small groups of villages or even individual villages has already been proved using the ATS-6 satellite.

INTELSAT-V, 25 years after the first communications satellite, provides several orders of magnitude higher capability. Compared with a total of 2,500 parts in INTELSAT-I, INTELSAT-V carries 35,000 parts. Despite this complexity, advances in modern technology have made it possible to reach reliabilities of the order of 99.999 percent.

The cost of INTELSAT research and development has been about 4 percent of the total revenue. Table 8.3 gives the costs of the space segment itself. The cost of an INTELSAT-I spacecraft in 1965 was about $4.5 million. INTELSAT-V spacecraft now cost approximately $33 million each. While the cost has gone

TABLE 8.3 INTELSAT Space Segment

INTELSAT satellite series	Operational date	No. of satellites procured	No. of satellites launched successfully	Incremental space segment costs (millions)
I	1965	2	1	$ 17
II	1966	5	3	$ 37
III	1968	8	5	$ 96
IV	1971	8	7	$270
IV-A	1975	6	5[a]	$280[a]
V	1979	7	5[a]	$470[a]

[a] Estimated.

up several times, the capacity of the satellite has increased even more. The launch costs of INTELSAT satellites have also been very high; for example, the launch of an INTELSAT-IV spacecraft costs about $30 million. Since the ground stations are owned and operated by each country, there are no available figures on total system cost. Total system costs are available for a proposed United States domestic system with 375 earth stations. The expected investment and operation costs are shown in Table 8.4.

In view of the large investments required and also the huge operating costs for the satellites, many countries are taking advantage of the INTELSAT operations and leasing satellite capacity. This is advantageous both for the IN-TELSAT organization and for the users. Table 8.5 shows the costs for use of the satellite by different countries.

New technology in the next decade will make it possible to construct and place in orbit very large satellites. This will make possible correspondingly smaller user terminals allowing the development of new applications which are neither feasible nor economically attractive with small satellites. In the past, the

TABLE 8.4 Costs of Proposed SBS System

Item	Investment (millions)	Operating Cost (millions)
Space segment	$117	$108
Earth Stations	176	121
System Management Facility	8	—
Space Ports	10	24
System Development	96	96
Headquarters	—	151
Maintenance	—	70
Total	$407	$570

TABLE 8.5 INTELSAT Space Segment Leases

Country	Date of service	Charges per year
Algeria	July 1975	$1,000,000
Brazil	July 1975	Regular charge for 360 units[a]
France	Aug. 1975	$1,000,000
Malaysia	Aug. 1975	$1,000,000
Nigeria	Dec. 1975	$1,000,000
Norway	Dec. 1975	$1,000,000
Spain	Mar. 1976	Regular charge for 180 units[a]
	Approved Contracts	
Colombia		$1,000,000
Nigeria		$1,000,000
Sudan		$1,000,000
Zaire		$1,000,000
Saudi Arabia		$1,000,000
Chad		$1,000,000
Libya		$1,000,000
Philippines		$1,000,000
Oman		$1,000,000
India		$1,000,000
Chile		$1,000,000

[a] The regular charge in 1976 was $8,280 per unit year; it is currently $7,380.

Note: The actual charges depend on whether a pre-emptible channel or a dedicated channel has been provided.

strategy has been to use inexpensive and small communications satellites and pay the price for very expensive ground stations. There is a reversal in this trend now and it is conceivable that in the next decade the operation of desktop ground terminals will become a reality. This in turn will make it possible for a larger number of users with low investment costs to use the satellite's capability.

In the future it may be possible to build very large—and very expensive—satellites to link up very large numbers of small inexpensive terminals. The very large investment required for the satellite may be compensated by the low cost per call if the number of terminals and the volume of communications are large enough.

8.3.2 Education

A natural outgrowth of the development of more powerful communications satellites requiring smaller earth stations has been the development of broadcasting satellites. In a broadcasting satellite system, one or more large earth stations can transmit radio or television programs to a satellite with a powerful transmitter which transmits the programs for reception by a large number of small terminals equipped for reception only. The first satellite with a broadcasting capability was the United States ATS-6 satellite launched in 1974.

The use of satellites for educational purposes has proved extremely advantageous in several situations, particularly in the United States. Several hundred experiments and demonstrations were conducted using ATS-6. For example, in the eastern United States, college-level courses were transmitted from the University of Kentucky to teachers in the Appalachian mountain region in order to explore the possibilities of upgrading teacher skills. In another experiment, ATS-6 was used to demonstrate the use of satellites for the improvement of the quality of primary education through an experiment in the West Indies for the transmission of course lectures between various islands. The ATS project also carried out an experiment on college curriculum sharing between Carleton University in Canada and Stanford University in California. This demonstrated the use of satellites in expanding the scope of curricula by sharing classes among universities and countries.

The use of broadcasting satellites for education has been experimental until now. Since terrestrial broadcasting or communications technology is little used in education today, it is not clear whether space technology could reduce the cost of education by replacing part of existing systems or would provide an additional resource which can be used by the existing system. A third possibility is that the most effective use of satellite educational broadcasting will be for programs directed at all ages and groups of people independently of the formal educational system, providing a parallel channel of mass education.

Given the uncertain role of satellite broadcasting in education, it would be difficult for a developing country to justify a satellite dedicated to educational broadcasting. India is resolving this problem by placing two broadcasting transponders on the INSAT satellites whose primary justification has been the communications and meteorological systems. Of a total space segment cost of about $100 million, perhaps about 25 percent could be attributed to the broadcasting system. On the ground, an antenna 3.7 m in diameter and the electronic equipment necessary to adapt a conventional television set to receive the signals may cost about $1,300. Receivers for 10,000 villages, for example, would cost about $13 million. Operation of the system, including program production and receiver maintenance, would also entail large expenditures.

8.3.3 Weather

Monitoring weather, particularly rainfall, can have substantial impact on economic activities. Of all the meteorological elements that influence man, rainfall is both the most significant and the most variable in space and time. Not surprisingly, therefore, it is also the most closely monitored. In most parts of the world additional rain guage networks are operated to supplement and extend the data coverage that would be provided by the general meteorological observation stations. However, it is beyond dispute that even these augmented rain gauge networks are still often inadequate to meet the needs of agencies, organizations, and individuals engaged in a very wide range of activities.

Water is vital to life. As its use grows, so we need to evaluate supplies better, both locally and globally, in order to plan the exploitation of water realistically, and so place the development of dependent operations such as domestic water supply, industrial expansion, and agricultural growth on firm foundations.

The monitoring of rainfall by conventional means is inadequate, especially in tropical and subtropical zones. In some cases, meteorological data disseminated via the Global Telecommunication System of the world weather watch are actually deteriorating. Meteorological stations in the international network are not maintained or operated to the specified standards, and some stations are closed. It is imperative that conventional methodologies for the evaluation of water resources be augmented by new approaches, so that we may obtain a better, not worse, view of rainfall distributions both in space and time. Remote sensing by satellites affords most hope that this may be achieved.

The network of conventional meteorological stations can be augmented by remote sensing techniques that are now becoming operational. The two technologies should be integrated to ensure reliable rainfall data. Remote sensing techniques, while useful, are not so well advanced that they could replace the venerable rain gauge. Furthermore, while rain gauge systems can be operated and maintained by each country, satellite rainfall estimation requires international cooperation and trained data interpreters.

Satellite-assisted rainfall monitoring has been used in irrigation design in Indonesia, water resource evaluation and management in the Arabian peninsula, river basin hydrology in the United Kingdom, and desert locust monitoring and control in northwest Africa. Results from the African studies, the most extensive so far, have been shown to be superior on some occasions to the existing rainfall information upon which crop predictions have been based, indicating something of the potential for satellite-improved rainfall assessment methods in agrometeorology and agroclimatology. A method, employing both rain gauge and satellite data has been used for desert locust survey and control in northwest Africa. This method is also being taught to northwest African nationals for use in a fully operational mode.

Satellite data were used in northeast Oman to assess the utility of a method for evaluating rainfall in a comparatively small area of rugged terrain and to provide information for subsequent rain gauge network design related to intended extensions of agriculture and coastal industry, both of which would increase the exploitation of local ground water reserves. Analysis of the monthly maps prepared as outputs from the exercise indicated that the method coped well with both spatial and temporal variations of rainfall.

It is in northwest Africa (Algeria, Libya, Morocco, and Tunisia) that the use of satellite data has had its widest application and is nearest to full operational implementation, in association with the activities of the Desert Locust

Control Commission for that region. Following successful early trials with the technique in the Ahaggar Massif of southern Algeria in 1976, it was applied to the whole desert locust recession area in those four countries (90 percent of their total surface area) during the winter and spring of 1976–1977 and to the same area in an operational training period for national personnel in the late winter and early spring of 1978–1979. Rainfall was mapped from conventional and satellite (NOAA–5 and METEOSAT) imagery.

For desert locust survey and control purposes, rainfall reports were required as soon as possible following the end of each week and aggregate reports following the end of each month. The monthly maps are used to help plan the movement of ground inspection teams, whose observations include the corroboration of evidence of rainfall, the development of associated vegetation, the level of moisture in the soil in potential locust breeding areas, and the identification and reporting of locust breeding areas and locust activity of different kinds. The monthly maps are also used as the basis for selection of high-resolution LANDSAT images to determine whether post-rainfall vegetation has in fact developed in the areas suggested by the rainfall mapping method. Surface characteristics and seasonal variations in vegetation growth have strong influences on rainfall and vegetation growth.

The advantages of satellite-assisted rainfall monitoring in remote desert areas were demonstrated during a brief test period in the Ahaggar Massif when special field data corroborated the satellite indication of a rainfall event north of Tamanrasset on 29 April 1976. During the more extensive applications in 1977, 1978, and 1979, there were many occasions when significant rainfall events in the desert were assessed using satellite data, which proved to be more accurate than the conventional data gathering methods using rain gauges.

For example, the satellite-improved map for the first week of April 1977 greatly clarified the extent of important events which yielded as much as 138 mm of rain at one station in northwest Libya—an event of considerable significance in an area so susceptible to flash flooding. The satellite image analyses indicated that the area of very heavy rain was relatively limited, but was surrounded by areas of moderate to heavy rain which were much more complex in pattern than the scant ground data had suggested. It is also worth observing that the combined use of conventional data and satellite imagery on occasion indicated very clearly those weather systems that had been responsible for the rains.

A further conclusion reached in the course of this work was that the extra rainfall information generated by the satellite data could be used in the assessment of crop conditions and potential yields, since this information was demonstrably more accurate than that on which crop assessments had been based in the FAO publication "Foodcrops and Shortages." The rainfall maps that will be prepared shortly on a fully operational basis for desert locust control support in north west Africa will represent the most detailed and accurate maps of rainfall available to anyone in that region for near real-time operations.

Rainfall monitoring can now be improved in many regions and for many purposes using satellite techniques. The cost of mounting such an operation depends on the size of the area in question, the climatic diversity of the area, the frequency and extent of rain-bearing cloud masses, and the purposes for which the monitoring program is intended.

Experience has shown that the cost of carrying out the analysis is small but data acquisition costs can be high, depending on the source of the satellite data. It is true that, for many purposes, operations can be designed to utilize satellite imagery supplied from outside a study region, but there are many potential applications of improved rainfall data for which daily or more frequent assessments may be required within hours rather than days of the completion of each period of time. For these needs to be met, it is necessary for satellite receiving equipment to be installed locally. The equipment should be capable of yielding good-quality images from the high-resolution visible and infrared radiometers on satellites of the TIROS–N family. A further capability to receive sector images of geostationary (e.g., METEOSAT, GMS, or GOES) satellite images could be invaluable in some cases. The capital outlay involved here would be in the range of $100,000 to $1 million. Clearly, for such a large investment to be justified, other possible uses of the satellite data beyond those in rainfall monitoring alone should be investigated.

8.3.4 Resources

LANDSAT–1, the first United States satellite dedicated to earth resources data experimentation, was launched in July 1972. Carrying a multispectral scanner, return beam vidicon cameras, tape recorders, and a data collection system, the satellite provides periodic imagery of the land surface of the earth. Nearly 80 developing countries have used remote sensing data from LANDSAT. This imagery, if it is to yield economic benefits, must be transformed into information which is useful for decision making. Managers normally are required to make decisions under conditions of less than perfect knowledge and information is sought to facilitate the decision process.

At the highest levels of planning, consideration is given to national goals and the "state of the world"; a national plan is formulated reflecting these considerations. Given the existing patterns of social and economic behavior, the national resource base, international considerations, and the political and economic feasibility of change via new technology, investment, and possible social and economic reorganization, the plan usually seeks to maximize public welfare.

The planning process requires knowledge of the national physical and cultural base. This includes the identification and distribution of natural resources (agricultural, geological, and hydrological) as well as existing patterns of land use. The traditional methods by which this information is acquired and stored is by means of cartographic and thematic maps. The production of these maps, by traditional means is usually costly and time-consuming; but without such a data base, a realistic national plan is difficult to achieve.

From the viewpoint of decision makers, information has economic value only in so far as it facilitates "better" decisions, broadly defined as those choices among alternatives which produce greater value. By evaluating projects in terms of present values, time is recognized as a major dimension in the analysis, and it is the time distribution of all costs and benefits which is relevant to the analysis.

The economic value of satellite imagery depends on its characteristics and the nature of the economy as well as on the cultural context into which it is introduced. To the extent that satellite data provide equivalent information to that acquired by conventional methods, but at lower cost, or permit "better" decisions (in terms of higher expected value) than would be possible in absence of the data, the data have economic significance.

During the next decade the technology of space remote sensing is expected to undergo steady growth and continuing change. Limitations and weaknesses of the first-generation technology will be reduced with systems being developed now.

Some advances have already come with the launching of LANDSAT–3 that has succeeded LANDSAT–1 and –2. The LANDSAT–3 multispectral scanner includes a fifth thermal sensing band, potentially valuable for estimating soil moisture; the return beam vidicon cameras have been modified into a two-camera panchromatic system with 40-m resolution. The enhanced capabilities have very much helped in some aspects of agriculture, in range management, in land use inventory, and in cartography.

The LANDSAT–D satellite will carry a second-generation scanner, the Thematic Mapper, equipped with seven spectral bands of 30-m resolution. The improved spatial and spectral resolution is especially significant for agriculture and range management—better crop identification, detection of soil moisture and crop stress, and assessment of range feed conditions—as well as for identification of geologic strata and fracture systems and recognition of numerous land-use classes. Remote sensing data have also been acquired by the SOYUZ–SALYUT (USSR), SKYLAB (United States), and BHASKARA (India) missions. Higher-resolution data will be provided by the SPOT (France) system to be launched in 1984.

Resource data analysis in this decade will also use data from a variety of special-purpose meteorological, oceanographic, and hydrological satellites. One experimental system employing imaging radar is of special interest to countries with persistent cloud cover which cannot be penetrated by sensors operating in the visible and infrared regions of the spectrum.

The multisource data, employed with plant growth, hydrological, and other dynamic models, should make possible the establishment of global and regional monitoring systems with important consequences for crop yield forecasting, water resource and range management, and observation of ecological conditions.

The analysis of satellite data ranges from relatively simple techniques, such as photo interpretation of visual imagery for identification of simple structures, areas, and broad classes, to very complex techniques, such as digital analysis of satellite data and computer enhancement of imagery of complex areas.

Many developing countries have begun to utilize LANDSAT imagery and have participated in training programs and in research projects. A shortage of trained personnel is the primary factor limiting the ability of developing countries to assimilate the technology. There is a need for programs to inform policy-makers, planners, and resource managers of the potentials of the technology, to provide short-term, in-depth training of scientists and resource specialists to enable them to analyze satellite data applicable to their resource sectors, and to provide longer-term academic training for those who will be involved with its technically more demanding aspects. To accommodate the larger number of people who will have to be trained in the next decade, educational institutions will have to modify their orientation and enlarge their capacities in this field.

Several countries have begun to take steps to institutionalize remote sensing activity within their bureaucracies; some have created special units for the purpose. The spread of receiving stations around the globe, each encompassing large areas, provides the opportunity for creation of cooperative, regional centers where basic facilities for processing can be concentrated.

Applications of satellite remote sensing have been developed largely in the developed countries. In general, these applications need to be adapted for use in developing countries. Some considerations based on experience in developing countries in some applications are presented below.

Crop Surveys: The use of LANDSAT data in developing countries has the following limitations:

1. The small irregular fields and the practice of intermingling crops, typical of many developing countries, will make it difficult to identify and measure the acreage of selected crops at current resolution levels.
2. Problems occur in the use of satellite data for the wet tropics because the same crop can be at many stages of growth throughout the year. In addition, current sensors cannot penetrate the seasonal cloud cover.

In spite of these limitations, if developing countries with inadequate crop acreage information can obtain crop acreage estimates of 75 to 80 percent accuracy from earth resources satellites, they will for the first time have a base upon which to design an agricultural statistical sampling system. Satellite data will provide improved crop acreage estimation in those countries that have sizable areas of large-scale, simple structure farming. It will also help in improved crop yield estimation through simultaneous use of meteorological satellites and other data and will provide better knowledge of soil capability and water availability.

Land Use: Satellite data can be used for land and soil reconnaissance surveys in the developing world. In Tanzania, digitally enhanced LANDSAT imagery, in conjunction with substantial small-area soil and plant community ground observations, with field and light aircraft observations, are providing a basic land system evaluation of the Arusha district at a scale of 1:250,000. In Mexico, soil maps at a scale of 1:1,000,000, intended to show the location and extent of the country's potentially arable soil resources, have been prepared largely with the aid of LANDSAT data. The sets (wet season and dry season) of LANDSAT color transparencies for most of the country provided the basis for the mapping project, which covered present land use as well as soil capability. The study of present land use covering the whole country (197 million hectares) took two years and cost $200,000 (0.1 cents per hectare). One significant finding was that 6.3 million hectares were in a state of advanced erosion. The study of soil capability covered 45 million hectares and was completed in one year. Satellite data were also used to indicate potential use for cultivated crops and irrigation.

Rangeland Management: Rangeland monitoring in developing countries can help with a number of important functions such as:

1. Timing of turnout and removal of livestock in grazing areas.
2. Preparation of plans to prevent overgrazing and to open up new areas for grazing.
3. Timely measures to reduce fire hazards.
4. Additional investment in range improvements by draining, irrigating, seeding, or fertilizing.

A study in the Arusha region of Tanzania employed LANDSAT data successfully in delineating boundaries for 550 distinct landscape units in an 8,300 square kilometer area on the basis of landform and vegetation characteristics. Fourteen grassland types of varying suitability for forage (three herbaceous, eight shrub/scrub, and three savannah) were recognized in the LANDSAT data. These delineations, fortified with detailed sampling information provided by aerial photography and on-site inspections, have identified promising areas for range, agricultural, and ground water development.

Forest resources: In most of the developing world, data on forest resources are generally limited in detail and extent. There are no comprehensive systems of forest inventory by aerial photography of the type found in the technologically advanced countries, but data requirements may also be substantially less stringent. In such a setting, LANDSAT data may be of significant benefit in providing basic information on the extent and location of forest resources and the changes occurring in the woodlands. Developing countries are increasingly aware of the need to manage their forest resources not only to meet their timber and energy needs but also to preserve the ecological balance and to prevent erosion, siltation of dams, and pollution of coastal waters. Mapping of forest vegetation, estimation of timber volume, and measurement of the rate of depletion, to which satellite data can make important contributions, are essential steps in planning control measures. In Brazil, LANDSAT imagery has been used to monitor a program for controlled development of large areas of the Amazonian forest for various purposes, especially cattle grazing. Landowners, with the help of government subsidies, are permitted to cut down trees up to a third of their land holdings. Routine and systematic use of LANDSAT imagery has proved to be the only economic way of enforcing the terms of the government-assistance contracts and of monitoring and controlling the volume of tree-cutting.

8.4 FUTURE ROLE OF DEVELOPING COUNTRIES IN SPACE

8.4.1 Operational Programs

The space technologies discussed in this chapter are being used, to a greater or lesser extent, in most developing countries. The majority of the countries have earth stations that communicate via INTELSAT or INTERSPUTNIK satellites, many have used remote sensing data from LANDSAT or from manned spacecraft, and many have established meteorological satellite receiving stations. A few countries have established LANDSAT receiving stations, and regional remote sensing programs are established or under consideration in several regions.

It seems clear that the established uses of space technology, such as international communications via satellite, will continue to grow. Additional countries will build ground stations, and existing stations will increase their volume. More recent developments, such as the use of leased transponders or national satellites for domestic communications networks, will develop rapidly in the next decade. Similarly, in the field of meteorology, satellite data and the network of receiving stations, using the operational system of geostationary meteorological satellites now being completed will continue to grow.

Experimental applications of space technology, such as remote sensing or educational broadcasting, will probably become operational before the end of the decade. National and regional ground stations and, in the case of broadcasting, satellites will be established, and the technology will be integrated into the resource management and education infrastructures in many developing countries.

8.4.2 Research and Development

Currently, most developing countries participate in space activities by using technology developed in the advanced countries. Only a few countries are actively engaged in research and development of new space technology. The lack of trained scientists and engineers and the limited financial resources make it very difficult for most developing countries to establish research and development programs.

There are various ways in which developing countries could participate in research and development activities without having complete space programs themselves. They could:

1. Participate in basic space research through the analysis and interpretation of data from research satellites.
2. Participate in the design and construction of satellites and ground receiving stations.
3. Research and develop new applications of communications satellites or of data from remote sensing and meteorological satellites.

Developing countries can undertake research and development primarily in cooperation with the developed countries. For problems of particular concern to developing countries, it may be desirable or necessary to establish cooperative programs between developing countries. This could be done through the following:

1. Bilateral programs with developing countries with space capabilities.
2. Regional or multilateral programs for research and development of applications for developing countries.
3. Regional or multilateral programs for designing and building satellites and ground receiving stations.
4. Regional or multilateral space agencies to develop a space capability.

8.4.3 International Cooperation

The resources required for space technology and the extensive coverage inherent in satellite orbits make regional or international cooperation desirable and, in many cases, even essential. If the infrastructure and personnel exist, the cost of a satellite system with ground support could be between $50 million and $500 million depending on the application. In many cases, this cost can be shared without reducing the benefits to any country.

Cooperative programs, particularly at the regional level, offer a number of advantages to developing countries:

1. Programs can be undertaken which surpass the capabilities of any single country.
2. The technology can be directed to the priorities of a group of countries with common needs.
3. The dissemination of data is controlled by the countries concerned.

8.4.4 Conclusion

Space activities will contribute to the economic and social development of developing countries to the extent that these countries take an active role in developing applications that meet their needs. This can and should be done through national programs, through cooperation between developing countries, and through cooperation with the technologically advanced countries. This cooperation should include financial and material assistance, but perhaps more important, it should include greatly expanded efforts to ensure that information on space activities is readily available and that people from developing countries will have access to a wide range of education and training programs in all relevant disciplines, at all technical levels, and for various periods.

ACKNOWLEDGEMENTS

This chapter was prepared with the assistance of a team of experts organized by the Committee on Science and Technology in Developing Countries (COSTED) of the International Council of Scientific Unions (ICSU). The team included the following members: Mr. S. Ramakrishna (India) (team leader), Mr. S. Ruttenberg (United States of America), Mr. J. Sahade (Argentina), and Mr. R. Sunaryo (Indonesia). The team consulted with other COSTED members and participants at COSTED meetings.

Additional assistance was received from the Committee on Space Research (COSPAR) of ICSU and the International Astronautical Federation (IAF). These contributions were compiled by Mr. J. Sahade (Argentina) and Mr. K. A. Ehricke (United States).

Chapter 9

TRAINING AND EDUCATION

OF USERS OF SPACE TECHNOLOGY

ABSTRACT

This chapter is the second of two on socioeconomic aspects of space activities. It was prepared with the assistance of an international team of experts organized by the International Institute for Aerial Survey and Earth Sciences (ITC) of Enschede, the Netherlands, and by the Committee on Space Research (COSPAR) (for Section 9.1), by the International Telecommunication Union (ITU) (for Section 9.2), and by the World Meteorological Organization (WMO) (for Section 9.3).

The purpose of this chapter is to examine the requirements and the opportunities for education and training in three major fields of space activities: remote sensing, communications, and meteorology. The current situation is reviewed, and to the extent possible, future requirements are considered. It is hoped that this chapter will assist governments in assessing their requirements and developing their national policies on education and training in space technology.

INTRODUCTION

An effective system of education is considered essential for all nations. Within this system, training programs are provided to develop specific skills. If a country is to use space technology effectively, it must have policy makers and administrators capable of assessing the political, social, and economic implications of the technology, scientists capable of developing and adapting the technology, engineers to design applications systems, technicians to construct and operate the systems, and teachers to teach the technology. A variety of education and training programs will be required, at various levels and with various orientations or specializations. It is important here to make a distinction between the terms education and training as used in this chapter.

The objectives of education are to bring the individual to an understanding of a subject, so that he may form independent opinions, establish priorities, understand, and discuss the methodology, the techniques used, and their applications. Education is concerned with the development of mental ability and of mental power, and thus with the attitude of persons.

The objectives of training are to teach individuals to carry out specific tasks based on an accepted methodology and for which known techniques are available. Understanding of the context is not always required; often only the ability to apply the technique is needed. Knowledge of the subject as a whole may not be necessary. Training brings the individual to a desired standard of efficiency. This is achieved by instruction and practice.

The specific requirements for education and training depend on the specific technology. Operational satellite systems in communications and meteorology require engineers and technicians with well-defined skills; experimental systems in remote sensing require a more flexible approach. The field of communications, with its large industrial infrastructure, may adopt satellite technology more easily than the fields of meteorology or remote sensing with their smaller infrastructure. The end users of the satellite technology may not be aware of the satellite as is usually the case in communications, they may receive interpreted data from meteorological satellites, or they may receive raw data from remote sensing satellites. In communications, therefore, only those directly involved with the satellite may need to be trained; in remote sensing a very wide range of people must be trained in various aspects of the new technology.

National policies on education and training in space technology should be part of an overall educational policy. Education should be regarded as a productive investment in human resources, resulting in personal growth and development, improved social satisfaction, higher efficiency, and better public services. Education and training are the indispensable complements of any investment in new technology and in expanded public services, and such investments are prime catalysts in socioeconomic development.

9.1 SATELLITE REMOTE SENSING

Remote sensing is here considered to refer in particular to data derived from satellite systems. In practice, the data are almost always used in association with data and information derived from aircraft and ground measurements and observations. It should be emphasized that the analysis and interpretation of such data only attain their full potential if they are integrated with existing methods of data analysis and classification for inventory and mapping.

If remote sensing data are to be obtained, interpreted, and applied, scientists and planners have to be educated in the properties and interactive processes of the natural environment. In addition, decision makers and politicians have to develop an awareness of the same so that the significance of the data obtained may be understood and may be applied in the context of economic, social, and political constraints. Men and women are required who are prepared to work across existing subject boundaries, to adjust to new techniques, and to apply these in their countries to socially relevant activities, as, for instance, environmental monitoring.

To create the awareness, flexibility, and motivation necessary to adjust to rapidly changing conditions, education is likely to be more effective than training for tasks which, because of the very nature of the new technologies, will alter quickly. To act as a catalyst to inject a new enthusiasm into the work of surveyors and earth scientists already using air photographs and fieldwork data, an understanding of the potential of resources satellite data is required. It should be recognized therefore that satellite remote sensing:

1. Provides information not previously available from aerial and/or ground survey procedures alone.
2. Provides new contexts for the reorganization and application of data already available.
3. Produces information more regularly and faster than has previously been economically possible.

The time available for the development and introduction of the necessary curricula is relatively short if full advantage is to be taken of the currently available satellite data. It appears, however, that the present experimental and the coming pre-operational phases of satellite remote sensing will continue through the 1980s. This will allow curricula development to be undertaken by both developing and industrialized countries. When, in due time, operational satellite systems are guaranteed, the educational and technical structure necessary for taking full advantage of their potential will then exist.

9.1.1 Remote Sensing Technology: Implications for Education and Training

9.1.1.1 Existing and planned remote sensing systems. Images of the earth's surface from space were acquired by early manned space missions of the USSR and the United States in the 1960s. These images indicated the potential of earth surveys from space, but the sporadic coverage in time and space and the variations in scale and geometry limited their usefulness. With the development in the 1970s of the automatic unmanned remote sensing satellites LAND-SAT 1, 2, and 3 (United States) and of the long-duration manned missions SOYUZ–SALYUT (USSR) and SKYLAB (United States), systematic acquisition and systematic application of remote sensing data became possible. In addition to the satellites specifically designed for earth resource surveys, meteorological satellites of the NOAA (United States) and METEOR (USSR) series were used for relatively low-resolution surveys, particularly for hydrological studies. In 1979, India launched a satellite for earth's observation (SEO), BHASKARA, a relatively low-resolution system which provided experience which is being used in planning for future high-resolution systems.

In the 1980s a variety of earth observation satellites will be launched and operated by a number of countries or groups of countries. These satellites will supply the user communities in various countries with images and digital data for research, for applications in existing and new fields, and for education. Because the design and construction of a new type earth observation satellite takes five to seven years, the current activities and plans with regard to these costly high-technology space vehicles will greatly determine the products users will get in the years to come. Moreover, the useful life of remote sensing satellites tends to be substantially longer than the design life, which itself has been improved from one year to three to five years. Hence it is possible to predict space technology activities over a period of up to ten years ahead.

The planning of education and training programs must therefore be based on these existing and planned systems. In addition to further missions in the SOYUZ–SALYUT and LANDSAT series, the following missions are planned for the early to mid-1980s:

1. The SPOT system (France, with Belgium and Sweden).
2. The marine observation satellites (MOS) and the earth resources satellites (ERS) (Japan).
3. The Land Application Satellite System (LASS) and Coastal Ocean Monitoring Satellite System (COMSS) [European Space Agency (ESA)].
4. The satellite for earth observation (SEO–II) (India).

Further information on these systems is provided in Chapter 5. Additional remote sensing satellite systems are being considered by Indonesia, Brazil,

China, and other countries. All of these systems are considered experimental or pre-operational, leading to operational systems in the late 1980s.

While most of the planned and proposed systems are designed to provide systematic coverage of most of the earth's surface, some systems are selective or specialized in their coverage.

1. A proposed Indonesian system, TERS would be placed in an equatorial orbit and would therefore only cover the equatorial zone.

2. SPOT will have a pointable sensor in order to increase frequency of coverage of selected areas.

3. A proposed STEREOSAT system would provide stereo pairs of images for topographic measurement.

4. A variety of short-duration experiments providing small quantities of data will be conducted during manned space missions.

These special features will provide useful data, but they will have limited impact on educational activities.

Other space activities relevant to this study include: meteorological programs, which include ocean observations; non-imaging remote sensing experiments (MAGSAT, GRAVSAT, microwave profiling and sounding techniques); conceptual studies for specific application satellites (ICESAT, MAPSAT, DESERTSAT, EARTHWATCH, etc.); and experimental programs (mainly designed to test remote sensing techniques in the spectral windows through which it is possible to observe the earth from space).

9.1.1.2 Technology and training.

Techniques for acquiring images of the earth's surface can be divided in five categories depending primarily on the region of the electromagnetic spectrum in which the observations are made. Before describing these techniques, however, two additional criteria must be introduced: spatial resolution and temporal resolution.

Spatial resolution is usually described as the size (or area) of the smallest meaningful picture element (pixel) on the earth surface. Orbital remote sensing pixel sizes range from 30 km to 0.3 m depending on the technique applied. Each order of magnitude has its specific applications, e.g., 30 to 3 km for meteorology and oceanography; 3,000 to 300 m for monitoring detailed meteorological processes; 300 to 30 m for observation of natural patterns, phenomena, and processes on land; 30 to 3 m for monitoring human activities and mapping man-made objects; 3 to 0.3 m for various local surveillance tasks.

Temporal resolution, the time interval between successive observations of the same area or objects, ranges with current orbital remote sensing from less than 20 min to a year or even to a decade. Satellites can be classified according to temporal resolution in a series of time intervals that increase by a factor about 3 (for instance, 20 min, 1 h, 3 h, 8 h, 1 day, 3 days, and then weekly, monthly, quarterly, yearly, etc.). The significance lies in the monitoring of processes visible on earth: A high temporal resolution, combined with a quick delivery of the data, is essential for following highly dynamic processes in nature or rapidly changing human activities.

Remote sensing of the earth's surface uses regions of the electromagnetic spectrum in which the atmosphere is transparent, the so-called atmospheric windows. The most frequently used windows are those in the visible and near infrared, the mid and far infrared (or thermal infrared), and the microwave regions of the electromagnetic spectrum. The techniques used in these windows are described below.

Multispectral scanning in the visible and near infrared windows will continue to be useful to developing and developed countries alike. Hence training and education have to continue to develop user communities in the 1980s. Data processing will gradually become standardized, providing high-quality products for visual analysis. Interactive automatic data analysis will be cost effective in special cases only. Preparations for monitoring, based on color imagery, will have to take place in this period.

Multispectral photography in the visible and near infrared with multilens cameras, with multiple cameras, or with multispectral imaging with TV-type cameras will continue for some time as the basis for experiments in certain applications. Education and training in these techniques should continue for those countries which participate in the experiments. Visual interpretation of color images, interactive image analysis on multispectral viewers/projectors, and interactive data analysis on electronic displays are among the fields of activities for which training is required. The slow and irregular dissemination of space photographs reduces their value for monitoring dynamic processes.

Panochromatic black and white imaging in the visible and near infrared with high spatial resolution (i.e., small pixel sizes) constitutes a practical perspective for those applications in which color, as such, supplies insufficient information. The trend toward smaller pixels from space started in the past decade and clearly continues, as shown by RBV images of LANDSAT-3, linear array "pushbroom techniques" in SPOT, metric camera experiments in SPACELAB, long focal length photographic cameras in other manned space programs, and specifications for STEREOSAT and MAPSAT programs. It is essential to give experienced photo interpreters additional education and training in the interpretation of these high-resolution black and white images, especially in stereoscopic image interpretation for mapping and inventory and for monitoring activities of people. Computer processing and automatic analysis of these images is hardly of interest; hence developing countries are here in the same position as developed countries. Change detection for updating maps and for production of small- and medium-scale map substitutes will be possible but may require an educational input to guide existing survey departments to adapt methods and infrastructure.

Thermographic remote sensing in the middle and far infrared windows as well as in the microwave window of the spectrum is useful for areawide appraisals of surface temperature and related parameters such as surface roughness. Uses are mainly in the fields of meteorology and oceanography, in coastal ocean monitoring, and in land applications where energy and temperature are direct and significant parameters. The intrinsic difficulties in interpreting thermographic records for earth science applications have not been substantially reduced by several thermographic earth observation experiments in the 1970s. The rather coarse spatial resolutions achievable in the 1980s will further limit the applications over land considerably. Therefore, it is likely that these remote sensing techniques will be of little value for worldwide applications in the earth sciences in the 1980s. Consequently, training and education can be limited to those research workers who have access to thermographic data from any meteorological satellite and are in a position to carry out fieldwork rather frequently.

Synthetic aperture radar (SAR) in the microwave window was demonstrated with SEASAT in 1978. Digital SAR data and images from several areas in North America and Western Europe are available for worldwide education. This side-looking radar observed the earth surface rather steeply for oceanographic applications; for land applications a more oblique viewing, for instance under 45°, is needed. SPACELAB experiments in the mid-1980s and experimental satellites with SAR, to be operated in the late 1980s by European and North American countries, will supply images for land applications. Due to satellite power restriction and limited receiving capabilities, only a reduced number of countries are going to be covered in this decade. Operational SAR satellite systems for civil applications are not yet foreseen. Although educational activities should give limited attention to satellite radar systems, research activities in certain countries may be backed up by specific educational programs in order to prepare for the late 1990s.

The more exotic imaging remote sensing techniques and the current and future non-imaging remote sensing techniques from space do not require a substantial educational effort in this decade, although some may be introduced and mentioned in regular courses aimed at the familiar techniques described above. It should be understood that the introduction of a completely new technology in society takes two to three decades. Proper familiarization, education, and training can ease the transition from exotic to conventional techniques but cannot actually shorten the introduction period. Moreover, these remote sensing techniques are usually applied in global programs and often are so specific that only a few research workers will get involved. Educational activities should primarily concentrate on those remote sensing techniques that are delivering a "user-friendly" product.

9.1.1.3 Applications.

The use of earth surveys from space may seem to be an extension of the use of airborne remote sensing for mapping and inventory. If this were true, education and training could follow the classical patterns and be mainly discipline-oriented in the earth sciences. However, a new technology will usually not overthrow a well-accepted and widely applied technology. Instead, the new technology will find its own users for new applications and will, together with the old techniques, improve the existing uses as to speed, efficiency, and economy and modify the classical applications and products. For this to happen the new technology must have certain advantages.

Satellite remote sensing has several properties not found in aerial photography. The data are acquired regularly and have a temporal resolution counted in days instead of years. Satellite imagery has a spatial resolution which is one or more orders of magnitude coarser than that of air photographs, which typically have equivalent pixel sizes from 1 to 0.1 m. The geometrical fidelity of a space image is usually high, that is, the planimetric properties compare well with medium- and small-scale maps, whereas terrain elevation and height of objects can be determined more accurately from air photographs by means of stereoscopic evaluation. These and other distinct and unconventional properties, like synoptic view/large area coverage, uniformity in greytone or color across the frame, possibility to acquire images under the same conditions (time of day or year, orbital position, pixel size, scanline orientation), make it clear that satellite remote sensing will lead to new surveying activities.

Applications of remote sensing satellite data can be divided into three categories: traditional surveying, mapping, and inventory activities; change detection and disaster surveying; and monitoring of dynamic phenomena. In a given discipline (e.g., geology, agriculture), one, two, or all three categories may be applicable.

Mapping and inventory are mainly concerned with the question of "what is where." Static situations and objects are being detected, identified, described, measured, and localized on earth and projected into a map. For the analysis of the earth's surface, many disciplinary fields have been developed: oceanography, hydrography, hydrology, geology, geomorphology, pedology, topography, forestry, other vegetation sciences as geobotany, and so on. In each field, maps at various scales are produced, often within the framework of systematic inventory of the whole territory of a country. Satellite imagery is useful for small- and medium-scale mapping activities in most disciplinary fields. However, the bulk of mapping activities lies in the production of large-scale maps, where the number of sheets is large and updating is needed frequently.

Satellite remote sensing data are also important for specialized or thematic surveys, in particular for multidisciplinary surveys: rural surveys, urban surveys, ecological surveys, integrated surveys for development planning, etc. The impact of satellite technology on these inventory and mapping activities will lead to new methods and products and will have to be accompanied by interdisciplinary and multidisciplinary education.

It is clear that in a rapidly developing society, change detection is of great importance. In modern society mapping suffers from the high rate of change:

change in land use in rural and in urban areas; change in the requirements for maps and inventories; change in concepts in the various disciplines of the earth and social sciences, leading to different interpretations of the same data; and change in the economical and technical factors on which the mapping methods are based. Inventory and mapping activities therefore should be based on this rule: "Do not collect or keep data that cannot be updated!"

Change detection can only develop as a regular surveying activity if data are required so often that the establishment of routines and structures for this dynamic activity is justified. Operational satellite remote sensing, therefore, will stimulate the development of change detection for:

1. Updating maps and other data records.
2. Producing map substitutes for evolving features (medium and small scale).
3. Comparing aged maps and new satellite images.
4. Studying processes by analyzing change patterns in a sequence of images or digital data, as a preparation for monitoring.
5. Rescue and relief planning, associated with the occurrence of disasters.

Modern remote sensing education and training should consider these activities.

Disaster surveying is a somewhat related activity concentrating on people, property, and livestock, but with a completely different setting. It is in some respect the ultimate form of change detection. It forms a challenging application of satellite data, requiring special change detection techniques with associated educational activities (possibly in the next decade by means of space technology) and well-trained *ad hoc* teams of surveyors, relief specialists, etc. Disasters can be natural (earthquake, cyclone, tsunami, flooding, extreme temperatures, excessive snowfall, etc.) but also man-made (fire, chemical pollution, accidental nuclear explosion, landslide, etc.). In all these cases, speedy information extraction, response, and action are essential. Planning disaster relief operations and subsequent monitoring, as well as damage assessment, are included. Disaster prediction and potential hazard mapping are related to those activities.

Monitoring is the active following of processes, directly or indirectly visible, in a series of satellite pictures, based on a thorough understanding of the nature of these processes and the factors influencing them. Knowledge about the way the various phases of a process manifest themselves in satellite data has to be acquired in advance by ground surveys. The objective of monitoring is to trigger actions if deviations observed are surpassing a set threshold. Monitoring, therefore, includes the establishment of criteria (which are often partly politically determined) and presumes that the will (and power) for action does exist. Monitoring programs will require extensive ground and air surveys in the early phase, relying more on the more economical satellite surveys as experience is gained.

Monitoring of moderately and highly dynamic processes in nature and of human outdoor activities will be feasible with operational satellite remote sensing systems and rapid delivery of data at low cost. Monitoring is highly desirable, for modern governments need information not only on "what is where" but also on "what is going on" and "what is wrong." Moreover, this surveying activity is also necessary for the overall economy of satellite earth observation, for monitoring has to be done frequently by many users.

Processes to be monitored are natural processes and man-induced processes. Natural processes are snowfall in the mountains; growth of natural vegetation; sedimentation in lakes and rivers; ocean circulation, coastal transport of pollutants by currents, and sea state; impact of drought. These are only to be monitored if they have an impact on society. The man-induced processes are of particular interest here. Examples are agricultural processes, shifting cultivation, deforestation, flooding, desertification, spreading of (vector) diseases, construction, air and water pollution, and urbanization.

Certain transborder processes and potential hazards (e.g., air and water pollution due to human activities elsewhere; spreading of vector diseases and locust swarms, flooding, desertification) may require a regional monitoring organization with national subcenters. However, as a general rule, monitoring and the related implementation of corrective and compensatory actions should be delegated to decentralized units and local authorities for reasons of efficiency (logistics and quick response) and in order to achieve participation by all concerned. There is no value in establishing a monitoring system for the observation of processes and activities affecting the environment unless a mechanism exists or can be established enabling action to be taken.

Monitoring of the environment should be carried out at grassroots level in a decentralized mode to stimulate participation of the local people. It is essential that each sizable community develop a capability to take action on the basis of adequate information about the environment. This can be achieved if so-called "barefoot surveyors" (local people responsible to the local authorities concerned) will become available to each community. These persons would require short periods of practical training in, among other things, satellite image reading, and in understanding those major environmental processes which have an every-day significance for their community. They could also form a link with professional surveyors in various governmental agencies and with national services.

The development of "geo-based information systems" is of considerable interest in the context of satellite surveying. Here data of various types, and from different sources, pertaining to particular areas on earth, are stored, manipulated, and transformed into data sheets, maps, and other products for planning and decision making. The flexibility of digital data processing combined with speedy input of new data (possibly from updating on the basis of satellite remote sensing records) offers new possibilities to the surveyor, cartographer, and planner. Because of the similarity between essential properties and processing methods of geo-based information systems and remote sensing systems, both should be treated together in education and training, if possible.

9.1.2 Remote Sensing Organizations and Training Opportunities

Since the first experimental observations from satellites were made in the mid-1960s, the communities involved in the application of space imagery have grown steadily. This trend is evident from the following:

1. The growing list of countries having earth observation satellite ground receiving stations.
2. The intensified exposure of the developing and developed countries to satellite remote sensing data applied in surveying projects.
3. The increasing use of space imagery data by international organizations in development assistance programs.

The existing structures of the community involved in the application of remote sensing consists of a number of groupings in which certain facilities for reception, distribution, user support, education, and training are found, either operated as regional facilities or as national facilities with international and/or regional functions.

Today's educational activities for satellite remote sensing consist of the following:

1. Universities and other educational institutions provide education as part of standard curricula or as separate programs.
2. Remote sensing centers, established for applied research in remote sensing and for its introduction to user agencies, often offer courses.
3. User agencies are conducting courses in the application of remote sensing or in digital data processing.
4. Countries are organizing irregular *ad hoc* educational courses in remote sensing on their own or in cooperation with other countries.
5. Private enterprises, independent research laboratories, and institutes are occasionally organizing workshops, seminars, and on-the-job training on a contract basis, often as part of development projects.
6. Owner-countries of earth observation satellites (so-called "space powers") by means of national or multinational space agencies are usually involved in activities creating awareness among potential users.

The total educational effort is of great value to countries in the position to have their people educated or trained, but the current situation has some disadvantages:

1. The various educational programs lack coordination.
2. Developing countries may have problems resulting from a temporary oversupply of educational opportunities offered in an incoherent manner.

3. Educational and training efforts tend to be directed to the top strata of the professional pyramid.

4. Current educational programs respond often solely to immediate needs, or are too strongly based on past experiences, instead of forming part of a longterm educational strategy or personnel development plan.

For countries at an early stage of setting up a particular structure for the use of satellite earth observation data, the cumulative effect of these disadvantages hampers their efforts.

Industrialized countries with negative demographic trends in the student age groups are confronted with an overcapacity of facilities for higher learning. In this context, international cooperation in remote sensing applied to environmental monitoring will stimulate reintegration of faculties and departments.

In view of the anticipated introduction of operational satellite remote sensing and the potential impact of this technology on society, and considering the educational capacity available for staff and personnel of the member states, there exists a structural shortage of educational potential as to facilities, curricula for modern types of applications, and teaching aids worldwide.

The following paragraphs list some of the organizations and training programs in the field of remote sensing with emphasis on regularly offered courses. The listing, by no means exhaustive, is meant to give an insight into the number of educational and training facilities regionally and worldwide.

The facts are based on information obtained from publications on those institutes that specialize in this field and on the results of questionnaires mailed to a number of institutes, organizations, government agencies, etc., all over the world. National educational institutions and those with a partly or wholly regional function, as well as institutions and organizations such as FAO and UNESCO, which provide courses, were specifically contacted.

9.1.2.1 Africa.

Africa is, in principle, characterized by five regional remote sensing centers under the umbrella of the African Remote Sensing Council (ARSC). The centers were created in accordance with resolution 313 (XIII) of the Council of Ministers of the United Nations Economic Commission for Africa (ECA). They are to receive, process, and distribute satellite data and to train and assist Africans in the use of these data for resources survey and development. The centers are located in Cairo, Egypt; Ile-Ife, Nigeria; Kinshasa, Zaire; Nairobi, Kenya; and Ouagadougou, Upper Volta. Of these, the Ouagadougou and Nairobi centers are the first to be operational. The LANDSAT ground receiving station of South Africa is not under the umbrella of ARSC.

Egypt. The Cairo Remote Sensing Center, under the aegis of the Egyptian Academy of Scientific Research, has a large collection of remote sensing equipment for research and production work, including laboratory and airborne instruments. Initially established as a national user support center, it is now designated as a regional training and user assistance center by ARSC. The center offers consultation and aid to other countries of the region in establishing their own remote sensing laboratories. It also plans to offer a three-week course on remote sensing every year.

Kenya. The Nairobi Regional Remote Sensing Facility has been attached to the already existing Regional Center for Services in Surveying and Mapping. Interpretation and analytical assistance are provided for users requiring facilities not available to them through their own agencies. The center provides seminars, short courses, and on-the-job training for users of remote sensing within the member states of ARSC (mainly for the countries of eastern Africa). Instruction is provided in English.

Nigeria. The Regional Center for Training in Aerial Surveys in Ile-Ife offers theoretical and practical courses in photogrammetry and elementary photo interpretation with a remote sensing component. Instruction is offered in both English and French.

Upper Volta. The Ouagadougou Regional Sensing Center is involved in training, user assistance, and the reproduction and distribution of satellite imagery. The training programs consist of practical introductory courses (three months), advanced courses (four months), and on-the-job training. Courses are geared for African attendance and instruction is provided in English and French. The first course was given in 1978.

Zaire. The Kinshasa Remote Sensing Center is still in the early stage of planning. Some training is already provided for nationals.

South Africa. South Africa has its own ground receiving station. The country's earth science institutes are involved in remote sensing education and training. Training and user assistance are also provided to other countries of the region.

9.1.2.2 Latin America.

According to the use of remote sensing data, various geographical groupings can be recognized. We may distinguish therefore Mexico, Central America and the Caribbean Islands, the Andean Pact countries (Bolivia, Colombia, Ecuador, Peru, Venezuela), Brazil and the countries of the south, or "los paises del Cono Sur" as they are locally known (Argentina, Chile, Paraguay, and Uruguay). Satellite remote sensing programs have been established by most of these groups, each approaching education and training somewhat differently in an attempt to deal with special needs and problems. Some countries have established a national remote sensing committee or coordinating body; others have no coordinating body, but a national lead agency can be identified. Some have no remote sensing infrastructure whatsoever. Currently, no overall coordination exists between the Latin American countries. However, at the meeting of the Society of Latin American Specialists in Remote

Sensing held in Quito, Ecuador, in November 1980, steps were taken to have better coordination in remote sensing education and training and to establish the Latin American Council on Remote Sensing. In addition, the Instituto Pan-Americano de Geografía e Historia (PAIGH) created an *ad hoc* working group to unify and define the criteria for education and training in remote sensing. The results of this working group were presented to the General Assembly of PAIGH in Bogota. The United Nations Economic Commission for Latin America (ECLA) and the Organization of American States (OAS) may also coordinate the various member states in the region in the future.

Three national educational centers with regional functions, which use Spanish as the language of instruction, and four mainly nationally oriented centers, to some extent open to foreigners, operate in Latin America.

Colombia. Centro Interamericano de Fotointerpretación (CIAF), a rather large institute, in 1968 started to offer courses in the use of aerial photography and remote sensing as applied to natural resources. The courses, offered annually, last 40 weeks and are oriented to civil engineers, geologists, soil scientists, and vegetation specialists. Short courses of 14 to 16 weeks are also conducted in the application of remote sensing to various disciplines. CIAF also provides consultation and research.

Panama. The Defense Mapping Agency, Inter-American Geodetic Survey, Cartographic School (DMA-IAGS), established in 1952, provides photogrammetry and cartography courses for countries of South and Central America. In addition, it annually offers a five-week course on the theory and application of remote sensing.

Venezuela. The Centro Interamericano de Desarrollo Integral de Aguas y Tierras (CIDIAT) is planning to offer two-week introductory courses in remote sensing.

Argentina. The Comisión Nacional de Investigaciones Espaciales (CNIE) operates its own ground receiving station and is responsible for processing and distributing LANDSAT data. It conducts its own seminars, workshops, and courses in remote sensing in collaboration with national educational institutes and organizations and with international organizations.

Brazil. The Instituto de Pesquisas Espaciais (INPE) operates a LANDSAT ground receiving station. It has ties with national educational institutes and runs its own seminars, workshops, and courses in remote sensing.

Ecuador. The basic objective of the Centro de Levantamientos Integrados de Recursos Naturales por Sensores Remotos (CLIRSEN) is to develop remote sensing technology in Ecuador. Its main projects are training and providing technical assistance on remote sensing to other agencies (public and private). CLIRSEN is planning to offer seminars at the national level.

Mexico. The IBM Centro Cientifico has a remote sensing program, mainly for research purposes. On request it offers courses in remote sensing (mainly in digital image processing) for participants from Latin America.

9.1.2.3 North America.   Due to the large numbers of individuals and agencies in North America that are engaged in the use of remotely sensed data and the existence of numerous educational facilities prior to the advent of satellite remote sensing, the community involved is rather diversified.

In both Canada and the United States, the initial processing and dissemination of satellite data occur at a central facility: the Canada Center for Remote Sensing (CCRS) in Canada and the National Aeronautics and Space Administration (NASA)/United States Geological Survey (USGS) in the United States. Education, training, and data analysis, however, occur at numerous federal, state, provincial, and private facilities. In the United States, formal education and training of federal users occurs primarily at one facility, the Earth Resources Observation System (EROS) Data Center, but many user agencies do provide some programs to their employees in both formal and on-the-job training. In addition, many universities provide opportunities for the training of practicing professionals through extension or continuing education courses. Many federal and state agencies possess analysis facilities geared to their own specific needs and, in addition, many private companies commercially offer analysis services. Canada has some coordination at a national level with federal and provincial agencies, institutions, and facilities, which is partly discipline and partly problem-oriented. CCRS, as a federal agency, provides a supportive role to Canada's provinces which have the responsibility for education. It has either funded, contributed to, or conducted training workshops on specific topics on remote sensing techniques.

Numerous universities, organizations, institutes, centers, etc., in Canada and the United States offer regularly scheduled courses in satellite remote sensing open to foreign participants. Four representative examples from Canada are described below. For the United States, a selection of three internationally oriented courses, in which USGS is involved, and a selection of courses with a significant international orientation given by a laboratory, an institute, and a university are described below. Three of these six offer on request custom-designed training programs under contract with national or international organizations. A list of educational possibilites in remote sensing can be obtained from the EROS Data Center, Sioux Fall, South Dakota.

9.1.2.3.1 Canada.   The CCRS in Ottawa sponsors digital workshops open to foreigners and offers postdoctoral fellowships and visiting scientist programs through the National Research Council.

The Alberta Remote Sensing Center and the Quebec Remote Sensing Center promote educational and professional development in remote sensing and conduct remote sensing courses.

The Geological Survey of Canada of the Department of Energy, Mines and Resources, Ottawa, gives short courses and on-the-job training in remote sensing.

Laval University in Quebec gives postgraduate courses in remote sensing and organizes short courses for participants from developing countries.

9.1.2.3.2 United States. The EROS Data Center (EDC) in Sioux Falls, South Dakota, a center of USGS, offers a variety of international training programs. Twice a year, five-week workshops are given in remote sensing with emphasis on LANDSAT data. The May workshop concentrates on vegetation assessment and land-use planning, the September workshop on geology and hydrology. Both include visual and digital analysis. EDC also offers numerous short courses, in particular in applications or analysis techniques.

North Arizona University and USGS, Arizona, jointly offer, on a regular basis, advanced international five-week courses in geological interpretation, land-use planning, environmental applications, and digital image processing.

The Office of International Geology, USGS, Virginia, regularly offers a one-week introductory course in data processing as a preparation for the digital image processing course conducted by USGS.

The Laboratory for Applications of Remote Sensing (LARS), Indiana, a research laboratory of Purdue University, offers training opportunities to foreign scientists in computer-aided analysis of multispectral remote sensing data. A one-week introductory short course is conducted monthly, and a one-week advanced course is taught annually, usually in the spring. LARS also offers a visiting scientists program.

The Environmental Research Institute of Michigan (ERIM) conducts training courses on a variety of remote sensing applications and analysis techniques. The training courses are not regularly scheduled but are often held in conjunction with development aid projects.

The Remote Sensing Institue (RSI) of South Dakota State University offers a 6 to 12 months' visiting international scientists educational program in remote sensing technology, with emphasis on LANDSAT data. The program provides training in the fundamentals of remote sensing and includes field and laboratory experiments.

9.1.2.4 West Asia (ECWA region).

Most countries have their own national remote sensing programs that are discipline oriented. The United Nations Economic Commission for Western Asia (ECWA) may play a role in the coordination of national programs.

9.1.2.5 Asia (ESCAP region).

Almost all countries within this region have national remote sensing programs established and are developing methodologies for mapping natural resources and monitoring environmental changes. The United Nations Economic and Social Commission for Asia and the Pacific (ESCAP) is already undertaking to provide coordination. Several countries have formed new organizational units to guide remote sensing research, provide services to user departments, conduct resources surveys, maintain data banks, organize education or training programs, and publish results of the use of remote sensing data and photography from satellites, aircraft, and balloons.

Countries with ground receiving stations for LANDSAT satellites are Australia, China (under construction), India, Iran (temporarily not functioning), Japan, and Thailand (under construction). The Indian Space Research Organization (ISRO) has a space program in which three satellites, including a remote sensing satellite, have been launched. Education and training within the region are available in many institutes of higher learning and also in some government departments which focus on on-the-job training and research programs oriented to their own needs. Some ESCAP countries are newcomers to the use of the technique and have no remote sensing infrastructure whatsoever. The only regional facility is the regional training center in Thailand.

Thailand. The Asian Regional Remote Sensing Training Center ARR-STC) at the Asian Institute of Technology (AIT) plans to conduct short courses in remote sensing (particularly on computer-aided analysis) to participants from the ESCAP region. Trainees for ARRSTC will be selected by the national remote sensing centers or programs.

Australia. Part-time education in remote sensing is mainly carried out at universities and institutes of technology, such as the South Australian Institute of Technology and the Footscray Institute of Technology, as components of undergraduate and postgraduate courses in earth sciences and surveying. Workshops and short courses in remote sensing, open to nationals of any country, are given by Footscray and the Australian Mineral Foundation.

Bangladesh. The Space Research and Remote Sensing Organization (SPARRSO) arranges in-service training, short courses, and package courses in remote sensing from time to time. Chittagong University (in cooperation with SPARRSO) and Dacca University include remote sensing in some of their postgraduate courses.

China. A large number of colleges and universities are dealing with remote sensing, mainly in research and education. The relevant institutions of higher learning and research have organized several remote sensing courses for over 2,000 national participants of different levels. Beijing University, in cooperation with other universities, has recently established a remote sensing training center.

India. The National Remote Sensing Agency (NRSA), Andhra Pradesh, has trained its own personnel. It has conducted several short-term special courses and on-the-job training in remote sensing tailored to user's requirements. Courses are primarily for national attendance but foreign participation is possible. NRSA also provides consultation on all aspects of remote sensing. The Indian Photo-interpretation Institute (IPI), Uttar Pradesh, is a part of NRSA. About 1,000 were graduated from the Institute between 1966 and 1980, and it is the largest educational institute in remote sensing in the ESCAP region. IPI provides a number of courses of up to one year in duration in airphoto-interpretation for earth sciences in which satellite remote sensing has lately been incorporated. These courses are primarily meant for national attendance, but some foreign participants are regularly accommodated. IPI also carries out research and is involved in consultation. Educational facilities in remote sensing are also available in some government departments and universities such as the Center of Studies in Resource Engineering (CSRE), Indian Institutes of Technology, Osmania University in Andhra Pradesh, Perarignar Anna University of Technology in Tamil Nadu, and Gujarat University.

Indonesia. At present this country undertakes its own remote sensing education and training programs through its respective national agencies and institutions. Short training courses are available at the National Co-ordination Board for Surveys and Mapping (BAKOSURTANAL). Facilities for education in remote sensing are also available at the Gadjah Mada University in cooperation with BAKOSURTANAL. Remote sensing on-the-job training is conducted at various government departments.

Iran. Education and training facilities on the application of satellite data to earth resources are organized by the Remote Sensing Center of Iran.

Japan. The Remote Sensing Technology Center of Japan (RESTEC) is involved in regional remote sensing education and training. RESTEC conducts educational programs including on-the-job training in computer-aided analysis of satellite data. It also organizes seminars and symposia in this field. Since 1978, one-month courses are organized annually for participants from the ESCAP region.

New Zealand. Education in remote sensing is included in the geography courses at the University of Auckland. The Physics and Engineering Laboratory (PEL) in the Department of Scientific and Industrial Research conducts short courses in remote sensing.

Pakistan. Through the Space and Upper Atmosphere Research Committee (SUPARCO) some training seminars are organized periodically for national insitutions in the applications of remote sensing. SUPARCO also organizes on-the-job training for a limited number of participants from the ESCAP region.

Sri Lanka. The Survey of Sri Lanka has a remote sensing section for resources evaluation. In-service training for a limited number of government officers is foreseen.

9.1.2.6 Western Europe (ESA, EEC, and the Council of Europe regions).

The European Space Agency (ESA) has been a major force in developing an awareness of the potential of satellite remote sensing by stimulating and sponsoring meetings and studies of satellite data applications. Recently it has established EARTHNET, a network for receiving, archiving, preprocessing, and distributing satellite imagery and data. The member states participating are Belgium, Denmark, the Federal Republic of Germany, France, Ireland, the Netherlands, Spain, Sweden, Switzerland, and the United Kingdom; Norway is considering participation. EARTHNET will be the ground segment in the European Remote Sensing Satellite Program. Receiving stations are at Fucino, Italy (LANDSAT); Kiruna, Sweden (LANDSAT); Lannion,

France (HCMM and NIMBUS G); Maspalomas, Spain (NIMBUS G). Additional receiving stations will be established, as for example, for SPOT at Toulouse, France. Preprocessing of SEASAT data occurs at Farnborough, United Kingdom, and Oberpfaffenhofen, Federal Republic of Germany. Each country maintains a national point of contact for providing data to users. Some of these are located within institutes of remote sensing research (Oberpfaffenhofen), some are developing user assistance services and their own applied research program (e.g., Sweden), and some are developing as national research centers (Farnborough).

The European Economic Community (EEC), through its Joint Research Center (JRC) at Ispra, Italy, in cooperation with European laboratories and research institutes, has initiated and coordinated education and training programs for Europe and the developing countries. A Demonstration Center has been established at Ispra. EEC cooperates with ESA by financing and organizing experimental remote sensing flights. JRC operates through various groups and laboratories to provide data, organizational support, and research and consulting advice.

The Council of Europe has been active in supporting discussions on applications of satellite data (Toulouse Conference) and in promoting the coordination of education in remote sensing. The Council of Europe maintains close relations with ESA and, through JRC, with EEC.

Some countries have developed national programs. For example, France has a well-organized national structure for the development and evaluation of satellite data for research objectives—Opération pilote interministérielle de télédétection (OPIT)—and has established its own satellite project: Satellite probatoire d'observation de la Terre (SPOT). In Italy, a commercial company, Telespazio, established with government support to develop and operate satellite earth stations, is now taking part in EARTHNET. Telespazio also provides support and consultant services in digital processing of remote sensing data.

Remote sensing is taught in Western Europe within the context of existing courses in airphoto-interpretation that are provided in geology, soil science, geography, agriculture, etc. These courses are designed for national participants but are generally open to foreign students. In some cases, a student may prepare a dissertation or thesis on a remote sensing subject as part of a remote sensing course associated with a research program.

Among the courses regularly offered in remote sensing applications with emphasis on satellite data are: in France, the courses provided by the Groupement pour le développement de la télédétection aérospatiale (GDTA), conducted in English and French, and those by the University of Strasbourg; in the Netherlands, the courses offered annually by the International Institute for Aerial Survey and Earth Sciences (ITC), ranging from 2 to 12 months' duration, and conducted in English for up to 200 students from the developing countries.

Short seminars, summer schools, symposia, etc., are provided by certain departments, laboratories, universities, institutes, etc., constituting either very general reviews of remote sensing or very specific training in the applications of techniques. Some national points of contact, for example, in Sweden and the United Kingdom, are now also providing such short courses. The Land Resources Division of the Ministry of Overseas Development in the United Kingdom provides short courses for students from developing countries.

The European Association of Remote Sensing Laboratories (EARSeL) is a nongovernmental grouping of European laboratories concerned with the initiation and development of pure and applied research, as well as educational programs. With the support of the Council of Europe, EARSeL has initiated a program to coordinate the development of remote sensing education in institutes of higher education. The objective is to promote higher degree courses in remote sensing and to encourage the development of a common syllabus appropriate for most institutes. The need to provide popular courses to the general public and informative demonstration courses/packages to politicians, decision makers, and managers has also been recognized.

9.1.2.7 Eastern Europe (CMEA and INTERCOSMOS regions).

In Eastern Europe, technical and scientific cooperation, including research and development (R and D), operations, applications, and education in remote sensing, is carried out through the Council for Mutual Economic Assistance (CMEA) and the Council on International Co-operation in the Study and Utilization of Outer Space (INTERCOSMOS). The second council forms an organizational setting for joint activities of nine socialist countries, in which some other countries participate (both nonsocialist and non-European). Joint exploration and experimental work is initiated and subsequently implemented by working groups of scientists within INTERCOSMOS.

Several universities and institutes within this region conduct education and training in satellite remote sensing. In the USSR, education and training in satellite remote sensing are carried out by several institutes, including the Moscow State University and the Moscow Institute of Engineers for Geology, Aerial Surveying, and Cartography. The Institute has been involved in educating engineers and researchers on natural resources remote sensing since 1976. The training of specialists covers planning, scientific and technical investigations and space experiments and equipment specification, design construction, and testing. Training is conducted in a combined effort by all the institutes giving training in satellite remote sensing techniques. The period of training is $5\frac{1}{2}$ years and the curriculum includes long practical and production training. United Nations-sponsored seminars are held in the region for participants from developing countries.

9.1.2.8 International programs.

The United Nations, through the Outer Space Affairs Division and the Division of Natural Resources and Energy, and the specialized agencies and programs, such as FAO, UNESCO and its Intergovernmental Oceanographic Commission (IOC), the United Nations Environmental Programme (UNEP), the United Nations Development Programme (UNDP), and the World Bank (IBRD), organize or support a variety of research activities, application projects, and training courses in remote sensing. Although many are tied to United Nations-sponsored projects, some are open to scientists from specific regions. The Interim Fund for Science and Technology for Development, administered and managed by UNDP under the policy guidelines of the Intergovernmental Committee on Science and Technology for Development (ICSTD), will also be used for funding research and educational projects in remote sensing.

At the recommendation of the United Nations Committee on the Peaceful Uses of Outer Space (COPUOS), with the endorsement of the United Nations General Assembly, two centers for remote sensing applications, education, and training are being established within the United Nations system. The first center, forming part of FAO in Rome, deals with renewable natural resources. It started regular education and training programs in 1976. It provides on-the-job training, problem-oriented workshops, regional and interregional training courses in cooperation with other United Nations members, ESA, and member states, and technical advice in specific projects. The second center, forming part of the Division of Natural Resources and Energy in New York, deals specifically with nonrenewable resources. It participates in seminars organized by the Outer Space Affairs Division.

The Inter-Agency Meetings on Outer Space Activities under the auspices of the United Nations Administrative Committee on Coordination (ACC), in which remote sensing education and training form a standard agenda item, are already concerned with the coordination of educational and training activities carried out through relevant international and/or regional programs. The coordinating task of the United Nations in this field is being hampered by the fact that COPUOS is lacking the necessary technical and practical advice of, e.g., a standing group of experts who should have remote sensing education as their terms of reference.

9.1.2.9 Summary.

Remote sensing organizations and training programs are evolving rapidly, and the information given above will need to be updated frequently on the basis of this currently available information. It is estimated that the present worldwide education and training capacity in remote sensing for participants from developing countries amounts to about 1,000 for courses exceeding six months and approximately 4,000 for seminars and courses of short duration—in total about 1,800 man-years annually. Whether or not this capacity will be used fully depends to a considerable extent on the funding of fellowships. Although many national, regional, and international sources for funding exist, none is committed solely to support education and training in satellite remote sensing.

9.1.3 Preparing for Operational Remote Sensing

9.1.3.1 Operational systems and personnel. In the decade 1991–2000 the introduction of operational remote sensing satellite systems will gradually take place. It is likely that each remote sensing technique will have its specific satellites operating in a particular orbit, with specific timing and periodicity. A trend in this direction is already visible in the observed changes from earlier experimental satellites, which were designed as multipurpose platforms with a variety of sensors, to dedicated satellites with a specific sensor package, moving in specific orbits.

All nations, irrespective of their degree of industrialization, will be faced with similar problems when confronted with the impact of fast technological change since this, with some delay, will result in structural changes in society. The fundamental problem lies in the inertia of society as a whole or of certain sectors of society. In the specific case of satellite remote sensing, the inertia of an already existing professional infrastructure for surveying and earth sciences has to be considered here. Additional problems will arise from the lack of equipment, slowness in transfer of knowledge, and limitations in means and manpower. Still, a new technology such as remote sensing also provides certain groups in society with the incentive for modernizing educational and professional structures, more or less independently of a country's level of development.

The main difference between industrialized and developing countries is that the industrialized countries can overcome many of the problems by joint action, but developing countries, in addition to joint action among themselves (in the framework of the Buenos Aires Plan of Action on Technical Cooperation among Developing Countries–TCDC), require initial material and technical support from industrialized countries before being able to achieve a certain degree of self-reliance. It should be noted here that certain countries, notably India and Brazil, while not being fully industrialized, nevertheless have developed competence in remote sensing comparable to that of the industrialized countries.

The cross section of society requiring education or training includes:

1. Decision-makers and planners, including politicians and senior officials, who should have a general awareness of satellite remote sensing and its practical and policy aspects.
2. Managerial persons in institutions, agencies, and private enterprises, who should have sufficient technical background to coordinate activities regarding specific applications of satellite data and to establish facilities for satellite remote sensing.
3. Personnel carrying out satellite surveying tasks at various levels, who should receive instructions for interpretation of imagery and digital data for mapping and monitoring in various disciplines and environment.

4. Technical support staff, from engineers to technicians, who should be responsible for construction, operation, and maintenance of facilities and equipment and who need manuals with instructions for performing technical tasks.

5. Research workers, who should develop interdisciplinary approaches in their work and possess in-depth knowledge on several aspects of satellite remote sensing.

6. Teachers, responsible for the education and training of the various groups of personnel, who should have insight in technical matters and in earth sciences and experience in educational technology and curriculum development.

These educational requirements should be specified in accordance with short-term and long-term objectives. For most categories of personnel, in-depth education is required. The objectives should be to provide scientists and technical support groups with the necessary background knowledge and skills required to process, interpret, and utilize remotely sensed data.

Industrialized nations that have consolidated organizations and infrastructures need training for specific skills and education for further development, improvement, or transformation of existing infrastructures.

Most developing countries are still at stages of institutional development in which a relatively large number of personnel at various levels and in different categories are to be trained in specific skills in a short period of time. In addition, education is required to establish or modify professional infrastructures. Marked differences in the level of socioeconomic and institutional development occur between regions and countries. This in turn is reflected in the number of personnel needed for the various categories and the levels of functions requiring education and training. It is also reflected in the degree of dependence on outside education or training facilities.

In the decade 1981–1990 education and training in remote sensing applications should be planned so as to pave the way for the introduction of operational satellite systems in the next decade, by making full use of the products provided by experimental and semi-operational remote sensing satellites.

For the operational systems of the 1990s, three scenarios can be conceived, each based on the way in which the data acquired by the satellites reach the user:

1. A fully centralized and more or less global system (I).
2. A partly centralized and regional system (II).
3. A fully decentralized, and in principle national, ground receiving station and dissemination system (III).

This logistic aspect is so significant for education and training of technical staff and for the fields of application of satellite remote sensing that the scenarios will be briefly described.

Scenario I consists of fully centralized data gathering and distribution systems operated by the owners of the satellite systems. Such systems, characterized by an almost worldwide coverage (including the oceans), large data streams, and high-level data processing, are well suited for certain research purposes and centralized global investigations. The logistics of interfacing with the dispersed user communities in the developing countries is complex and the delivery of data requested by the user takes a relatively long time. For static applications (as systematic mapping), in which the time factor does not play a substantial role, countries can use the data, provided that the pre-processing suits their particular purpose.

Scenario II includes global systems of earth observation satellites and a network of ground receiving, pre-processing, and user support stations. These stations would be located in countries selected to provide maximum coverage of the continents. The stations are regional, except in case of very large countries, and supply national and local remote sensing centers with imagery and digital data. Survey departments can use the satellite images for small-scale mapping and as a tool for planning the use of other sensors. Since monitoring of moderately slow processes is becoming possible, the network provides for distribution to a dispersed user community.

Scenario III foresees a maximum degree of decentralization of ground receiving stations with pre-processing and a mainly national-oriented data dissemination network. National stations will establish priorities as coverage of areas for which user departments request images and will allocate the data acquisition to that satellite which best suits the purpose. Simple computer-assisted interpretation methods will become operational. Speedy delivery of images and digital data to the local authorities becomes possible by means of modern communications networks including communications satellites. Selective remote sensing from space fits well in this scenario and will allow the exploitation of the full potential of satellite earth observation technology. The introduction of the necessarily advanced but "user-friendly" technology (as, for instance, microelectronics) on a selective base constitutes a complement to the socio-economic development policies.

The three scenarios will occur simultaneously, for technical, practical, economical, and political reasons. Similar scenarios have to some degree materalized already for meteorological satellite ground stations. In the case of satellite communications, one can notice the same development from regional/centralized to national/decentralized modes of operation. In fact, the LANDSAT series of satellites has stimulated the transition from scenario I to II already. Although the experimental character of new sensors (for instance, the thematic mapper in the next LANDSAT) may be a reason to centralize data reception again, one may expect that in due time scenario III will also develop for remote sensing.

The availability of satellite remote sensing systems for each of the three scenarios will have a significant impact on education and training requirements.

In scenario I emphasis will be on the ability to operate sophisticated automated and interactive systems working mainly with digital data. In scenario II, both automatic, semi-automatic, and interactive digital processing and information extraction as well as visual interpretation are applied, calling for education differentiated with regard to the techniques used either in regional or national centers. In scenario III, most countries, except the rather small ones (which may cooperate with neighbor countries), will need to train the operators for their ground receiving stations. As to the users of the data and their education, emphasis is on the "user-friendliness" of subsystems and on products designed to suit different user groups, each working with techniques suited to their conditions. This facilitates non-expert usage of imagery, and it permits a progression from classical applications in the earth sciences to new uses.

Hard figures on personnel requirements for the 1990s are difficult to give, but three principles can be used by individual countries or groups of countries to determine a number or at least to make an educated guess about it. The principles are based on the idea of critical mass, on across-border cooperation, and on comparison with existing infrastructures in surveying.

Critical mass is a term from nuclear physics and refers to the amount of nuclear fuel needed to sustain a chain reaction. The idea applies here in particular to the creation of a small team of experts, engineers, and technicians for remote sensing applications in a country. Two or three persons are usually insufficient to have an impact on a new field or on a new technology in an established department or institute. Based on experiences gained in India, teams of about 30 persons at different levels and in diverse disciplines (the critical mass) are required for local remote sensing centers to support surveying departments in each of the federal states of that large country.

Many developing and industrialized countries are too small to adhere to the idea of critical mass in each of the survey organizations in their country. In such cases, multinational cooperation may offer a solution, for example:

1. A regional center providing all necessary functions for a group of nations (the regional remote sensing centers in Africa) or an international center providing functions in certain fields of application (the FAO Remote Sensing Center in the field of resources).
2. A combination of a number of small national facilities serviced by a regional support facility (the Regional Center for Services in Surveying and Mapping in Kenya).
3. National teams strengthened by across-border exchange of experience (under programs for bilateral or multilateral cooperation, as at present in certain parts of Asia, Latin America, and Europe).

A comparison with the infrastructure in surveying can be made on the basis of a United Nations-sponsored study, "Inventory of Needs in Surveying and

Mapping: A Global Inventory."[1] In this study, the number of professional cartographers and topographic surveyors, including photogrammetrists, was found to correlate with the number of inhabitants of a country. Countries with a well-established infrastructure in this domain of surveying employ on the average 5.2 "cartographers" per 10,000 inhabitants. If one includes all surveyors in earth science disciplines, the proportion becomes one surveyor per 1,000 inhabitants. It appears that per survey discipline, one professional per 10,000 inhabitants is reasonable for a surveying infrastructure serving the national authorities of a well-developed country.

Inventory and mapping infrastructures, based on the use of air photographs and field data, exist in many countries. Operational use of satellite data in central departments, in local offices, and by field parties will affect the work of a small percentage of the total staff, because small- and medium-scale mapping forms a limited fraction of the overall national mapping activities. This percentage is put tentatively at ten; thus one person per 10,000 inhabitants might use satellite data. In the case of a single discipline, one employee per 100,000 inhabitants will be needed.

Tempting as it may be to apply these numbers to determine the amount of personnel any country, region, or continent might need for satellite surveying activities within the domain of mapping and inventory, it is advisable to consider three additional factors. The first is the structural change the new technology of satellite remote sensing will have on established survey organizations, methods, and products. The second is the relation between the total population of a country and the critical mass idea. Several European countries, Japan, and the United States fall in the population range of 300 to 30 million. However, half the world's countries fall in the 30 to 3 million, and one-fifth in the 3 to 0.3 million range, while one-seventh have less than 300,000 inhabitants. It is clear that the number of 5.2 per 10,000 and the numbers derived therefrom should be applied with the greatest care. The third factor is the difference in the population structure (population pyramid): The countries for which the 5.2 per 10,000 rule holds have a low percentage of young people, whereas in many developing countries 45 percent of the population are under the age of 16.

Monitoring is a different activity and it should have its own infrastructure and be organized with a central core and many decentralized offices, including personnel working at the grass-roots level. Here again it is fair to assume that the number of personnel required is proportional to the population in a country, province, district, or community, because, in essence, the effects of man's activities and their impact on the environment, and subsequently the impact of a deteriorating environment on the living conditions of the people, are to be monitored. No strict rule can be given for the number of personnel for monitoring, but it is suggested that it will be about the same as for a single mapping

[1] A. J. Brandenberger in *World Cartography*, XVI (UN E.80.I.12) New York, 1981, 3–72 (ST/ESA/SER.L/16).

organization, say, one person per 10,000 inhabitants. All employees in a monitoring organization should be familiar with the use of satellite data, in contrast to the situation in the mapping organizations, where only 10 percent of the personnel will be involved with space imagery. Because of fundamentally different structure and activity patterns, the critical mass idea applies less to monitoring than to mapping.

Disaster surveying does not require a large number of personnel, the permanent, highly centralized core of the organization is rather small, and the teams of surveyors are available on an *ad hoc* basis from the mapping and monitoring organizations.

For all surveying activities using satellite data, personnel are required who bring in the support sciences to the extent needed (mathematics and physics, in particular, statistics, data processing, communication theory, electronics) and the supporting techniques to operate the data reception and data processing systems (including maintenance of instruments and software and management of the facilities). These staff members are included in the estimated numbers given above.

9.1.3.2 Education and training requirements.

By using the above estimates for personnel as a basis, it is possible to estimate the required capacity and the programs for education and training in satellite remote sensing for the short-term and long-term surveying activities for mapping and monitoring.

Three rules of thumb have to be used in this context. The first rule is that, in case of an established department or organization with a constant number of employees, the educational program, including refresher and updating courses ("recurrent education"), should be sufficient to cope with 10 percent of the personnel annually in order to compensate for attrition due to departure, retirement, or transfer.

The second rule is that, in order to establish a new department or organization, the annual educational program should be large enough to cope with about 30 percent of the personnel to be employed for full operation, depending to some extent on the number of years (5, 7, 10?) it will take to arrive at this stage. The effective capacity needs to be large because not all trained persons will eventually be holding jobs involving their subject of study. Moreover, as long as the critical mass has not been reached, the number of dropouts is relatively large and the percentage may even have to be raised to 40 in certain situations.

The third rule applies to cases that lie between the situations described above: the rapid introduction of a promising technological innovation in an existing department or organization by means of education and training of personnel already employed. In such a case, the educational program should allow for an annual throughput of about 20 percent of the number of employees who will ultimately apply the innovative technology in their conventional work. The introduction of satellite remote sensing in a mapping and inventory organization forms such a case. In total, it affects one person per 10,000 inhabitants, and consequently the annual number of persons to be trained is one per 50,000 inhabitants.

For a correct appreciation of the last figure (1:50,000), it is illustrative to apply it to a country with five million inhabitants (i.e., halfway on the world list of population per country; 50 percent of the world's nations have a smaller population). If this model country had a topographic survey department and mapping departments for all of the conventional earth science disciplines, 100 persons would have to be educated or trained annually. The required capacity depends directly on the duration of training per person. Assuming that the use of satellite imagery and digital data for mapping and for updating requires an average training of three months, then the educational capacity should be 300 man-months, or 30 man-years annually, and the facility should have 30 trainees at a time for the first five years.

An independent estimate comes from Latin America, where CIAF has made a survey of education and training needs in remote sensing for mapping and inventory activities.[2] Out of a total population of about 350 million, about 35,000 persons are needed in the established departments in the various countries to introduce satellite remote sensing in their programs. This corresponds with the 1:10,000 ratio indicated above. To reach that stage within ten years, the third rule (one per 50,000 inhabitants) should be applied. The number of trainees will thus be about 7,000 annually. Again, with an average duration of three months, this amounts to 1,750 man-months, or 150 man-years, per year during the ten years the buildup phase is going to last.

Comparison with another estimate pertaining to about 100 developing countries of the world is rewarding too. The National Academy of Sciences (NAS) of the United States has estimated training needs over the period 1975–1985 in the report *Resource Sensing from Space.*[3] These amounted to 27,000 to 40,000 persons for nondegree training (3 to 12 months) and 3,000 to 9,000 persons for degree programs (one to four years). The ten years cumulative educational need is about 30,000 man-years for an estimated total population of 2,000 million. To derive the annual demand, it is necessary to apply a correction for dropouts during the ten years. Consequently, the capacity required is not 3,000 but, say, 5,000 man-years per year. This falls short of the estimate of 12,000 man-years which results if the second rule is applied and an average training of three months assumed. The difference may be explained by the following factors: in 1975 only small-scale reconnaissance mapping was foreseen; a limited number of survey departments in the earth sciences were considered; the lower strata of the personnel pyramid were not included, as indicated by the long duration of the programs.

[2] "Informe del Centro Interamericano de Fotointerpretación (CIAF) Sobre Entrenamiento y Educación de Los Usuarios de La Technología Espacial," Serie 4, CIAF, Bogota, Colombia, October 1980.

[3] *Report Sensing from Space: Prospects for Developing Countries* (Washington, D.C.: National Academy of Sciences, 1977).

The various educational programs have to be directed to the different strata of the personnel pyramid, and they may have to be provided at various locations. The largest number of personnel need specific types of training at rather low levels; the required educational facilities should be available in every country and can even be located at provincial or local centers. A smaller number need medium-level education, preferably with multidisciplinary inputs; the required educational facilities could be national or regional when this offers advantages. The smallest number need high-level education aiming at both specialization and integration, with emphasis on the consequences of technological innovations. The required educational facilities should, for reasons of efficiency, be regional or international, especially for countries having a relatively small population.

The introduction of satellite surveying in classical mapping and inventory activities is a short-term program which could well take place in this decade. After that, the first rule applies and the capacity has to be halved. If, however, structural changes occur in the mapping departments (under the influence of pressure from within society or of space technology, as, for instance, geodetic satellites with Doppler positioning techniques), additional educational effort may be required.

For the long-term planning of education facilities, one also has to anticipate the establishment of a new organization for monitoring, for which a new personnel pyramid has to be erected, with one person to be employed per 10,000 inhabitants. Here the second rule is valid; so 30 percent of the pyramid's contents has to be trained annually during the buildup phase: one person per 33,000 inhabitants. Because monitoring has to be carried out in a highly decentralized mode of operation, an average training of two months is assumed. Consequently, the educational capacity for monitoring is equal to that for mapping: 30 man-years annually per 5 million inhabitants.

From these short- and long-term considerations it appears that a part of the educational capacity used in this decade for the introduction of satellite remote sensing in the mapping organizations can later be used for training of personnel for a new monitoring infrastructure. Training needs for disaster surveying can probably be satisfied within the same facilities, backed up by local seminars, because the number of personnel is small. Altogether one may expect a gradual transition from education for "static" surveying to training for "dynamic" surveying.

9.1.3.3 Educational curricula and materials.

Curricula and teaching aids have to be developed for remote sensing personnel, from planners and managers to research workers and teachers, production personnel, and technical support staff. The wide spectrum of these groups makes it impossible to design standard curricula. Curricula for the principal surveying and mapping questions have been developed in one form or another in the facilities that at present provide courses in satellite remote sensing. It is most effective to continue using existing programs, which prove to have a rather great similarity. However, changes in course contents and the aims of educational programs are already needed in this decade.

Monitoring calls for different curricula. Initially, politicians and planners should be reached; later, research workers in institutes and departments should become involved. Teachers have to be taught, and learning methods and teaching aids should be prepared in order to reach the lower strata of the personnel pyramid. Penetration down to the grass-roots level and training of "barefoot surveyors" require a different approach than educating postgraduate students.

Disaster surveying with the help of satellite remote sensing data calls for still another style of education and training. Much depends on the structure of disaster relief organizations and on the potential disasters and hazards in the region or country. Local seminars may have to be held regularly for maintaining the capabilities of the members of the relief teams and for keeping them abreast of new developments. Teaching materials, learning packages, and new learning methods are badly needed for training in this "dynamic" surveying activity.

Handbooks with detailed instructions are probably needed for all three satellite surveying activities: mapping, monitoring, and disaster surveying. Practical examples, pertaining to the region or country, have to be included for instruction. Important aspects are climate, ecology, local environment, sociocultural structure of society, etc. Monitoring should specifically be treated for characteristic environments or areas (e.g., semi-arid, wet tropical, coastal, mountainous, agricultural, urban). Space imagery has a double function here: It serves as demonstration material for all groups and as actual "worksheets" for the users.

Modern educational methods are certainly needed for the various levels of trainees, because the current methods do not respond adequately to the demands. Programmed learning packages and "tele-education" by means of the new media become sheer necessities. Both also offer openings for new educational activities. Direct television broadcasting satellites, video cassettes and video discs, recording and playback equipment for office and home, and simulation of digital image processing on personal microprocessors all deserve an active role in modern remote sensing education. Color television screens and computer displays are especially valuable because color is an essential element for the application of multispectral remote sensing techniques.

Teaching the teachers (the last but not the least group of trainees) is another pressing necessity for realizing the potential of satellite remote sensing, but here much has already been achieved. Many alumni of the courses described above possess adequate knowledge to teach and train personnel in their own country. But the lack of teaching aids, demonstration equipment, and learning packages is hampering the educational activities to some degree. When these shortcomings can be removed, the potential resources of national instructors, teachers, and professors could be better exploited. The introduction of new media and methods would then also be possible and could well act as a multiplication factor.

The total population of the developing countries is now about 3,000 million; the short-term educational capacity demand for remote sensing applications is estimated at about 18,000 man-years annually. This is only for the introduction of satellite remote sensing data in ongoing mapping and inventory activities in this decade. The available capacity at present falls a factor 10 short of this demand.

The long-term educational capacity required for the introduction of operational satellite earth observation for monitoring and disaster surveying is equally large, and mapping will continue to generate a certain demand. These demands in this and the next decade justify considerable efforts of all countries and international organizations. Such efforts should be directed, among others, to the full use of existing facilities, the development of new, practical curricula, the introduction of efficient modern teaching aids, the creative use of the new media, and teaching of the teachers. No quantitative assessment of the effect of such efforts can yet be given, but together they certainly can narrow the capacity gap considerably.

9.1.4 Conclusion

There are many activities that can be undertaken immediately by national and international administrations to start the development of the education and training programs that will be required for operational remote sensing programs. For example:

1. Education and training in satellite remote sensing can be integrated into established educational activities in airborne remote sensing and aerial photography.
2. Existing centers and facilities should be supported in their operations and their development.
3. The national educational infrastructure should be developed in each country in a way that best meets that country's social and educational applications and technological needs.
4. Coordination between programs should be maintained and improved to ensure the widest availability of new knowledge in the rapidly changing field of satellite remote sensing.
5. Long-term planning of remote sensing education should be supported within an overall development strategy.
6. Educational systems for the lower strata of professions must be developed using modern educational methods, instructional material, and instructor training.

7. Specialized teams for disaster surveying can be created and trained by means of local workshops and seminars.

8. Packages of instructional material can be prepared and distributed widely through television and other mass media.

9.2 SATELLITE COMMUNICATIONS

As of early 1981, some 500 earth station antennae of various types were in operation in 120 countries. Within five years, this number will double. Earth stations are the essential link between the national telecommunications network of each country and the space segment—the satellite. They can be of various types, depending on the grade of service offered. However, all are extremely complex communications stations that require continuous attendance by highly qualified technical personnel for operation and maintenance. According to the present estimation of the manpower requirements, some 15,000 technical staff will have to be trained over the next five years to ensure efficient operation.

Training needs cannot be expressed in numbers alone. Consideration must also be given to the required job performance level, the entry level of the trainees, and the training facilities required. The educational system must train large numbers of trainees with wide variations in sociocultural background and experience and train them in a field in which very few organizations have the required know-how and experience. Furthermore, it should be recognized that training must be part of and based on the fundamental policies of each government with regard to the telecommunications network:

1. Network management policy.
2. Maintenance policy.
3. Personnel policy.

Training in satellite communications, and particularly in the operation and maintenance of satellite earth stations, will be considered here in the context of a personnel program since training cannot be separated from other aspects of personnel management. If an employee is to perform a job, he must learn the job from an experienced instructor, he must be motivated, and he must have the proper tools.

The following model for developing a training program based on the above principles is described in detail in ITU Training Development Guidelines.[4]

---

[4] These guidelines were produced (1979) as a result of the UNDP-sponsored project Course Development in Telecommunications—CODEVTEL.

9.2.1 Training Programs and Methods

In order to plan a training program, it is necessary to carry out the following activities:

1. Review or prepare the organization chart.
2. Review the existing and planned technical systems.
3. Prepare a list of jobs for which training is required.
4. Determine the numbers of staff in each job.
5. Conduct a personnel survey to determine likely sources for recruitment, education and training received, experience, and language ability in order to identify any required remedial training.
6. Conduct a job analysis for each of the jobs identified above. For this critical step, the cooperation of equipment manufacturers, administrators, and experts in task analysis is usually required; international cooperation can be extremely useful since the experience of other administrations in the operation and maintenance of similar equipment can be taken into account.

The organization of the training program will depend on the job analysis carried out as indicated above and on the entry level of the trainees. Normally, a technician receiving training for earth station operation or maintenance should have completed training in general telecommunications and in microwave techniques. Such training is carried out in many national telecommunications training centers and is usually part of any UNDP/ITU training project. In addition to this training, practical experience in the maintenance of terrestrial microwave systems is highly desirable.

Training methods can be many, including lecturing, self-study, tutoring, and laboratory exercises with model equipment. However, training cannot be complete without practical on-the-job training under close supervision of experienced staff. This requirement and the low numbers of staff to be trained for any one location (no more than a few per year after the first staff group has been trained) make specific demands on the training program. Individualized training is highly desirable, including programmed learning, audiovisual learning aids, and simulated practice at the real earth station itself.

The organization of such a training program is very expensive and may be beyond the reach of many countries. However, this problem can be effectively solved by international cooperation.

In the framework of the UNDP/ITU project CODEVTEL, individualized training such as described above was developed by a national team of course developers in Singapore, with the assistance of international staff, following the methods recommended in the ITU Training Development Guidelines. This training covered one area of particular importance to the administration, namely, the lineup procedures according to the INTELSAT Satellite System Operating Guide.

The course was tested in Singapore and, having proved its value, was translated into French and Spanish. Because of the very widespread use of standard procedures in satellite communications, 23 countries have expressed their interest in using this course with minor modifications. The widespread use that can result from such international cooperation makes it cost effective to develop very sophisticated training materials and job aids which can ensure that all essential jobs are carried out according to the standards necessary for continuous operation and a very high grade of service. In this respect, it should be noted that exceptionally high standards are used in space communications; except in special circumstances, the reliability of an earth station should not be less than 99.8 percent, corresponding to no more than 18 hours of service interruption in a whole year. To achieve such service standards, precise job standards and highly trained staff are indispensable, and the need for international cooperation is evident.

9.2.2 Staff Motivation and Working Environment

Personnel must not only be trained to do a job but also motivated to perform it to the highest standards. A few factors likely to have a major influence on staff motivation are listed below:

1. Perceived status of job. This may depend on many factors, including the salary level when compared to other opportunities, the level of technology used, the physical conditions of the work, any publicity given to the job location, a smooth organization, etc.
2. Perceived career opportunity.
3. Feedback on performance. The staff member should receive clear indications on the difference between a good and a poor performance and on the value that is attached to the highest work standards.
4. Opportunity for self-achievement. If the administration aims at high interest among its most promising staff members, it should provide opportunities for further individual study and discussions with national and international specialists.

It should be recognized that the labor market in space communications is highly competitive; skilled staff who are dissatisfied with their employment conditions are very likely to find better opportunities in some other country. International cooperation, leading to more uniformity in skill levels and employment conditions, could result in more staff stability and improved performance of the worldwide telecommunications network.

In many cases, maintenance deficiencies can be traced to the lack of proper tools, measuring instruments, materials, or equipment. It should be stressed here that training is only one condition for proper operation and maintenance in any technical organization. It should be complemented with proper management procedures ensuring that the necessary means for carrying out each job are continuously available. With respect to the continuity of service, the management of spare parts also plays a critical role. Proper attention has to be given to these aspects at an early stage during the planning phase of any installation.

9.2.3 The United Nations and Training in Satellite Communications

The United Nations has played a major role in promoting worldwide cooperation for training in satellite communications. The United Nations Outer Space Affairs Division, ITU, UNESCO, and UNDP have all participated in this effort. ITU has had the primary responsibility for coordinating those efforts.

The ITU technical cooperation activities have included the implementation of numerous projects. The total expenditure for these projects approached 30 million United States dollars in 1980, financed for the main part by UNDP. Almost two-thirds of the total field expenditure was disbursed for the training of staff to meet the manpower demand in various sectors of telecommunications in developing countries. This assistance consisted of the establishment or improvement of national or multinational training institutions, as well as in-service and on-the-job training, the organization of short-term specialist meetings and seminars, and the administration of fellowships. Fifty-four percent (315 out of 584) of all the expert missions in 1979 dealt directly with the development of human resources in telecommunications, the experts serving as instructors, lecturers, training experts, or project managers. On a regional basis, the percentage of missions devoted to the field of training was 54 percent in Africa, 47 percent in the Americas, 73 percent in Asia and the Pacific, and 41 percent in the Middle East.

Examples of ITU activities in which assistance was provided to member countries are listed below:

1. Organization of or participation in international seminars:
   1965 Tokyo, Japan: Seminar on Satellite Communications
   1966 Washington, United States: United States Seminar on Communication Satellite Earth Station Technology
   1968 London, United Kingdom: United Kingdom Seminar on Communication Satellite Earth Station Planning and Operation
   1969 Geneva, Switzerland: Seminar on Recent Progress in Telecommunication Technique—Integration of Satellite Communications into the General Telecommunication Network
   1970 Rabat, Morocco: Seminar on Space Communications in the Service of Progress and Co-operation

1976 Kyoto, Japan: Seminar on Satellite Broadcasting
1976 Khartoum, Sudan: Seminar on Satellite Broadcasting in the 12 GHz band
1976 Rio de Janeiro, Brazil: Seminar on Satellite Broadcasting

2. Provision of expert services, training curricula, and training programs in member countries, mostly in the framework of projects financed by UNDP. In 1979, 38 countries received assistance in telecommunications training under various national ITU projects, and 18 experts in satellite communications provided advice and training to developing countries.

3. Organization of fellowships. Fellowships are offered each year to trainees from developing countries, enabling them to profit from the experience acquired in technically more advanced countries. During 1979, 618 trainees commenced, continued, or terminated programs. Of these, 11 dealt specifically with satellite communications.

4. Exchange of training materials. The training program developed under the CODEVTEL project produced seven individualized training modules, each comprising a slide and tape presentation, a workbook, learning aids and job aids, and tests. These materials have been disseminated among the countries participating in the CODEVTEL project and are in English, French, and Spanish.

9.3 SATELLITE METEOROLOGY

Education and training programs in satellite meteorology can be divided into three groups according to their objectives:

1. Training of users of satellite data.
2. Training of operating and maintenance personnel for receiving stations.
3. Education of engineers and scientists for research and development in reception, processing, and applications of satellite data.

Training of users is the largest group, both in terms of demand and availability. Training seminars have been held in many countries, and satellite meteorology forms part of many training programs in general meteorology. The potential demand for training in applications is large since satellite data can be used in a number of disciplines: weather forecasting, storm warning, climatology, hydrology, agrometeorology, and others. Because of the large numbers of users to be trained, such training can be organized nationally as well as regionally or internationally.

Training of receiving station personnel might most efficiently be organized regionally or internationally, since all but the largest countries would have only one receiving station. Ideally, arrangements for initial and ongoing training should be part of the equipment procurement program and should be organized in cooperation with the manufacturer. Training should include practice on the equipment to be used.

Education of engineers and scientists is of longer term and more general, and it would normally be part of a university program in engineering or the natural sciences. Scientific study would generally be most productive in a context of active ongoing research programs. Since such programs exist in a rather small number of countries, education of scientists in the field of satellite meteorology might best be organized on an international basis.

9.3.1 The United Nations and Training in Satellite Meteorology

The United Nations and its specialized agencies and programs have made a major effort to make the technology of satellite meteorology accessible to developing countries. The United Nations Outer Space Affairs Division, WMO, UNEP, FAO, UNESCO, and UNDP have all participated in this effort. WMO has the primary role in coordinating United Nations activities relating to satellite meteorology.

The Eighth Congress of WMO, held in April 1979, decided that the Education and Training Program should be regarded as a matter of very high priority. During the eighth financial period (1980–1983) every effort will be made to provide assistance and advice in education and training to enable member countries to meet the increasing demands for qualified personnel for the application of meteorology and operational hydrology to various fields of economic and social development. The activities under this program include, among others, the following:

1. The preparation of syllabuses and training publications.
2. The establishment of new and the strengthening of existing regional and national meteorological training centers.
3. The organization of training courses, seminars, and workshops at regional and national levels.
4. Collaboration with the United Nations and its specialized agencies in the field of education and training in areas of WMO competence.
5. The award of long-term and short-term fellowships for training purposes.
6. Maintenance of a training library in the WMO secretariat for the purpose of advising members on availability of training materials.

In view of the need for publishing compendia of lecture notes on meteorological satellites, a compendium for training class I meteorological personnel is now under preparation. In addition to the WMO training publications, some 400 slides on meteorological satellites are available from the

Regional Seminar on the Interpretation, Analysis and Use of Meteorological Satellite Data which was organized by WMO and held in Tokyo in November 1978 and from the training course on satellite data interpretation and application, which was organized by WMO in cooperation with Colorado State University from September to December 1979. These slides are available on request in the WMO Training Library for reproduction of member countries to serve as training aids in the field of satellite meteorology.

Training activities in the field of satellite meteorology that were sponsored or co-sponsored by WMO from 1978 to 1980 include:

1. FAO/WMO/ESA Training Course (in English) on the Application of Satellite Remote Sensing to Agrometeorology and Agroclimatology, Rome, 2 to 13 October 1978; 20 participants attended.
2. Regional Training Seminar (in English) on the Interpretation, Analysis and Use of Meteorological Satellite Data, Tokyo, 23 October to 2 November 1978; 32 participants from 20 member countries attended. (Co-sponsored by the United Nations space applications program).
3. Training Course (in English) on Satellite Data Interpretation and Analysis, Colorado, United States, 15 September to 14 December 1979; 15 participants from 15 member countries attended.
4. Technical Conference (in English, French, and Russian) on the Use of Data from Meteorological Satellites, Lannion-Trégastel, France, 17 to 21 September 1979; over 200 experts from 40 countries participated.
5. FAO/WMO/ESA Training Course (in French) on the Application of Satellite Remote Sensing to Agrometeorology and Agroclimatology, Rome, 3 to 12 October 1979; 20 participants attended.
6. Workshop (in English and Spanish) on the Use of Satellite Data for Hurricane Detection and Prediction, Mexico City, Mexico, 17 to 22 March 1980; 42 participants from 15 member countries attended.
7. FAO/WMO/ESA Training Course (in English) on the Application of Satellite Remote Sensing to Rural Disasters, Rome, 27 October to 7 November 1980; 22 participants from 19 countries attended.
8. Course on Satellite Meteorology, Erice, Italy, 12 to 22 November 1980; this course was not sponsored but announced to member countries by WMO.

The training activities for 1981 were:

1. A one-week seminar on the Application of Satellite Data to Cyclone Forecasting.
2. Fourth FAO/WMO/ESA Course (in French) on the Application of Remote Sensing to Rural Disasters, Rome, 12 to 23 October 1981.

Fellowships are awarded from the following sources of funds:

1. United Nations Development Programme (UNDP).
2. Trust Funds (TF).
3. Voluntary Cooperation Programme (VCP).
4. WMO Regular Budget (RB).

The fellowships are either long-term (more than one year) or short-term (less than one year). In 1979, 28 such fellowships were awarded for training in meteorological satellites (two under UNDP, 12 under VCP, and 14 under WMO RB). Twelve fellowships were requested for 1980–1981.

9.3.2 Requirements and Opportunities for Training

An indication of the continuing training needs of the developing countries may be obtained by noting their responses to the WMO training activities listed above. For the Colorado (1979) training course, which was limited to 15 places, there were 24 applicants, and for the FAO/WMO/ESA course (1980), there were over 150 applicants for the 22 places. The large demand for training courses and fellowships and the normally expected training needs for new staff and the retraining of serving staff provide an indication of substantial training needs in this field.

It should also be noted that five of the six regional associations of WMO have indicated that they require training courses, seminars, or workshops in satellite meteorology and that the WMO Congress at its eighth session (1979) agreed that there is a need to organize specialized training courses in the field of satellite meteorology.

In the developing countries there are limited opportunites for training in the field of satellite meteorology. As indicated in the WMO publication No. 240, *Compendium of Training Facilities,* there are two developing countries which offer courses dedicated to the interpretation of meteorological satellite imagery. There are also, however, training institutions in 12 other countries which include the subject of satellite meteorology in thier syllabuses for the training of meteorologists.

In the developed countries, training opportunities are available in the national meteorological services and in several universities and institutes. In the universities, courses in satellite meteorology are offered as units in various degree programs, and the subject is also available for research specialization in some institutions. National meteorological services organize special training programs for overseas students or fellows, for example, the programs in English at the National Environmental Satellite Service in the United States, in French at the Centre de météorologie spatiale in France, and in German at the Central Office of the Meteorological Service of the Federal Republic of Germany.

9.3.3 Organization of Training Programs

The WMO publication *Guidelines for Education and Training of Personnel in Meteorology and Operational Hydrology* (WMO-No. 258) contains detailed syllabuses for training in the field of satellite meteorology. These syllabuses provide recommended curricula for the training of class I and class II meteorological personnel specializing in satellite applications.

Training courses can be organized in developing countries in two ways. First, nationals who have attended courses and seminars abroad can train other staff in their respective services or regions. Such international courses provide a multiplier effect through the training of trainers, and this in turn contributes to the improvement of the developing countries' self-reliance. Second, the organization of seminars or workshops in the developing countries through technical cooperation programs carries the training to the developing countries. Training conducted in the developing countries generally requires the acquisition from abroad of suitable materials and training aids such as data, photographs, documented case studies, etc., and may also be limited by the degree of sophistication of the satellite equipment in the country.

Training which includes recent technological advances and equipment and new techniques in data reception, handling, and interpretation can only be acquired in those institutions or services where the developments are taking place and where the scientists are constantly updating their knowledge and expertise in the subject.

9.3.4 Conclusion

Training in the developing countries is either secondhand from indigenous staff who were trained abroad and who remain isolated from the centers and institutions in the developed countries where new developments are constantly taking place, or firsthand through the medium of specially organized events in the developing countries utilizing overseas expertise. Training is available abroad, but it is almost entirely dependent on the availability of fellowship awards from one source or another. Together, these three avenues of training are unable to cope with the training needs of the developing countries in the fields of satellite meteorology and applications. This situation is becoming even more adverse as a result of the rapid rate of development in space technology and the slow rate of transfer of this technology to the developing countries.

ACKNOWLEDGEMENTS

The ITC international team included the following contributors: W. C. Draeger (United States of America), Prof. A. P. Kapitsa (Union of Soviet Socialist Republics), Mr. J. P. Ouedraogo (Upper Volta), Wing Commander K. R. Rao

(India), Dr. Hernan Rivera (Colombia), Prof. R. Savigear (European Economic Community), and Mr. Michel Yergeau (Canada). In addition, the following ITC staff assisted with the preparation of the paper: Mr. F. C. d'Audretsch, Dr. E. S. Bos, Prof. S. A. Hempenius, Prof. J. J. Nossin, Mr. J. Richardson, Prof. C. Voûte and Mr. Tsehaie Woldai.

As part of the preparation of this chapter, a questionnaire was sent to a number of organizations and institutions involved in education and training in remote sensing, and an outline of Section 9.1 was sent to a number of international and regional organizations for comments. In particular, useful information was provided by the Food and Agriculture Organization of the United Nations (FAO) and the United Nations Educational, Scientific and Cultural Organization (UNESCO).

The COSPAR contribution was prepared by Prof. T. Sahade of Argentina.

Part IV
INTERNATIONAL
COOPERATION

Chapter 10

MULTILATERAL

INTERGOVERNMENTAL COOPERATION

IN SPACE ACTIVITIES

ABSTRACT

This chapter, which is one of three chapters on international cooperation, is aimed at providing information on the mandates and programs of various intergovernmental organizations active in space science and technology. It was prepared by the conference secretariat based on information available to it from the organizations concerned.

10.1 INTERNATIONAL TELECOMMUNICATIONS SATELLITE ORGANIZATION (INTELSAT)

10.1.1 Origin

On 20 August 1964, representatives of 11 nations signed interim agreements establishing the International Telecommunications Satellite Consortium, whose purpose was the design, development, construction, establishment, operation, and maintenance of a global commercial telecommunications satellite system. Soon afterward, as global satellite communications became a reality and commercial viability for satellite communications was proven, many other nations joined this consortium. During the period 1969–1971, a plenipotentiary conference, in which the governments of the consortium member countries participated, held a series of meetings in Washington, D.C., to determine a permanent structure for the consortium.

The conference culminated in the conclusion of two agreements: the intergovernmental "agreement" and the "operating agreement." These agreements entered into force on 12 February 1973. The parties to the agreement are the governments of the member states, and the signatories of the operating agreement are either the governments which are parties to the agreement or their designated telecommunications entities, public or private. These two agreements superseded the interim agreements under which INTELSAT had operated as the International Telecommunications Satellite Consortium since 1964.

10.1.2 Purpose and goals

The main purpose of INTELSAT, under its definitive agreements, is to "continue to carry forward on a definitive basis the design, development, construction, establishment, operation and maintenance of the global commercial telecommunications satellite system as established under the provisions of the Interim Agreement and the Special Agreement."

10.1.3 Membership

The agreement specifies that the government of any state party to the interim agreement or the government of any other state member of the International Telecommunication Union (ITU) may accede to the INTELSAT Agreement.

The preamble to the INTELSAT agreement states that satellite communications should be available to the nations of the world "on a global and nondiscriminatory basis." With this in mind, the INTELSAT system is also available for use by countries that are not members. Utilization charges to these users are the same as to member countries.

As of December 1981, membership stood at 106 countries, as follows: Afghanistan, Algeria, Angola, Argentina, Australia, Austria, Bangladesh, Barbados, Belgium, Bolivia, Brazil, Canada, Central African Republic, Chad, Chile, China, Colombia, Congo, Costa Rica, Cyprus, Denmark, Dominican Republic, Ecuador, Egypt, El Salvador, Ethiopia, Fiji, Finland, France, Gabon, Germany, Federal Republic of, Ghana, Greece, Guatemala, Guinea, Haiti, Holy See, Honduras, Iceland, India, Indonesia, Iran, Iraq, Ireland, Israel, Italy, Ivory Coast, Jamaica, Japan, Jordan, Kenya, Kuwait, Lebanon, Libya, Liechtenstein, Luxembourg, Madagascar, Malaysia, Mali, Mauritania, Mexico,

Monaco, Morocco, Netherlands, New Zealand, Nicaragua, Niger, Nigeria, Norway, Oman, Pakistan, Panama, Paraguay, Peru, Philippines, Portugal, Qatar, Republic of Korea, Saudi Arabia, Senegal, Singapore, Somalia, South Africa, Spain, Sri Lanka, Sudan, Sweden, Switzerland, Syria, Thailand, Trinidad and Tobago, Tunisia, Turkey, Uganda, United Arab Emirates, United Kingdom of Great Britain and Northern Ireland, United Republic of Cameroon, United Republic of Tanzania, United States of America, Upper Volta, Venezuela, Viet Nam, Yemen, Yugoslavia, Zaire, and Zambia.

10.1.4 Financial Aspects and Shareholding

INTELSAT operates as a financial cooperative and its ownership is shared by all of its member countries. Members contribute to the capital requirements of the organization in proportion to investment shares based on their use of the system, as measured by their payment of utilization charges. The capital requirements are subject to a ceiling, consisting of the cumulative capital contributions made by the members less the cumulative capital repaid to them plus the outstanding amount of contractual capital commitments of INTELSAT.

Initially, the capital ceiling was set to be 500 million United States dollars when the operating agreement entered into force. Since then, the total has been increased to keep pace with the growth of the system, and it currently is 1,200 million United States dollars.

INTELSAT assets include primarily the cost of satellite construction and launch services. Satellites are generally procured under long-term contracts which provide for payments by INTELSAT over the contract period.

The revenues of the INTELSAT system, which are received from the lease of the satellite capacity to the system users, are calculated to cover the operating expenses, the depreciation of the system, and a return of 14 percent on the signatories' net investment. The revenues received, after subtracting the operating expenses, are shared out among the member countries. These same countries, however, as users, will have paid in most of those revenues in the first place. In so doing, they establish their percentage shareholding in the organization and they are called upon to contribute accordingly to the capital investment in the system.

A country as user will, in effect, pay itself back as owner, and an important purpose of the utilization charge is to provide a means of measuring the use, and so arrive at an equitable way of sharing the cost of the system among the members. The investment share of each member is recalculated every year, based on the total utilization charges paid by that member for the previous six months, and there are arrangements for adjusting the shares of countries who wish to have a lower or higher percentage shareholding than their utilization.

As well as setting the percentage of investment capital which each signatory contributes to the system and the revenue which it receives back, the share holding also governs the voting rights in the Board of Governors.

10.1.5 Organizational Structure

The organizational structure of INTELSAT consists of four units: an Assembly of Parties, composed of representatives of all governments that are parties to the INTELSAT agreement: a Meeting of Signatories, composed of representatives of all signatories (governments or their designated telecommunications entities) to the INTELSAT operating agreement; a Board of Governors, composed largely of representatives of those signatories whose investment shares, individually or in groups, are not less than a specified amount; and an Executive Organ.

The Assembly of Parties considers those aspects of INTELSAT that are primarily of interest to the parties as sovereign states, as well as the resolutions, recommendations or views put to it by the other bodies of INTELSAT. It formulates policies and long-term objectives consistent with the principles, purposes, and scope of the INTELSAT activities. Each party has one vote.

The Meeting of Signatories considers resolutions, recommendations, or views put to it by the other bodies of INTELSAT and also considers matters relating to the financial, technical, and operational aspects of the INTELSAT system.

Among its various responsibilities, the Meeting of Signatories considers and decides upon any recommendation by the Board of Governors concerning increase in the capital ceiling; determines annually the minimum investment share for representation on the Board of Governors; and establishes general rules and policies, upon the recommendation of, and for the guidance of the Board of Governors, concerning the operations and management of the INTELSAT satellite system on a nondiscriminatory basis. Each signatory has one vote.

The Board of Governors is responsible for all decisions concerned with the design, development, construction, establishment, operation, and maintenance of the INTELSAT space segment and for the decisions necessary to carry out any other activities undertaken by INTELSAT. The Board considers all resolutions, recommendations, and views addressed to it by the other bodies of INTELSAT. It is assisted by Advisory Committees on Technical Matters on Planning and by a Budget and Accounts Review Committee.

The Board endeavors, usually successfully, to take decisions unanimously. Otherwise, decisions are taken by a weighted vote based on investment shares.

As of December 1980, the Board consisted of 27 governors representing 91 signatories. In accordance with the provisions of the INTELSAT agreement, 22 governors are serving on the Board on the basis of having investment shares equal to or in excess of the minimum established value for this purpose. Five other governors represent ITU regional groups. These are: Africa Group I, Africa Group II, the Caribbean Group, the Central American Group, and the Nordic Group.

The Executive Organ is headquartered in Washington, D.C., and is headed by a director general, who is the chief executive and legal representative of IN-TELSAT, and who is responsible to the Board of Governors for the day-to-day management and operation of INTELSAT.

10.1.6 Program and Activities

The INTELSAT global satellite system is made up of two elements: the space segment, consisting of satellites and associated facilities owned by IN-TELSAT, and the ground segment, consisting of the earth stations owned by telecommunications entities in the countries in which they are located.

10.1.6.1 Space segment.   INTELSAT operations started when the INTELSAT-I (EARLY BIRD) satellite was launched on 6 April 1965. It had a capacity of 240 telephone circuits or one television channel. This satellite was used for transoceanic communications between one station in North America and one of the many stations in Europe. The system expanded rapidly providing a global coverage when INTELSAT-III satellites came into use during the period 1968–1969, carrying 1,200 simultaneous telephone circuits plus television. IN-TELSAT-IV satellites introduced in 1971 had a capacity of 4,000 telephone circuits plus television. Further international traffic growth required the introduction of INTELSAT-IV-A satellites in 1975, with a capacity of 6,000 circuits plus television.

In 1980 the first satellite of the INTELSAT-V series, the largest and most sophisticated commercial satellites ever built, was launched. These satellites, capable of handling 12,000 circuits plus television, utilize a number of new technologies, among them, the use of the 11/14 GHz bands, frequency re-use in the 4/6 GHz bands through spatial separation of antenna beams as well as use of dual polarization capability, and frequency re-use in the 11/14 GHz bands through spatial separation of antenna beams. INTELSAT-V will utilize new modulation/multiple access techniques, viz., Time Division Multiple Access/Digital Speech Interpolation (TDMA/DSI), as a means to enhance the capacity of these satellites.

INTELSAT currently keeps 13 satellites in orbit, including five satellites for international operations—three satellites over the Atlantic Ocean region (AOR), one satellite over the Indian Ocean region (IOR), and one satellite over the Pacific Ocean region (POR). In addition to international operational satellites, each ocean region is provided with space satellites for traffic restoration in case of an emergency and with satellites providing domestic services. The satellites in orbit often represent two or three generations of satellite technology.

The 1981 Atlantic Ocean configuration for operational satellites, an INTELSAT-V satellite, an INTELSAT-IV-A, and an INTELSAT-IV as Primary Major Path I and Major Path II satellites, respectively, is an example of this. In such a configuration, the Primary satellites act as the major medium of transmission carrying most small traffic links and diversity circuits for most major routes. As their name implies, the major path satellites carry diversity traffic among the heavy traffic routes. They also carry traffic between limited "community of interest" earth stations.

All these are geosynchronous satellites, located at suitable longitudinal locations on the geosynchronous satellite orbit, at an altitude of approximately 35,780 km above the earth.

Table 10.1 shows the deployment plan of operational INTELSAT satellites for the international system in the latter part of 1981, and Table 10.2 shows the orbital locations intended for INTELSAT satellites during the period 1980–1984 for all satellites (international, spare, and domestic leases).

10.1.6.2 Ground segment.
In the ground segment, growth has kept pace with the development of the space segment. Since operations were begun in 1965, the number of countries and territories operating in the system has increased from 5 to more than 130, the number of earth stations from 5 to more than 240, and the number of antennae from 5 to more than 300.

These earth stations are owned and operated by the international telecommunications entities of the country in which they are located. INTELSAT currently authorizes three standards for earth stations that operate international services through its satellites. The standards govern, in the main, antenna performance, acceptable interference noise levels, and transmission parameters related to the operation of earth stations.

Standard earth stations are defined as those earth stations which conform to the mandatory performance characteristics for operation in the specified fre-

TABLE 10.1   Deployment Plan of Operational INTELSAT Satellites for the International System in the Latter Part of 1981

	Satellite Type	Location (° longitude east)
(a) *Atlantic Ocean region*		
Primary satellites (most earth stations in the region have access)	V (F-2)	335.5
Major Path I satellite (to which Atlantic region users with a second antenna have access, plus a few users with one antenna but having traffic requirements that are at present completely met on this satellite)	IV-A (F-4)	325.5
Major Path II satellite (to which users with three antennas have access, plus a few users with one antenna but having traffic requirements that are at present completely met on this satellite)	IV (F-1)	341.5
(b) *Indian Ocean region*	IV-A (F-3)	63.0
(c) *Pacific Ocean region*	IV (F-8)	174.0

TABLE 10.2   Orbital Locations for INTELSAT Satellites 1980–1984

INTELSAT Space Station Type(s)	Location ° E (longitude)	Intended Use
IV	307.0	Domestic leases
IV, IV-A, V	325.5	AOR MP 1
IV-A	329.0	AOR spare
IV-A, V, MCS	332.5	AOR spare
IV-A, V	335.5	AOR primary
IV, IV-A, V	338.5	AOR spare
IV, IV-A, V, MCS	341.5	AOR MP 2
IV, IV-A	356.0	AOR residual
IV, IV-A	359.0	AOR leases
IV, IV-A, V, MCS	57.0	IOR spare/leases
IV, IV-A, V, MCS	60.0	IOR spare/MP
IV, IV-A, V, MCS	63.0	IOR primary
IV, IV-A, V, MCS	66.0	IOR spare
IV, IV-A, V	174.0	POR primary
IV, IV-A, V	179.0	POR spare

quency band and which have an elevation angle to the satellite with which they operate of not less than the values given below:

1. Five degrees for earth stations operating in the 6/4 GHz band.
2. Ten degrees for earth stations operating in the 14/11 GHz band.

Earth stations which do not conform to any one of these standards are considered nonstandard earth stations. The three standards are briefly described below:

1. *Standard A*. This standard, which generally calls for large, parabolic antennae (about 30 m in diameter) is the most widely used in the system. Their figure of merit (G/T) is nominally 40.7 dB/K, and they operate in the 6/4 GHz frequency bands. The standard A stations allow highly efficient use of the satellite orbit and the frequency bands allocated to the Fixed Satellite Service.
2. *Standard B*. This standard was developed by INTELSAT for situations that called for a more economical alternative to standard A. Standard B earth stations, which have parabolic antennae of about 11 m in diameter and operate in the 6/4 GHz frequency bands, are particularly suitable for countries with small traffic demands. Their figure of merit (G/T) is nominally 31.7 dB/K. Some countries use a standard B station as an initial entry into the INTELSAT system, later upgrading it to standard A when traffic growth warrants a larger investment. Because standard B earth stations are less efficient users of the space segment resources, charges for

satellite circuits through them are 50 percent higher than those for standard A earth stations.

3. *Standard C.* These stations, with antennae of between 14 and 19 m in diameter, are designed specifically to operate within the 14/11 GHz frequency bands. Their figure of merit (G/T) is nominally about 40 dB/K.

There are also a number of nonstandard earth stations operating with the INTELSAT system, but these are largely for domestic leased transponder services, for experimental purposes, for the INTELSAT tracking, telemetry, command, and monitoring (TTCM) network, or for the specialized high-performance test antennae used to check the satellites in orbit, especially just after launch.

Because of the varied technical problems that could arise in the operation of nonstandard earth stations with INTELSAT satellites, INTELSAT considers these stations for approval on a case-by-case basis subject to space segment availability and compatibility with various intersystem operational criteria. Normally, INTELSAT will approve nonstandard earth stations for operation in the INTELSAT global network only on a temporary basis.

Standard or nonstandard earth stations may be approved for operation with leased transponder space segment capacity on a case-by-case basis.

10.1.6.3 Tracking, telemetry, command, and monitoring (TTCM).
INTELSAT, through contracts with some signatories, maintains a system of TTCM stations around the world. Through these stations, as their name implies, commands are sent to the satellites for changing their position or operational configuration, and information is collected on their position, orientation, and "vital health signs." The stations also monitor the technical characteristics of communications carrier transmissions to ensure that they meet standards for power, frequency, frequency band occupancy, noise and other characteristics. In addition, some of these stations incorporate specialized high-performance test antennas and equipment to check the INTELSAT satellites in orbit.

TTCM stations are located in specific regions of the world to ensure adequate coverage and visibility of the INTELSAT satellites, not only during their operating lifetime but also during launch.

10.1.7 Requirements and Performance of the INTELSAT System

10.1.7.1 International services.
INTELSAT provides bulk international telecommunications capacity to countries' telecommunications authorities, which, in turn, utilize this capacity for telephony service, leased circuits, telegram and data facilities, and satellite television channels.

For preassigned service, INTELSAT generally leases satellite capacity to telecommunications authorities on a permanent basis by "unit of utilization." A unit of utilization, or a half circuit, can be described as the amount of capacity required to provide a two-way voice circuit to the satellite which, when matched with a half circuit at the other end, provides a full circuit between two standard A earth stations. So it can be seen that two half circuits or units would be required for each two-way telephone conversation between two standard A earth stations.

For telegraphy and record service, the telephone channel can be used as a whole or subdivided between various telegraph channels, depending on the transmission rate and bandwidth requirements. The telephony channels can also be used for data transmission, facsimile, etc. In addition, some telephony channels are used alternatively for both voice and data.

For wideband data transmissions, either groups of telephony channels or digital single channels per carrier (SCPC) are utilized, depending on the availability of digital transmission and interface equipment at the earth stations. INTELSAT currently offers wideband data services up to 1.5 megabits per second. With the introduction of TDMA in the INTELSAT system in the near future, and the general proliferation of digital terrestrial networks, satellite communications will become increasingly better suited, technically and economically, for signals or digital origin. But since INTELSAT must serve as a global common denominator for all telecommunications services, the INTELSAT system will continue to provide analog FM services as well in the foreseeable future.

In addition, each INTELSAT satellite has a certain amount of its capacity set aside for "on demand" services, such as television, or restoration of services carried on transoceanic telecommunications cables. Intercontinental television is a unique service introduced by INTELSAT in international network operations. Two channels (and in special cases more) are available on each INTELSAT satellite for television. Television is noted for the complex interface arrangements necessary for the multidestination transmissions, particularly with video networks having different transmission standards. Orders for television programs are collected by the international telecommunications carriers in the various countries and passed on to the INTELSAT operations center in Washington, D.C. Here, orders are coordinated to ensure that the satellite capacity is free at the time desired.

Further, INTELSAT has, since 1971, operated a demand-assignment service in the Atlantic Ocean region called single channel per carrier PCM multiple access demand assignment equipment (SPADE), which enables individual telephone circuits to be provided on demand for communications between a large number of countries. This service is particularly suited to those routes where traffic would not warrant the provision of full-time circuits.

10.1.7.2 International traffic growth within the INTELSAT system.

Figure 10.1 and Table 10.3 show the continuous and vigorous growth of the traffic in the INTELSAT system for full-time telephony channels over the period

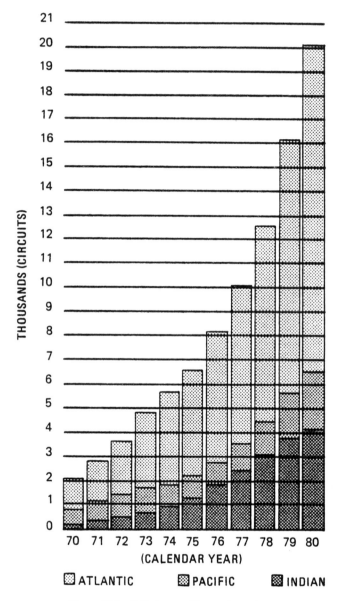

Figure 10.1 Full-time satellite use by region.

TABLE 10.3 Full-Time Traffic Growth in the INTELSAT System

	Circuits			
	Atlantic	Pacific	Indian	Total
1965	75	—	—	75
1966	86	—	—	86
1967	209	135	—	344
1968	360	211	—	571
1969	914	452	51	1,417
1970	1,316	656	157	2,129
1971	1,757	827	327	2,911
1972	2,374	924	450	3,748
1973	3,145	1,126	636	4,907
1974	3,847	930	976	5,753
1975	4,431	963	1,290	6,684
1976	5,391	986	1,883	8,260
1977	6,502	1,115	2,483	10,100
1978	8,127	1,470	3,050	12,647
1979	10,454	1,950	3,807	16,211
1980	13,741	2,310	4,208	20,259

1965 through 1980. In the last three years the growth rate has been 25 percent, 28 percent, and 25 percent, respectively. Other nonfull-time traffic in 1980 is given below.

In 1980, 31 SPADE terminals (used only in the Atlantic Ocean region) were in operation, with an average of about 38,500 access minutes per day.

Also, about 15,600 television transmissions were relayed by INTELSAT in 1980, with total usage of about 30,000 billable half-channel hours.

Finally, total temporary service for the INTELSAT space segment for 1980 was approximately 87,500 unit days. This included restoration of services carried on transoceanic cable systems on 11 occasions, with a total of about 61,000 unit days. This traffic, like television, however, tends to fluctuate yearly.

10.1.7.3. International via-routing and diversity. The World Routing Plan, as originally conceived by ITU, has undergone a conceptual change largely because of the availability of communications satellites. This is due to the large numbers of direct circuit paths established in the INTELSAT system. The prevailing mode for the bulk of telephone calls is a direct one-link connection, part of which is derived through the use of dedicated via-routed[1] circuits.

Global planners have always considered terrestrial and satellite systems to be complementary and have established many combined terrestrial/satellite paths as a means of establishing direct access circuits. In a few instances, where

[1] The term "via-routed" means that a satellite circuit is established between two earth stations, at least one of which is not in the end-user country.

this seemed impracticable, double-hop satellite circuits have also been established.

Another feature of the INTELSAT system is diversity routing for purposes of service protection. The INTELSAT network frequently offers diversity of two or more paths either through the use of satellites operating in the same region or by the use of satellites operating in two regions with common coverage areas such as in the Atlantic/Indian Ocean or Indian/Pacific Ocean regions.

10.1.7.4 International system availability.
Each satellite is designed with hardware redundancy (Table 10.4) so that high service reliability can be achieved by switching to redundant elements by ground command in the event of subsystem failures affecting parts of the communications payload.

The continuity of service provided by the space segment for 1980 was 99.996 percent; the average continuity of service achieved by circuits through the earth stations in the INTELSAT system was 99.950 percent; and the overall average continuity of service for the complete system on the pathway-by-pathway basis was 99.890 percent.

10.1.7.5 Satellite capacity for domestic communications.
A growing number of countries use INTELSAT satellites for communications within their borders as well as for international traffic. These countries utilize spare capacity on INTELSAT satellites, which is leased to them for their domestic communications. Many countries in the world, both developed and developing, face urgent requirements to expand their internal communications networks. To do this using conventional terrestrial links can be a drawn out, expensive process and in some cases, because of terrain, climate, etc., can be virtually impossible.

Recognizing this, INTELSAT agreed to lease spare capacity on its satellites for domestic purposes. This gave countries the possibility of creating almost "instant" extensive networks with links capable of carrying quality communication—even television—to anywhere an earth station could be constructed.

Originally the lease cost of a transponder with global coverage for domestic purposes was 1 million United States dollars per year, on a pre-emptible basis. In 1980 this cost was reduced to 800,000 United States dollars per year. If a fraction

TABLE 10.4

| | INTELSAT | | |
	IV	IV-A	V
Receivers on board	4	6	15
Receivers in service at one time	1	3	7
Transmitters (TWTAs) on board	24	32	43
Maximum number of transmitters (TWTAs) in operation at one time	12	20	27

of a transponder is needed (one-half or one-quarter), the cost is reduced proportionately.

Table 10.5 gives the long-term allotments of space segment capacity for domestic service as of December 1980.

10.1.8 Cooperation with System Users and International Organizations

10.1.8.1 INTELSAT assistance and development program (IADP).

The scope of this program, established in 1978 is to assist INTELSAT signatories and users and potential signatories and users toward more effective utilization of INTELSAT facilities. This includes the following specific activities:

1. Performing preliminary feasibility and/or viability studies for earth segment facilities and equipment intended to work via the INTELSAT space segment.

2. Providing assistance with regard to the following: defining specifications; providing reports to institutions in support of applications for financing; establishing criteria or guidelines for the technical evaluation of proposals; monitoring the implementation of, and undertaking supervision of, construction and acceptance testing of earth segment facilities intended to work via the INTELSAT space segment.

3. Providing assistance with regard to the planning and performance of tests, experiments, and demonstrations concerning the technical and operational development of new services that could be provided on the INTELSAT space segment.

TABLE 10.5   Transponder Leases in Service (December 1980)

Country	Number of Transponders	Lease Conditions
1. Algeria	one	pre-emptible
2. Australia	one and one-half	pre-emptible
3. Brazil	three and one-half	pre-emptible
4. Chile	one-quarter	pre-emptible
5. Colombia	one-quarter	pre-emptible
6. France	three-quarters	pre-emptible
France	one	full-rate
7. India	one-quarter	pre-emptible
8. Nigeria	three	pre-emptible
9. Norway	one-half	pre-emptible
10. Oman	one	pre-emptible
11. Peru	one-quarter	pre-emptible
12. Saudi Arabia	two and one-quarter	pre-emptible
13. Spain	one-half	full-rate
14. Sudan	one	pre-emptible
15. Zaire	one	pre-emptible

4. Providing technical assistance and advice on the upgrading, modification, or modernization of INTELSAT earth segment facilities in order to allow for more efficient utilization of the INTELSAT network.

5. Organizing INTELSAT training programs and seminars related to the planning, procurement, implementation, or operation and maintenance of earth segment facilities intended to work or currently utilized with the IN-TELSAT space segment. Such activities would include conducting earth station technology seminars.

6. Providing assistance and/or advice on telecommunications systems planning and development studies when such studies would involve increased or more efficient utilization of the INTELSAT space segment.

7. Assisting in providing information required for frequency coordination of potential or planned INTELSAT earth stations through ITU.

The assistance is normally provided free of charge and is generally on a first-come, first-served basis. However, if necessary, the Director-General would take into account the need for the service, its importance to INTELSAT, and the amount of work to be carried out. Where the assistance requested would require more than two man-months of effort or extensive outside consultant services, the request would be considered on the basis of a cost reimbursement contract, or, alternatively, would be presented to the Board of Governors for approval.

This program has proven very effective since its initiation, and a number of important projects have been completed and others undertaken. These projects covered international as well as domestic traffic situations. As of 1980, the program has been utilized since its inception in 1978 by 24 INTELSAT system users. This number is expected to grow in the future.

10.1.8.2 Effect of the INTELSAT system on international traffic tariffs.

Since its establishment, INTELSAT has maintained a history of reducing rates for telecommunications charges despite the continuous tide of world inflation. Table 10.6 shows the chronological history of INTELSAT space segment charges during the period 1965–1981 for full-time voice circuits in terms of units of utilization (a unit of utilization is equivalent to half a voice circuit when standard A stations are utilized at both ends as defined above, as well as the charges for non-full-time services, viz., SPADE, television, and cable restoration).

Several factors have contributed to this reduction in rates:

1. Technological advances and innovations over the years, resulting mainly from research and development activities sponsored by INTELSAT, which made possible spacecraft components of improved performance and lighter weight. This factor results in less expensive satellites for a specific purpose and capacity.

TABLE 10.6   Chronological History of INTELSAT Space Segment Charges

	Full-Time Voice Grade Units Per Year	Per Month
27 June 1965	$32,000	$2,666.66
1 Jan. 1966	20,000	1,666.66
1 Jan. 1971	15,000	1,250.00
1 Jan. 1972	12,960	1,080.00
1 Jan. 1973	11,160	930.00
1 Jan. 1974	9,000	750.00
1 Jan. 1975	8,460	705.00
1 Jan. 1976	8,280	690.00
1 Jan. 1977	7,380	615.00
1 Jan. 1978	6,840	570.00
1 Jan. 1979	5,760	480.00
1 Jan. 1980	5,040	420.00
1 Jan. 1981	4,680	390.00

	Spade	
1 Nov. 1973	$0.15 per minute of holding time at each end	
1 Mar. 1975	0.10	"
1 Jan. 1977	0.09	"
1 Jan. 1978	0.08	"
1 Jan. 1979	0.07	"
1 Jan. 1980	0.06	"
1 Jan. 1981	0.005	"

	Television	
		Channel Per Minute
27 June 1965		$14.80
1 Jan. 1966		9.60
18 July 1968		8.75
1 Jan. 1981		8.00

	Cable Restoration
27 June 1965	$ 3.70 per hour per unit (minimum 24 hours)
1 Jan. 1966	0.04 per minute per unit (minimum 24 hours)
1 Apr. 1966	58.00 per day per unit
1 Jan. 1971	43.50 per day per unit
1 Jan. 1972	36.00 per day per unit
1 Jan. 1973	31.00 per day per unit
1 Jan. 1981	28.00 per day per unit

2. Increased capacity of INTELSAT satellites over the years, through the implementation of state-of-the-art transmission techniques, such as frequency re-use of the allocated frequency bands several times and the introduction of modern modulation and multiple access techniques.

3. Through careful planning of the INTELSAT satellite system, a higher fill factor is achieved in utilizing the space segment.

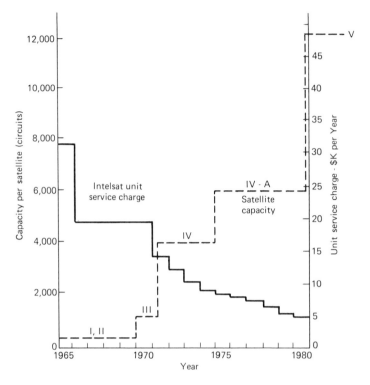

Figure 10.2 INTELSAT satellite capacity and service charge.

Figure 10.2 shows the change in capacity of INTELSAT satellites over the years and the resulting annual charges for a full-time unit of utilization over the period 1965–1981. It can be seen that the unit of utilization service charge in 1981, in real terms, is less than one-fifteenth of its value in 1965.

10.1.8.3 Rural and thin route communications.

Telecommunications have long been recognized as a vital part of the infrastructure of rural areas. It is not surprising, then, that extensive programs for modernizing and building telecommunications facilities are now under way throughout the world. Many countries are turning to INTELSAT for satellite services to enable them to meet their ever-increasing international and domestic communications and requirements.

INTELSAT has reduced its space segment charges for domestic lease services as well as for full-time circuits. This has made it easy for large, sparsely populated countries, marked by inhospitable terrain which renders the building of conventional terrestrial systems extremely difficult if not impossible, to lease capacity in INTELSAT satellites to overcome nature's barriers and provide up-to-date, reliable domestic telecommunications services to their outposts. An IN-

TELSAT satellite provides a wide area of coverage, largely independent of distance. Thus, INTELSAT has enabled developing countries to leapfrog over generations of communications technology without having to invest a great deal of time and money in a telecommunications satellite system of their own that would usually have capacity far beyond their real needs.

10.1.8.4 International Telecommunication Union (ITU). INTEL-

SAT considers that the establishment of formal relations with ITU is desirable, and for some time it has been involved in discussions with the Secretary-General of ITU toward this objective.

These discussions culminated in July 1980, when the Director-General of INTELSAT and the Secretary-General of ITU signed a document concluding arrangements for the collaboration between the two organizations in the field of space communications. The major points of these arrangements are:

1. Exchange of publications and documents of mutual interest.
2. Exchange of information on conferences and meetings, in accordance with the statutes and rules of each organization.
3. Consultation on questions deemed to be of common interest for the development of telecommunications.

Any additional agreement for the establishment of formal relations between ITU and INTELSAT would require the approval of the ITU Plenipotentiary Conference (the next conference is scheduled for the last quarter of 1982). INTELSAT intends to explore with ITU the most appropriate means to conclude such an agreement.

In the area of the studies performed by the International Consultative Committees for Radio (CCIR) and for Telephony and Telegraphy (CCITT), INTELSAT actively participates in several study groups of both committees, the terms of reference of which include subjects that are related to the activities of INTELSAT. INTELSAT utilized the CCIR/CCITT recommendations in the design and operations of the INTELSAT network.

INTELSAT participates in almost all ITU administrative conferences as an observer and in an advisory capacity, in accordance with the ITU convention.

10.1.8.5. International Maritime Satellite Organization (INMAR-

SAT). Based upon the discussions with the pre-INMARSAT International Joint Venture and the INMARSAT Preparatory Committee, an amendment to the INTELSAT–V contract was made to include a maritime communications subsystem (MCS) on several of the INTELSAT–V satellites.

After the establishment of INMARSAT, INTELSAT provided INMARSAT with details about the technical and operational aspects of the maritime communications subsystem on INTELSAT–V satellites, as well as a description of the contractual terms that INTELSAT would expect to conclude with IN-MARSAT.

As of December 1980, INTELSAT and INMARSAT were in the final stages of concluding the contractual arrangements for the provision of three or more maritime communications satellite subsystems to INMARSAT on a lease basis.

10.1.8.6 Asia Pacific Telecommunity (APT).

Considering the scope of APT and its emphasis on strengthening the regional aspects of international telecommunications, it was felt that both APT and INTELSAT would benefit from the establishment of formal relations between the two organizations. Consequently, a memorandum of formal relations between the two organizations. Consequently, a memorandum of understanding between INTELSAT and APT was approved by the INTELSAT Board of Governors in December 1980, and it will take provisional effect when it becomes approved by the Management Committee of APT, pending approval by the 1982 Assembly of Parties of IN-TELSAT.

10.1.8.7 Intergovernmental Maritime Consultative Organization (IMCO).

IMCO and INTELSAT reached agreement on formal relations in 1976. The major points of the agreement are:

1. Regular consultation on matters of common interest.
2. Participation by each organization in the meetings of the other, if they are of interest, on an observer basis.
3. Exchange of scientific and technical information, subject to such arrangements as may be necessary to withhold confidential information.

10.1.9 Future Outlook: INTELSAT-V Follow-on Systems

The demand for international services via the INTELSAT system is growing at a rapid rate of about 25 percent a year. This is clear from the forecasts provided by the INTELSAT users at their Annual Global Traffic Meetings. Additionally, INTELSAT is at the present time considering offering capacity for domestic lease services on a planned basis. The forecast growth of this service is also significant. Other new services which can be economically provided by communications satellites are appearing on the horizon, and therefore capacity for them has to be planned. Examples of such new services include computer communications, real-time information systems, electronic mail, videoconferencing, data collection, etc.

Ever since the introduction of the INTELSAT-V series, INTELSAT has been studying its requirements for additional satellite capacity. If the capacity per satellite for the next generation satellite system were to remain the same as or close to those of the INTELSAT-V satellites, the number of orbit slots needed for INTELSAT satellites in each of the ocean regions would grow fast and multiple antennas at the INTELSAT earth stations would be required in order to restore the flexibility and connectivity of the current network. Clearly, this is not a desirable situation, and therefore the successive generations of satellites have to accommodate higher and higher volumes of traffic. This solution is not only desirable from the point of view of the system users but also from the point of view of efficiency of utilization of the geostationary satellite orbit.

The increase in capacity per satellite can be effected through one or more of the following techniques: (a) use of advanced technology for modulation and multiple access, (b) use of increased frequency bandwidth per satellite, and (c) multi-re-use of the currently used frequency bands.

In fact, a compromise has to be made among the various options for meeting the traffic requirements because of the limitations on the payload that can be accommodated on the launch vehicles that will be available in the near future.

The results of the studies made by INTELSAT have shown that the optimum approach for the coming decade is as described in the following paragraphs.

1. *Satellite deployment plan.* In the Atlantic Ocean region three operational satellites will be needed: a primary and two major path satellites. In the Indian Ocean region one operational satellite is sufficient initially, but later two satellites will be needed: a primary and a major path satellite. In the Pacific Ocean region one operational satellite will be needed.

2. *INTELSAT-VI design.* The performance specification of the INTELSAT VI has been prepared and the request for proposals (RFP) issued in March 1981. The satellites will have the following characteristics: (a) compatibility with launch by either the United States Space Transportation System (STS) or the European Ariane-4 expendable launch vehicle; (b) sixfold multiple frequency re-use in the 6/4 GHz frequency bands; (c) re-use of the 14/11 frequency bands; (d) TDMA operating in a portion of the 14/11 GHz band and satellite-switched TDMA operating in a portion of the 6/4 GHz band.

    Studies are being conducted of an STS optimized INTELSAT-VI spacecraft concept that would provide INTELSAT with increased capacity as well as allow the incorporation of additional capabilities in the INTELSAT-VI series of satellites.

3. *Interim solution.* In light of the rapidly increasing traffic requirements for both international and domestic services which are expected to create a need for additional capacity in orbit before the INTELSAT-VI series becomes available, INTELSAT is investigating alternatives to fulfill this need. INTELSAT intends to procure modified INTELSAT-V satellites (V-A satellites) with a modest increase in capacity and modified configura-

tion that are tailored to the traffic patterns forecast in the various ocean regions during the lifetime of this satellite series. The extra capacity will be derived mainly by making improvements to the basic INTELSAT-V, incorporating late technological developments and utilizing more complex antenna systems. These satellites are expected to be operational by the end of 1983, and their weight and size will be compatible with the currently available expendable launch vehicles.

Beyond INTELSAT-VI, the INTELSAT planners are faced with a number of exciting concepts: multipurpose space platforms, satellite clusters, and a string of satellites interconnected by intersatellite links. The multipurpose platforms could be assembled in space and could include communications packages that would provide services of different types. These packages would share the number of common facilities on the platform, such as power supplies, stabilization, etc. The cluster concept could consist of a central switching satellite around which would be grouped individual satellites providing different services that would communicate to the central satellite by means of very short radio links.

Finally, a string of satellites that would communicate with each other through intersatellite links could provide distributed coverage over the whole globe and provide different types of services.

10.2 INTERCOSMOS

10.2.1 Foundation and Membership

INTERCOSMOS is a program of comprehensive cooperation among the socialist countries for the peaceful exploration and use of outer space. Ten countries (Bulgaria, Cuba, Czechoslovakia, the German Democratic Republic, Hungary, Mongolia, Poland, Romania, the USSR, and Viet Nam) take part.

In April 1965 the Soviet government sent a letter to the governments of the other socialist countries proposing that they should discuss specific action aimed at combining their efforts in the peaceful exploration and use of outer space, with due regard for each country's scientific and technological capabilities and interests. In accordance with an understanding reached following an exchange of communications among the heads of government, meetings of representatives of Bulgaria, Cuba, Czechoslovakia, the German Democratic Republic, Hungary, Mongolia, Poland, Romania, and the USSR were held at Moscow in November 1965 and April 1967 to discuss the nature and form of such cooperation. The second meeting adopted a complex program of joint space activities which, at a 1970 meeting of the directors of the national coordinating bodies of the countries participating in the cooperative effort, was given the name INTERCOSMOS.

Seeking to consolidate their favorable experience in the field of cooperation and promote its further development, on 13 July 1976, representatives of the nine socialist countries taking part in the INTERCOSMOS program signed in Moscow an intergovernmental "Agreement on cooperation in the peaceful exploration and use of outer space" (which entered into force on 25 March 1977). The agreement is open to accession by other countries interested in cooperating in the investigation of space, subject to the consent of the contracting parties. Viet Nam acceded to the agreement on 17 May 1979, thus becoming the tenth country participating in the INTERCOSMOS program.

10.2.2 Organizational Structure

The INTERCOSMOS program embraces joint activities in the following areas:

1. Space physics.
2. Space meteorology.
3. Space biology and medicine.
4. Space communication.
5. Investigation of the earth by means of aerospace techniques.

National coordinating bodies responsible for organizing and arranging for the implementation of both the overall program and the bilateral and multilateral individual projects and studies have been set up in each member country.

The highest coordinating body in INTERCOSMOS is the Conference of Directors of the National Co-ordinating Bodies, which discusses and makes decisions on matters of principle relating to the organization and execution of joint activities in space. Sessions of the Conference take place at least once a year and generally rotate among the member countries. Decisions and recommendations adopted by the Conference are binding on the countries whose directors vote in favor of their adoption.

Permanent mixed working groups, comprising scientists and specialists from the member countries, have been set up to coordinate the execution of programs and work plans in the five areas indicated. They direct the execution of joint space investigations and experiments, conduct regular reviews of progress in the projects undertaken, and consider proposals for further joint activities, new types and methods of cooperation, the creation and production of scientific instruments and devices, and the scientific and practical uses to which the results should be put. As necessary, they draw up recommendations and submit them to the Conference for consideration.

These organizational arrangements for cooperation under the INTER-COSMOS program have worked successfully for almost 15 years and make it possible to solve routinely any problem that arises, with due regard for each country's interests.

Overall coordination of the activities of national bodies between Conference sessions, pursuant to the Agreement of 13 July 1976, is entrusted to the national body of the depositary country, the INTERCOSMOS Council of the USSR Academy of Sciences.

The countries participating in the INTERCOSMOS program have no common source of financing. Each country finances the experiments and the development and production of the scientific instruments in which it is interested. The Soviet Union provides, free of charge to its partners in cooperation, rocket technology facilities and launch services. The scientific results of joint experiments are the collective property of all participants in the project.

10.2.3 Goals and Achievements of the INTERCOSMOS Program

Twenty INTERCOSMOS satellites, eight high-altitude VERTIKAL research rockets, and a large number of meteorological rockets designed to investigate the ionosphere and magnetosphere and study the sun, solar activity, cosmic rays, and the parameters of the upper atmosphere were launched between October 1969 and November 1980. More than 200 scientific instruments and devices were developed and manufactured for experiments carried by these satellites and rockets. A number of space vehicles launched by the USSR as part of its national program (METEOR and COSMOS satellites, PROGNOZ automatic stations, SOYUZ spacecraft, and the SALYUT-6 orbital station) carried instruments designed by scientists and specialists from the socialist countries as a part of the INTERCOSMOS program.

Joint research in space biology and medicine is designed to study the influence of various spaceflight factors on the human body and devise the necessary preventive measures.

Work in the field of space communications has led to the creation of the INTERSPUTNIK international space communications system and organization, which transmits television programs, telephone messages, and other information.

Joint work in the investigation of the earth by means of aerospace techniques is being carried out with a view to solving problems relating to scientific methodology and developing experimental equipment for remote sensing of the earth.

The INTERCOSMOS program has produced valuable scientific results, some of which constitute major contributions to the space sciences and to the practical application of space research in various fields. The joint scientific publications that have been prepared on the basis of these results number more than 500; the results of joint space research have been regularly reported at COSPAR meetings, IAF congresses, and other international symposia and conferences and were very favorably received by the international scientific community.

In 1976 the Soviet Union proposed the participation of citizens of socialist countries in international manned space flights. During consultations held in July and September 1976 between representatives of Bulgaria, Cuba, Czechoslovakia, the German Democratic Republic, Hungary, Mongolia, Poland, Romania, and the USSR, it was agreed that citizens of all countries participating in the INTERCOSMOS program would take part in flights on board Soviet spacecraft and space stations together with Soviet cosmonauts between 1978 and 1983.

The spaceflights of seven international crews were carried out between March 1978 and September 1980; these crews consisted of Soviet cosmonauts and cosmonauts who were citizens of Czechoslovakia, Poland, the German Democratic Republic, Bulgaria, Hungary, Viet Nam, and Cuba. The flights took place on the following dates:

1. 2–10 March 1978: The flight of cosmonauts A. A. Gubarev (USSR) and V. Remek (Czechoslovakia).

2. 27 June–5 July 1978: The flight of cosmonauts P. I. Klimuk (USSR) and M. Hermaszewski (Poland).

3. 26 August–3 September 1978: The flight of cosmonauts V. F. Bykovsky (USSR) and S. Jaehn (German Democratic Republic).

4. 10–12 April 1979: The flight of cosmonauts N. N. Rukavishnikov (USSR) and G. I. Ivanov (Bulgaria).

5. 26 May–3 June 1980: The flight of cosmonauts V. N. Kubasov (USSR) and B. Farkas (Hungary).

6. 23–31 July 1980: The flight of cosmonauts V. V. Gorbatko (USSR) and Pham Tuan (Viet Nam).

7. 18–26 September 1980: The flight of cosmonauts Y. V. Romanenko (USSR) and A. Tamayo Méndez (Cuba).

At the present time, cosmonaut candidates from Mongolia and Romania are being trained for spaceflights at the Y. A. Gagarin Cosmonaut Training Center near Moscow.

Approximately 100 scientific experiments jointly prepared by scientists and specialists of the socialist countries taking part in the INTERCOSMOS program were carried out during these flights. These experiments follow the INTERCOSMOS program's traditional lines of research (space biology and medicine, materials processing in space, study of the physical properties of outer space and of the earth's atmosphere and surface). They are a natural outgrowth of research in these fields and reflect the increasingly practical trend of joint work in space with a view to producing economic benefits and meeting the practical needs of people.

These experiments aid in improving the medical and biological care provided on manned spaceflights, advancing medical services for people on earth, understanding the underlying principles of technological processes in space, studying the natural environment in which man lives, and studying the earth's natural resources in order to make their consumption more rational, provide the necessary monitoring of environmental pollution, and solve other global ecological problems.

Over the next five years, in accordance with jointly elaborated long-term plans, cooperation within the INTERCOSMOS program will be developed further along all the lines mentioned above, using various automatic and manned space vehicles.

10.3 INTERNATIONAL SYSTEM AND ORGANIZATION OF SPACE COMMUNICATIONS (INTERSPUTNIK)

10.3.1 General

INTERSPUTNIK is an intergovernmental international organization with membership open to all states. The Agreement on the Establishment of the "INTERSPUTNIK" International System and Organization of Space Communications was signed on 15 November 1971; after ratification, it entered into force on 12 July 1972 and was registered with the United Nations.

INTERSPUTNIK was created to meet the need for exchanges of radio and television programs, telephone and telegraph links, and other information among different countries.

The legal and technical principles on which INTERSPUTNIK is based were developed within the context of cooperation between the socialist countries in the exploration and use of outer space for peaceful purposes under the INTERCOSMOS program. INTERSPUTNIK operates on the basis of respect for the sovereignty and independence of states, equality, non-interference in internal affairs, mutual assistance, and mutual benefit.

Any state that shares the goals and principles of INTERSPUTNIK activities and assumes the obligations under the agreement may become a member.

The current members of INTERSPUTNIK are Afghanistan, Bulgaria, Cuba, Czechoslovakia, Democratic Yemen, the German Democratic Republic, Hungary, Mongolia, Poland, Romania, the USSR, and Viet Nam.

10.3.2 Structure

The governing body of INTERSPUTNIK is the Board, comprising one representative from each member of the organization. Each board member has one vote. The Board considers general policy issues, approves plans for the establishment, operation, and development of the communications system, determines the technical specifications for the earth stations, authorizes the inclusion of earth stations in the communications system, approves the plan for the distribution of channels and the rates for transmitting a unit of information, elects the director-general, and determines the structure and staff of the Directorate, in addition to considering other matters related to INTERSPUTNIK activities. Sessions of the Board are held at least once a year.

The permanent executive and administrative body of INTERSPUTNIK is the Directorate, which is situated in Moscow. It is headed by the director-general, who is empowered by the Board to represent INTERSPUTNIK in all matters arising out of the agreement. The Directorate staff is recruited from among nationals of the states whose governments are members of INTERSPUTNIK, with due regard for their professional qualifications and equitable geographical representation.

An Auditing Commission, whose membership is determined by the Board, has been established to supervise the financial and economic activities of INTERSPUTNIK.

10.3.3 The Communications System

The INTERSPUTNIK communications system comprises a space segment and earth stations. The space segment, which includes communications satellites and control systems, is the property of the organization or is leased by it from its members. The earth stations are the property of the countries that build them or the organizations that operate them.

INTERSPUTNIK currently operates using Soviet satellites on a lease basis. The system employs two STATSIONAR satellites in geostationary orbit at longitudes of 14° W (Atlantic region) and 53° E (Indian region). Two relay units on board each satellite are used for telephone and telegraph links and for the exchange of radio and television programs.

Thirteen ground stations operate in the INTERSPUTNIK system: seven in Europe (Bulgaria, Czechoslovakia, German Democratic Republic, Hungary, Poland, and two in the USSR), four in Asia (Afghanistan, Laos, Mongolia, and Viet Nam) and one each in Central America (Cuba) and North Africa (Algeria).

There are plans to build earth stations in Syria, Democratic Yemen, Guinea, and a number of other countries.

In addition to the members of the organization, other countries (for example, France, Italy, Spain, Yugoslavia) also use the channels of the INTERSPUTNIK system.

10.3.4 Level of Utilization

The INTERSPUTNIK communications system is used mainly for exchanges of television programs with broadcasts lasting four to eight hours daily. Some 40 percent of the television program exchanges made by the member countries of INTERVIDENIE is arranged through the INTERSPUTNIK system. The programs cover major political, cultural, and sporting events and other happenings around the world.

More than 20 countries participate in exchanges of television programs through the INTERSPUTNIK system. About the same number use its channels for international telephone and telegraph links.

Plans for the development of the system in the next few years include bringing additional channels into use on board the STATSIONAR satellites and introducing new equipment in the earth stations, so as to make possible an increase in the amount of information transmitted and to improve the quality and reliability of the communications channels.

10.3.5 International Cooperation

INTERSPUTNIK coordinates its activities with ITU and other international organizations in connection with the use of the frequency spectrum and the application of standards for communications channels, and in other areas as well. Relations between INTERSPUTNIK and the Pan-African Telecommunication Network (PANAFTEL) are being expanded and consolidated. INTERSPUTNIK has concluded a number of international agreements, [for example, with the Council for Mutual Economic Assistance (CMEA) and with the International Radio and Television Organization (OIRT)], and it participates in the INTERCOSMOS program, identifying specific areas and forms of cooperation on matters of common interest.

10.3.6 Conclusion

The work of INTERSPUTNIK and other international organizations serves to confirm the effectiveness of using space for communications purposes. In recent years a large number of international, regional, and national communication systems using earth satellites have been set up.

A consolidation, within the specialized agencies of the United Nations and elsewhere, of the cooperation and interaction among countries and international organizations in solving the interrelated problems in this area will promote the development of international exchanges of information and ties of friendship around the world.

10.4 EUROPEAN SPACE AGENCY (ESA)

10.4.1 General Information

The European Space Agency (ESA) is an intergovernmental international organization which fosters cooperation among European countries, thereby enabling them to develop, carry out, and finance European space projects in common.

ESA was established on 31 May 1975. It combines the activities previously conducted by the European Space Research Organization (ESRO) in the construction of scientific satellites and of the European Organization for the Development and Construction of Space Vehicle Launchers (ELDO) in the construction of launchers and the building of applications satellites. The establishment of a single organization reflected the need not only for Europe to pool its efforts within a single body but also to achieve a more adequate balance in European programs. The agency's priorities are set through the joint political will of member states. Such organizational unity has provided Europe with an effective instrument for ensuring its presence in space and thereby securing a political independence that no member state could have attained on its own.

10.4.2 Major Objectives

ESA's major objectives are listed in its Convention. Its purpose is to provide for and to promote, for exclusively peaceful purposes, cooperation among European states in space research and technology. In this connection, it elaborates and implements a long-term European space policy and coordinates and integrates national and European programs. The formulation of its programs is based on the principles of independence, balance, and cooperation.

Europe's concern for independence is reflected in particular by the development of a launcher program—Ariane—for placing satellites into orbit and thereby securing greater freedom of action.

The concern for equity in the distribution of the previous programs is shown in a balanced sharing of the budget among scientific programs (ESA has placed 12 scientific satellites into orbit) and applications programs (remote sensing, meteorology, and telecommunications).

The concern for cooperating with other countries that have space programs has been reflected particularly in the development of the manned European space laboratory (SPACELAB), which will be launched by the United States Space Shuttle. This will also enable Europe to participate in the development of manned spaceflights and to benefit from the many scientific investigations that will be carried out on man in space.

In order to develop a coherent industrial policy through the pooling of the bulk of the funds allocated to space activities in Europe, the industrial work involved must be shared in a manner commensurate with the financial contributions of each member state. This helps the ESA to attain one of its goals, that of improving the competitiveness of European industry.

10.4.3 Member States

ESA has 11 member states: Belgium, Denmark, France, the Federal Republic of Germany, Ireland, Italy, the Netherlands, Spain, Sweden, Switzerland, and the United Kingdom. Austria participates in some ESA programs as an associate member state, Norway as an observer, and Canada through a cooperation agreement.

10.4.4 Cooperation Arrangements

10.4.4.1 ESA policy.   ESA's policy is formulated by the Council, which is composed of representatives of member states. Depending on their importance, decisions are taken either unanimously, by a two-thirds majority of member states, or by a simple majority. In certain important, but not all, cases, this system avoids vetoes that would impede the smooth running of the agency.

Various program councils and committees are empowered to make decisions or submit recommendations to the Council. These councils and committees are also composed of representatives of member states chosen, in some cases, on the basis of their participation in programs.

The director general is the chief executive of the agency and its legal representative. He is responsible for managing the agency, conducting programs and implementing policy, and also for proposing activities and programs that are submitted to member states for approval. He is assisted in his duties by directors responsible for various areas of activity and a staff of about 1,400.

Under the Convention, ESA makes maximum use of existing installations and equipment in member states. Accordingly, national space agencies are closely involved in carrying out a number of the agency's programs. The outstanding example is the Ariane program, for which the French National Center for Space Research (CNES) is the prime contractor.

10.4.4.2. Financing.   ESA's budget is financed by its member states. The level of the 1981 budget is some 600 million accounting units, or nearly $720 million. Member states contribute to mandatory activities according to their national income. These activities include the general budget and the scientific programs. Financial support for the optional programs is given on a flexible basis; each state participating in a particular program determines its percentage contribution. Nonmember states which cooperate in individual programs also contribute at a rate chosen by them (for example, Austria participates in the SPACELAB and L-SAT programs, Canada in the L-SAT program and in the definition phase of the remote sensing program, and Norway in the MARECS and remote sensing programs).

The bulk of the contributions of European governments has been paid out by the agency to European manufacturers of satellites, launchers, and ground stations and suppliers of other equipment and services required by the joint space programs of the member states. The implementation of the agency's programs by European industry assures the latter, through the granting of contracts, of a high return on the investments of participating states.

10.4.4.3 Cooperation with nonmember states.
Cooperation with nonmember states is encouraged by the Convention that established the agency. The agency is developing its relations with nonmember states and international organizations. Its essential role is to publicize European space achievements and thereby to promote cooperative activities. Efforts are directed especially toward countries or groups of countries interested is availing themselves of the space applications programs. Many developing countries have been given demonstrations of space applications systems that could be made available in order to enable them to accelerate their development, especially in the fields of meteorology and remote sensing.

10.4.5 Results

The European programs include scientific satellites that are helping to expand man's knowledge of the universe, applications satellites used for long-range telecommunications, meteorology, and remote sensing, and rocket development culminating in the production of Ariane, the first European heavy launcher capable of putting into orbit large geostationary telecommunications satellites (including broadcast satellites). The programs also include SPACE-LAB, which will enable European astronauts to do research in space.

10.4.5.1 Scientific programs.
ESA's scientific programs cover a range of disciplines centering on the study of the earth's environment and nearby or distant celestial objects.

In 1980 four scientific satellites of ESA were in orbit around the earth:

1. COS-B, launched in 1975, is investigating gamma-ray astronomy.
2. ISEE-2, launched in 1977, is investigating the magnetosphere and sun–earth relationships. ISEE-2 is one of the three satellites of the international program which ESA has undertaken in cooperation with NASA.
3. IUE, launched in January 1978, is an example of cooperation between ESA, NASA, and the United Kingdom's Science Research Council and is doing astronomical research in the ultraviolet range.
4. GEOS-2, launched in 1978, is continuing the magnetospheric research begun by GEOS-1.

Among the ongoing scientific projects, mention must be made of the following:

1. EXOSAT, to be launched by Ariane in July 1982, is designed to identify and investigate celestial sources of X-rays with a degree of accuracy heretofore unattained.

2. The Space Telescope is another example of cooperation between Europe and the United States; the program calls for the operation for some 15 years of a telescope 2.4 m in diameter that will be put into orbit in 1983 by the Space Shuttle. The European contribution to the program is providing the faint-object camera and the solar array.

3. The International Solar Polar Mission is another joint ESA and NASA program that involves the operation of a European satellite to be launched in 1985 by the Space Shuttle. The Mission will explore regions far from the plane of the earth's orbit and, in particular, space above the solar pole.

4. There are two programs involving devices carried on board SPACELAB: lidar, an active instrument for investigation of the atmosphere, and the space sled, which will investigate human behavior in space and the space sickness experienced by astronauts.

5. HIPPARCOS, to be launched by Ariane, is designed to collect astrometric data on a large number of stars.

6. GIOTTO, expected to be launched in 1985, is designed to encounter Halley's comet one year later.

10.4.5.2 Applications programs.

10.4.5.2.1 Earth Observation Program.   Europe has developed weather satellites to investigate scientific phenomena in the earth's atmosphere and disseminate to users images and meteorological data obtained by processing. METEOSAT-1, launched in 1977, is Europe's contribution to the World Weather Watch and to the Global Atmospheric Research Program (GARP), which comprises five satellites. There are four others: three United States and one Japanese. METEOSAT-2 was launched in December 1981 on the third Ariane development flight.

METEOSAT has a number of missions: photography, image dissemination, and data collection. The processing of the data yields a number of meteorological parameters, such as wind velocities and fields, sea surface temperatures, and cloudtop height. Such data are coded and entered into the ground communications network of the World Meteorological Organization (WMO).

The SIRIO-2 satellite, which is to be launched in 1982 by Ariane, will carry out two experimental missions relating to meteorology and geodesy. The objective of the first mission, known as MDD (Meteorological Data Distribution), is to improve the flow of information in Africa. The objective of the second, called LASSO (Laser Synchronization from Stationary Orbit), is to improve world synchronization of atomic clocks by means of laser techniques.

The missions of the remote sensing satellites cover a great variety of applications: crop inventories and production forecasts, water resources management, and monitoring of river banks and coastal areas, fisheries; marine currents, and pollution. Such applications can benefit not only the industrialized countries but also the developing countries, which, for the most part, do not have the necessary information at their disposal for the effective management of their natural resources. Accordingly, ESA has established a three-faceted program:

1. The establishment of a network of data reception stations (EARTHNET program), to receive, pre-process, and disseminate data from the United States remote-sensing satellites LANDSAT, HCMM, NIMBUS-7, and SEASAT. The network comprises five stations in Sweden, Italy, and the United Kingdom, France, and Spain.
2. The design of a European remote sensing payload to be carried on the first mission of SPACELAB, for the purpose of assessing the operational possibilities and the capabilities of two instruments, a metric camera and a microwave radar.
3. Preparation of European remote sensing satellite projects. The objective of this preparatory program is to study and develop European remote sensing satellites to be launched by Ariane from 1986 to 1988.

This system is designed with a view to establishing:

1. An ocean observation satellite system for coastal and ocean areas (1987).
2. A satellite system for land applications (1988).

10.4.5.2.2 Telecommunications.   One of Europe's main objectives is to set up a network of links for both traditional services (telephone and television) and new, specialized services (teleconferences, data transmission, etc.) and make them available to European postal, and telecommunications, and radio administrations.

In 1971 research led to the development of an orbital test satellite (OTS). OTS is the third communications satellite developed in Europe. It follows the French–German SYMPHONIE and the Italian SIRIO. Launched in 1978, its missions have focused on demonstrating the performance of the equipment on board (payload, space vehicle systems and subsystems) in preparation for a later preoperational mission that will offer a traffic capacity of 6,000 telephone circuits, or 4,500 telephone circuits and two television channels. OTS operates in the 14 and 11 GHz frequency bands.

The next generation, that of the European Communications Satellites (ECS), is designed for a European regional operational system that will retransmit telephone or television signals and transmit data between off-shore oil rigs and coastal stations. The European Telecommunications Satellite Organization (EUTELSAT) is to be supplied with five ECS satellites to help to meet the needs of European users during the 1980s.

One type of ECS satellite, MARECS, using the ECS platform, will provide long-range links between ships and land stations. MARECS-A was launched in December 1981, and MARECS-B is planned for mid-1982. The two satellites will be operated by INMARSAT, a world maritime communications organization founded in 1979. INMARSAT will have a network of six satellites, payloads on three INTELSAT-V and three MARECS satellites.

Since 1976, ESA has been studying plans for a heavy platform adapted to Ariane's capacities that would enable it to carry telecommunications payloads for direct broadcasting and for new services. Broadcasts could be received directly in individual homes through the use of a parabolic antenna (less than 1 m in diameter) and a converter adaptable to conventional television sets. The L–SAT program responds to the need for versatility in providing for the full range of future applications.

10.4.5.3.1 Resources

10.4.5.3.1 The Ariane Launcher.   The Ariane heavy launcher will serve to provide Europe with a launching capability for its own scientific or applications satellites, enabling it to have a share in the extensive market for launchings expected in the 1980s, estimated at about 200 geostationary satellites. Ariane, designed, among other things, to place satellites weighing over 1,000 kg into geostationary orbit, became operational in 1982. After its four flight-qualification tests, the European launcher was declared operational and will be available to European and non-European users.

Projections of the potential market for the Ariane launcher, assuming a rate of four launches per year and the possibility of launching two satellites with the same launcher, point to an estimate of some 30 to 50 launches during the 1980s. A program for further development of Ariane aimed at improving the launcher's performance, is already under way (Ariane-2 and -3).

10.4.5.3.2 SPACELAB.   SPACELAB, the reusable manned space laboratory developed for ESA by European industry, has been designed to be placed into orbit by the NASA Space Shuttle. The first flight model was delivered to NASA in 1982. As many as four persons can work in the pressurized SPACE-LAB module for periods of about one week. Two payload specialists, one from the United States and one from Europe, will take part in the first mission.

The Spacelab program, carried out under an agreement signed in 1973, is the most important program of cooperation between ESA and NASA. In all, 38 scientific and technological experiments (24 European, 13 United States, and one Japanese) have been scheduled for this first mission, which was planned jointly by ESA and NASA. The disciplines involved in this first mission include astronomy, atmospheric physics, solar physics, plasma physics, remote sensing, the life sciences, and materials processing.

ESA is currently studying programs for the utilization and further development of SPACELAB.

10.4.6 Conclusion

The European Space Agency is a novel achievement of noncommercial international cooperation in the field of space. It ensures greater efficiency of most of the efforts of its member states in both scientific and applications programs. The agency offers ample opportunities in international cooperation for the progress of the exploration of space for peaceful purposes and in the interest of all mankind, in accordance with the Outer Space Treaty, the cornerstone of space law adopted by the United Nations in 1967.

10.5 INTERNATIONAL MARITIME SATELLITE ORGANIZATION (INMARSAT)

10.5.1 The Establishment of INMARSAT

By resolution A.305 (VIII) of 23 November 1973, the Assembly of the Inter-Governmental Maritime Consultative Organization (IMCO) decided to convene an international conference to decide on the principle of setting up an international maritime satellite system and to conclude agreements to give effect to this decision.

Pursuant to this decision, the International Conference on the Establishment of an International Maritime Satellite System was convened in London on 23 April 1975 for its first session and concluded its work at its third session on 3 September 1976.

The following instruments were adopted by the Conference:

1. Convention on the International Maritime Satellite Organization (INMARSAT).
2. Operating Agreement on the International Maritime Satellite Organization (INMARSAT).

Both agreements entered into force on 16 July 1979, i.e., 60 days after the date on which states representing 95 percent of the initial investment shares had become parties to them. The headquarters of the organization are in London.

10.5.2 Purpose

The main purpose of INMARSAT is to make provision for the space segment necessary for improving maritime communications, thereby assisting in improving distress and safety of life at sea communications, efficiency and management of ships, maritime public correspondence services, and radio determination capabilities. The organization seeks to serve all areas where there is need for maritime communications and acts exclusively for peaceful purposes.

10.5.3 Membership and Structure

Thirty-two states had become parties to the INMARSAT Convention by 15 October 1980 and several states are in the process of acceding.

The members of INMARSAT are: Algeria, Argentina, Australia, Belgium, Brazil, Bulgaria, Byelorussian Soviet Socialist Republic, Canada, China, Denmark, Egypt, Federal Republic of Germany, Finland, France, Greece, India, Italy, Iraq, Japan, Kuwait, Netherlands, New Zealand, Norway, Poland, Portugal, Singapore, Spain, Sweden, Ukrainian Soviet Socialist Republic, USSR, United Kingdom, and United States.

INMARSAT is open to all states for membership. In addition, nonmember countries may use the space segment for navigation services on conditions determined by INMARSAT.

INMARSAT has three main organs: the Assembly, where all member states are represented and have one vote each: the Council, which has 22 signatories and voting power in relation to investment shares; and the Directorate, headed by a director-general. Substantive decisions are taken with a two-thirds majority, both in the Assembly and in the Council. The functions of the Assembly are to consider and review the general policy and lay-term objectives of the Organization, express views, and make recommendations thereon to the Council. All other decisions are taken by the Council. The Council, in this way, decides on all financial, operational, technical, and administrative questions.

In accordance with the Convention, INMARSAT operates on a sound economic and financial basis, having regard for accepted commercial principles. It is financed by contributions from its members in proportion to their use of the space segment. Member states receive capital repayment and compensation for use of capital when revenues allow such repayment.

Article 6 of the Convention provides that the Organization may own or lease the space segment. The charges for use of the space segment have the objective of earning sufficient revenues for INMARSAT to cover its operation, maintenance, and administration, the provision of such operating funds as are necessary, amortization of investments made by signatories, and compensation for use of capital.

10.5.4 The Scope of INMARSAT Activities

The following operational requirements, which needed to be satisfied by a maritime satellite system were laid down at the International Conference on the Establishment of INMARSAT:

1. Handling of distress cases, including distress messages, and search and rescue control communications, as well as determination by land-based stations of the position of any ship in distress and of search and rescue units involved.

2. Distribution of urgent safety messages, including requests for medical assistance.

3. Interrogation of the land-based station by the ship or other mobile craft to obtain position information (this could possibly be followed by environmental, meteorological, and oceanographic information) or regular interrogation of the ship at appropriate time intervals by the land-based station, together with the transmission of position and other information.

4. Initial position determination, providing accuracies of the order of one to two nautical miles. As technology develops accuracy may be improved until the system is suitable for navigation near coasts and in narrow passages and fairways.

5. Transmission of highly accurate standard frequencies and time signals.

6. Selective calling and multiple access technique to facilitate communications.

7. Selective calling of ships by coast stations for establishing public correspondence through terrestrial communications.

8. Public correspondence, including ship's and company operational business carried out by telegraphy and telephony.

9. Data transmission, including facsimile, teleprinter, and wideband systems.

10. Automatic warning of ships which are continuously tracked by the system that they are approaching shallow waters, underwater constructions, drilling and production platforms, etc.

11. Advising ships which are continuously tracked by the system on anti-collision actions and on avoidance of continuously tracked navigational hazards, for example, icebergs.

12. Automation of the position-reporting system based on position information as mentioned in (3).

13. Traffic control, including collision warnings, especially in converging areas, subject to the radio determination system providing sufficient relative accuracy [see (4)].

14. Distribution of meteorological, hydrographic, and oceanographic information to ships.

15. Individual meteorological and oceanographic advice to ships by land-based stations, for example, weather routing, navigation through ice, etc.

16. Collection of meteorological, hydrographic, and oceanographic observations from ships.

The International Conference on the Establishment of an International Maritime Satellite System further recommended that at an early date arrangements should be made to undertake the study of the institutional, financial technical, and operating consequences of the use by INMARSAT of multipurpose satellites providing both a maritime mobile and an aeronautical mobile capability.

10.5.5 Commencement of Satellite Communications Services

The multitude of operational requirements can only be met step by step, contingent in particular upon the cooperation of the international maritime community and the further buildup of INMARSAT. As a first step, INMARSAT is establishing a space segment to provide L- and C-band maritime mobile satellite communications capacity over the Atlantic, Pacific, and Indian Ocean regions. Operations began in 1982 using MARISAT and MARECS satellites.

10.6 ARAB SATELLITE COMMUNICATION ORGANIZATION (ARABSAT)

10.6.1 Origin and Membership

The Arab interests in telecommunications are taken care of by the national post and telecommunications administrations which coordinate their efforts in the Arab Telecommunication Union (ATU). The members of the Arab League founded ARABSAT by means of an ARABSAT Charter signed on 14 April 1976. Twenty-one Arab States are members of the ARABSAT Communication Service: Algeria, Bahrain, Democratic Yemen, Egypt, Iraq, Jordan, Kuwait, Lebanon, Libya, Mauritania, Morocco, Oman, Palestine, Qatar, Saudi Arabia, Somalia, Sudan, Syria, Tunisia, United Arab Emirates, and Yemen. There are plans for integrated telecommunications networks by the mid-1980s.

10.6.2 Organizational Structure

ARABSAT consists of three main organs:

1. The General Assembly, composed of all ARABSAT members, and which meets once a year.
2. The Board of Directors, composed of representatives of nine member states, five of which are permanent: Kuwait, Lebanon, Libya, Saudi Arabia, and United Arab Emirates, the four others are elected by the General Assembly.
3. The Executive Body, composed of a number of administrative units or sections which are defined in the ARABSAT's internal regulations.

ARABSAT headquarters is in Riyadh, Saudi Arabia.

10.6.3 Goals

The main objective of ARABSAT is to establish, operate, and maintain a regional telecommunications system for the Arab region. The ARABSAT system will serve as a complement to the terrestrial network for routing interregional public telecommunications traffic between main international switching centers and provide new possibilities for television program exchange among Arab countries.

The users of the ARABSAT system will be the members of the Arab League, and the ARABSAT system design has been based on the requirements expressed in traffic meetings and on multilateral consultations as well as in surveys made by ARABSAT teams. The ARABSAT system will be equipped to furnish the following services:

1. Regional and domestic telephony, telegraphy, telex, and data transmission.
2. Regional and domestic television.
3. Community television.

10.6.4 System Component

The ARABSAT space segment is composed of two satellites that will be located on the geostationary orbit at 19° E and 26° E. A third one will serve as spare.

Ground facilities including earth stations and terrestrial transmission systems will have to be set up and will have to be compatible with the ARABSAT space segment. A control network will consist of control stations and one control center at Riyadh.

The ARABSAT satellites will be designed for launch on either the European Ariane or the American Space Shuttle.

The ARABSAT system will be brought into use by 1984 and will provide services for the immediate requirements of the Arab member states as well as for their needs for telecommunications up to 1990.

Chapter 11

ROLE OF THE

UNITED NATIONS SYSTEM

IN SPACE ACTIVITIES

ABSTRACT

This chapter, the second of two on international cooperation, is aimed at providing information on the mandates and programs of various organizations within the United Nations system which are active in space science and technology. It was prepared by the Conference secretariat, based on information available to it from the organizations concerned.

INTRODUCTION

In the decade of the 1960s the international community had already come to accept the results made possible by outer-space technology in areas such as communications, meteorology, and management of the earth's environment and its resources. As a response to the expression of this new interest by a growing number of countries, the United Nations convened in Vienna, in 1968, the First

United Nations Conference on the Exploration and Peaceful Uses of Outer Space to inform member states of the results and the potential of space applications, to discuss how nations might benefit from active participation, and to consider a possible role for the United Nations in this effort. The conference provided a stimulus within the United Nations for new initiatives to enable as many countries as possible to take part in the opportunities available for applying space technology to the needs of economic, social, and cultural development. The United Nations system directed its attention to the area of space applications and each of the organizations within the system undertook programs to promote outer-space applications in the developing countries.

Those programs were intended to promote an awareness in the developing countries, particularly among policy-makers, to facilitate education and training of experts in these countries in the area of space applications, and to provide assistance in establishing space applications activities.

Since 1968, the United Nations system has held a large number of panel meetings, seminars, and training workshops relating to space applications in all regions of the world. A program of fellowships has been administered under the programs and a modest level of activity in providing expert advice to countries has also been undertaken. Many developing countries have already benefited and in some cases have, as a result, started national programs varying from the simple use of automatic picture transmission (APT) machines for receiving satellite weather information to more complex uses of digital analysis and computer enhancement of satellite sensed data of earth resources such as agriculture, minerals, and forests. Assistance to member states of a more substantial nature, i.e., the provision of hardware, technical assistance, etc., has been provided by the United Nations system through its international funding agencies such as the World Bank and UNDP. The procedure followed in such cases is given in Appendix 11.1.

The programs have thus been successful, although relatively limited in scope. Recognizing the increasing activity in the area of space applications and benefits to be derived therefrom, the developing countries have already called for their expansion.[1]

As outlined in this study, space technology clearly offers a variety of applications that can benefit member states; choices need only be made as to which applications would be most desirable and effective. In realizing the benefits that space technology can provide, the United Nations and its agencies can play an integral role in helping any member state to recognize and develop its own intrinsic capabilities. As needs are expressed by member states, appropriate resources and efforts will have to be provided to respond to them effectively.

---

[1] See, for example, the recent reports of the United Nations Committee on the Peaceful Uses of Outer Space and of its Scientific and Technical Sub-Committee.

11.1 ORGANIZATIONS WITHIN THE UNITED NATIONS SYSTEM DEALING WITH SPACE ACTIVITIES

11.1.1 United Nations

11.1.1.1 Outer Space Affairs Division. With the establishment of the United Nations Committee on the Peaceful Uses of Outer Space in 1961, the Outer Space Affairs Division of the United Nations Secretariat was assigned the responsibility of implementing decisions of the Committee and its subsidiary bodies related to the promotion of international cooperation and peaceful uses of outer space. Among the primary tasks carried out by the Committee, in which the secretariat assists, are the development of international treaties, conventions or legal principles governing the activities of states in the peaceful exploration and use of outer space, and the implementation of global systems or programs sponsored by the international community.

Following the decision of the Committee on the Peaceful Uses of Outer Space to take upon itself the promotion of international cooperation in the practical applications of space technology, the United Nations program of space applications was initiated in 1969 with the following objectives: (a) to create an awareness among the relevant policy-makers and government agencies of the benefits from space applications technology and (b) to provide training and education programs to enable officials from developing countries to gain practical experience in areas of space technology applications. The activities of the program are subject to the annual review and approval of the Committee.

The space applications program organizes, sponsors, and conducts a variety of seminars, panels, and workshops in fields of practical applications of space technology, particularly in space communications, space meteorology, and remote sensing as applied in various disciplines including cartography, agriculture, forestry, geology, oceanography, and other related earth sciences. As interest is expressed, various seminars/panels/workshops relating to specific problems or regions are organized together with specialized agencies and/or concerned member states. Over 30 such international seminars/panels/ workshops (with some 1,100 participants from developing countries) have been held in various parts of the world, including developing countries that have undertaken programs aimed at integrating space technology in their economic and social development.

Under the space applications program, the United Nations can, upon request by member states or organizations within the United Nations system, provide within the limits of resources available technical advisory services in the applications of space technology for development. The program has also coordinated several surveys on the needs of developing countries in the utilization of space technology for development, including visiting missions to countries in the

Middle East and Africa. The program is also responsible for the administration of fellowships for advanced training in fields of space science and technology which are offered by interested member states to individuals of developing countries. For example, 37 fellowships were awarded.

The space applications program will continue its efforts toward increasing the awareness of decision-makers, policy planners, and scientific and technical personnel regarding the benefits of space technology applications. Increased emphasis can be expected on the applications of such technology for social and economic development and on the promotion of discussions and information exchange between developed and developing countries on space research and its applications. Regional meetings and seminars will focus increasingly on presenting the latest techniques in remote sensing and on providing instruction on their use.

11.1.1.2 Natural Resources and Energy Division.   Within the Department of Technical Co-operation for Development, the Natural Resources and Energy Division is responsible for a broad program of activities in the fields of cartography (surveying and mapping), energy, geology and mining, and water resources. The program consists of two basic activities: operational or field projects for technical assistance and non-operational projects including studies, seminars, and conferences. Both aspects of the program focus on meeting the needs of developing countries.

Of primary interest to the Natural Resources and Energy Division is the use of remote sensing as a tool for resources exploration, particularly as it can benefit the developing countries. In this connection, the Remote Sensing Unit within the Division cooperates with those sectors of the United Nations and its agencies responsible for project execution, particularly in the field of natural resources development. This cooperation involves such activities as providing substantive assistance in the design and evaluation of space-related remote sensing projects; facilitating access to data remotely sensed by satellites; advising on the feasibility of particular remote sensing applications, and providing support for remote sensing centers being established in various regions. In a non-operational context, the Division participates, together with other United Nations bodies, in sponsoring various seminars and training programs and in providing fellowships to individuals from developing countries.

The Division will continue the activities contained in its current program, placing particular emphasis on the uses of remote sensing in natural resources development. Attention is focused in particular on the need to increase access to remote sensing data and to promote knowledge of the latest techniques of data collection and analysis. As a result of this, the Division may expand its activities in the area of training and fellowships. In addition, it will provide increased support to member states and regions in the establishment of remote sensing centers, as interest in such centers grows.

11.1.1.3 Regional commissions. The Economic Commission for Africa (ECA) serves the aspirations of member states by contributing to their development process. In accordance with the United Nations General Assembly resolutions 2955 (XXVII) and 3182 (XVIII) on the peaceful uses of outer space and the need for international cooperation in space activities, ECA has endeavored to create, among member states, an awareness of the potential of satellite communications and space remote sensing technology. The aim is to promote the application of satellite communications and remote sensing technology in the survey, evaluation and planning for the rational utilization of the natural resources of Africa, and the monitoring of the impact of human activities on the environment. Representatives of member states attending the ECA Conference of Ministers held in February 1975 in Nairobi, Kenya, unanimously adopted resolution 280 (XII) requesting the executive secretary to establish a remote sensing program for Africa and endorsing the decision to establish in Africa a regional ground station for receiving and processing data transmitted by remote sensing satellites.

A mission of experts, led by an official of ECA, visited several African states with planned or ongoing remote sensing programs in order to see what facilities each had to offer for setting up the station. The report submitted by the mission proposed that three ground receiving and processing stations should be established at Kinshasa, Zaire, Ouagadougou, Upper Volta, and Nairobi, Kenya, and that five training and user assistance centers should be established at Kinshasa, Ouagadougou, Nairobi, Cairo, Egypt, and Ile-Ife, Nigeria. Of these five centers, those in Nairobi and Ouagadougou are now regionally active and the one in Cairo is now functional.

These centers are, first, to train African personnel to use satellite data and imagery in economic development planning in their countries and, second, to provide African cadres that would eventually run the receiving stations and training and user assistance centers.

The first meeting of the Conference of Plenipotentiaries of the African Remote Sensing Council (ARSC) was held in Ouagadougou in September 1978 to adopt the constitution and formally establish the African Remote Sensing Program. Seven states ratified the constitution at that time.

At the inaugural meeting of ARSC (Ouagadougou, October 1979), convened by the ECA after ten signatories had been registered, 13 states became members of the Council. In accordance with article II of the constitution, the major objectives of the Council are to:

1. Harmonize the policies of member states with regard to remote sensing.
2. Provide member states with an effective machinery for implementing a comprehensive remote sensing policy.
3. Promote the development of remote sensing activities and coordinate such activities with a view to improving the exploitation and development of natural resources of economic interest to more than one member state.

4. Foster the development of closer relations between member states in the application of remote sensing.
5. Use remote sensing techniques to monitor the ecological impact of the exploitation of natural resources.
6. Promote the establishment of receiving, processing, training, and user assistance centers in member states and coordinate the activities of such centers.
7. Ensure that all member states have access to the benefits of remote sensing.
8. Promote and encourage training and exchange of personnel as well as of remote sensing ideas and experience among member states.

The Economic and Social Commission for Asia and the Pacific (ESCAP) and the Economic Commission for Latin American (ECLA) have also recognized the potential for the use of remote sensing applications in their respective regions. Plans for the establishment of centers on a regional scale, however, remain the subject of further study and discussion.

11.1.1.4 Office of the United Nations Disaster Relief Co-ordinator (UNDRO).

UNDRO was established on 1 March 1972. The mandate of UNDRO is to mobilize, direct, and coordinate relief activities of the United Nations system on behalf of the secretary-general, in response to requests from stricken countries, to promote the study, prevention, control, and prediction of natural disasters, and to advise on pre-disaster planning.

The activities of UNDRO focus on the development and use of data-gathering techniques for forecasting and predicting natural phenomena likely to cause disasters. In cooperation with the United Nations, various specialized agencies, and regional economic bodies, UNDRO sponsors training programs in the use of such techniques and in disseminating information on technological developments with disaster implications for personnel from disaster-prone developing countries. UNDRO is also studying the use of satellites for obtaining imagery for post-disaster relief coordination purposes.

Within its limited resources, UNDRO cooperates in sponsoring training programs in the use of remote sensing for disaster forecasting and in dissemination of information on relevant technological developments.

11.1.1.5 United Nations Environment Program (UNEP).

UNEP was created in 1972 in response to a recommendation of the Conference on the Human Environment. The UNEP secretariat, under the guidance of a 58-nation governing council, was entrusted by the General Assembly with a number of responsibilities which include coordinating environmental programs within the United Nations system, keeping their implementation under review, and assessing their effectiveness. It is also to advise, as appropriate, intergovernmental bodies of the United Nations system on the formulation and implementation of environmental programs.

Of primary interest to UNEP is the use of remote sensing as a tool for systematic data collection on environmental variables. The Global Environmental Monitoring System (GEMS), developed and coordinated within the United Nations system by UNEP, represents a major and long-term effort in this field. Based on data acquired through GEMS, a quantitative picture of the natural and man-made global and regional trends undergone by critical environmental variables and renewable natural resources is being obtained. Thus, within the GEMS program, a variety of projects are being executed in such fields as anti-desertification, rangeland development and management, and earth resources assessment, including forests and soils.

UNEP will continue its coordination of GEMS and, at the same time, increase its activities within the scope of that system. Among the primary areas for further work are the uses of remote sensing in monitoring problems related to agriculture and land-use practices together with climate monitoring by satellite, global ozone monitoring, and studies of climatic change. Emphasis will also be placed on training, particularly as relevant for individuals from the developing countries.

Because a main role of UNEP is to stimulate and coordinate environmental activities within the United Nations system, it therefore works very closely with the specialized agencies so that many UNEP-sponsored and supported programs are described under the activities of the individual agencies, the most notable being those with WMO and FAO.

11.1.1.6 United Nations Development Program (UNDP).   UNDP is a funding and coordinating organization that works with 150 governments and more than a score of international agencies to promote improved living standards and economic growth in developing countries throughout the world. UNDP provides assistance to developing countries through technical cooperation activities covering virtually the entire economic and social spectrum of development—including agriculture, education, industry, power production, transport, communication, public administration, health, housing, and trade—reflecting the diverse needs of recipient governments. UNDP-supported activities focus on five main areas: surveying and assessing natural resources and other potential development assets; stimulating capital investment to help realize these possibilities; training in a wide range of vocational and professional skills; transferring appropriate technologies and stimulating the growth of local technological capabilities; and economic and social planning. UNDP-supported programs give emphasis to assisting national personnel to assume full responsibility when the program's support phases out, to promoting technical cooperation among developing countries, and to aiding the least developed countries and the poorest population groups of any developing country. A governing council of 48 member states is responsible for monitoring the content and conduct of operations and for allocating funds.

The majority of UNDP-funded projects relating to space science and technology are in the three fields of natural resources surveys, technology transfer, and planning. Activities of special interest to UNDP include those in communications, weather and pest forecasting, radio and television broadcasting for development, and a variety of resource surveys, including crops, forestry, minerals, soil, and water. Among the projects that have been or are being carried out are: development of an experimental satellite communications earth station and the organization of relevant on-site training courses; establishment of a regional meteorological center; preparation of a variety of feasibility studies; development of a telecommunications system; regional integration of telecommunications networks; establishment, together with FAO, of a training program on the uses of remote sensing in crop production; and a UNDP/WMO project for a hydrometeorological survey.

In conformity with government priorities, UNDP will maintain its present variety of projects in the future, possibly increasing its emphasis on the provision of training facilities and expert services. In the field of space science and technology, remote sensing and satellite communications are likely to remain important areas of interest to UNDP. Additional emphasis may be placed on regional projects such as the establishment of specialized research development and training centers, the development of telecommunications networks, and the establishment of remote sensing stations based on requests by governments and evaluation in light of other priorities and requests for assistance. (See Appendix 11.1 for a description of the procedure for the establishment of a UNDP-assisted space applications project in a member state.)

11.1.1.7 United Nations Industrial Development Organization (UNIDO).

The institutional mandate of UNIDO is to contribute and assist in accelerating the industrialization of the developing countries. On the global scene, UNIDO is committed to achieving the goals and objectives set forth in the Lima Declaration and Plan of Action, namely, to increase the share of the developing countries to the maximum possible extent and, as far as possible, to at least 25 percent of total world industrial production by the year 2000. This goal was reconfirmed in the New Delhi Declaration and Plan of Action.

As the preparation and development of outer space activities and related technical ground-based operations concern, among other things, research, development, application and use of technology as well as testing, training, and supporting technical operations, the involvement and participation of a large number of industries and industrial services are evident. In view of its coordinating role within the United Nations system for all matters relating to industrialization, UNIDO accordingly takes considerable interest in the industrial aspect of the development and application of space technologies. The industries which are actually involved in space activities are either traditional (mechanical, metallurgical, electrical and electronic, chemical, engineering and building industries) or involve technologically advanced industries of the telecommunications and electronic branch. These are industrial areas in which UNIDO has experience, and in which it intends to develop its activities and operations, as may be required by the international community, in general, and by the developing countries, in particular, in terms of technical cooperation, surveys or studies, collection and dissemination of industrial and technological information and data, training, consultations, and expert group meetings.

11.1.2 Specialized Agencies and Other Organizations

11.1.2.1 International Telecommunication Union (ITU).
Within ITU (which has 154 member countries) the development of space technology has not given rise to any entirely new activities requiring the establishment of space structures. On the contrary, already in 1959 at a World Administrative Radio Conference, frequency allocations were made to the space research service, and in 1963 the world "Extraordinary Administrative Radio Conference to allocate frequency bands for space radiocommunication purposes" was held in Geneva to allocate frequencies to each radiocommunication service using space techniques. Obligatory procedures for the coordination, notification, and registration of the use of radio frequencies and the geostationary satellite orbit were developed, adopted, and included in the international treaty entitled "International Telecommunication Convention and Radio Regulations." These allocations and procedures were reviewed, developed, and revised in world radio conferences at Geneva in 1971 and 1979.

The objectives of ITU are as follows: (a) to maintain and extend international cooperation for the improvement and rational use of telecommunications of all kinds; (b) to promote the development of technical facilities and their most efficient operation; and (c) to harmonize the actions of nations in the attainment of those ends.

Pursuant to those objectives, ITU is responsible for the management and the application, on a day-to-day basis, of the regulations governing the use of the radio-frequency spectrum and the geostationary satellite orbit. Any new developments in space technology, in so far as telecommunications are concerned, can be studied and solved within existing structures. All constituent elements of ITU are concerned with space communications because space radiocommunication means "any radiocommunication involving the use of one or more space stations or the use of one or more reflecting satellites or other objects in space," in other words, radiocommunication services using space techniques. These include space research service, earth exploration-satellite service, fixed-satellite service, aeronautical mobile-satellite service, maritime mobile-satellite service, radionavigation-satellite service, meteorological-satellite service, broadcasting-satellite service, etc. The International Radio Consultative Committee (CCIR) studies the technical and operating questions for all these space services and their use of the spectrum and issues recommendations. The International Telegraph and Telephone Consultative Committee (CCITT) studies the technical aspects of interconnecting satellite systems with land systems and their operating principles, and it establishes recommendations on this subject.

ITU also conducts an extensive technical assistance program which has assisted countries in carrying out feasibility studies and training experts in the use of space technology for communications.

11.1.2.2 World Meteorological Organization (WMO).

Artificial satellites have had a considerable impact on the activities of WMO and have resulted in very great benefits to national meteorological services throughout the world. The effect on WMO has in fact been so important that almost all the constituent bodies of the organization are engaged directly or indirectly in outer space activities.

The purposes of WMO include:

1. Facilitating international cooperation in the establishment of networks of stations and centers to provide meteorological and hydrological services and observations.
2. Promoting the establishment and maintenance of systems for the rapid exchange of meteorological and related information.
3. Promoting standardization of meteorological and related observations and ensuring the uniform publication of observations and statistics.
4. Furthering the application of meteorology to aviation, shipping, water problems, agriculture, and other human activities.
5. Promoting activities in operational hydrology and furthering close cooperation between meteorological and hydrological services.
6. Encouraging research and training in meteorology and, as appropriate, in related fields.

The role of satellites has considerably increased, not only for obtaining various kinds of observational data, in particular quantitative data, but also for the collection and distribution of information in support of various WMO programs. Indeed, it is now generally recognized that satellites are indispensable for the success of the World Weather Watch (WWW), the Global Atmospheric Research Program (GARP), the Tropical Cyclone Program, the Integrated Global Ocean Services System (IGOSS—jointly with IOC), the Hydrology and Water Resources Program, agrometeorological programs, research, and other WMO programs.

WMO has also undertaken extensive education and training programs for experts from developing countries in order to train them in the use of satellite data for meteorological purposes.

11.1.2.3 United Nations Educational, Scientific and Cultural Organization (UNESCO). Within UNESCO, the Sector of Culture and Communication (especially the Divisions of Free Flow of Information and Communications Policies and of Development of Communication Systems) is responsible for space communication activities. These activities are intended to study and provide advice to member states on the use of space communications for the furtherance of UNESCO's aims, taking into account the Declaration of Guiding Principles on the Use of Satellite Broadcasting. The Sector also has long-standing programs for the training of radio and television staff for program production, an activity that has equal relevance to terrestrial or satellite broadcasting systems.

The promotion of international arrangements and conventions is carried out in close cooperation with the Office of International Norms and Legal Affairs. Assistance to member states, principally in the form of missions to provide expert advice on the use of space communications for education and national development, is carried out through the Division of Methods, Materials and Techniques in the sector of education.

The interests of UNESCO in remote sensing of the earth by satellites and other spacecraft are connected mainly with studies of the natural environment and its resources. The use of remote sensing, together with conventional (airborne) remote sensing techniques, is being dealt with by the following units of the UNESCO Science Sector: Division of Ecological Sciences, Division of Earth Sciences, Division of Water Sciences, Division of Marine Sciences, and the secretariat of the Intergovernmental Oceanographic Commission.

In the sector of culture and communication, the Division of Cultural Heritage could use remote sensing techniques related to the location and subsequent restoration of important historical and cultural monuments.

11.1.2.4 Food and Agriculture Organization of the United Nations (FAO). In 1973 FAO appointed a remote sensing officer and in 1976 a remote sensing unit was established in the Agriculture Department to coordinate and implement the objectives set forth by the FAO conferences. In 1980 further steps were taken toward implementing those objectives and FAO converted its Remote Sensing Unit, based on recommendations from the Committee on the Peaceful Uses of Outer Space, into a remote sensing center for renewable resources within its Agriculture Department.

The Remote Sensing Center acts as the organization's focal point for activities in this field. It provides technical backstopping to a large number of field projects whose work involves the application of remote sensing and develops projects where such applications predominate. It is closely involved in the formulation and execution of regular program activities with a remote sensing component, including a range of activities carried out in collaboration with other United Nations agencies. It has developed facilities at FAO headquarters, among which are a global index of satellite imagery, including the LANDSAT 16-mm browse file, a library of such imagery for developing countries and of remote sensing literature, and a laboratory for the interpretation and analysis of aerial photographs and satellite imagery. The major instruments available include a color-additive viewer, a density slicer, and a zoom transferscope. Photographic facilities for producing hard copy from the display instruments are currently being acquired and a capability for selective computer processing of multispectral data is being established.

FAO also conducts a range of training activities both in developing countries and by using the facilities of its Remote Sensing Center. Much of its work is devoted to advising and assisting developing countries and other international organizations in relation to the applications of remote sensing and the development of programs and facilities for this purpose. The Center also represents FAO in a range of bodies within the United Nations system and in international scientific organizations concerned with remote sensing.

11.1.2.5 World Health Organization (WHO).

WHO has made every effort to keep abreast of developments in space research, for potential results of such research can have a bearing on certain of its programs, particularly when they relate to man's relationship with his environment.

The main fields of interest of WHO in relation to its present and future activities are: communicable diseases, environmental health, epidemiology and communication sciences, occupational health, cardiovascular diseases, radiation health, nutrition, mental health, human genetics, organization of medical care, training and education in medicine, and health education. The particular fields of interest comprise the techniques employed to determine vector habitats in relation to malaria, schistosomiasis, and trypanosomiasis.

Remote sensing of air and water pollution is an attractive alternative to land-based monitoring systems because large areas may be covered in a short time. In addition, the use of remote sensors may be cheaper in the long run and give quicker results than the conventional methods of sampling and laboratory analysis. However, many of these techniques are at present in various stages of development and only a few can be used routinely.

11.1.2.6 International Civil Aviation Organization (ICAO).

The general function of ICAO is to ensure that international civil aviation may be developed in a safe and orderly manner and that international air transport services may be established on the basis of equality of opportunity and operated soundly and economically.

The economic application of satellite services to international civil aviation is an important objective of ICAO, and a great deal of activity has been devoted to the definition of appropriate operational requirements, and the best means of satisfying those requirements. The aeronautical fixed service, which interconnects the ground infrastructure responsible for the safety of flight, already utilizes satellite communications links where available and feasible. The aeronautical mobile service, which provides communications between aircraft in flight and the ground communications network, could for the first time benefit from the provision by satellites of instant, static-free communications between pilots and controllers over all parts of the earth's surface. The aeronautical radio navigation service, utilizing a variety of navigation techniques, could benefit from the application of satellites to provide radio determination of aircraft position during flight over any part of the earth's surface. In each of the two latter applications, the desirability of providing service over the polar regions adds to the complexity of possible solutions, and it extends considerably the range of technical problems that must be considered by ICAO.

In addition to the above, ICAO is interested in and is working on a number of other aspects concerning the peaceful use of outer space, including the use of satellites by search and rescue services, the definition of outer space, transport to and from, and through, outer space, and more mundane problems associated with safety. An example of such a problem is the return to earth of space debris, which could constitute a hazard, particularly to civil aircraft in flight.

In each of these areas of activity, ICA pays particular attention to the potential of the peaceful application of space technology to provide essential aeronautical safety services in developing countries as part of its function to establish international air transport services on the basis of equality of opportunity, and to ensure that they are operated safely and economically. Particular attention is given to the coordination of ICAO activities with the related activities of other international organizations in the interest of minimizing wasteful duplication of international effort.

11.1.2.7 Inter-Governmental Maritime Consultative Organization

(IMCO).   Since 1966, IMCO has taken a considerable interest in the development of space techniques for maritime purposes. Until 1971, its activities in the utilization of space techniques were largely related to the development of maritime requirements in preparation for the World Administrative Radio Conference for Space Telecommunication. The technical bodies of IMCO prepared two recommendations which specified, among other things, that space techniques could be used for alerting and locating ships in cases of distress and emergency, for facilitating search and rescue operations through more effective communications, and for the promulgation of urgency and safety messages. Ancillary functions could be automation of a position-reporting system, position determination and information, traffic guidance, automatic navigational warning system, weather routing, direct printing, and facsimile.

A general outline emerged for a maritime satellite system that would satisfy primarily telecommunications requirements (safety, distress, and public correspondence) and, in addition, would perform a number of other functions to be determined on a priority basis according to their merit, compatibility, and consistency with the optimum use of the frequency spectrum available and the organizational and financial factors involved in the development and eventual operation of the system.

In 1975–1976 IMCO convened the International Conference on the Establishment of an International Maritime Satellite System which adopted two instruments: the Convention and the Operating Agreement on the International Maritime Satellite Organization (INMARSAT). The Conference also established a preparatory committee which carried out a comprehensive study to facilitate the effective operation of INMARSAT. INMARSAT was established on 16 July 1979 when the convention and the operating agreement entered into force.

In October 1980 IMCO developed requirements for a future global maritime distress and safety system which is expected to be implemented by 1990 and which will replace the present radiotelegraph morse system. Ship-to-shore long-range distress alerting in the future maritime system will use low-power distress transmitters operating through geostationary and polar orbiting satellites which will also be used for shore-to-ship long-range alerting. Low-power distress transmitters will also be carried in ships' survival craft. The possibility of using a combination of active and passive alerting, including positioning, is under consideration. The future global maritime distress and safety system will be developed in close cooperation with ITU and INMARSAT.

11.1.2.8 International Labour Organization (ILO).

The ILO does not have activities at this stage which relate specifically to outer space. However, developments in this field have an impact on ILO activities or call for supportive action in areas within its competence. Thus outer space activities and related ground-based operations involve specific hazards to life and health, the prevention of which is obviously essential. The preparation of outer space activities involves the participation of a large number of industries with considerable manpower. These industries employ various traditional and advanced technologies, with specific occupational hazards that need to be controlled through collective agreements as well as legislative, administrative, and technical measures. This is a field in which the ILO has a long-established experience, and in which it intends to continue and increase its activities, in particular, through the preparation of international standards and technical manuals, the collection and dissemination of information, the organization of congresses and symposia, and technical cooperation with developing countries. The ILO is following the development of satellite-based air navigation, surveillance (radar), and communications systems so as to determine their possible impact on cockpit and air traffic control personnel. It is also concerned with the impact of new communications techniques (e.g., direct satellite broadcasting) on the employment and conditions of work of performers and audiovisual workers.

11.1.2.9 World Intellectual Property Organization (WIPO). In 1969 WIPO and UNESCO engaged in a joint effort aimed at outlawing the "piracy" of radio and television signals where such signals pass through an artificial communications satellite and carry a "program." "Program" is taken to mean what is intended for, and can be picked up on, domestic radio or television receiving sets. This effort resulted in the adoption in 1974 of a new convention on the protection of program-carrying signals transmitted via satellite. The provisions of this convention are not applicable where distribution of signals is made from a direct broadcast satellite.

11.1.2.10 International Atomic Energy Agency (IAEA). The IAEA has no activities or plans directly concerned with the peaceful uses of outer space. The activity most directly related to this subject is the continuing subprogram "Emergency assistance to any Member State following an accident or natural disaster involving radioactive materials."

11.1.2.11 The World Bank (International Bank for Reconstruction and Development or IBRD). As an international financial institution, the World Bank is indeed interested in the peaceful uses of outer space and in knowing how developments in this field might contribute to the development process in its developing member countries. The World Bank is particularly interested in satellite data analysis, education, and telecommunications.

Two sections within the World Bank deal with satellite imagery: the Economic and Resource Division of the Agriculture and Rural Development Department and the Cartographic Section of Administrative Services; in addition, the Education Projects Department retains the services of a mass media specialist to assist developing countries in formulating plans for using communications media, including satellites, to upgrade and expand education.

11.2 PROGRAM AREAS OF ORGANIZATIONS

11.2.1 Remote Sensing

11.2.1.1 International and regional centers for remote sensing activities

11.2.1.1.1 Established centers for remote sensing activities. Two international remote sensing centers have been established within the United Nations system, as a result of recommendations made by the Committee on the Peaceful Uses of Outer Space. These centers are located in the United Nations Natural Resources and Energy Division of the Department of Technical Co-operation for Development and at FAO in Rome.

The Remote Sensing Center of FAO, established in January 1980, provides remote sensing advisory services and technical assistance, including training, to member states; it also provides remote sensing support to FAO headquarters and field programs, coordinates remote sensing activities at headquarters and in the field, and serves as the liaison point between FAO and other major organizations concerned with space applications. These activities, covering renewable resources, include agriculture, fisheries, forestry, land use, soil survey, water management, wildlife, and the conversion of natural resources.

The Remote Sensing Unit of the Natural Resources and Energy Division carries out operational field projects for technical assistance and nonoperational projects, including studies, seminars, and conferences in the area of nonagricultural resources.

On a regional level, an important source of remote sensing information is the African Remote Sensing Programme established by the Economic Commission for Africa (ECA) in 1977. This program is designed to assist member states in applying current techniques for the exploration of natural resources and for monitoring environmental changes. In order to ensure substantial and continuing support for the operations of the program, an African Remote Sensing Council (ARSC) was also established, which consists of national ministers responsible for activities relating to remote sensing.

Five training and user assistance centers are eventually to be established under the program; two are currently functioning as planned. These centers are the Regional Remote Sensing Center at Ouagadougou, Upper Volta, and the Regional Remote Sensing Facility co-located with the Regional Center for Services in Surveying and Mapping (RCSSM) in Nairobi, Kenya.

At Ouagadougou, training and user assistance services are to be housed in the same premises as the receiving and processing facilities. The center has set up a lecture hall, library, and photographic laboratory. It performs three main functions: training, user assistance, and the reproduction and distribution of LANDSAT imagery. To date, there have been eleven introductory courses attended by some 110 students from 18 countries in Africa. For its user assistance service, the center has the following facilities available for use: (a) an index to existing remote sensing imagery covering Africa and an archive of selected imagery for the period, (b) a photographic laboratory for reproducing imagery with special processing when necessary, (c) a reference library of books, maps, technical reports and journals, and (d) equipment for ground surveys to aid interpretation.

At the Nairobi Center, regional remote sensing activities are conducted under the auspices of the "Regional Center for Services in Surveying and Mapping." In 1979 and 1980 the center organized ten short courses, each lasting between two to three weeks in Nairobi (Kenya), Dar es Salaam (Tanzania), and Gaborone (Botswana). The topics covered included water resources, vegetation and soils, cartography, forest and range monitoring, agriculture and land use, resource assessment, and transportation engineering. In addition, the center has organized three four-day information seminars in Dar es Salaam, Nairobi, and Mbabane (Swaziland) for government agencies, university and college staff, and others. For its user assistance services, the center has the following facilities available for use: (a) LANDSAT imagery browse file covering the whole of the eastern and southern Africa region, (b) map files containing small-scale topographic map series for Kenya and Botswana, and (c) image analysis facilities including multispectral color-additive viewers, density slicer, zoom transfer scope, and reflecting projector.

Since 1977 these two training and user assistance centers have become valuable sources of remote sensing information and have periodically organized training courses, seminars, and workshops in remote sensing.

11.2.1.1.2 Potential centers for remote sensing activities.   Plans are progressing toward the establishment of two regional remote sensing centers for the regions of the Economic and Social Commission for Asia and the Pacific (ESCAP) and of the Economic Commission for Latin America (ECLA).

In the case of ESCAP, the Natural Resources and Energy Division, together with UNDP, has prepared the terms of reference for a regional remote sensing program having two basic objectives: (a) to enable participating countries to upgrade their national capabilities for gathering and analyzing data on their natural resources and (b) to enable participating countries to monitor environmental changes using data provided by airborne and spaceborne remote sensing systems. The program as envisioned would provide training and technical assistance, conduct research, assist in management and development, and promote the general exchange of information. If funding is secured by 1982, ESCAP is expected to begin the initiation of the program in association with FAO and the Department of Technical Co-operation for Development.

A similar project formulation for regional technical co-operation has been prepared for Latin America. Argentina has offered to host the establishment of a regional remote sensing center that would assist countries in Latin America in the use of remote sensing technology for resource development. The project could also take advantage of existing Argentinian receiving and data processing facilities.

11.2.1.2 Sources of information on the status of remote sensing technology (publications, seminars, and projects).   In addition to the established centers described above, a variety of other United Nations organizations and agencies can provide information on specific aspects of remote sensing technology. Brief descriptions of publications available from United Nations sources are provided below.

As part of GEMS, UNEP is preparing a handbook of methods to be used by ecological monitoring units (EMU) for habitat monitoring in arid and semi-arid areas. The book will cover the three functional levels of EMU data collection (ground survey, low-level systematic reconnaissance flights, earth resource assessment satellites) as well as methods for data analysis and data handling.

A comprehensive glossary of remote sensing terms in English, French, and Spanish will be available from FAO. Remote sensing documents are available on soil degradation mapping, forest cover monitoring, rural disaster monitoring, desert locust survey, high-altitude aerial photography, and side-looking airborne radar (SLAR). There are also publications on several problem-oriented training courses for developing countries as well as a world index of space imagery. Ten international and regional training courses and several country-based workshops are planned for 1982–1983.

Reports of Study Group 2 (space research and radio astronomy) of the International Radio Consultative Committee (CCIR) of ITU reflect that group's continuous examination of the technical aspects of utilizing the Earth Remote Sensing Service.

Remote sensing techniques are very often utilized individually in research projects related to science programs of UNESCO. For instance, remote sensing imagery of natural resources is used in a number of national research projects of the Man and Biosphere Program (MAB) and remotely sensed data have been analyzed for research in some projects of the International Geological Correlation Program (IGCP) as well as in the International Hydrological Program (IHP). These activities are carried out primarily through the participation of member states in UNESCO programs.

Additional information is available from UNESCO in the form of studies such as an introductory monograph to be published in 1982 on earth science data handling, with particular application to mineral resource data. A section of the monograph will describe how to utilize satellite data obtained and how to interpret such data for mineral exploration purposes.

As a contribution to the International Hydrological Program, UNESCO published a report (1979) on remote sensing of snow and ice" (Technical Paper in Hydrology No. 19). Under the current phase (1981–1983) of IHP, a research project on the "applications of remote sensing to hydrology, including groundwater" is included. Under this project, a panel of experts will prepare a report on the current applications and future activities to be undertaken within the framework of IHP.

Information on remote sensing and hydrology can be obtained from WMO, which has published a number of reports on various aspects of the subject and plans to publish additional reports, in particular one on data transmission methods, including satellite systems and meteoburst. The WMO Commission for Hydrology (CHY) continues to follow developments in the use of remote sensing techniques in all aspects of operational hydrology and, at the same time, detailed guidelines are being prepared on the use of satellites for hydrological data transmission, including maintenance and operation of data collection platforms hardware. A workshop on the application of space technology to hydrology has been included in the WMO program of meetings for 1982–1983.

Remotely sensed observations from satellites are considered an important component of the IOC/WMO Integrated Global Ocean Services System. The IOC Working Committee for the Global Investigation of Pollution in the Marine Environment is actively evaluating the remote sensing of machine pollutants, and the need for related ground truth data, for calibration purposes. The maintenance of archives and the exchange of remotely sensed oceanographic data represent one of the major tasks of the IOC Working Committee on International Oceanographic Data Exchange and, in particular, of its Task Team on the Exchange of Airborne and Satellite Remotely Sensed Data.

In the area of disaster prevention, UNESCO is developing single data-gathering techniques for the forecasting and prediction of natural phenomena, such as earthquakes and volcanoes which are likely to cause disasters. FAO and WMO are developing such techniques for surveys of food shortages and disasters, including drought, floods, and tropical storms. UNDRO and UNEP are assisting in these activities and, along with various regional economic bodies, cooperating in the training of personnel from disaster-prone developing countries in the use of monitoring techniques and in the dissemination of information on technological developments with disaster implications. Also in the field of disaster prevention, FAO has prepared a paper summarizing its remote sensing activities related to agricultural disaster.

As part of GEMS, UNEP and FAO have established a project to assess the present state of world tropical forest and woodland resources. Three regional assessment reports (Latin America, Africa, and Asia) covering 76 countries were released in 1981, with a global assessment based on the regional reports released in 1982. Data have been extrapolated to December 1980 to provide a common base against which to set future assessments. UNEP and FAO intend to expand this forest assessment activity to include subtropical and temperate forests, thereby enabling regular global forest assessments to be made.

Also as part of GEMS, UNEP and FAO have developed simple tropical forest cover inventory and monitoring methods. The results of the initial study (Benin, Togo, and United Republic of Cameroon) are available. Tropical forest cover programs will be developed in Asia, Latin America, and Africa over the next decade. Data from the programs will form an important input to the forest assessments described above.

11.2.1.3 Education and training programs through fellowships, training workshops, and summer schools.

Work has been undertaken by various organizations, including the United Nations, UNESCO, WMO, FAO, and UNEP, to train specialists from member states, particularly those from developing nations, in the various applications of remote sensing technology.

The United Nations program on space applications, in cooperation with member states, specialized agencies, and other organizations, organizes and sponsors four to five training courses and seminars each year relating to applications of space technology. Since its inception, the program has trained more than

1,000 participants and over 40 courses have been held for the benefit of the African, Asian, European, Latin American, and North American regions. Areas of study have included the uses of remote sensing in hydrology, agriculture, meteorology, forestry, and natural resources management as well as satellite communications for education and development.

The space applications program also administers fellowships offered by member states to candidates from developing countries for advanced training in fields related to space science and technology. Among the member states that have provided such fellowships are the governments of Austria, Belgium, Brazil, India, Italy, and the Netherlands.

The following postgraduate training courses dealing with the application of remote sensing in integrated natural resources research, management, and development are available through UNESCO:

A postgraduate training course in integrated surveys at the International Institute for Aerial Survey and Earth Sciences and in remote sensing applications at the ITC/UNESCO Center in Enschede, Netherlands.

A postgraduate course in integrated study and rational use of natural resources at the universities of Paris, Montpellier, and Toulouse, France.

A postgraduate course in natural resources research and land evaluation at the University of Sheffield, United Kingdom of Great Britain and Northern Ireland.

A postgraduate training course on ecosystem management at the Technical University in Dresden, German Democratic Republic, with the support of UNEP.

The FAO Remote Sensing Center in Rome and the Remote Sensing Unit of the Natural Resources and Energy Division in New York provide training (both at their respective headquarters and in the field) which focuses on the applications of particular subjects or subject groups and is conducted by experts in such fields. The training concentrates on the analysis of remotely sensed data using techniques that have proven applications. For example, the FAO Center provided on-the-job training in Rome in 1980 for selected specialists from Benin, Liberia, Malawi, the Netherlands, the United Kingdom of Great Britain and Northern Ireland, and Venezuela.

FAO and UNDP have collaborated in providing training in remote sensing, either at selected academic centers in developed countries or with training partly provided by the FAO Remote Sensing Center in Rome. In addition, FAO, under the FAO-funded Technical Co-operation Program (TCP), has provided training courses in China. Exemplary of the training courses sponsored by FAO in cooperation with other organizations is a 1981 international training course, planned together with WMO, the European Space Agency (ESA), and UNDRO, on the applications of remote sensing techniques to rural disasters.

The Natural Resources and Energy Division participates in several training programs and is instrumental in granting training fellowships for work in advanced remote sensing laboratories. One aspect of its program for 1981 was a one-week meeting held at headquarters and attended by remote sensing experts from seven countries to evaluate the use of the most recent remote sensing technology in the development process. This program is, among other things, intended to identify technical cooperation projects in which remote sensing can be used as a tool for development on national and regional levels.

The two regional remote sensing centers of the African Remote Sensing Program provide training for individuals from their member states. Over 100 students from 18 countries of the west and central African subregions have already been trained at these centers. Some fellowships are offered by donor countries, and this has enabled the centers to successfully organize training workshops and seminars since 1977. With the aid of technical assistance, future operators for the Ouagadougou receiving station are being trained in the countries of the fellowship donors. Training facilities are also available through bilateral agreements with the European Economic Community (EEC), France, Canada, and the United States of America. An example of the programs provided is a seminar on the utilization of remote sensing in highway engineering held in 1981 in Nairobi, Kenya. In addition, discussions with EEC have reached an advanced stage on the organization of a seminar on yield forecasting early that same year.

11.2.1.4 Expert services and survey missions to identify specific areas of applications relevant to a given country, or group of countries and to carry out special studies on pilot projects. In the area of resource development, the Natural Resources and Energy Division provides expert services in project preparation and execution and has participated in survey missions to identify the need for remote sensing techniques in water, mineral, energy projects, cartography, hydrography, and geodesy. The missions have been fielded at government request and the necessary expertise drawn from available staff in the Minerals, Water Resources and Cartography and Information Branches of the Division. Several pilot projects have also been jointly designed by the Division and UNDP in order to strengthen the capacity of national laboratories for the assessment, selection, acquisition, and adoption of foreign remote sensing technology and expertise.

In areas of environmental concern, UNEP has sponsored several projects, both independently and together with other organizations of the United Nations system. Among these projects are:

1. A four-year project, together with WMO and ESCAP, to improve tropical storm forecasting for the Bay of Bengal and Arabian Sea regions.
2. A pilot project, together with FAO and as part of GEMS, to inventory and monitor changes in tropical forest cover; inventories were completed for

Benin, Togo, and the United Republic of Cameroon, and other inventories are planned.

3. Potential projects on marine pollution detection, monitoring of mangrove forests, and hydrological monitoring of changes in the African Sahelian region.

4. Several projects, as a result of the United Nations Conference on Desertification, to monitor desertification processes and natural resources; these include independent national projects using the Ecological Monitoring EMU approach and dealing with rangeland development and management and four long-term transnational anti-desertification projects in Africa.

As part of GEMS, UNEP and FAO, in cooperation with the government of Senegal, have established a pilot project for the inventory and monitoring of Sudano-Sahelian pastoral ecosystems. This four-year project will involve the establishment of an EMU which uses three functional levels of data collection, viz. ground survey, low-level systematic reconnaissance flights, and earth resource satellites. Data provided by the EMU will be used primarily for rangeland development and management purposes and to monitor possible rangeland degradation and desertification processes. The project became operational in late 1979 to form part of a global network of rangeland EMUs within GEMS.

Also as part of GEMS, a UNEP and FAO project has developed methodologies for assessing soil degradation, covering such factors as water and wind erosion, salinization, and alkalinization. The methodologies include the use of remote sensing information derived from LANDSAT images. Provisional maps (1:5,000,000) for Africa north of the equator and for the Near East are presently available which show the present rate and the present state of soil degradation and the risk of soil degradation. The system will undergo field testing for use at other scales during the second phase of the project.

FAO, in cooperation with UNEP, UNESCO, and WMO, prepared for the United Nations Conference on Desertification a global map of desertification at a scale of 1:25,000,000. Consequently, a new cooperative project between UNEP and FAO is now starting which will field test a methodology for the assessment and cartographic presentation of desertification for areas affected or likely to be affected by desertification. The methodology will be suitable for development and management purposes, at national and regional levels, and in the preparation of appropriate and useful desertification maps.

A number of FAO-executed UNDP projects are concerned with remote sensing applications to agriculture (including land use, forestry, fisheries, and the conservation of natural resources). Among these projects are the completion of vegetation maps at 1:250,000 in Nigeria (based on SLAR imagery) and the increasing use of LANDSAT imagery for land resources evaluation in Indonesia. In Sierra Leone, nationwide reconnaissance surveys of vegetation, land-use physiography, and soils were completed with the aid of high-altitude infrared color photography and panchromatic black and white photography. This photography is now to be used in cooperation with the United States Agency for International Development (USAID) as a base for forecasting and monitoring agricultural production. Other important FAO projects using remote sensing include crop production monitoring in Argentina, anti-flood management in India, resource assessment in Viet Nam, the strengthening of national agricultural training in China, and desert locust monitoring. In the development of remote sensing applications for desert locust survey and control, initial funding by USAID (1979–1980) will assist toward operational satellite monitoring of precipitation and vegetation conditions in the desert locust recession areas of northwest Africa and in developing automated digital techniques. The Regional Desert Locust Office in Algeria has used METEOSAT, TIROS-N, and LANDSAT imagery to extract information on rainfall and precipitation and will be constructing a regional map, based on LANDSAT, to show potential desert locust habitats. The program is being extended to India and Pakistan in cooperation with the national agencies (NRSA, SUPARCO).

Remote sensing is a frequent component of minerals, forestry, cartography, agricultural, and other survey projects and related training activities that are funded by UNDP and executed by the United Nations, the specialized agencies, IAEA, and the World Bank. There are, however, a number of projects relating to remote sensing techniques for which UNDP/Office of Projects Execution (OPE) is executing agency, including the following: strengthening of the Afghan Cartographic and Cadastral Survey Institute, Afghanistan; techniques for exploration of mineral resources at the National Geological Research Institute (NGRI), Hyderabad, India; études sur l'industrie pétrolière et les ressources minières, Madagascar; enquête géologique et géophysique sur l'anomalie magnétique de Djidian-Kéniéba, Mali; emergency rehabilitation of the source of water supply of the city of Tete, Mozambique; occupation des sols de l'arrondissement de Say: prises de vues aériennes et photo-interprétation (régions africaines); aerial survey of northern Thailand watersheds, Thailand.

Current World Bank projects[2] utilizing satellite remote sensing data include those carried out in Bangladesh, Indonesia, Nepal, Nigeria, Pakistan, and Yemen. The applications of remote sensing technology in these projects include their use in the preparation of base maps and categorization of land use/land cover, flood damage assessment and rehabilitation programs, hydrological development projects, and work related to agricultural, forestry, and settlement programs. More than a dozen other minor projects utilizing satellite data are in

---

[2] Two sections within the World Bank utilize satellite imagery: the Resource Planning Unit of the Agriculture and Rural Development Department; the Cartographic Division of Administrative Services. The Resource Planning Unit is engaged in a variety of analytical work utilizing satellite tape data for use in compiling feasibility studies for Bank-financed projects for its developing member countries and the Cartographic Division which primarily uses the satellite film products for cartographic applications (producing or bringing up to date small-scale maps for Bank reports).

various stages of operation: India (forestry), Somalia (livestock), Zaire (agricultural development), Paraguay (forestry), and Pakistan (forestry).

11.2.1.5 Provision of equipment to facilitate the undertaking of specific projects and the establishment of national centers.   FAO provides assistance by making recommendations on equipment, on relevant training, and on the development of necessary infrastructure (e.g., Argentina, China, Liberia, Nigeria, Sudan, Viet Nam). Recommendations on equipment have ranged from the choice of the simplest needs of field teams to centralized interactive systems for LANDSAT imagery and equipment associated with cartographic needs of governmental departments.

Within the terms of reference of providing impartial advice and assistance to technical projects, member states, and United Nations bodies, the FAO Remote Sensing Center is providing assistance in the development of remote sensing programs on national and regional levels. In several projects, provision has been made for the purchase of equipment to enable the analysis and interpretation of remote sensing data in national laboratories.

11.2.2 Communications

11.2.2.1 Coordination of the use of the radio frequency spectrum and geostationary satellite orbit.   The ITU World Administrative Radio Conference (WARC), Geneva, 1979, has further developed and revised the coordination, notification, and registration procedures for the use of radio frequencies and the geostationary satellite orbit included in the international treaty entitled "International Telecommunication Convention and Radio Regulations." These obligatory procedures are applied by the telecommunications administration of each country member of the union and the International Frequency Registration Board (IFRB), a permanent organ of ITU. The 1979 revision and the procedures will enter into force on 1 January 1982.

At its May 1980 session, the ITU Administrative Council established the following calendar of conferences related to space radiocommunications, subject to the necessary consultations and approval processes of article 54 of the Convention:

A five-week Region 2 broadcasting-satellite planning conference beginning 13 June 1983.

A six-week first session of a WARC on the use of the geostationary–satellite orbit and the planning of corresponding space services beginning in June 1985.

A six-week second session of a WARC on the use of the geostationary–satellite orbit and the planning of corresponding space services in September 1987.

11.2.2.2 Dissemination of information on the status of technology. Meetings of the ITU International (Radio) Consultative Committee (CCIR) and the International Telegraph and Telephone Consultative Committee (CCITT) consider various matters relating to the work of ITU, including standardization and studies in the field of space communications.

The work of CCIR encompasses different space disciplines, thus making it possible to evolve coordination procedures, the technical bases to be used in the obligations, and the basic planning parameters for the orderly, efficient, and rapid exploitation of space radiocommunications techniques. Several meetings of CCIR concerning space radiocommunications service were held in 1981.

CCITT study groups will meet in 1981 and 1982 to program the necessary modifications to existing recommendations concerning the integration of satellite telecommunications in the international telecommunications network (transmission characteristics, signaling, and operation).

Other studies available from CCITT relate to the parameters to be defined in the development of an international maritime system for communication via satellite with ships at sea, in order to permit the integration of these new services in the existing public international telephone or telegraphic services.

Meetings of the ITU World Plan and Regional Plan Committees develop plans for the international telecommunications network to facilitate coordination in the development of an international telecommunications service including satellite communication systems.

In addition to the activities of ITU in this area, both UNESCO and the World Bank have prepared useful information on satellite communications. A comparative study of planning processes for satellite communications will be published by UNESCO in this biennium. The Education Department of the World Bank has completed two relevant papers—one entitled "Radio for Education and Development" and the other entitled "Economics of Educational Radio." Both World Bank papers include a brief examination of the potential of educational radio broadcasting via satellite.

11.2.2.3 Education and training programs through fellowships, training workshops, and summer schools. The United Nations program on space applications, which has held several seminars and workshops in this field, has scheduled four regional seminars relating to satellite communications for education and development in Argentina, Ethiopia, France, and Indonesia during the 1981–1982 period as part of the preparatory work of the UNISPACE 82 Conference.

Following are examples of past and present programs sponsored by ITU, some of these in cooperation with UNDP: provision of a small earth station in India for training as part of the second phase of the ITU/UNDP Advanced Level Telecommunication Training Center (ALTTC) project; development of a complementary training program in satellite communications which will be used to train over 600 technical personnel required to install, operate, and maintain the domestic Indian National Satellite System for Television and Telecommunications (INSAT), scheduled to become operational in 1982; training in Central America on line-up procedures–satellite systems operation guidelines.

In August 1980, ITU and the Federal Republic of Germany agreed to conduct a pre-feasibility study which would utilize advanced technology including a domestic satellite system specifically designed to operate with battery-powered inexpensive earth stations to provide telecommunications in the rural and isolated areas of participating countries.

A two-week seminar by the IFRB of ITU on the management of the radio-frequency spectrum and the optimum use of the geostationary orbit was held from 21 September to 3 October 1980 with another similar seminar scheduled for 1982.

11.2.2.4 Expert services and survey missions to identify specific areas of applications relevant to a given country or group of countries and to carry out special studies in pilot projects. Among the services and expertise provided by ITU are the following:

1. Planning assistance which has been instrumental in arriving at a consensus among the island countries of the South Pacific for the establishment of regional satellite communications facilities to meet their national telecommunications requirements. During a meeting of the prime ministers of the region in July 1980, island member countries expressed their wish to join in a suitable satellite consortium for their national telecommunications and noted the possibility that the Australian government would take account of this when finalizing its own national satellite system.

2. Support for space communications research activity at the Space Applications Center, Ahmedabad, India, as part of an ITU/UNDP project that includes propagation measurements and development of space communications hardware at frequencies above 10 GHz.

3. Assistance to Samoa through the South Pacific Regional Project for the management of its earth station project (financed through EEC funding) in the form of planning, supervision of the installation, as well as final acceptance testing.

4. Planning and assistance for the implementation of a pilot data network in Indonesia using PALAPA (the Indonesian domestic satellite system) and small earth stations. The novel form of communications architecture employed is called "Packsatnet" (Packet Satellite Data Network). Packsatnet is being evolved as part of an ITU/UNDP project aimed at providing an appropriate and well-tested concept for a unified nationwide public data network. In addition, it will serve as a test bed for future research work.

5. Technical expertise for the administrations of Benin, Cape Verde, and

Equatorial Guinea to assist them in preparing specifications and issuing tenders for the installation of INTELSAT earth stations in their countries.

6. Provision of the services of an expert in power sources to participate in the maintenance of equipment and in the training of technicians in power sources for earth stations in the Sudan.

Also active in this field, UNESCO will provide advisory services to the Gulf States on the use of a satellite for program exchanges and to the Arab States Broadcasting Union on the planning of the Arab satellite network. Further, expert services will continue to be provided by UNESCO to member states on request to advise on the program and management aspects of the establishment of satellite communications systems for broadcasting.

In connection with the improvement and expansion of the developing countries' public telecommunications systems, the World Bank has helped finance a number of earth stations for domestic and international communications via satellite, when this has been shown to be the least-cost solution for handling the expected traffic. Both large type A stations and, more recently, smaller terminals have been included.

Since most of the countries where the World Bank is likely to finance telecommunications projects over the next few years already have at least one earth station, emphasis in 1981–1982 will be on the expansion of exisiting systems. Satellite communications systems which are being tried in other sectors (e.g., education) with Bank support may also lead to the construction of fairly sizable networks of small stations likely to be utilized for extending public telephone service to rural and remote sites for the first time.

The World Bank has assisted in the introduction of solar power supplies for use in microwave repeater stations and other applications. The utilization of these sources is likely to increase in the next few years, because the cost of this solution has become competitive in a number of situations and the Bank continues to support the extension of telecommunications services to remote and isolated areas.

In addition, the World Bank's Departments of Education and Population, Health, and Nutrition retain the services of mass-media specialists who assist developing countries in formulating plans for using communications media, including satellites, to upgrade and expand education, training, and information dissemination. The basic policy of the Bank is that countries should choose the medium most suitable for their requirements. The use of communications or broadcasting satellites has not yet been included in any Bank-financed projects or FAO-executed projects, but an increasing number of countries have recently been interested in using satellites for educational broadcasting including agricultural programs. An experimental small ground station linked with a satellite for educational radio broadcasting and administration communications is now being installed in the Philippines and a technical study will be carried out to review its feasibility. This and other trials may lead to the extensive use of communications or broadcasting satellites in future Bank projects.

11.2.3 Meteorology

11.2.3.1 Dissemination of information on the status of technology.   Under the auspices of WMO, the World Weather Watch (WWW) plan for 1980–1983, (WMO publication No. 535) approved by the Eighth World Meteorological Congress, relies substantially on satellites for the operation of all three elements essential to the WWW, namely, the Global Observing System (GOS), the Global Data-Processing System (GDPS), and the Global Telecommunication System (GTS).

Details for the GOS satellite subsystem are given in WMO publication No. 411, *Information on Meteorological Satellite Programmes Operated by Members and Organizations,* which is brought up to date as required. A description of the types of information derived from meteorological operational satellites is given in the *Manual* and in the *Guide on the Global Observing System,* WMO publications Nos. 544 and 488, respectively.

The quantitative data which have been prepared and are being distributed in real-time on the GTS of the WWW include: vertical temperature profiles of temperature and humidity of the atmosphere, wind field derived from cloud displacements for the tropics and middle latitudes, sea surface temperature fields, and information on snow and ice cover. In accordance with the WWW concept, the space-based subsystem is designed to meet the need for satellite data on three levels: global, regional, and national. Apart from the information obtained through the satellite direct read-out system, which is mostly used operationally for national and regional purposes, the above-mentioned quantitative data have been operationally used to meet global and, in some cases, regional needs. As stated in the WWW plan for 1980–1983, the two subsystems of the GOS, one based on the surface and the other on space, will complement each other.

In the field of aviation meteorology, the common WMO/ICAO regulatory material on meteorological service for international air navigation requires, as a standard practice, the display in meteorological offices of meteorological satellite photos or mosaics and/or nephanalyses (which are largely based on satellite imagery) to assist flight crew members and others concerned with the preparation of flights, and for use in briefing and consultation. In light of this requirement, and in the knowledge that the analysis of satellite images can be of value in establishing upper winds and in improving aerodrome and landing forecasts, the Eighth WMO Congress included studies on the increased use of data from meteorological satellites in the program of aviation meteorology for 1980–1983.

The reports of rapporteurs of the Commission for Agricultural Meteorology on the Application of Satellite and Remote Sensing Techniques in Agricultural Meteorology are under review for possible publication as a WMO technical note in 1981-1982.

The Eighth Congress of WMO has approved the holding of a workshop on the applications of satellites to agrometeorological activities and the preparation of a technical report on the results of this workshop. These projects are planned for 1982-1983.

The seventh session of the WMO Technical Commission for Special Applications of Meteorology and Climatology [now the Commission for Climatology and Applications of Meteorology (CCAM)] appointed, in early 1978, a rapporteur on requirements for satellite data for special applications of meteorology and climatology. The rapporteur will report on his studies at the 1982 session of CCAM.

In the field of research, the WMO Commission for Atmospheric Sciences is continuing its work through its Working Group on Satellite Meteorology, which in 1979 published a WMO *Technical Note on Quantitative Meteorological Data from Satellites* (WMO publication No. 531).

As regards the implementation of the WMO Global Ozone Research and Monitoring Project, satellites are expected to play an increasingly important role, especially for determining vertical ozone profiles. A comprehensive review of ozone measurements from satellites, including data available, was published in January 1981.

The WMO World Climate Program (WCP) is being carried out with the collaboration of the International Council of Scientific Unions (ICSU) and with the cooperation of UNEP and JOC and many other international organizations [which co-sponsor the SCOR (ICSU)/IOC Committee on Climate Changes and the Ocean (CCCO)] and will require a comprehensive three-dimensional observational description of the complete global climate system (atmosphere, oceans, land, cryosphere). Satellites will play a major role in this endeavor to observe and monitor the climate system.

In the field of information services, WMO continues to issue and bring up to date the following publications: *Information on Meteorological Satellite Programmes Operated by Members and Organizations* (WMO No. 411); *Information on the Applications of Meteorological Satellite Data in Routine Operations and Research* (abstracts, annual summaries, and bibliographies) (WMO No. 475); *The Role of Satellites in WMO Programmes in the 1980s* (WMO No. 494); *Operational Techniques for Forecasting Tropical Cyclone Intensity and Movement* (WMO No. 528); *Satellite Data Requirements for Marine Meteorological Services, Marine Service Affairs Report No. 14* (WMO No. 548).

UNEP has now published five issues of the annual *Ozone Layer Bulletin,* which contains information on the ongoing and planned ozone layer investigation programs and the results of research on the ozone problem. The most recent issue (January 1980) contains an assessment of ozone depletion and its impact as of November 1979, together with recommendations for future work. Recent research results and ongoing and planned research programs relevant to the World Plan of Action on the Ozone Layer are also included.

UNEP, the International Livestock Center for Africa, and the government of Kenya sponsored an International Workshop on Aerial Surveys in Nairobi from 6 to 12 November 1979. The workshop formed part of the GEMS renewable natural resource program and was organized in order to discuss progress in the use of very low-level aerial surveys in the natural resource monitoring of arid lands. Consideration was also given to the integration of satellite and other remotely sensed data into ecological monitoring programs. Proceedings of the workshop were published in January 1981.

The Arab Development Institute, the Royal Scientific Society of Jordan, and UNEP sponsored a seminar on environment monitoring for the Arab world, in Amman, Jordan, from 26 to 29 October 1980. This meeting, part of GEMS, informed participants of the latest appropriate technologies available for the monitoring of soils, erosion, desertification, agriculture, pastoralism, meteorology, and marine and terrestrial pollution. Particular attention was given to developments in ecological monitoring with emphasis on the use of the EMU concept. The seminar was intended for both decision-makers and senior technical staff and practical benefits for both development planning and management were stressed throughout. Special consideration was given to the use of remote sensing techniques. The seminar proceedings were published in 1981.

11.2.3.2 Education and training programs through fellowships, training workshops, and summer schools.

WMO, both independently and together with other United Nations agencies, governments, and national organizations, sponsors a variety of seminars and training programs relating to space applications and meteorology. Among past programs were:

1. A technical conference on the use of data from meteorological satellites, which was jointly organized by WMO, ESA, and the Société météorologique de France (SMF). Over 200 experts from 40 countries attended the conference which reviewed the use of satellite data for weather forecasting and analysis, climatology, agrometeorology, and ocean–surface phenomena.

2. A training workshop on the use of satellite data, under the auspices of the WMO Tropical Cyclone Programme, with the objective of providing senior hurricane forecasters from the region with up-to-date information and practical training in the use of satellite data for hurricane detection and prediction.

WMO and FAO are planning to hold a Spanish-language training course on the application of remote sensing techniques in agrometeorology for Latin America in 1981–1982. This follows similar courses for English and French-speaking participants held in 1978 and 1979. Also under consideration is a follow-up course in late 1981 to be organized by FAO in cooperation with WMO, ESA, and UNDRO on the application of remote sensing techniques to rural disasters.

WMO has also concluded that there is a need to publish a compendium of lecture notes from regional seminars concerning meteorological satellites. Consequently, WMO has begun preparing such publications and a compendium of lecture notes on meteorological satellites for training class 1 meteorological personnel is under preparation.

In addition to the WMO training publications, some 400 slides on meteorological satellites and satellite data interpretation and application are available in the WMO training Library for reproduction for member countries, on request, to serve as training aids in the field of satellite meteorology.

WMO also awards a number of fellowships in this field under the different sources of its technical cooperation programs. During 1979, 28 fellowships were awarded for training in meteorological satellites—two under UNDP, 12 under the WMO Voluntary Co-operation Programme (VCP), and 14 under the WMO regular budget. Twelve fellowships were requested for 1980–1981 and a similar number may be needed for 1982.

In addition to the fellowships of WMO, UNDP has provided fellowships for training in satellite meteorology to individuals from the following countries: Algeria, Bangladesh, Burma, Portugal, Rwanda, and Thailand. Some courses conducted at the WMO Training Center in Niamey, Niger, include the interpretation of satellite data and maintenance of APT equipment.

A training program is also provided by the United States under the WMO VCP in the interpretation of meteorological data from satellites. The program is for meteorological personnel of the developing countries.

11.2.3.3 Expert services and survey missions to identify specific areas of applications relevant to a given country or group of countries and to carry out special studies on pilot projects.

WMO, often together with UNDP, is providing the services of experts in the field of meteorology to many countries. Among the countries which have benefited from such expert services are Bangladesh, China, and Yemen. Similar missions are planned for Indonesia and Portugal. The purpose of such missions has generally been related to the establishment, operation, or maintenance of APT equipment. Future projects under UNDP auspices are planned for Brazil, Chile, Colombia, and Mongolia.

Assistance in the field of satellite meteorology is also expected to be provided under trust-fund arrangements in two countries. An APT station is to be installed at the WMO Training Center in Niamey, Niger, and a trust-fund agreement has been concluded with Saudi Arabia for the operation and maintenance of APT equipment in Yemen.

Under the WMO VCP, new APT stations including Weather Facsimile

Service (WEFAX) capability will be provided to Antigua, Belize, Egypt, Kenya, Mauritania, Mozambique, Nicaragua, Sudan, United Republic of Tanzania, and Upper Volta. Equipment for bringing up to date or upgrading existing APT stations will be provided for Colombia and Pakistan. It is expected that additional equipment for the reception of WEFAX broadcasts will be installed in Burundi, Cape Verde, Djibouti, Nepal, Thailand, Tunisia, Zaire, and Zambia.

11.2.4 Maritime Communications

Although only recently established, INMARSAT has the potential for becoming a major source of information, training, and assistance on the use of satellites in maritime activity. A number of operational capabilities have been identified for INMARSAT; their realization is contingent upon the cooperation of the international maritime community and the further buildup of INMARSAT. As a first step, INMARSAT will establish a space segment to provide L-and C-band maritime mobile satellite communications capacity over the Atlantic, Pacific, and Indian Ocean regions.

11.2.5 Air Navigation

ICAO interest in the application of satellite technology to civil aviation is concentrated in the three general areas of fixed telecommunications, mobile telecommunications, and radio navigation. Its international participation in related current activity has consisted mainly of the provision of operational information, the collection and dissemination of statistical data, and the coordination of supporting activity among relevant international bodies.

ICAO has, for a number of years, been using satellite links in its aeronautical fixed service for the distribution, between fixed points on the earth, of safety communications dealing with air traffic control, aeronautical meteorology, aeronautical information exchanges, and operational control messages. Such links often form part of the worldwide aeronautical fixed telecommunications service and, where implemented, have eliminated communications deficiencies caused chiefly by unreliable high-frequency radio links. Satellite links are also being used successfully in some cases to provide voice communications between area control centers to assist in the transfer of responsibility for air traffic transiting their areas of responsibility.

Throughout 1980, the ICAO secretariat participated in an international study which was developing, by computer simulation, a comprehensive assessment of the needs of international civil aviation for mobile communications, radio navigation, and surveillance safety services over the next 25 years. The study was to assess the ultimate capacity of existing systems to provide these safety services, taking into account expected technological and economic developments, and it includes various combinations of satellite and other technologies, as well as advanced all-satellite solutions.

The above study was done by member states, and the ICAO participation consisted mainly of the provision of data and of assistance with technical and operational details. When the computer simulation is completed the ICAO Air Navigation Commission and Council will review the situation and develop further the organization's future work program in the use of satellites for international civil aviation.

11.3 COORDINATION OF OUTER SPACE ACTIVITIES WITHIN THE UNITED NATIONS

In view of the variety of activities and programs that relate to applications of space technology, the need for close cooperation among the organizations of the United Nations system was recognized at an early stage by the United Nations Committee on the Peaceful Uses of Outer Space.

Acting upon recommendations of the Committee, the General Assembly has consequently called upon organizations concerned to consider undertaking programs jointly or in close consultation or cooperation with one another. Such recommendations have included, for instance, those calling for cooperation in the implementation of the World Weather Watch in the 1960s as well as in subsequent programs for the strengthening of meteorological services and research and for the expansion of training and educational opportunities in these fields, and those calling for cooperation in technical and other assistance to help meet communications needs of member states and for the effective development of domestic communications.

The implementation of these and subsequent legislative actions has required close consultations at the secretariat level, and as programs and activities began to encompass such general areas as training and education to meet the needs of developing countries, coordination of efforts has proved increasingly necessary.

Following the recommendation of the Committee on the Peaceful Uses of Outer Space, a Sub-Committee on Outer Space Activities has been established under the Administrative Committee on Co-ordination (ACC) of the United Nations, to coordinate the activities of the various organizations involved in space activities. This interagency subcommittee, whose meetings are attended by representatives of the organizations concerned in the United Nations system, meets annually to work out joint programs and coordinate their related activities. The results of its work are reported annually to the Committee on the Peaceful Uses of Outer Space and its Scientific and Technical Sub-Committee.[3] Thus, this ACC subcommittee provides an effective mechanism for the conduct of activities and for promoting concerted action, especially with regard to the application of space technology for the development projects.

[3] The latest report is to be found in document ACC/1980/30.

APPENDIX 11.1: PROCEDURE FOR THE ESTABLISHMENT OF A UNDP-ASSISTED SPACE APPLICATIONS PROJECT IN A MEMBER STATE

Member states of the United Nations are conversant with the five-year indicative period used in budgeting UNDP assistance. Under this procedure, UNDP informs each member state of the financial assistance it may count on from UNDP if pledges to UNDP are met. Thereafter, it is the prerogative of each government to assign its own priority to the projects it wants UNDP to fund.

The procedure to be followed in requesting UNDP assistance for a space application project is similar to the one cited above. The government concerned assigns a high priority to the scheme and communicates its desire for the project to UNDP through its UNDP resident representative. At this stage, the UNDP appoints a relevant executing agency (EA) of the United Nations to act on its behalf and to advise the recipient country's government at every stage of the project. The suggestion, for example, to utilize remote sensing satellite techniques in the process of executing any project may come either from the recipient country or from the project personnel representing the executing agency. In the case of a remote sensing project, the executing agency could be FAO for renewable resources projects, or WMO for projects involving meteorological and water resource studies, or the Natural Resources and Energy Division for nonrenewable resources projects.

The EA arranges for the preparation of the country's program that will indicate how the government proposes to utilize the funds requested in a space applications/remote sensing project. Its actual preparation is undertaken through the collaborative efforts of the EA and the recipient country. Once the program is approved by UNDP, it then becomes the working document of the project.

Where the EA has the expertise and facilities needed for executing the project, it may choose to do so itself. In many instances, however, such projects are often awarded as contracts to private organizations that specialize in remote sensing technology applications. Irrespective of the approach employed, the EA monitors the project continuously; it also prepares interim reports as may be necessary, as well as the final report of the project for UNDP.

It might be appropriate here to give an example of how a remote sensing project in a member state has been funded within the United Nations system.

A11.1.1 A Case Study: UNDP-Funded Remote Sensing Project in a Member State

Following the launching of the LANDSAT I satellite by the United States in July 1972, the decision-maker in this country approved his country's participation in the LANDSAT program on 26 January 1974. Thereafter, his government constituted the National LANDSAT Committee, a policy-making and advisory group, and mandated it to:

1. Coordinate the activities of various sectors (such as hydrology, agriculture, forestry, fisheries, geology, and cartography) involved in the country's LANDSAT program and to provide overall policy guidance to the LANDSAT Task Force in implementing the program.
2. Examine and approve the country's LANDSAT program proposal before submitting it to NASA.
3. Evolve a suitable program for the training of local personnel in satellite data technology and to arrange for their training.
4. Communicate with NASA and other external agencies in connection with LANDSAT and other satellite programs.

This Committee was made up of heads of government ministries affiliated with agriculture, defense, water resources and flood control, forestry, fisheries and livestock science and technology, and home affairs.

A LANDSAT Task Force was also established as the executive body responsible to the National LANDSAT Committee. The Task Force was mandated to:

1. Prepare sectoral projects and an overall project proposal for the LAND-SAT program of the country.
2. Set up and maintain ground observation stations with the assistance of concerned government agencies.
3. Collect and analyze ground station observation for coordination with LANDSAT data.
4. Collect, store, and process all incoming data and disseminate relevant data to the departments and agencies concerned.
5. Select persons for the ground truth program and provide in-service training.
6. Prepare and submit quarterly progress reports to the National LANDSAT Committee in order to implement this program.

Eight main fields of investigations were identified: agriculture, forestry, oceanography and fisheries, water resources, cartography, interpretation techniques development, geology, and meteorology.

Project areas and test sites, as well as investigators and co-investigators for each sector were selected, and a proposal for the country's participation in the LANDSAT II program was submitted to NASA in May 1974. In order to gain a better understanding of this technology, the principal investigator undertook a mission to Europe and North America for the express purpose of familiarizing himself with the remote sensing activities in these regions. Another external organization was appointed by the government to assist it in the initial implementation of its program. A LANDSAT center was established at its new premises and a subcontractor was selected to assist the country in the application of remote sensing for land inventory. By the time these activities were concluded, the country was in a position to take the next step of involving its nationals on a comprehensive remote sensing application program implementation. This involvement required external assistance which was provided by UNDP with FAO as the executing agency. The assistance consisted of appointing a remote sensing adviser or project manager for the country's remote sensing program. This necessitated the preparation of the submission to the government and UNDP of a detailed work plan for the remote sensing program. The manager required for the project was to:

1. Assist with all aspects of work in the development and management of the Remote Sensing Center. This includes advising on the establishment of the Center, its policy, and its functions. As required, advise the Planning Commission and the National ERTS Committee.
2. Advise on all technical aspects of remote sensing, including overseas training and local training and covering data collection, storage, retrieval, and distribution. Assist in setting up a library of LANDSAT imagery covering all of the country.
3. Assist in participation by the Remote Sensing Center in ongoing field programs and research related to remote sensing.
4. As required, advise Center officers in applying remote sensing techniques to field programs.
5. Determine the need for and prepare, as necessary, a project proposal for follow-up activities to this project.
6. Advise the director of the country's Remote Sensing Center on integrating bilateral aid programs with the operational activities of the Center.
7. Establish and participate in local training courses in remote sensing.

In addition, funds were sought from UNDP to train local scientists abroad under a UNDP/FAO program. Orientation courses for national investigators or co-investigators on satellite data were undertaken. Following these exercises, the local investigators worked side by side with the contractor to collect ground truth data for the cooperative investigative program on land survey. The first phase of the country's remote sensing project under the project was supported by UNDP contributions amounting to $469,150. The country contributed more than ten times as much, namely, $5,039,276 in local currency.

The assignments to be executed in the UNDP-funded project included:

1. *Agriculture:* To inventory dry and wet land during the winter crop season; to resolve and develop identification keys for major crops, e.g., rice, jute,

sugar cane, wheat, and potatoes; and to estimate yields and production of the major crops, particularly of the winter (boro) crop. Additional objectives were to determine crop calendars for the major crops; to locate and estimate crop damage by floods, drought, disease, pests, and weeds; to investigate the application of satellite data, with appropriate inputs from other sources; and to determine soil moisture content during different seasons.

2. *Forestry:* To inventory forest resources in certain forest areas and to identify changes in the coastal areas requiring afforestation measures to stabilize land formations.

3. *Water resources:* To prepare dry season water and wet lands inventory and to map inundated areas and soil products on cultivable land. Additional objectives were to undertake studies on salinity distribution and to study changes in river courses and coastal morphology.

4. *Cartography:* To assist in the revision of small-scale maps of the country and to assist in the preparation of various types of thematic maps.

5. *Oceanography and fisheries:* To determine the effectiveness and reliability of satellite data for marine and inland fisheries purposes and to determine feasibility of locating schools of fish and prawn by use of remote sensing techniques.

6. *Interpretation techniques development:* To screen, select, reproduce, store, and retrieve LANDSAT data made available by NASA; to correct atmospheric effects on LANDSAT data; to coordinate the development of classification and automatic pattern recognition techniques through identification of spectral signatures; to develop sampling techniques and systems models for programs based on the application of satellite data; and to operate, maintain, repair, and calibrate the instruments supplied under this project.

7. *Meteorology:* To investigate the utilization of satellite imagery for meteorological observation, with the objective of augmenting existing cyclone studies; to study cloud dynamics using satellite imagery; and to forecast crops using meteorological and satellite data.

8. *Geology:* To identify gross lithological/formational divisions; to delineate broad structures in the northern and eastern parts of the country; and to study the changes in the river channels in selected areas.

Chapter 12

ROLE OF NONGOVERNMENTAL

ORGANIZATIONS

IN SPACE ACTIVITIES

ABSTRACT

This chapter, the third of three on international cooperation, is aimed at providing information on the mandates and programs of various nongovernmental organizations active in space science and technology. It was prepared by the conference secretariat, based on information available to it from the organizations concerned.

12.1 INTERNATIONAL COUNCIL OF SCIENTIFIC UNIONS (ICSU)

12.1.1 Origin

"In view of the great importance of observations during extended periods of time of extraterrestrial radiations and geophysical phenomena in the upper atmosphere, and in view of the advanced state of present rocket techniques, the

CSAGI[1] recommends that thought be given to the launching of small satellite vehicles, to their scientific instrumentation, and to the new problems associated with satellite experiments, such as power supply, telemetry, and orientation of the vehicles.'' With this recommendation, made in 1954 by its Special Committee on the International Geophysical Year, ICSU first became involved in research with satellites. However, the Council's interest in space research goes back to 1919 when its predecessor, the International Research Council, accepted a proposal from one of its Scientific Union members, the International Union of Geodesy and Geophysics (IUGG) to cooperate with the International Astronomical Union (IAU) in order to investigate the relations between solar phenomena and terrestrial magnetism and electricity. One result of this cooperation was the creation in 1925 of a Commission for the Study of the Relations between Solar and Terrestrial Phenomena. The annual grant received by the Commission at this time amounted to £40 ($160).

The experience gained in developing the International Geophysical Year (IGY) program showed the value of continued cooperative rocket and satellite research and, in October 1958, ICSU created the Committee on Space Research (COSPAR) to provide the scientific community throughout the world with a means of exploiting the capabilities made available by the new "space" techniques, and to stimulate the participation of scientists not actively engaged in the IGY rocket and satellite program.

Currently, 13 of the 18 international scientific unions within ICSU are actively participating in the work of COSPAR. Those with a wide range of space research activities are described in detail below. In addition to the unions and COSPAR, there are eight committees with space research activities, in particular the Scientific Committee on Antarctic Research (SCAR) (see below) and the Special Committee on Problems of the Environment (SCOPE). The latter is primarily concerned with the use of space techniques for monitoring and for surveying natural resources. In addition, SCOPE has carried out some studies on the possibility of monitoring a number of environmental parameters from space platforms.

The Inter-Union Commissions for Frequency Allocations for Radio Astronomy and Space Sciences (IUCAF), Radio Meteorology (IUCRM), and Solar-Terrestrial Physics (IUCSTP) (described below) are active in encouraging studies using a variety of space techniques. A new Inter-Union Commission for Studies of the Moon was created by ICSU in 1970. The Commission organizes cooperation between the unions of astronomy, geodesy and geophysics, and geological sciences, as well as COSPAR, for the study of the moon.

ICSU was established in 1931, as a direct successor to the International Research Council (founded in 1919) to provide a central body through which the world scientific community can deal with problems of common interest and can encourage international scientific cooperation.

---

[1] Comité spécial de l'année géophysique internationale.

12.1.2 Purpose and Objectives

The objects of the Council are:

1. To encourage international scientific activity for the benefit of mankind, and so promote the cause of peace and international security throughout the world.
2. To facilitate and coordinate the activities of the international scientific unions.
3. To stimulate, design, and coordinate the international interdisciplinary scientific research projects.
4. To facilitate the coordination of the international scientific activities of its members.

12.1.3 Membership

ICSU is composed of two categories of members:

1. International Scientific Union members.
2. National members.

There are currently 18 International Scientific Union members:

International Astronomical Union (IAU)
International Union of Geodesy and Geophysics (IUGG)
International Union of Pure and Applied Chemistry (IUPAC)
International Union of Radio Science (URSI)
International Union of Pure and Applied Physics (IUPAP)
International Union of Biological Sciences (IUBS)
International Geographical Union (IGU)
International Union of Crystallography (IUCr)
International Union of Theoretical and Applied Mechanics (IUTAM)
International Union of the History and Philosophy of Science (IUHPS)
International Mathematical Union (IMU)
International Union of Physiological Sciences (IUPS)
International Union of Biochemistry (IUB)
International Union of Geological Sciences (IUGS)
International Union for Pure and Applied Biophysics (IUPAB)
International Union of Nutritional Sciences (IUNS)
International Union of Pharmacology (IUPhar)
International Union of Immunological Societies (IUIS)

The National members are academies of science, national research councils, etc., in the following countries: Argentina, Australia, Austria, Belgium, Bolivia, Brazil, Bulgaria, Canada, Chile, Cuba, Czechoslovakia, Denmark, Egypt (Arab Republic), Finland, France, German Democratic Republic, Germany (Federal Republic of), Ghana, Greece, Hungary, India, Indonesia, Iran, Iraq, Ireland, Israel, Italy, Japan, Jordan, Kenya, Korea (Democratic People's Republic), Korea (Republic of), Lebanon, Madagascar, Mexico, Monaco (Principauté de), Morocco, Netherlands, New Zealand, Nigeria, Norway, Pakistan, Philippines, Poland, Portugal, Romania, South Africa, Spain, Sri Lanka, Sudan (Democratic Republic), Sweden, Switzerland, Taiwan, Thailand, Tunisia, Turkey, USSR, United Kingdom, United States of America, Uruguay, Vatican City State, Venezuela, Vietnam Socialist Republic, Yugoslavia.

12.1.4 Associates

The Council has three National Associates in Jamaica, Nepal, Malaysia, and 16 scientific associates:

International Federation for Documentation (FID)
International Federation for Information Processing (IFIP)
Pacific Science Association (PSA)
International Statistical Institute (ISI)
International Society of Soil Science (ISSS)
International Association for Water Pollution Research (IAWPR)
International Federation of Library Associations and Institutions (IFLA)
International Union for Quaternary Research (INQUA)
International Brain Research Organization (IBRO)
International Council for Laboratory Animal Science (ICLAS)
International Federation of Societies for Election Microscopy (IFSEM)
International Radiation Protection Association (IRPA)
International Federation for Automatic Control (IFAC)
International Union Against Cancer (IUCC)
International Union of Forestry Research Organizations (IUFRO)
International Union of Psychological Sciences (IUPS)

12.1.5 Committees, Commissions, and Services

Scientific Committee on Oceanic Research (SCOR)
Scientific Committee on Antarctic Research (SCAR)
Committee on Space Research (COSPAR)
Scientific Committee on Water Research (COWAR)

Committee on Science and Technology in Developing Countries (COSTED)

Committee on Data for Science and Technology (CODATA)

Committee on the Teaching of Science (CTS)

Scientific Committee on Problems of the Environment (SCOPE)

Scientific Committee on Solar Terrestrial Physics (SCOSTEP)

Scientific Committee on Genetic Experimentation (COGENE)

Inter-Union Commission on Frequency Allocations for Radio Astronomy and Space (IUCAF)

Inter-Union Commission on Radio Meteorology (IUCRM)

Inter-Union Commission on Spectroscopy (IUCS)

Inter-Union Commission on the Application of Science to Agriculture, Forestry and Aquaculture (CASAFA)

Inter-Union Commission on the Lithosphere (ICL)

ICSU Abstracting Board (ICSU   AB)

Federation of Astronomical and Geophysical Services (FAGS)

World Data Centers (WDC)

Of the above-mentioned organizations, 14 of the unions, eight of the committees, four of the commissions, and the two services are involved in activities relating to the peaceful uses of outer space. Each organization involved in space research is described briefly below.

12.1.6 The International Astronomical Union (IAU)

The Union was established in 1919 "to promote the study of astronomy in all its aspects." Various international groups such as the Union of Solar Studies, International Latitude Service, International Time Association, International Committee for the "Carte du Ciel," etc., had been responsible for various aspects of astronomical research prior to this.

IAU is interested in the peaceful exploration of outer space from the point of view of obtaining the maximum amount of information on the positions of, the physical structure of, and the radiations from all natural celestial bodies, including the material that pervades interplanetary space, normally referred to as the interplanetary medium. It is concerned that the above should not be contaminated in any avoidable way by man-made interferences. IAU is also interested in the use of artificial celestial bodies either as such or as astronomical observatories. IAU's activities are limited to acting as the aegis under which the astronomers of all countries may conduct astronomical research. It also acts as a forum for the discussion, correlation, and coordination of results and programs. One of the principal ways in which the IAU helps to promote the study of astronomy is by the organization of triennial general assemblies and meetings, symposia, colloquia, and workshops. The last General Assembly was held in 1979: the next will be in 1982 in Greece. In addition to meetings of the 39 scientific commissions, the 1979 Assembly included discussions on subjects such as exploration of the solar system, ultraviolet astronomy from recent space experiments, very hot plasmas in circumstellar, interstellar, and intergalactic space, etc. The discussions, discourses, and reports of the Commission meetings will be published in "Highlights of Astronomy."

The IAU has also organized symposia or colloquia on subjects such as dynamics of the solar system, refractional influences in astrometry and geodesy, physics of solar prominences, and scientific research with the space telescope.

The Union organizes (approximately annually) Schools for Young Astronomers. The last three were held in Yugoslavia in 1980, Spain in 1979, and Nigeria in 1978. These are intended especially for young astronomers from developing countries.

The Union is participating in two new major international efforts of significance in the peaceful uses of outer space:

1. The Giant Equatorial Radio Telescope (GERT) in Africa.
2. A project to monitor the earth's rotation and intercompare the techniques of observation and analysis (MERIT).

12.1.7 The International Union of Geodesy and Geophysics (IUGG)

In 1919 the Union was formed to promote the study of problems relating to the shape and physics of the earth, including the oceans and the atmosphere, that require international collaboration. The Union has seven international associations, four of which have specific interests in space science (the International Association of Geomagnetism and Aeronomy, the International Association of Meteorology and Atmospheric Physics, the International Association of Geodesy, the International Association for the Physical Sciences of the Ocean) and two which use data from remote sensors (International Association of Hydrological Sciences and International Association of Volcanology and Chemistry of the Earth's Interior).

12.1.7.1 The International Association of Geomagnetism and Aeronomy (IAGA).   IAGA focuses its studies on those regions of planetary atmospheres where ionization and dissociation play a role, their dynamics and relevant chemical processes, on the sources and sinks of energy and particles, and associated mechanisms of interaction with the neutral atmosphere and the local electromagnetic field; on the origin, configuration, and variations of the planetary magnetic field and its interaction with the planetary body and the plasma envelope; and on the dynamics of magnetospheric and solar wind

plasmas and their mutual interaction. The Association has been particularly active in developing with SCOSTEP the Middle Atmosphere Programme (MAP) and will organize a symposium on Middle Atmosphere Sciences in Hamburg in 1981 and jointly with IAMAP on radiation problems in the middle atmosphere and troposphere–stratosphere interactions.

12.1.7.2 The International Association of Meteorology and Atmospheric Physics (IAMAP).

The increasingly important role of space observations in scientific studies of atmospheric physics and meteorology have had a major impact on the development of IAMAP over the past 30 years, largely because of the effective way in which satellite observations complement ground-based ones.

At present IAMAP, in cooperation with COSPAR and the JSC, is concentrating on two problems whose solution is of particular importance in improving global models of the atmosphere and is vital in any attempts to improve climate predictions: radiation processes in the atmosphere and the atmospheric composition.

12.1.7.3 The International Association of Geodesy (IAG).

The International Association of Geodesy (IAG) helps coordinate international programs involving geodetic links and promotes the study of scientific problems of geodesy.

The Association has played a primary role in the creation of the Satellite Geodesy Central Bureau which ensures maximum use of new techniques and geodetic liaisons between continents and regions, such as the Europe–Africa Geodetic liaison, the Arctic–Antarctic Arc, etc. It also brings together research workers involved in various disciplines using space geodesy techniques.

12.1.7.4 The International Association for the Physical Sciences of the Ocean (IAPSO).

The International Association for the Physical Sciences of the Ocean makes full use of satellite measurements of sea surface temperatures, wind speed and direction, wave height, etc. The advent of satellites has introduced a new precision in the location of drifting buoys and of research vessels. The Association works closely with SCOR and COSPAR.

Information about the activities of the Union are published in the *IUGG Chronicle* and in the Association publications.

12.1.8 The International Union of Radio Science (URSI)

URSI was created at the beginning of the twentieth century to study the new radiotelegraph communications system. Studies now include the scientific bases of radiocommunications and applications of radio methods to research in general, for example, radio telescopes carried by spacecraft have been especially effective in elucidating the low-frequency characteristics of solar radiation and its extension into interplanetary space. The development of satellites helped provide a major advance in telecommunications, and it should not be overlooked that radio in one form or another is involved in the control and command, telemetry, and tracking of space vehicles.

Information about the Union is published in the *URSI Information Bulletin.*

12.1.9 The International Geographical Union (IGU)

The principal bodies in the International Geographical Union that make use of remote sensing in their studies are the Commissions on Geographical Data Sensing and Processing, Geomorphological Survey and Mapping, Agricultural Productivity and World Supplies, Desertification in and around Arid Lands, and Environmental Atlases.

News about the International Geographical Union are published in the *IGU Newsletter.*

Several other unions such as the International Union of Geological Sciences (IUGS) and the International Union of Biological Sciences (IUBS) make use of remote sensing in their studies; the International Union on Physiological Sciences, the International Union of Biological Sciences, the International Union of Biochemistry, and the International Union of Pure and Applied Biophysics are involved in biological experiments in space, and the International Union of Pure and Applied Physics and the International Union of Pure and Applied Chemistry carry out studies involving the physics and the chemistry of the earth's atmosphere and of the effects of influences from outside it. Certain other unions, including the International Union of Theoretical and Applied Mechanics and the International Union of Crystallography, are involved in experiments on materials in space.

12.1.10 Scientific Committee
on Oceanic Research (SCOR)

The Scientific Committee on Oceanic Research was established in 1957 by ICSU to continue the work in oceanography begun during the IGY and to further international scientific activity in all branches of oceanic research.

The Committee has the following working groups: Ocean–Atmosphere Materials Exchanges, River Inputs to Ocean Systems, Living Resources of the Southern Ocean, Oceanographic Programs during the First GARP Global Experiment, etc., which make use of data obtained by remote sensing. The recently created Committee on Climate Change and the Ocean (jointly sponsored by SCOR and the Intergovernmental Oceanographic Commission) is active in many fields which involve the use of satellites for remote sensing, for interrogation of buoys and sondes, and for position finding. The Committee was fortunate in being able to make use of SEASAT during its limited period of operation.

News about the Committee's activities is published in *SCOR Proceedings*.

12.1.11 The Scientific Committee
on Antarctic Research (SCAR)

The Scientific Committee on Antarctic Research was set up in 1958 by ICSU and is charged with furthering the coordination of scientific activity in the Antarctic. The Committee immediately assumed responsibility for the international cooperative scientific programs in the Antarctic initiated by the IGY. In fostering programs of circumpolar scope and significance in all appropriate scientific disciplines, SCAR works closely through ICSU bodies and, in the field of space research, maintains liaison with COSPAR.

The Working Group on Meteorology played an important role in designing experiments using satellite-borne sensors and ground measurements to study the Antarctic ice caps during the Global Atmospheric Research Programme. These are continuing within the framework of the World Climate Research Programme.

Remote sensing is also being used to provide information on the extent of and changes in the Antarctic ice cap.

News about the Committee is published in the *SCAR Bulletin*.

12.1.12 The Committee on Space Research (COSPAR)

The activities of COSPAR are discussed in Section 12.2.

12.1.13 The Committee on Science and Technology
for Development (COSTED)

The Committee on Science and Technology for Development was established by ICSU in 1966 in order to coordinate and encourage efforts by the members of the ICSU family to assist the developing countries in the fields of science and technology.

The Committee has organized jointly with COSPAR and other ICSU bodies a number of meetings and workshops such as those on Space and Development in Bangalore in 1979, Applications of Space Observations to Marine Resources in Developing Countries in Washington in 1980, and is planning meetings on Satellite Communication—Prospects for Developing Countries in Brazil and Kenya and on Remote Sensing for Resource Evaluation in the United States.

12.1.14 The Scientific Committee on Problems of the Environment (SCOPE)

Since its creation in 1969 one of SCOPE's tasks has been to assess and evaluate the methodologies of measurement of environmental parameters, and the first SCOPE report, on Global Environmental Monitoring, was prepared for the UN Conference on the Environment in 1972. SCOPE continues to work in close cooperation with the Monitoring and Assessment Research Centre (MARC) which it was instrumental in creating in 1975, whose current program includes the development of assessment methodologies for use in environmental management which involves remote sensing.

Remote sensing also plays an important role in SCOPE's principal study on biogeochemical cycles and also in the two newest projects on the processes of land and soil transformation and ecosystem dynamics in freshwater wetlands and shallow water bodies.

Information about SCOPE's activities is published in *SCOPE Newsletter.*

12.1.15 The Scientific Committee on Solar-Terrestrial Physics (SCOSTEP)

SCOSTEP will organize its next major quadrennial inter-Union Symposium on selected topics in solar–terrestrial physics in 1982 in association with the next COSPAR meeting in Ottawa, Canada.

Other activities include the International Magnetospheric Study (IMS); the Middle Atmosphere Programme (MAP); Solar Maximum Year (SMY); Monitoring the Sun–Earth Environment, etc.

The International Magnetospheric Study (IMS). The globally coordinated observational phase of the IMS, 1976–1979, closed on 31 December 1979, and the IMS Data Analysis Phase has now begun. The IMS Steering Committee met in 1979 and prepared a final report on the Observational Phase (1976–1979), on measures to put the Data Analysis Phase (1980–1985) on the firmest possible footing, and on recommendations for a long-range program (beginning about 1985) of coordinated observations that will complement prospective magnetospheric and interplanetary spacecraft missions late in this decade.

Plans are proceeding for the establishment of a second International Computerized Data Analysis Workshop Center (DAWOC) similar to that at the IMS Satellite Situation Center (IMS-SCC) at the Goddard Space Flight Center, U.S.A. Such centers are equipped with small specialized computers capable of taking in data of innumerable different types, of digesting the data, and of displaying them as graphical functions of time on closed-circuit television screens with great flexibility as to scales, time-resolution, etc.

Middle-Atmosphere Programme (1982–1985). A series of meetings has been organized to develop and extend the program based on the preliminary

MAP projects. These are projects that are already being organized where international interest and sufficient facilities exist: besides scientific results, the meetings will provide experience in how to reduce the total program to manageable proportions without compromising its interdisciplinary character.

Solar Maximum Year (SMY), August 1979–February 1981. The activities within the SMY are now completed. The SMY Steering Committee coordinates three groups with slightly different but compatible interests, namely, the Flare Buildup Study (FBS), the Study of Energy Release in Flares (SERF), and the Study of Travelling Interplanetary Phenomena (STIP), interested, respectively, in physical conditions occurring before, during, and after flares. Detailed plans have now been worked out for the close coordination of observations of many different types involving some scores of ground-based observatories and spacecraft, in particular the United States "Solar Maximum Mission" (SMM) spacecraft, which was designed for this purpose.

Monitoring the Sun–Earth Environment (MONSEE). MONSEE is a continuing activity which serves all the others. It is concerned with the problems of collecting, safekeeping, cataloging, and distributing nearly 100 types of data obtained on regular schedules at more than 400 stations and of data sets from special sources. This large body of organized data, managed by the ICSU World Data Centres system, constitutes invaluable background or correlative material for all SCOSTEP programs, for whose specialized data-management problems the MONSEE Committee also renders assistance. The Committee is increasingly directing its attention to improving the efficiency of data handling methods through the increased use of computer-compatible techniques, and it encourages data centers to produce data digests (indices, charts) that are more useful than raw data to the nonspecialists.

SCOSTEP issues several publications including the following: *STP Notes, IMS Bulletin, MONSEE Bulletin.*

12.1.16 The Inter-Union Commission on Frequency Allocations for Radio Astronomy and Space Science (IUCAF)

The Commission was formed in 1960 to take all appropriate steps to secure protection from interference for a number of channels of radio frequencies which are required for the pursuit of radio astronomy and research in space science.

The Commission, which comprises representatives of URSI, IAU, and COSPAR, with some independent consultants and advisers from the International Radio Consultative Committee (CCIR) and the International Frequency Registration Board (IFRB), meets periodically. At such meetings and by correspondence, the Commission discusses the technical merits of the requirements of the scientific community and submits proposals to the International Telecommunication Union (ITU) with a view to securing an adequate allocation of frequency channels suitably distributed throughout the spectrum.

For space research, the requirements are for frequencies suitable for transmission from satellites to the earth in order to communicate the results obtained with appropriate measuring equipment. In addition, full protection from interference is required for the frequencies used to determine the position of the satellite accurately.

For radio astronomy, a series of frequencies at approximately harmonic intervals is sought, with, in addition, a number of discrete frequency channels where atomic or molecular constituents of the earth's environment are known to resonate.

The Commission met in April 1979 to prepare for the World Administrative Radio Conference (WARC). During the WARC, which was called from 24 September to 6 December 1979 to revise radio regulations, scientific interests were strongly represented, with IUCAF playing a coordinating role. During the whole of the conference there was at least one observer with the undivided responsibility for representing IUCAF, and 13 scientists, with close connections with the Commission, from ten countries participated as members of their national delegations. Close contact was maintained by the IUCAF representatives with others concerned, partly by informal meetings of all those able to attend, but mainly through personal contacts as the need arose. A close link was established between IUCAF, national bodies representing space science activities, and the European Space Agency, which also organized discussion meetings.

There is reason to believe that the results are regarded as generally satisfactory, but with some remaining problems that will need to be solved by detailed national or international negotiations within the framework of the regulations. Increased demand for frequencies by most services has resulted in more frequency sharing and, in particular, the extent to which the passive services (radio astronomy and earth sensing by the use of radiometers) will be affected by the other services will need to be assessed. IUCAF will be identifying the most serious problems and deciding what action should be taken.

A meeting will be arranged soon to assess the role of IUCAF in the post-WARC period and to plan the detailed activities.

12.1.17 The Inter-Union Commission on Radio Meteorology (IUCRM)

The Commission, which was set up in 1959, is currently concerned with aspects of meteorology that affect the propagation of electromagnetic waves in the earth's atmosphere and through planetary atmospheres and with the application of electromagnetic techniques of meteorology.

The Commission recently sponsored or cosponsored a number of colloquia or symposia on topics associated with research in or from space, for example, Air-Sea Interaction and Its Effects on Electromagnetic Wave Propagation, Loftus, Norway; Wave Dynamics and Radio Probing of the Ocean Surface, Miami, Florida, United States; Oceanography from Space, Venice, Italy; and Middle Atmospheric Dynamics and Transport, Urbana, Illinois, United States.

12.1.18 The Inter-Union Commission on the Application of Science to Agriculature, Forestry and Aquaculture (CASAFA)

CASAFA, which was created by ICSU in 1978, has decided to concentrate its activities on the semi-arid tropics. A number of its projected activities will include studies involving observation, assessment, and monitoring using remote sensing techniques.

12.1.19 The Federation of Astronomical and Geophysical Services (FAGS)

FAGS was established by ICSU in 1956. Each of the 11 permanent services in the Federation is sponsored by one or more of the scientific unions of ICSU. Each service is under the authority of a board whose membership is decided by the interested unions. The day-to-day activities of a service are controlled by a director appointed by the board.

FAGS is administered by a Council which includes representatives of the unions; UNESCO is invited to send an observer to meetings of the Council.

The essential role of a service is to act as an international center for the collection and preliminary processing, on a long-term basis, of many kinds of geophysical and astronomical data. The processed data are published and are used for research and for record purposes.

FAGS provides for:

1. The processing, in a central location, of data received from observatories in many countries, ensures a high degree of homogeneity in the published data. In consequence, subsequent utilization of the data is greatly facilitated by the work of the services.

2. Uninterrupted series of homogeneous data extending over long periods of time are required for certain types of scientific research. The services help to preserve the desired continuity even though it is inevitable that, from time to time, particular stations in the various networks may close down and others may commence operation.

The following list of services gives a brief indication of the broad fields in which they work:

(a) *The Earth*

Bureau Gravimétrique International (precise measures of gravity)

Service Permanent des Marées Terrestres (tidal deformation of the solid earth)

International Polar Motion Service (movements of the axis of rotation of the earth)

Permanent Service on Fluctuation of Glaciers (long-term changes in glaciers)

Permanent Service for Mean Sea Level

Bureau International de l'Heure (precise measure of time; rotation of the earth)

(b) *The Earth's Environment*

International Ursigram and World Days Service (collation of data on the physical characteristics of space; prediction of disturbances due to solar eruptions, etc.)

Permanent Service of Geomagnetic Indices (collation of geomagnetic data and compilation of planetary indices of activity)

Quarterly Bulletin on Solar Activity (collation of data on solar activity based on optical and radio observations)

12.1.20 The Panel on World Data Centres (WDCs)

The WDC Panel was created by ICSU in 1968 to advise the officers of ICSU on the management of the WDCs, which were created on the initiative of ICSU to handle data obtained during the International Geophysical Year.

In June 1979 the ICSU Panel on WDCs published the "Fourth Consolidated Guide to International Data Exchange through the World Data Centres." This fourth edition supersedes the "Third Consolidated Guide" of December 1973. The introductory chapter, "General Information and Principles," which is a synthesis of material that has accumulated since the publication of the first "Guide" in 1957, has been completely rewritten in the interest of simplicity, clarity, and brevity. The principles by which the WDC system operates, which were adopted when it was first established to serve the International Geophysical Year of 1957–1958, have been adjusted only slightly in order to accommodate the handling of the large sets of reduced data from such data-intensive programs as the GARP Atlantic Tropical Experiment (GATE) and the First GARP Global Experiment (FGGE). For such programs, it is simply not practicable to archive all the original (so-called "raw" or "Level I") data.

Almost all the guides for the individual disciplines or groups of disciplines have been revised since 1973 and some new guides have been added (volcanology, heat flow, and two in meteorology).

12.1.21 The Inter-Union Commission on the Lithosphere (ICL)

The Inter-Union Commission on the Lithosphere was created by ICSU, IUGG, and IUGS in 1980 to help in the elucidation of the nature, dynamics, origin, and evolution of the lithosphere, with special attention to the continents and their margins. Plans are proceeding for a wide range of studies, including Recent Plate Movements and Deformation, Subduction, Collision, Aceretion, Mineral and Energy Resources, Environmental Geology and Geophysics which will make use of the observations from space.

12.1.22 The Joint WMO-ICSU World Climate Research Program (WCRP)

The WMO-ICSU Agreement on the World Climate Research Programme (WCRP) came into effect on 1 January 1980 and replaced the earlier Agreement on the Global Atmospheric Research Programme (GARP), which had been so effective in organizing a number of significant efforts in meteorological research directed toward the improvement of weather forecasting and the understanding of the physical processes of climate. ICSU and WMO have agreed that there should be only one World Climate Research Programme and it should be sponsored jointly by both organizations. Both organizations have approved a transition from GARP to WCRP that will take place gradually from the inception of the WCRP agreement and a transition in membership from JOC to JSC that will reflect this change in programmatic emphasis.

With the inception of WCRP, the Joint Organizing Committee for GARP has been converted to the WMO-ICSU Joint Scientific Committee (JSC) for the World Climate Research Programme, which will be responsible for the planning and overall coordination of WCRP. It will also be responsible for the completion of the planning and studies initiated within the framework of GARP.

12.1.22.1 The World Climate Research Programme Plan. The first task achieved by JSC was to develop a detailed plan for WCRP based on the following strategy:

1. To make an inventory of all the problem areas or research elements of concern in climate research.
2. To review the status of research within these elements and thereby establish what work needs to be done to accelerate progress under each heading.
3. To identify the organizations and scientific bodies which should organize the work in each case and to make appropriate recommendations.
4. To define the data requirements of WCRP and to give guidance for the long-term global monitoring of climate.

In the implementation of this strategy JSC considered that in WCRP there will be requirements for:

1. Extensive experimentation with climate models for the evaluation of model performance and for the conduct of predictability and sensitivity studies.
2. The investigation of several climatologically significant processes both to gain improved understanding and to aid the design of improved parameterization schemes for models of:
   Cloudiness and radiation
   Ocean processes
   Ocean cryosphere processes
   Hydrology and land surface processes
   Radiatively important gases (including carbon dioxide)
   Aerosols
   Solar–terrestrial relationships
3. A full three-dimensional observational description of the entire climate system (including a need for many special forms of data) to support the whole program and for observational and diagnostic studies.

In preparing the plan, JSC endorsed the conclusion of its predecessor, JOC, that two elements amongst the climatologically significant processes (namely, cloudiness/radiation and oceanic processes) are critical and call for special attention because of their overriding scientific importance and because there is an immediate need to organize technical programs for their solution.

JSC has recognized that in a program with such wide implications several other international scientific bodies with WMO and ICSU must necessarily be involved. Indeed, for many of the research elements, vigorous and appropriate research programs are already in progress. Therefore, JSC intends to make maximum use of existing bodies and organizations, e.g., IOC-SCOR/CCCO, COSPAR, various IAMAP Commissions, WMO/CAS, etc.

12.1.22.2 The Global Atmospheric Research Programme (GARP) Global Weather Experiment.   The FGGE Operational Year was successfully completed on 30 November 1979, and it appears that the final data set will allow the scientific objectives of the FGGE to be realized. The production of the FGGE data sets is proceeding as planned, although somewhat behind schedule.

Virtually all nations in the world participated in this experiment. In addition to the operational World Weather Watch observational system, it included several special observing systems:

5 geostationary meteorological satellites
4 polar orbiting satellites

47 research ships

9 specially dedicated long-range aircraft

313 balloons in the tropical stratosphere

368 buoys in the southern hemisphere

97 commercial aircraft equipped with special meteorological instruments

The operational phase is thus completed but the research phase is expected to last for several years.

JSC noted with pleasure that the number of FGGE research programs being carried out exceeds 500. Within the Visiting Scientist Programme more than 30 institutions have expressed their willingness to accept scientists wishing to participate. The Committee supported the measures recommended by the WMO Executive Committee Inter-Governmental Panel on FGGE concerning the designation of WMO fellowships to further MONEX and WAMEX research and for the funding of the program.

Because of the urgent need for sound scientific advice, to guide the design of the future composite observing system, JSC attached great importance to a program of Observing System Experiments which had been planned under the auspices of its predecessor, JOC. The JSC arranged for this work, which will concentrate on investigations of the impact of observations by the FGGE Special Observing Systems on the large scale weather analyses and forecasts, to be given high priority.

12.1.22.3 Other GARP subprograms.   *The Tropical Subprogram.* Six years after the field phase of GATE, an International Conference on Scientific Results of GATE was held in September 1980 in Kiev, USSR. The outcome of the conference has been published as a monograph entitled "A Synthesis of GATE Scientific Results."

The GATE Oceanographic Atlas ("Physical Oceanography of the Tropical Atlantic during GATE") has also been published.

JSC considered that the tropical subprogram had been highly successful as evidenced by the extensive contribution to the literature listed in the GATE Bibliography (more than 1,000 scientific papers arising directly or indirectly from GATE have now been published).

The Monsoon Subprogram. The field phases of the Regional Experiments [Winter and Summer Monsoon Experiments (MONEX) and West African Monsoon Experiments (WAMEX)] were successfully completed. Research in MONEX began during the field phase and the preliminary scientific results were presented at the Sixth Planning Meeting for MONEX (Singapore, November 1979).

The Mountain Subprogram. A primary goal of the mountain subprogram, particularly its Alpine Experiment (ALPEX), is to provide a basis for designing improved procedures for quantitatively representing in numerical models the dynamical and physical effects of orography on various scales.

In order to satisfy this primary goal, the ALPEX Observational program (September 1981–September 1982) must provide data on both large- and small-scale mountain phenomena, including local wind effects; this will be essential for a better understanding of the mutual interactions between the small-scale mountain effects and synoptic-scale events. The scientific basis for the experiment and other details are set forth in the ALPEX Experiment Design Proposal (Geneva, July 1980).

12.1.23 Commemoration of the Anniversaries of the First International Polar Year (1882–1883), the Second Polar Year (1932–1933), and the International Geophysical Year (1957–1958)

ICSU has decided to commemorate in 1982–1983 the anniversaries of the First International Polar Year, the Second International Polar Year, and the International Geophysical Year. In addition to special presentations at the ICSU General Assembly in Cambridge, United Kingdom in September 1982 there will be commemorations at the COSPAR and SCOSTEP meetings, at the IAU and IUGG General Assemblies, etc.

12.1.24 ICSU Publications

The ICSU publishes a Year Book which provides information about ICSU, its members, associates, committees, commissions, services, etc. It gives the list of officers and principal committees and commissions and working groups of the members of the ICSU family. ICSU also publishes a quarterly Newsletter.

ICSU Organization and Activities was published in 1976 and the second edition of *ICSU: A Brief Survey* was published in 1979.

12.1.25 Relations with the United Nations Organization

ICSU was granted Category A Status by ECOSOC in 1971 after having been on the Roster for a number of years. The Council worked closely with the UN Advisory Committee on the Application of Science and Technology to Development (ACAST), the UN Office of Science and Technology and the Secretariat for the UN Conference on Science and Technology for Development. Discussions have already taken place about cooperation with the Director of the new UN Centre on Science and Technology for Development.

ICSU concluded a formal agreement with UNESCO in 1946 shortly after its formation, and in 1961 was admitted to consultative and associate status.

In addition to the agreement with the World Meteorological Organization for the World Climate Research Programme, ICSU has a working arrangement that was approved in 1960.

ICSU also has consultative status with the International Atomic Energy Agency, a working agreement with the International Radio Consultative Committee of the International Telecommunication Union, specialized consultative status with the Food and Agriculture Organization, official relations with the World Health Organization, and has recently concluded an agreement with the UN Development Programme.

ICSU is currently cooperating with all the above organizations in a wide range of activities, some of which involve scientific space research activities.

12.2 COMMITTEE ON SPACE RESEARCH (COSPAR)

12.2.1 Origin

At the International Geophysical Year (IGY) Rockets and Satellites Conference, held in Washington in October 1957 following the successful launching of Sputnik I, the first artificial satellite, a resolution was passed which drew attention to the importance of continued scientific research utilizing instrumented rockets and earth satellites, and it was recommended that the international scientific unions federated in ICSU develop suitable means for continuing this work.

Following up the final recommendation of the Fifth Assembly of the IGY Committee, held in Moscow in August 1958, ICSU, at its General Assembly of October 1958, set up the Committee on Space Research (COSPAR) as a special ICSU committee. The present charter of COSPAR was approved by ICSU in November 1959.

In 1961 the ICSU Executive Board, recognizing the continuing responsibilities of COSPAR, classed it as a scientific committee with unlimited duration.

12.2.2 Purpose and Objectives

The purpose of COSPAR, under its charter approved by ICSU in 1959, is to further, on an international scale, the progress of all kinds of scientific investigations which are carried out with the use of rockets or rocket-propelled vehicles. In September 1975, ICSU decided to extend the terms of reference of COSPAR to include also space research experiments carried out with the use of balloons. COSPAR is concerned with fundamental research and it will not normally concern itself with technological problems such as propulsion, construction of rockets, guidance, and control. The COSPAR objectives are achieved through the maximum development of space research programs by the international community of scientists working through ICSU and its national academies and adhering unions. Operating under the rules of ICSU, COSPAR ignores political considerations and considers all questions from solely the scientific viewpoint.

12.2.3 Organization and Membership

Reflecting the dual nature of the membership of ICSU itself, COSPAR is composed of representatives of national academies of sciences (or equivalent institutions) and of representatives of the international scientific unions.

At present 35 national academies of sciences (or equivalent institutions) of the following countries are members of COSPAR: Argentina, Australia, Austria, Belgium, Brazil, Bulgaria, Canada, Czechoslovakia, Denmark, Finland, Federal Republic of Germany, France, German Democratic Republic, Greece, Hungary, India, Indonesia, Iran, Iraq, Israel, Italy, Japan, Mexico, Netherlands, Norway, Pakistan, Poland, Romania, South Africa, Spain, Sweden, Switzerland, Union of Soviet Socialist Republics, United Kingdom of Great Britain and Northern Ireland, and United States of America.

The following 13 international scientific unions retain membership in COSPAR: International Astronomical Union (IAU), International Union of Biochemistry (IUB), International Union of Biological Sciences (IUBS), International Union of Crystallography (IUCr), International Union of Geodesy and Geophysics (IUGG), International Union of Geological Sciences (IUGS), International Mathematical Union (IMU), International Union of Physiological Sciences (IUPS), International Union for Pure and Applied Biophysics (IUPAB), International Union for Pure and Applied Chemistry (IUPAC), International Union of Pure and Applied Physics (IUPAP), International Union of Theoretical and Applied Mechanics (IUTAM), International Union of Radio Science (URSI).

12.2.4 COSPAR Interdisciplinary Scientific Commissions and Other COSPAR Bodies

Following a reorganization completed in 1980, the internal structure of COSPAR is based on interdisciplinary scientific commissions (ISCs). Among the responsibilities of the ISCs are the following: (a) to discuss, formulate, and coordinate internationally cooperative experimental investigations in space; (b) to encourage interactions between experimenters and theoreticians in order to maximize space science results, especially interpretation arising out of analyses of the observations; (c) to stimulate and coordinate the exchange of scientific results; (d) to plan symposia and topical meetings for discussion of the results of space research; (e) to carry out these tasks in the closest possible association with other organizations interested in these and related tasks; and (f) to prepare a statement on recent scientific developments in the area of interest to the Commission for the COSPAR report to the United Nations.

The terms of reference of COSPAR ISCs and other bodies are provided below:

1. ISC A on Space Studies of the Earth's Surface, Meteorology and Climate: To promote and enhance effective international coordination, discussion, and cooperation in areas of studies of the lower atmosphere-ocean–land system, where space observations can make unique and useful contributions.

2. ISC B on Space Studies of the Earth–Moon System, Planets and Small Bodies of the Solar System: To study the planetary bodies of the solar system (including the earth), especially evolutionary, dynamic, and structural aspects with planetary atmospheres included in so far as these are essential attributes of their main body and smaller bodies including satellites, planetary rings, asteroids, comets, meteorites, and cosmic dust.

3. ISC C on Space Studies of the Upper Atmospheres of the Earth and Planets, including Reference Atmospheres: To stimulate planning of cooperative research programs, to investigate specified aspects of the properties and structure of the upper atmospheres of the earth and planets, to plan symposia and topical meetings in which new results are presented and discussed, and to develop comprehensive reference atmospheres and ionospheres for the earth and planets.

4. ISC D on Space Plasmas in the Solar System, including Planetary Magnetospheres: To study various topics involving space plasmas by establishing subcommissions devoted to specific areas.

5. ISC E on Research in Astrophysics from Space, Sub-Commission E.1., Galactic and Extragalactic Astrophysics: To Co-operate closely with IAU Commission 44 and 48 and help establish close contacts between physicists and astronomers in cosmic space research. Sub-Commission E.2, Solar Physics: To coordinate problems associated with the sun as a star, including energetic particles in the heliosphere, which are of interest to the Commission.

6. ISC F on Life Sciences as Related to Space: To study, among other things, the effects of extraterrestrial environments on living systems and closed ecosystems, including biological effects of changes in radiation and in gravitational forces; biological and medical studies of human beings in space flight, life support systems, and nutritional problems in space; and earth-based studies of effects of extreme environments on biological systems.

7. ISC G on Materials Sciences in Space: To review fundamental theoretical and experimental investigations which will yield significant new understanding in the utilization of the physical conditions of space for scientific objectives; to recommend promising new avenues for future research; and to coordinate exchanges of information on scientific objectives.

8. Advisory Panel on Space Research and Developing Countries: To advise COSPAR on means of assisting developing countries in benefiting from

the possibilities of space research; to help COSPAR provide relative assessments of the benefits of space science and technology for countries at different levels of development; to maintain close contacts with the national committees for COSPAR in developing countries and to advise COSPAR as to ways that could help those countries strengthen their space programs, to provide a forum for all developing countries to discuss the undertaking of cooperative space programs and the possibilities of participation in space research; to facilitate the participation of developing countries in cooperative space programs; and to foster affiliations of developing countries with COSPAR.

9. Panel on Technical Problems related to Scientific Ballooning: To provide a forum for exchange of ideas on technical and operation aspects of, on instrumentation specific to, and on areas of research that could be supported by, scientific ballooning; to arrange symposia and disseminate information on technical advances and other information of interest to groups using balloons for space science; and to promote stratospheric balloon flights at different latitudes involving balloon drifts over thousands of kilometers by establishing cooperative efforts and appropriate chains of stations for data collecting.

10. Panel on Potentially Environmentally Detrimental Activities in Space: To discuss the different aspects of potentially environmentally detrimental activities in space, to produce a collection of survey papers for presentation to the COSPAR community and to the United Nations Committee on the Peaceful Uses of Outer Space, and to cooperate with that Committee in preparing for the Second United Nations Conference on the Exploration and Peaceful Uses of Outer Space.

11. Technical Panel on Dynamics of Artificial Satellites and Space Probes: To support and coordinate all activities aimed at the detailed description of the motion of artificial celestial bodies, and to promote the unification of procedures for the adjustment of satellite orbits from very precise tracking data, particularly the comparison and possible unification of definitions of orbital elements.

12. Advisory Committee on Data Problems and Publications: To elaborate rules concerning rocket and satellite information and data exchange in the frame of international services such as SPACEWARN, world data centers, and satellite warning centers and to formulate proposals on COSPAR publication policy.

12.2.4.1 The SPACEWARN system.

An international mechanism has been established so that space scientists of all countries participating in COSPAR, and also of other countries wishing to use SPACEWARN services, can have prompt access to information on new launchings of spacecraft, satellites, and space probes and to continual orbital information on the scientific satellites. The mechanism also serves to distribute widely and promptly the international designations of these artificial satellites.

This distribution of information started on an *ad hoc* basis in 1958 as part of the IGY World Days program and has continued as the SPACEWARN system. SPACEWARN is managed for COSPAR by the International Ursigram and World Days Service (IUWDS), a permanent service of URSI, in association with IAU and IUGG, and in close liaison with other ICSU bodies such as COSPAR.

The Greek letter satellite designation scheme was initiated by the Harvard College Observatory in 1957 for Sputnik I and was based on the long-standing IAU mechanism for designating newly discovered astronomical objects. COSPAR adopted and continued this system at its third meeting in Nice in 1960. IAU transferred to COSPAR the responsibility for designating artificial astronomical objects (satellites), and at its fifth meeting (Washington, 1962), COSPAR decided to change from Greek letters to an Arabic number system. A single letter follows the Arabic numbers to distinguish the various objects placed in orbit by a single launch. The general principle remained the same: satellites and space probes with a lifetime of more then 90 minutes, which require a designation for scientific purposes, are numbered according to the order in which they are launched. The COSPAR *Guide to Rocket and Satellite Information and Data Exchange,* adopted at the Washington meeting, assigned to the COSPAR secretariat the responsibility for making the final assignment of scientific designation according to the universal time of launch.

The guide was incorporated into the *Guide to International Data Exchange through the World Data Centres,* published in 1963, and with slight modifications into the *Fourth Consolidated Guide to International Data Exchange through the World Data Centres* (1979).

IUWDS, on behalf of COSPAR, coordinated a SPACEWARN network for rapid communication of satellite information. There are four satellite regional warning centers: (1) Western Pacific—Radio Research Laboratories, Japan; (2) Western Europe—Fernmeldetechnisches Zentralamt der Bundespost, Federal Republic of Germany, and Appleton Laboratory, England; (3) Eurasia—Astronomicheskiy Soviet Akademii Nauk, USSR and Soviet Geophysical Committee, USSR; and (4) the Western Hemisphere—also IUWDS World Warning Agency for Satellites (WWAS)-Goddard Space Flight Center, United States.

Associate centers which also handle satellite message traffic are located at Nederhorst den Berg, Netherlands, and at Salisbury, Australia.

The main types of messages transmitted by the SPACEWARN network are launching announcements and, when requested, current orbital elements. The messages are sent from one continent to another by telegram, and further distribution to national centers and individual laboratories is achieved by the most effective rapid means available.

The *SPACEWARN Bulletin,* a publication for the rapid distribution of information on satellites and space probes, is issued to the COSPAR national SPACEWARN contacts for satellite information, to satellite regional warning centers, and to various leaders and participants in COSPAR activities. Recipients are requested to arrange for any appropriate further distribution of this bulletin to interested individuals and institutions in their regions or countries. To improve the effectiveness of international distribution of satellite and space probe information via the SPACEWARN system, spacecraft are identified, according to the urgency and detail of information needed by the scientific community, by the following categories.

Category I: Spacecraft particularly suited for international participation, especially those for which prior arrangements have been circulated through COSPAR channels; essentially continuous satellite radio beacons, usually on frequencies less than 150 MHz, designed for cooperative ionospheric experiments; satellites with continuous telemetry of scientific experiments; balloon satellites; flashing light satellites; objects for laser tracking; and satellites in orbits of particular interest for which optical observations from the ground constitute a scientific experiment.

Category II: Space experiments of unusual general scientific interest or popular interest and manned spaceflights and space probes not included in category I.

Category III: All other space experiments, satellites with command telemetry only, test vehicles, etc., not included in category I or II.

Each month summaries of the launches received during the preceding reporting period are published. Official confirmation of the international identification occurs when it appears in the *COSPAR Information Bulletin* (three times a year). When possible, the *SPACEWARN Bulletin* contains more detailed information on scientific satellites of particular interest, such as a list of recent international designations, the texts of launching announcements, data on spacecraft particularly suited for international participation, and launching reports.

12.2.5 COSPAR Meetings and Symposia

Until 1980, every year COSPAR held its Plenary Meeting, comprising, in addition to business activities, scientific sessions on chosen topics and specialized symposia and workshops organized in collaboration with international scientific unions and other international nongovernmental and intergovernmental bodies. In addition, in the interval between annual COSPAR meetings, COSPAR organized its own symposia in various locations or co-sponsored the meetings of other bodies. Beginning in 1980, COSPAR Plenary meetings are to be held biennially; the 1982 Plenary and scientific events are to take place in Ottawa, Canada.

COSPAR meetings and specialized symposia serve as a useful forum for exchange of information on the results of investigations using space technology, and in the case of general symposia, for comparing the results of space activities with results obtained by other techniques.

COSPAR gatherings, open to any scientists of the world, also give an opportunity for the elaboration of plans for coordinated efforts in specific disciplines, as well as in undertakings of an interdisciplinary character.

A number of workshops and scientific sessions organized by COSPAR are of particular interest to developing countries. The contribution of space research to the problems of national development, the practical use of information obtained by space means by scientists from developing countries, and the transfer of know-how is among the important tasks of COSPAR.

12.2.6 COSPAR Publications

From 1960 on, the proceedings of scientific sessions of COSPAR Plenary meetings, symposia, and workshops were published in three regular series: (a) *Space Research;* (b) *Life Sciences and Space Research;* and (c) *Advances in Space Exploration.*

Beginning with 1980 the above series was discontinued and the proceedings of meetings and symposia appeared in the COSPAR journal *Advances in Space Research.*

In addition to the regular series indicated above, irregular proceedings such as the *COSPAR International Reference Atmosphere (CIRA)* and other volumes on specific subjects were published.

COSPAR is also producing COSPAR technique manuals. The last issues in this series dealt with remote sensing techniques and were of particular interest to the developing countries.

The *COSPAR Information Bulletin* appears three times a year. About 1,600 copies of it are being distributed free of charge to national committees for COSPAR, individual scientists participating actively in COSPAR, and interested international organizations.

12.2.7 Relations with the United Nations, Its Specialized Agencies, and Other International Organizations

Since 1961 COSPAR has maintained very close contact with the United Nations. As required by the United Nations Scientific and Technical Sub-Committee of the Committee on the Peaceful Uses of Outer Space, COSPAR submits an annual report entitled "Progress of Space Research" and also prepares specific studies on certain subjects at the Sub-Committee's request. COSPAR has observer status with the United Nations Committee on the Peaceful Uses of Outer Space and is regularly represented at its meetings and at those of its Scientific and Technical Sub-Committee.

Close collaboration also exists between WMO and COSPAR. On the request of WMO, COSPAR has carried out a number of studies on space observation systems for GARP and is currently involved in similar studies on the WMO/ICSU World Climate Research Program. Numerous symposia organized by COSPAR, dealing with meteorological topics, are being co-sponsored by WMO.

COSPAR, through participation in the Inter-Union Commission on Frequency Allocations, is cooperating with ITU and especially with its CCIR, presenting the needs for frequency allocations in specific radio bands, for scientific space research purposes.

Contacts also exist between UNESCO and COSPAR. UNESCO has, on several occasions, co-sponsored COSPAR-organized symposia or workshops and offered some financial support in the form of contracts for the organization of these events.

Occasional contacts exist also with the United Nations Environmental Programme (UNEP), FAO, and WHO in the areas of COSPAR activity of interest to them (monitoring of the environment from space, crop estimation, and planetary quarantine problems, for example).

12.3 INTERNATIONAL ASTRONAUTICAL FEDERATION (IAF)

12.3.1 Origin

In September 1950, the first International Astronautical Congress was held in Paris, on the initiative of astronautical societies in France, the Federal Republic of Germany, and the United Kingdom. Delegates of these societies, joined by delegates of societies in Argentina, Austria, Denmark, Spain, and Sweden, met with the intention of forming an international association or federation "to promote the development of interplanetary travels."

At the second Congress in September 1951, the original eight participants were joined by delegates of astronautical societies in Italy, Switzerland, and the United States, and in 1952, at the third Congress held in Stuttgart, the delegates of all 11 societies formally adopted the first Constitution of the International Astronautical Federation. Since 1956, three members from the USSR and socialist countries in Eastern Europe have joined.

In 1960, the Federation created the International Academy of Astronautics (IAA) and the International Institute of Space Law (IISL), which operate autonomously and co-operate closely with the Federation.

Congresses have been held annually since 1950, generally in September or October. The evolution of the Congress programs can be seen in the subjects studied since the first Congress in 1950, which was limited to reports on activities in the different countries forming the Federation at that time. In 1951 (London), a technical program was set up to discuss three subjects: satellites, rockets for satellites, and space stations. Then, gradually, new topics appeared in the programs. In Washington in 1961, for example, the program dealt with space propulsion, astrodynamics, energy conversion, research on combustion, bioastronautics, solar system exploration, space vehicles, structures, and instrumentation. Starting with the Congress in Baku in 1973, in addition to the various topics regularly discussed, it was decided to choose a particular theme for each Congress. The themes so far chosen are: Baku, 1973, Space activity—impact on science and technology; Amsterdam, 1974, Space stations, present and future; Lisbon, 1975, Space and energy; Anaheim, 1976, A new era of space transportation; Prague, 1977, Using space—today and tomorrow; Dubrovnik, 1978, Astronautics for peace and human progress; Munich, 1979, Space development for the future of mankind; Tokyo, 1980, Applications of space developments; Rome, 1981, Mankind's fourth environment.

The scientific and technological subjects that appear in the program most frequently are: fluid mechanics in space applications; propulsion; energy problems for space vehicles; materials and structures; bioastronautics; space transportation; reliability of space systems; application satellites: meteorology, communications, earth resources, geodesy and geodynamics, manned stations, and space laboratories; solar system exploration; supervised youth rocket experiments; education in astronautics.

Over the years, committees have been set up to study specific problems in areas of current concern as, for example, the Bioastronautics Committee, the Education Committee, or the Committee on Space Applications. The committees contribute to the preparation of the congresses in cooperation with the International Programme Committee.

The Committee on Space Applications has set up working groups to analyze possibilities for making the benefits obtained from space technology more accessible to mankind. These activities have an international scope that reaches to all parts of the world.

12.3.2 Purposes and Objectives

IAF is a nongovernmental association of national societies, institutions, and bodies that share the objectives described in the Federation's Constitution.

The objective of the Federation is to encourage the development of astronautics for peaceful purposes and to ensure widespread dissemination of scientific and technical information related to space.

One of the functions of the Federation is to encourage astronautical research, which it does through its congresses and other meetings on specific subjects or problems and which provide a neutral ground for the world's space experts to become acquainted and to exchange ideas in a friendly and informal atmosphere.

The Federation cooperates with other worldwide organizations that are interested in different aspects of astronautics or that are involved in the peaceful use of outer space.

The Federation is international, interdisciplinary, and nongovernmental. These three essential characteristics make it a unique organization for promoting the international development of astronautics at the scientific and technological levels and the exploration of outer space.

Administered by a bureau composed of seven members headed by a president, the policies of the Federation are decided by the General Assembly during its annual meeting. These policies are not influenced by any government or interested group.

The purposes of IAF are set forth as follows:

1. To foster the development of astronautics for peaceful purposes.
2. To encourage the widespread dissemination of technical and other information concerning astronautics.
3. To stimulate public interest in and support for the development of all aspects of astronautics through the various media of mass communication.
4. To encourage participation in astronautical research of other relevant projects by international and national research institutions, universities, commercial firms, and individual experts.
5. To create and foster as activities of the Federation academies, institutes, and commissions dedicated to continuing research in, and the fostering of, all aspects of the natural and social sciences relating to astronautics and the peaceful use of outer space.
6. To convoke and organize—with the support of its respective academies, institutes, and commissions—international astronautical congresses, symposia, colloquia, and other scientific meetings.
7. To cooperate with appropriate international and national governmental and nongovernmental organizations and institutions and advise on all aspects of the natural, engineering, and social sciences related to astronautics and the peaceful uses of outer space.

12.3.3 Organization and Membership

Founded in 1950, the Federation is at present composed of 60 members from 36 countries. These members consist of three categories:

1. National members: Several astronautical societies from the same country can be admitted as national members but there can be only one member who is entitled to vote on all matters discussed by the General Assembly.
2. Institution members: Universities, schools, institutes, or laboratories in-

volved in education or research in the field of astronautics can be admitted as institution members.

3. Associated members: International organizations whose purposes and activities are in accordance with the objectives of IAF can be admitted as associated members.

The General Assembly is the supreme governing body of the Federation. The day-to-day activities of the organization are carried out by the Bureau which consists of a president and five vice-presidents elected each year by the General Assembly, and the last-retired president; the presidents of IAA and IISL and the General Counsel are also members of the Bureau but do not vote.

12.3.4 Congresses of IAF

Each year the IAF Congress is organized and hosted by one of the national societies. The congresses are open to anyone who wishes to attend, regardless of whether one is affiliated to an IAF member society or not.

Since 1973, a theme is selected for presentation and discussed at the opening session of the Congress. Other sessions are also directed to the theme. The thirty-first Congress, held in Tokyo in September 1980, consisted, for example, of 47 sessions and almost 400 papers. The thirty-second IAF Congress was held in Rome in September 1981.

In addition, IAF organizes conferences for students and awards prizes for the best papers. The Congress is also the occasion of the presentation of the Allan D. Emil Memorial Award. This award is presented annually for an outstanding contribution in space science, space technology, space medicine, or space law which involved the participation of more than one nation and/or which furthered the possibility of greater international cooperation in astronautics.

The proceedings of the congresses are published, and special issues of *Acta Astronautica* (see Section 12.3.7) are devoted to them.

12.3.5 IAF Committees

The principal committees are of two types: administrative and substantive. Committees falling within the first category are the International Programme Committee, the Publications Committee, the Committee on Financing, the Committee for the Promotion of Activities and Membership, the Student Activities Committee, the SYRE Study Group, and the Committee for Liaison with the United Nations.

The substantive committees of the IAF are as follows:

1. The Education Committee, concerned not only with the educational programs relating to astronautical disciplines but also to the utilization of

space techniques in teaching methods, both by the transmission of courses by the best teachers in a certain discipline and by mass education, particular in the developing countries.

2. The Bioastronautics Committee, created to discuss and promote all activities relating to this discipline, especially man's role in them.

3. The Committee on Space Applications, covering a broad field of activity and directly connected with living conditions on the earth: earth resources, meteorology, ecology, telecommunications.

4. The Committee on Lighter-than-Air Systems, devoted to exploring civilian uses of balloons.

5. The Space Energy and Power Working Group, whose role is to call attention to new space concepts and developments to meet energy needs on earth from an international standpoint and to ways of coordinating national and international activities in that area.

12.3.6 Relations with the United Nations, Its Specialized Agencies, and Other International Organizations

An important aspect of the Federation's activity has been to establish a close relationship with various relevant international intergovernmental and nongovernmental organizations concerned with astronautics or related fields.

Among the governmental organizations should first be mentioned relations with the United Nations Committee on the Peaceful Uses of Outer Space, to which the Federation was granted observer status in 1976. IAF participates in meetings of this committee and its subcommittees, regularly submits reports on current advances and trends, and makes useful contributions, for example, as regards the evaluation of earth resources observed by satellites and the utilization of other data and systems, especially in developing countries. The Federation also took part, through international teams of scientists and experts, in contributing, within its area of expertise, to the series of background papers for UNISPACE-82.

The Federation also has relations with UNESCO, WHO, WMO, ITU, and IAEA.

Among the nongovernmental organizations, there is a particular need for coordination and agreement with COSPAR, the "Committee on Space Research" of ICSU. In some areas, the activities of IAF and COSPAR are so closely related that collaboration and coordination of activities are both desired and possible. COSPAR has recently decided to hold its plenary meetings every second year and IAF has decided to do likewise.

COSPAR will be given the possibility of organizing or co-sponsoring some meetings at the IAF congresses and the same opportunity will be open to IAF at COSPAR plenary meetings.

12.3.7 International Academy of Astronautics (IAA)

The International Academy of Astronautics was founded by IAF in 1960 with a Founding Committee chaired by Theodore von Karman, assisted by A. G. Haley, J. C. Cooper, and F. J. Malina. The current president is Dr. C. Stark Draper (United States).

IAA is composed of persons elected by the members of the Academy to three sections: basic sciences, engineering sciences, and life sciences. (In 1980, there were 560 members and corresponding members in more than 30 countries and 11 honorary members.)

The governing body of the Academy is the Board of Trustees, comprising the president, four vice-presidents, the last-retired president, and 12 trustees (four from each section).

The Academy cooperates very closely with IAF on and participates regularly in IAF activities, but it retains full autonomy in scientific activities and in matters of administration and finances.

The Academy's activities are oriented toward advanced ideas on space exploration and subjects of importance for the future.

The areas of interest of its committees are indicative of this orientation, for example: communication with extraterrestrial intelligence (CETI); space economics and benefits; energy and space; space rescue and safety studies; man in space studies; gasdynamics of explosions and reactive systems; space relativity; scientific and legal liaison; manned research on celestial bodies (MARECEBO); history of the development of rockets and astronautics.

The committees organize symposia and other scientific meetings in their respective subject areas annually or at longer intervals. Many of these meetings are held within the framework of the IAF congresses. Proceedings of the meetings are published under the auspices of the Academy.

The Academy has received support in or for its meetings from several international organizations: UNESCO, WHO, WMO, IAEA.

Acta Astronautica is the official journal of the Academy. It not only reflects the scientific activities of the Academy but also publishes articles from outside sources relating to astronautics.

A section of the journal (ex Mundo Astronautico) is reserved for general information on the operation of the Academy and the Federation, including an annual activities report of the IAA president.

In 1970 the Academy published the *Astronautical Multilingual Dictionary* in seven languages. It is out of print, but a limited number of copies are available at IAA headquarters.

The Academy has made an annual award, The Daniel and Florence Guggenheim International Astronautics Award, in recognition of an outstanding contribution to the development of astronautics.

12.3.8 International Institute of Space Law (IISL)

The international character of space activities quickly showed the need for organizing international cooperation in the studies of legal problems of outer space. For this reason, a permanent Committee on Space Law was set up in 1958 by IAF under the chairmanship of A. G. Haley.

This committee was replaced in 1960 by the International Institute of Space Law. It now has 400 individual members from 48 countries and is governed by a Board of Directors comprising the honorary president, the president, two vice-presidents, a secretary, a treasurer, and 11 members. The current president is Professor I. H. Ph. Diederiks-Verschoor (Netherlands). Prof. E. Pépin (France), who led the IISL between 1963 and 1973, has been elected honorary president.

Aside from a number of working groups that have been set up, the main activity of the Institute is its annual Colloquium on the Law of Outer Space held within the framework of the IAF Congress. Eminent jurists specializing in the law of outer space participate regularly in the colloquia and discuss the new situations resulting from the progress of space exploration and legal measures taken or proposed by the international community. Proceedings of the colloquia on the law of outer space are published regularly by IISL.

The Institute undertook a survey on the teaching of space law throughout the world and has organized symposia on this subject. It has published an annual worldwide bibliography on space law and related matters.

IISL assists IAF in its representation at meetings of the Legal Sub-Committee of the United Nations Committee on the Peaceful Uses of Outer Space.

12.3.9 Future of IAF

As an international, interdisciplinary, nongovernmental organization, the International Astronautical Federation has an increasingly important role to play in most aspects of space activity, particularly when programs need coordination at an international level.

The Federation is proud to state that since 1980 it has member societies from all countries that launch artificial satellites. The Federation, however, does not have member societies in all countries which can draw significant benefits from space activities. At this stage, participation in the IAF activities by institutions or individuals from all countries not represented in IAF, in particular from developing countries, is to be warmly encouraged.

Action inside the Federation will be continued at three levels: the annual

Congress, the specialized committees, and the relations with international organizations:

1. The quality of presentations must be continually heightened by a careful selection of the papers to be presented at the congresses because the success of the congresses is essential for the life and the professional standing of the Federation.
2. The committees and their activities must evolve in line with current problems and programs.
3. Close contacts must be maintained with international organizations concerned with specific problems of astronautics, and agreements should be reached between the Federation and those organizations in order to achieve the maximum benefit from their efforts.

INDEX